# Commemorating Coal Mining Worldwide

*International museums, heritage centres and sites related to coal mining*

AF898170

Margaret Lindsay Faull

BAR INTERNATIONAL SERIES 3076 | 2022

BAR
PUBLISHING

Published in 2022 by
BAR Publishing, Oxford

BAR International Series 3076

*Commemorating Coal Mining Worldwide*

ISBN 978 1 4073 5812 3 paperback
ISBN 978 1 4073 5813 0 e-format

DOI https://doi.org/10.30861/9781407358123

A catalogue record for this book is available from the British Library



COVER IMAGE *Headgear of the Grande Miniera di Serbariu at the Centro Italiano della Cultura del Carbone in Carbonia, Sardinia, Italy (© Centro Italiano della Cultura del Carbone).*



BAR
PUBLISHING

BAR titles are available from:

BAR Publishing
122 Banbury Rd, Oxford, OX2 7BP, UK
EMAIL info@barpublishing.com
PHONE +44 (0)1865 310431
FAX +44 (0)1865 316916
www.barpublishing.com

## Of Related Interest

**Visitor Experiences and Audiences for the Roman Frontiers**
*Developing good practice in presenting World Heritage*
Edited by Nigel Mills

Oxford, BAR Publishing, 2021 BAR International Series **3066**

**Le miniere delle Alpi Apuane (Toscana)**
*Storia dello sfruttamento minerario dall'antichità al XX secolo*
Marco Baldi, a cura di Deborah Giannessi

Oxford, BAR Publishing, 2021 BAR International Series **3034**

**Paisajes mineros y modos de vida en el norte de Mendoza, Argentina (S. XIX-XX)**
Osvaldo H. Sironi

Oxford, BAR Publishing, 2020 BAR International Series **2982**

**A Gazetteer of the British Iron Industry, 1490–1815**
*Volumes I and II*
Peter King

Oxford, BAR Publishing, 2020 BAR British Series **652**

## Acknowledgements

I am particularly grateful to Ray Christison, of the City of Greater Lithgow Mining Museum Incorporated (**5**), for information about Australian sites; Filip Depuydt, guide-animator at the Site du Grand-Hornu (**37**), for information about Belgian sites; Robert Wolensky, professor emeritus at the University of Wisconsin for contacting sites in America; and the International Committee of Museums representatives of various countries. I should like to express my gratitude for their advice to Sir Neil Cossons and my colleagues at the Network of European Coal Mining Museums (**29, 108, 244, 297, 327**), especially Jill Clapham and Rosemary Preece (**386**) and Virginie Malolepszy (**78**). I am also grateful to the Ambassador of Colombia, the Royal Thai Embassy and the Russian Embassy for assistance in contacting museums in their countries. I have had assistance and information from too many people to be acknowledged individually here, but to all of whom I am extremely grateful. I would especially like to express my thanks to those people who have sent me information and images about their sites, all of whom are acknowledged in the footnotes to the individual entries. I should also like to thank the three anonymous reviewers of this book for their very helpful suggestions and to BAR for agreeing to publish what is a very specialised book with a possibly limited audience.

There are some people without whom this book could not have been written. Firstly Maggie Heaton, the Executive Administrator at the NCMME, put all the captions together with the images and gave immeasurable other assistance with the project. Secondly Darran Cowd, formerly Collections Officer at the NCMME and now Curator, Folkestone Town Council, who identified the majority of the machinery shown in the various images and provided enormous amounts of technical information. Technical advice was provided by various other people who are acknowledged individually where they assisted, especially Mark Carlyle, curator of industry at the NCMME. Where no specific reference is made to advice from an individual, then the technical information can be assumed to have come from Darran Cowd. Finally thanks are due to Tom Arents, a postgraduate doctoral student at the time in Tokyo, for information and advice about Japanese sites and translation of Japanese documents (**248-77**); Eusebi Casanelles, former director of the Museu de la Ciènca i de la Tècnica de Catalunya, for information about museums in Spain (**322-36**); and Margarete Merz, the Personal Assistant to the Director of the Deutsches Bergbau Museum (**108**), for advice on the translation of German terms. I should like to acknowledge JSTOR for allowing free access to up to 100 articles a month during the pandemic.

Finally I am extremely grateful to David Anderson, Director-General, Amgueddfa Cymru-National Museum Wales (**388-9**), the International Committee of Museums and Collections of Decorative Arts and Design (ICDAD ICOM) and the Marc Fitch Fund for grants to cover the cost of publishing all the images in colour.


I am all too conscious that there may well be other coal-mining sites of which I am not aware and that details in the entries in this book may also be incorrect; in many cases I could not obtain information and had to rely on facts set out on the internet. Errors in this book are entirely my responsibility.

# Contents

# Preface

> In the metabolism of the Western world the coal-miner is second in importance only to the man who ploughs the soil. He is a sort of grimy caryatid upon whose shoulders nearly everything … is supported (Orwell 1937, p.18).

Britain was the world's first coal-burning nation, coal that heated London, provided steam for mills and factories and powered the locomotives of the railway revolution, coal that drove the first industrial economy. Wherever in the world there was accessible coal it was mined, the most prolific fuel of the modern age, supplanting wood as a source of heat and firing the boilers of steamships and whole navies, of power stations, the furnaces of great steelworks, the cupolas of small foundries and millions of domestic fires. Today, in a world preoccupied with creating a carbon-neutral future, we can for the first time see the boundaries of the Coal Age, from its hesitant inception through a triumphant universal crescendo to a demise being forced as much by societal coercion as any reservations about coal's efficacy as an efficient source of heat.

But the triumph of coal was accompanied by trauma, issues of safety, the dangers of deep-pit mining being an ever-present accompaniment of an industry that employed hundreds, and in many countries thousands, of miners. Almost without exception, the important coal-mining areas of the world suffered catastrophic disasters, from fires and explosions, collapsed workings or flooding. Invariably, there were harrowing scenes of families clustered around the pithead waiting for news of their loved ones – and their breadwinners. Coal too became a social and political force, through the growth of trade unions and powerful movements to assert that, if the price of coal was high, it was not high enough.

The mining of coal in the traditional coalfields of the Old World is now a political impracticality, while elsewhere – in Australia and China, for example – no such scruples constrain the sinking of bigger and more efficient pits. Here Margaret Faull charts the history of this global industry through the material evidence of its remains, in museums and preserved sites. Hers has been a mammoth task, unique not just in the comprehensiveness of its content, but in the eminence of the author's experience, understanding and perception. The author has brought her unique knowledge and understanding to bear on a subject of global significance. This book could not have been written by anyone else.

And the timeliness of its publication, in the aftermath of COP 26, adds a special poignancy to the archaeology of an industry that, as no other, has dominated the world.

Sir Neil Cossons OBE FSA FMA
Rushbury
Shropshire

# Abbreviations

| | |
|---|---|
| a.m. | *ante meridiem*, 'before midday' |
| BCE | Before the Christian era, the equivalent of B.C. ('before Christ') |
| CE | The Christian era, the equivalent of A.D. (*anno domini*, 'the year of our Lord') |
| EU | European Union |
| NCB | National Coal Board |
| NCMME | National Coal Mining Museum for England (**386**) |
| p.m. | *post meridiem*, 'after midday' |
| SIG | Stiftung Industriedenkmalplfege und Geschichtskultur (Foundation for the Preservation of Industrial Monuments and Historical Culture: **228**) |
| UK | United Kingdom |
| UNESCO | United Nations Educational, Scientific and Cultural Organisation |
| WWI | World War I (1914-18) |
| WWII | World War II (1939-45 in Asia and Europe, except for 1941-5 in Russia: **307**) |

# Geological periods

| | |
|---|---|
| Carboniferous | from 358.9 million years ago |
| Permian | from 298.9 million years ago |
| Triassic | from 251.9 million years ago |
| | from 201.3 million years ago |
| Cretaceous | from 145.5 million years ago |
| Cenozoic | from 65.5 million years ago |
| Tertiary | from 65.5 million years ago |
| Miocene | from 23 million years ago to 5.3 million years ago |
| Quaternary | from 2.6 million years ago |
| Holocene | from 11,700 years ago to the present day |

# Abstract

Coal is still a dominant fuel in America, Australia, China, Mongolia and Russia, but more and more countries are abandoning coal and turning to alternative sources of energy, increasingly renewable energy provided by wind, wave or solar power. So now is an opportune time to take stock of the impact that coal and coal mining have had on countries around the world. Some 48 countries have institutions recording or commemorating that history in the form of museums, heritage centres or preserved colliery sites, of which this book considers 432 examples. It is not a straightforward account of the history of coal mining but rather an examination of how that history has been presented by different institutions and cultures.

The introduction discusses some common themes found in many different institutions, such as the history of the development and subsequent spread worldwide of technology, the problems encountered in mining, how coal was transported underground, up the shaft and onto the markets, the workforce, including training, women and children working underground, slavery and forced labour. Mining is a dangerous occupation so miners sought to protect themselves with invocations of gods or saints, or by joining unions to bring pressure to bear on the owners. Disasters dogged coal mining, often with political consequences or leading to protective legislation and the erection of memorials to the victims. There are many different approaches to presenting this history, dependent on the financial resources available and the political approach of the country concerned, and many museums also consider topics such as art and fossils.

At the start of the section on each country there is a summary of the coal resources of that country and of its coal-mining history. Each institution is then discussed in detail, including the nature of the site, the history of the institution, including staff and visitor numbers and form of governance, and especially how it portrays coal mining, coal miners and coal-mining communities. Many have statutory protection, especially where they occupy a site with substantial colliery remains, some have been recognised by UNESCO as being of international significance, and some are on designated routes of coal mining.

Some topics are so universal that they are covered at too many sites to be able to be discussed in detail in the introduction. These include those museums that deal specifically with the history of coal mining, that have archives or a library, or display or have in storage large machinery, such as the massive machines used on opencast sites, Also frequent are displays about social history and living conditions of coal-mining communities. Others have an underground, either genuine or a simulation, while some opencast sites allow guided visits and also demonstrate how areas that have been degraded by mining can be reclaimed afterwards by landscaping. After opencasting, the voids left are often allowed to fill with water to provide recreational areas. The institutions with these features are listed in the concordance, which also details a few items not covered in the individual entries, such as whether there are visitor facilities.

# Introduction

Coal, in its various forms, is found in all the continents of the world and has been valued for hundreds of years for the heat and power that it can generate. Coal fuelled the Industrial Revolution (**379**), starting in the UK and then spreading to continental Europe and subsequently around the world. During the eighteenth century industrial goods were moved mainly by canal. In the nineteenth and first half of the twentieth century, coal was moved within countries by locomotives on the newly created railways and from country to country by ships. The steam locomotives and steam ships in their turn increased the requirements for the coal with which they were powered.

Producing coal from underground mines is a dangerous enterprise and technology was developed gradually to combat dangers such as gas explosions and roof falls. But such technology was often expensive and only adopted when enforced by legislation or when it enabled the coal to be produced at a greater profit. The social status of miners echoed the social conditions of other workers in each country, ranging from slaves, serfs and convicts to miners with better conditions and pay than other industrial employees.

Coal is, however, bulky and dirty, so from just before WWI it began to be replaced by oil (**404**), less of which was required to produce the same amount of power. Throughout the twentieth century coal mines began to be closed, either because of the use of the cleaner and less bulky oil or natural gas or because less expensive coal could be imported from countries where it could be produced more cheaply. The dangers of climate change and global warming became increasingly apparent, caused partly by the burning of coal. The Cop 21 Paris Agreement, signed in 2016, saw many countries pledge to phase out all use of coal, although the largest users or producers, America, Australia, China, India and Mongolia, continue to rely heavily on coal.

Museums devoted to the history of coal mining and coal miners are to be found on every continent. This book deals with coal-mining museums, heritage centres and preserved colliery sites throughout the world. It is not a history of coal mining worldwide, but rather the depiction of that history by the places discussed in this publication. This introduction deals with some aspects that go towards illustrating the development of coal mining and coal-mining cultures around the world. The concordance at the end of the volume deals with 18 aspects of coal-mining history that are to be found at a large number of sites. Certain aspects that are covered in the entries are not included in the introduction or the concordance, such as governance or methods of funding, but can be found by going through the entries in the volume. It is hoped that this publication can be used to illustrate the different ways in which museums and heritage centres present the subject; to provide a data base for those interested in pursuing a particular line of research or interest; and to compare and contrast approaches in different parts of the world – for example, museums in communist or former communist countries may display details of senior company staff or awards to miners for high productivity (*e.g.* **348.1**).

Coal occurs in various grades. Anthracite, known as black or hard coal, is the highest grade of coal, because it has the highest carbon content, the fewest impurities and the highest energy density of all types of coal. It produces little ash or smoke, so was especially valued for domestic heating. Bituminous coal, also classified as black coal, contains asphalt or bitumen, a tar-like substance; it releases more methane than other types of coal, so stringent safety precautions, such as good ventilation and gas monitoring, are essential during mining to prevent explosions. It is especially used in electricity generation. Dry bituminous coal was particularly used to power steam ships. Coking bituminous coal, known as metallurgical coal, is used to make coke for smelting iron ore (**379**). Falling between anthracite and bituminous coal is thermal coal, known as steam coal, which was especially used for powering steam locomotives. Lignite, or brown coal, has a carbon content around 25 to 35 *per cent*. Also known as brown coal is sub-bituminous coal, which contains 35 to 45 *per cent* carbon. Brown coal is the least valued, because it produces a lot of smoke with relatively little heat, and is difficult to store or transport, being liable to self-combust. Many eastern European countries have large lignite deposits, which they had to use to produce electricity during the communist era (1947-91), leading to major pollution and serious health issues amongst the population. Coal was not necessarily used just as it was mined (known as run-of-mine), but often had to be mixed with other types of coal to obtain the best results; for example, in the UK after 1972 semi-bituminous Kentish coal used in coke manufacture for foundry usage was increasingly added to higher-volatile coals from West Durham (Trueman 1954, pp. 160-1).

## Early history and development of technology

Coal deposits are found widely throughout the world. Each country where coal is found developed its own mining culture, although developments in technology tended to spread rapidly around the world.

### *Development of technology*

The first use of coal appears to have been in Czechia in the late Palaeolithic, before 8,000 BCE (**70**), and continued in

Europe with collection on the surface from outcropping seams. Coal seems to have been exploited in the Late Iron Age and used extensively for industrial processes in the Roman period (Travis 2008, pp. 159-73). There are documentary references to Roman coal mining in France and Italy (**introductions to France and Italy**) and evidence for Roman coal mining in the Saarland (**226**), Serbia (**314**), Spain (**331**) and the UK (**352**). There is then a break in evidence for coal mining in Europe during the Early Middle Ages (CE 478 to about 1000), although coal was probably still collected by local people where it outcropped (**103**), as also happened in Japan (**261**). Folk stories are often told about how coal was discovered (*e.g.* **10, 261, 314**).

The earliest documented use of coal in medieval Europe (Benoit and Verna 1999 for pre-Industrial Revolution use of coal in Europe) comes from Bulgaria in the tenth century (**45**) and from Germany in 1113 (**97**), with examples in France in the twelfth and thirteenth centuries (**introduction to France**), the UK in the later-thirteenth century (**361, 368**) and then in Spain in the late-sixteenth century (**introduction to Spain**). Coal mining was often initiated by religious houses and artisan mining went on in these early areas until the nineteenth century. Surprisingly very little information about coal mining and coal-mining sites has come from archaeological excavations, although landscape archaeologists have contributed considerably to the study of coal mines in the landscape (for examples see [Debrabant] 2009). There have been very few excavations by archaeologists. Professional excavation at Le-Molay Littry uncovered the mid-nineteenth-century foundations of the heapstead, the top of the shaft and the foundations of the steam boilers (**91.1**) and at Caphouse Colliery the foundations of a miner's cottage (**386**). In some cases clearance and restoration of abandoned sites has included a limited degree of excavation to identify elements on the site (**393.2**).

Most of the evidence for early coal mining has been obtained instead from opencast sites cutting through bell pits by workings close to the surface, showing the typical bell shape in profile (**241, 395.3**). Mining by bell pits went on until the nineteenth century, although there is evidence in England for shaft mining from the early-fourteenth century (**395**). In north-western Germany the first documented case of mining with a vertical shaft, as opposed to an adit, is from the early-eighteenth century (**128**).

In the High and Late Middle Ages (1000 to 1500 CE) Germany was the leading country for mining, both in technological developments and publications (**108**). In the seventeenth century Sweden took over as the leader in mining technology and mining education. The longwall system, however, is believed to have been developed in England in the first quarter of the seventeenth century (**379**) and in the eighteenth and nineteenth centuries it was the UK that had become the leading coal-mining country and the major exporter of coal overseas.

The UK and Germany developed various new technologies, materials and techniques in the nineteenth and twentieth centuries, when coal became essential to industrial development for powering machinery, for trains as railways spread and replaced horse-drawn vehicles, and fast steam ships displaced windjammers, as well as providing coal gas for heating and cooking. At the other end of the history of coal mining, the availability of gas and oil began the decline in the use of coal. This started in the UK in 1913 (**404**) and accelerated throughout the twentieth century and into the twenty-first, including in competition with clean energy such as windfarms and photovoltaic panels. In Europe the European Coal and Steel Community (ECSC) programme started pit closures as early as the 1950s (**37**). The last of the small mines of the Ruhr shut in 1976 (**148**). The last deep coal mine closed in the UK in 2015 and in Germany in 2018. In 2016 Belgium became the first country in Europe to phase out coal-fired electricity supplies completely, followed by Austria in 2020. Despite the ECSC programme, many people would not accept the likely demise of coal (**386**) and some mine operators still went ahead with modernisation, only to have to shut down with the new installations barely completed (**190**). Steam-powered winding engines were still being manufactured as late as the 1960s (**200**).

***Problems***

The major problems that arose in coal mining were water, subsidence and gas. Major exploitation of coal in the UK and then elsewhere was triggered by the discovery of how to smelt iron ore with coke made from coal (**379**) and how to deal with water underground. Groundwater moves downwards and enters underground areas, which are then in danger of flooding. The water produced in coal-bearing strata (**327**) can be either dangerous or beneficial. The dangers of water were illustrated graphically by nineteenth-century disasters (**30, 384**) and then by the Lofthouse Colliery disaster (**383**), while water could impede shaft sinking (**97, 370**). A number of museums have displays about methods of removing water underground (*e.g.* **327**). The earliest method was to drive a channel underground, although this did not work where the conditions were very wet (**84**), and some mines were designed with a slope so that water could drain away naturally, which also allowed tubs to be moved more easily by gravity (**122, 148**). Some mines used a pump operated by a horse-driven whim gin to dewater the pit (**235**).

In the UK Thomas Newcomen invented the atmospheric steam engine in 1712, both to remove minewater and to raise coal to the surface. One was installed the same year at Dudley (**360**) in England, while the 1795 engine at Elsecar survives *in-situ* (**371.1-2**). There are a working replica at Dudley (**360.2**) and models in Austria (**21.1**) and Hungary, where two Newcomen engines were installed in 1722 (**235**). A Newcomen engine was installed in France in the middle of the eighteenth century (**91**). Three later rotative Newcomen engines, which were operated manually, also survive (**398**). In 1783 in the UK James Watt patented an improved double-acting engine. An example, probably from 1802, is preserved in France (**91.2**). Sometimes pits were so wet that the conditions defeated modern methods of removing water and had to close (**130**). Once mines shut, they continue to produce water, which can cause surface flooding if not controlled. This has especially been a problem in Germany (**110, 143, 214**) and the UK (**384, 386**).

On the other hand, minewater has been used to provide good-quality potable drinking water to communities in Poland (**304**) and Sweden (**338**), and to water livestock (**386**). The deeper mines went, the greater the heat; hot water pumped to the surface can be used for heating buildings (**113, 386**) or for mineral-rich *onsen* (**250**) or hydrotherapy baths (**154**). In Poland water underground was so plentiful that it was used to form canals on which coal could be moved (**298, 301**).

The need for ventilation to supply fresh air to the miners and remove toxic, explosive or suffocating gases was another major problem. Methane gas, known as firedamp, was the greatest threat to life underground, as it is highly explosive (**88, 323, 332**); small animals were used to detect carbon monoxide after explosions (**327**). Compressed-air, hydraulic and pneumatic systems were used to power machines safely underground without risking explosions (**112**). In the nineteenth century, furnace shafts were used to draw fresh air into the pit for the miners to breathe (**284, 386**), but with the risk of igniting any methane gas present. Fresh air was moved around the mine by small children opening and closing air doors underground, known as trappers in England (**378.1**) and doorboys or nippers in America (**426**). Another method was to use bellows, mechanically driven by a waterwheel or an animal (**327**). All modern mines, including demonstration mines that come under mining legislation (**96, 112, 239, 329, 386, 388**), have to have effective ventilation, either natural (**56, 291**) or provided by fans on the surface (**239, 388**). Fire underground, whether from an explosion or the ignition of methane, could cause devastating loss of life (*e.g.* **234**). In one case fire spread to the underground from a forest fire (**314**).

The realisation that coal dust in the air could also result in explosions (**80**) meant that it became essential to suppress the dust. This was usually done by stone dusting (**80**) and later also water spray (**370**), as had been recommended in the UK in 1901 (**350**). In Russia and Japan blasting the coal out of the seam with a high-pressure jet of water was used to reduce coal dust (**251**), although this technique failed in Germany (**179**).

To deal with the results of these dangers, countries established mines-rescue stations with training facilities (**68, 354.2**) in surface galleries (**263**). Many museums display equipment used by mines-rescue teams (**264, 309, 398**), sometimes displayed on models (**274, 321**). Amongst the equipment of each mines-rescue station was an emergency winder, or winch, that could be moved to any colliery where the winding engine was out of action and miners were trapped underground. With the closure of so many mines, the emergency winder is now sometimes housed at a museum (**381, 386, 388**).

One problem in mining areas is subsidence on the surface (Orwell 1927, pl. [5]). In the past this happened slowly as roadways collapsed underground. As roof supports were removed with retreat mining (**382**) during the twentieth century, subsidence occurred almost immediately. On occasion buildings or headframes had to be demolished because of subsidence (**231, 232, 391**) or were affected by opencasting (**402**). One mine in Czechia was filled underground to avoid subsidence on the surface (**69**) and in Japan areas of subsidence on the surface were filled with waste rock (**258**). Both subsidence and opencasting resulted in voids on the surface, which are now often filled with water to provide recreational tourist areas (**72, 77, 111, 119, 121, 123.1, 230, 245**).

A large proportion of museums deal with the development of technology in sinking shafts, driving roadways, cutting coal and raising it to the surface. Shaft sinking was hazardous work, as was the subsequent maintenance of shafts, which was done by men suspended in a kibble (**33.2, 241, 328, 355.2, 386**; see also image Henriksson 1996, p. 188). An alternative was to drive a sloping roadway, known as an adit, or drift, from the surface down to the level of the coalface.

Early mining worked the coal by pillar-and-stall (**47**) in Europe, Japan and northern America (**47, 48, 49, 238, 250, 253, 307, 388, 398.2, 429, 430**), with fires and later gunpowder and then other explosives often used to bring down the coal (**324, 339**). The pillar-and-stall system continued up to the twentieth century in some areas, such as America, as it suited working by individuals, but was generally replaced by the more efficient longwall working by a group of miners. Originally this was done by advance mining, until during the twentieth century retreat mining was adopted as more cost-effective (**382**). Both methods involved various problems, such as flowblow, where the roadway ballooned upwards (**250**), with the floor sometimes even reaching the roof. Other problems included that of working in very confined areas, for example, where the roadways went out under the sea (**248, 254.2, 260, 324, 370**).

### *Transport underground*

Effective movement of the coal underground became a necessity with the development of large-scale mining. Initially coal had been carried in baskets on the backs of mineworkers or dragged along on wicker corves (**386.4**). In Scotland the baskets were transported to the surface by women climbing long steep ladders (*cf.* **250, 273**). At first horses were used only on the surface to transport coal for local use and to provide power for winding gins (**372.1-3**). Rails for moving tubs of coal underground were developed in the eighteenth century. Ponies underground are first recorded in the mid-eighteenth century, although they could not work in very low seams (**91**), where boys still moved the coal in tubs, one pushing and one pulling (**378.1-2**). In the UK the 1842 Act, and even more so the 1872 Act (**378**), made ponies indispensable in moving coal from the coalface to the shaft, and ponies were used in very large numbers (**18**). In the Ruhr pit ponies were first used in 1853 (**212**). Mules were used rather than ponies in Spain (**325.1**) and America (**419, 424, 429**), as well as oxen (**424**).

Ponies were trussed to be lowered down the shaft (**33, 356**), as otherwise some would have panicked in the shaft or refused to go underground (**370**), although others came to love it (**18**). Many museums have displays or collections about pit ponies/horses or display stables where they were kept (**33, 45, 90, 107, 112, 132, 277, 299, 309, 314, 360, 362, 369, 370.4, 385, 388, 395**). One aspect not usually dealt with in museums is that, where ponies were kept underground, the oats for their feed attracted mice or rats (Orwell 1937, p. 22). The two were never found together, as rats would always kill or drive out the mice. Miners took their food underground in vermin-proof metal containers, known as snap tins in the UK. Miners' meals underground usually consisted of bread and dripping or lard (**90**), never butter which would go rancid in the hot conditions, and a bottle of cold coffee or tea (*ibid.*, p. 20).

The heavy work meant that pit ponies often had a low life expectancy, although legislation was passed in attempts to protect them. Miners believed that ponies were regarded as more important than them, because pit ponies had to be bought. In Japan ponies were eaten at the end of their lives (**276**), while after disasters trapped miners on occasion ate either one of the ponies (**80, 384**) or else the corn stored to feed them (**375**).

Ponies went on being used in mines that had not mechanised or where it was difficult to supply electricity because there were very tight spaces, such as where the roadways went out under the sea (**370**). The last pit pony in Germany retired in 1951 (**232**), in Australia in 1990 (**9**) and in England in 1994 (**370**). In Australia monuments have been erected to the memory of the ponies (**10, 18**).

Ponies were gradually replaced by mechanised transport underground, such as mine cars and locomotives. The first electric locomotive was introduced underground in 1882 (**227.1**) and eventually diesel-powered locomotives were developed. This led to armoured face conveyors, although locomotives were still used to transport miners from the shaft to the coalface and back.

Various countries in Europe, especially Germany and the UK, and also northern America and Japan, developed machinery such as roadheaders and heliminers for driving roadways, shearers for cutting coal and conveyor systems for transferring it to the shaft or drift bottom and then on to coal-sorting facilities on the surface. Each new development was usually adopted quickly by mines (**19, 255, 355**) and by companies producing mining machinery, and then improved further. So there was a constant movement of development, modification and transfer of technology around the world. Many museums highlight specific innovations developed in their area; for example, in Silesia (then part of Germany, now in Poland) the first powered roof supports and many other technological innovations (**298**), or in Germany the first mine theodolite in 1836 (**105**). Similarly the first electrically driven main-shaft winding engines in Europe (**129**), as opposed to the oldest European working steam-winding engines using the traditional system of winding with two wheels (**298, 314**).

***Raising coal to the surface***

Headgears, or headframes, were similar throughout the world, although there was experimentation, especially in Germany. The monumental Malakoff headframe was developed in the 1850s (**127.1, 149, 163, 168, 181, 184, 197, 209, 231**) and in the early-twentieth century a headgear was designed with a third leg, known as the *Zschetzsche* design (**207, 210, 218**). In 1901 a *Hammerkopfturm*, an enclosed tower winder, was designed with the winding engine on the top (**147**). This form of headgear was used at a number of other German mines until the end of the 1920s (**147, 189**), when a new form – a high rectangular cube – was designed, which became the model for modern oil rigs (**144, 205**).

Also invented in Germany in 1887 was the *Koepe*, or friction-winding, system, which did not require any headgear (**127.2**). This was widely adopted, first in Germany (**105, 113, 147, 153, 165, 200**) and then throughout Europe (**68, 298, 303, 366, 381, 396**). In Turkey it was modified to work with a headgear and a winding engine on the ground (**347**). The system was expensive, but more effective for deeper mines (**366**). The use of concrete for colliery construction was pioneered in Germany, the first concrete headgear being constructed in 1910 (**104.1**). The first concrete headgear in Canada appears to date to 1913 (**52.1-2**), the same year as one in England (**104**), and in America to 1923 (**425**).

***Other aspects***

Miners sometimes resisted the introduction of new techniques, which often had unforeseen consequences; for example, rails laid underground to make it easier for miners to move coal tubs, then enabled the owners to use younger children and women to move the tubs (**378.1-2**). Similarly much work was done to devise a safer light for use underground, resulting in the Davy lamp (**331.1, 355**), but miners found it gave less light by which to work. Also the greater protection meant more-gassy mines being worked, resulting in a succession of devastating explosions in the UK, of which the Oaks Colliery (**355**) was just the worst example. Miners preferred the older wooden pit props, which creaked when a roof collapse was imminent, to the stronger metal ones (Orwell 1937, pp. 41-1).

Conflict and coal mining have also often been intertwined ([Debrabant] 2016) and the bitterness from the sufferings of war had an effect on mining communities. So the offer of assistance by German mines rescuers after the Courrières disaster was at first rejected because of the bitterness from the Franco-Prussian War, 36 years earlier (**80.2**). Indeed on being told in 2018 that the Germans appeared to have the earliest use of concrete for headgears, a Belgian responded that he was not surprised because the Germans were always keen on the use of concrete, "we still have some of their concrete constructions along our border" (*i.e.* from the pre-WWII Siegfried Line). One of the major reasons for creating the EU was to decrease this level of bitterness and to avoid yet another pan-European war, such as had been fought in 1914-18 with the loss of 20 million lives and 1939-45 with the deaths of 75 million people.

Sometimes systems developed independently in parallel. For example, in Taiwan blue, white and yellow helmets designated different jobs (**344**). In the UK miners wore white helmets, trainees yellow and mines-rescue service red; in Germany managers wore white, coalface workers yellow and mines-rescue service red helmets. This sort of colour-coding system does not seem to have been used in other countries. Some countries developed specific greetings between miners. In German-speaking countries this was *Glückauf*, 'good luck' (**110, 210, 226**), including Poland until 1945, after which it became *Szęść Boże*, 'God bless' (**introduction to Poland**), while in Japan the common greeting was *go anzen ni*, 'be safe' (**248**).

**Surface features**

Surface facilities were essential to the running of any mine. The winding-engine room with a tiled floor was always kept scrupulously clean and only the winding enginemen were allowed to enter. In addition to the winding-engine house and whichever winding system was in use, there were administrative buildings and various workshops. Particularly important were the joiners' and blacksmiths' workshops, in which all mine machinery was repaired and many items were produced, including shoes for the pit ponies (**68**). Many of these buildings have been preserved for use as museums and, where suitable, other purposes such as community centres. Some museums still display the early form of powering workshop machinery with line shafting (**132, 255.3, 328.2, 386**).

Deep mines are extremely hot, so that miners usually worked naked except for helmet with cap lamp, belt to carry the cap-lamp battery and boots (Orwell 1937, p. 20). This meant that the coal dust in the air coated their entire bodies, already wet with perspiration. Originally colliers had to go home 'in their dirt', as washing facilities were not provided (*ibid.*, pp. 33-4, pl. [6]; **344**). The first pithead baths in the UK that were constructed to a recommended standard opened in 1928 (**391**). These had an individual stall for each miner, as also in later pits in Belgium (**32**), whereas in Taiwan bathing facilities were communal (**344**). Miners changed from their everyday clothes before going underground and, in the UK and the Campine coalfield in Belgium, their work clothes were stored in heated lockers after they finished work to allow them to dry. In America, Japan and continental Europe, clothes were stored on either hooks (**424**) or in birdcage lockers, which were open baskets suspended from the ceiling by ropes with a padlock at the bottom of each, to which only the individual miner had the key (**63.1, 70, 276.2, 334**).

Coal comes to the surface either by conveyor belt up the drift or in tubs up the shaft. The building housing the shaft top, with the wheels above, is known as the heapstead (**87.1, 99.3, 280.2, 304.2, 387.4**). Material coming out of a coal mine consists of coal, stones and dirt, and so needs to be sorted on the surface in what is known as a screens building (**378**). In recent years this has been done mechanically, but in the past it was sorted by women (**244.2**) or by miners either too young, too old or too disabled to work underground (**426;** Orwell 1937, p. 4). In the UK it used to be said that every miner was a child twice in his career – at the very beginning and very end of his working life. Once sorted, the coal was usually cleaned in a washery; where the sorting and washing facilities were in the same building, it was known as a coal-preparation plant. The coal then dropped through hoppers into waggons or tubs underneath (**46.1, 52.1, 64, 78.2, 334.2, 342.2, 344.2, 364, 386.1**), with the waste material deposited on spoil heaps. These could reach massive proportions and in some areas, such as Wales, looked remarkably like Iron Age hill forts. Because of the inefficient way in which coal was sorted from waste by hand, the spoil heaps still contained small pieces of coal; miners and their wives struggled to recover this broken coal during times of unemployment or strike action, despite the dangers involved (*ibid.*, pp. 93-6, pls [1, 3]).

In many cases the spoil heaps have now disappeared. In the Netherlands the spoil heaps were removed (**280**). In some countries, such as the UK, the heaps often were washed in the 1970s and 1980s to retrieve the small coal, after which the remaining waste was spread out so that it no longer looked like a traditional spoil heap. In other cases spoil heaps were levelled to provide a base for a building (**327, 380**), the shape used for the design of an office building (**189**), turned into a dry rock garden (**113**), grassed (**226**), planted with flowers (**345**) or turned into a country park (**390, 391.1, 395**). In Japan the spoil from mines on small adjacent islands was dumped between the islands to create a single large island (**260, 261**), while in Germany tunnels were dug into one spoil heap to provide an air-raid shelter during WWII and later a training tunnel and testing area for new underground machines (**161**). The largest spoil heap in the Puertollano coalfield has been landscaped to form an urban park (**326**), while spoil heaps in Germany and Sweden have their own name (**109, 226, 338.2**).

Many spoil heaps have developed their own microclimates and are rich in animal and plant life (**28, 130**). In some cases guided tours are offered of the heaps to describe the birds, reptiles and mining heritage (**81, 90**). Access is sometimes provided by steps cut into the spoil heap (**22, 269**), so that visitors can reach the vantage viewing point at the top (**115, 215**). The Hoheward and Hoppenbruch spoil heaps, which together form the largest spoil-heap landscape in Europe, have two arches which form the 'horizon observatory' (**231.3**).

A number of impressive spoil heaps are preserved in Japan (**269, 271**; see also **252, 258**), but it was in northern Europe that spoil heaps really came into their own. The north European plain, where WWI was predominantly fought (**79, 81**), is extremely flat without distinguishing features. There the spoil heaps form a major part of the landscape (**28, 35, 81, 97, 160**) and have contributed to the Nord-Pas-de-Calais being granted World Heritage Site status (**77**). Two organisations, the Chaînes des Terrils, were set up in 1986 to protect and display the spoil heaps in Belgium (**28.1-2**) and France (**81.1-2**). Where spoil heaps were threatened; for example, in Belgium where the 20 *per cent* coal remaining in the heaps could have been reclaimed by washing, the authorities moved in to protect them (**38**). Spoil heaps are used for a variety of purposes: paragliding (**38**), mountain biking, orienteering, skiing, luge (**81**), skating and hiking (**113**) or as a base for works of art (**22**). These spoil heaps are all safeguarded by regulating what activities are permitted on them and regular biological monitoring.

Once brought to the surface and sorted from the waste, coal had then to be moved to where it was to be used. Transport, coal and the uses of coal are inextricably linked. Coal is dirty, heavy and needs to be supplied in bulk. In medieval Europe, movement of goods was usually by packhorse or water-borne, so working of coal tended to be on a small scale and usage strictly local. Yet in the Roman period there is proof of coal being moved considerable distances (Travis 2008, pp. 133-58). In the Middle Ages coal tended to be moved further afield only from those places that were close to the sea, such as Bristol and County Durham in the UK from where coal was shipped to London (**361, 368, 375, 399**). Use of coal locally increased once chimneys had been invented, which allowed more salubrious domestic consumption of coal. Chimneys began to be used on the European Continent as early as the twelfth century, but were regarded with suspicion in England, where widespread adoption did not really come until the seventeenth century and only then with the appointment of wardens in cities to control chimney fires (https://ultimatehistoryproject.com/chimneys.html).

From the early-eighteenth century it became possible to move bulky goods by canal (**concordance**) and in the nineteenth century by railway (**concordance**), sometimes constructed specifically for moving the coal (**271**). The locomotives were powered by coal, so increasing demand, while permitting movement of coal to areas where it could be used both for steam engines in factories and for domestic heating. The final stimulus to coal mining came from the switch from wind to steam power for shipping (**261**); indeed the cost of the journey from Chicago to Liverpool dropped by two-thirds between 1873, when using sail, and 1884, using steam (Mount 2021, p. 18). In Japan Nagasaki, with its coal supply, became important as the closest port to China and a stopover where foreign commercial ships and naval vessels could refuel (**261**). The coal mines of Zonguldak made it an essential supply base for the Ottoman navy (**introduction to Turkey**). The British Navy established a network of international coaling stations to maintain the fleet at sea (**278**), while the American navy favoured Pocahontas coal, because it did not produce smoke (**428**).

## The workforce

The workforce used in the mines differed greatly from country to country. In some areas and at certain times miners were well-paid and regarded as highly skilled, whereas in others they endured appalling working and living conditions and were often trapped by their status in society.

### *Training and dissemination of knowledge*

Once invented, mining technologies tended to spread widely, partly by publications and training in schools of mines and partly by movement of miners. Mining is a skilled occupation and training is required for miners to be effective, as well as safe. Originally miners learned on the job and there are still echoes of this in the Forest of Dean, where legally anyone meeting certain conditions and experience is entitled to become a Free Miner (**367**), although the mines inspectorate now ensures that those working underground are properly qualified. In Czechia and Slovenia miners were apprenticed for training underground and were initiated into the trade by 'jumping the leather' (**67, 316**). In the nineteenth century mining companies in the Nord-Pas-de-Calais provided vocational-training centres, including training for work in the mines (**77**). In Germany earlier galleries and tunnels dug into spoil heaps have been used for training (**112, 161**), as well as a tunnel created in 1932 under the university in Cologne specifically to train mining engineers (**96**).

The facilities or collections of some museums are used in countries still dependent on coal for energy to train and encourage young people to enter the mining profession (**349**). In Ukraine large machines belonging to the museum are kept in the grounds of the adjacent technical college (**349**). The museum's simulated roadways in India were specifically designed both for the public and as a training facility for mining engineers in an area producing much of India's coal (**238.2-3**). The drift mine of the Zonguldak Maden Müzesi (**347**) is still used as a training centre for mineworkers. In America the underground roadways of the Portal 31 Mine are used by universities to train mining engineers (**417**). The eyesight and reflexes of winding enginemen have to be tested regularly and miners who make up mines-rescue teams have to be extremely fit and undergo specialist training in rescue techniques. In Japan a simulated underground roadway on the surface (**263**) and in Taiwan the underground drift (**344**) are used for mines-rescue training and practice. Conversely some museums began life as mines-rescue training facilities (**68, 347**).

As mining became more technical, independent schools of mines were established. Apart from the dangers of gas and explosions involved in coal mining, there was little difference between the techniques of mining coal, copper, gold, lead and silver. So schools of mines taught general mining techniques, perhaps the earliest being the eighteenth-century mechanical laboratory at the Falun copper mine (**introduction to Sweden**), with the earliest tertiary institution being at *Selmecbánya* in Hungary (**316**). These schools had aids such as models and geological collections, and some are now open to the public as museums. The earliest is probably the museum of the school of mining in Russia (**311**), while in 1807 the École Pratique des Mines provided an impetus to the development of mining in the Saarland in Germany (**105**). In Japan the Kaijima Company provided vocational training for unskilled workers (**258**), while in New Zealand, where coal and gold mining are still important, some 30 schools of mines were established in the nineteenth century. Two of these are preserved, together with their collections (**289, 290**). In Poland a school of mines established in 1889 is now a major museum, with the training roadways created in 1927 now providing an underground experience (**300**). In Sweden an early-twentieth-century mining school is now a museum (**338.1**) and in Germany a museum was created in 2007 in a 1929 mining-vocational school (**232**), as in France in 1986 (**90**).

Knowledge of mining technology was also disseminated by publications, the earliest and most famous of which is Agricola's 1556 *De Re Metallica* (**108.3**). In France in 1777 the eminent chemist, Lavoisier, published material about early mining from outcropping coal seams (**92**) and Thury produced a manuscript in 1800 on the techniques of the Compagnie des Mines de Littry (**91**).

### *Women and children*

Women and children worked underground in a number of countries. Generally legislation was passed to outlaw this practice, at different times and in different places. It appears that children still work in coal mines in Colombia, India and North Korea. It is usually assumed that the use of children was a practice that only developed in Europe and Asia with the Industrial Revolution (**379**). The Spanish National Archaeological Museum in Madrid, however, has the Roman gravestone from Cástulo of a child miner, four years old, carrying a pick and a basket (inventory number 16744), who would have worked in a lead/silver mine. So it is quite likely that children would similarly have worked underground in Roman coal mines.

In many pits during the nineteenth century women worked half-naked alongside completely naked men, to whom the women were not necessarily related. Impetus for legislation came from this immorality with its danger of sexual assault (**378, 386**), combined with the appalling conditions underground and the heavy work involved, especially for children (**378.1-2, 387, 394, 397**). The earliest legislation in Europe appears to have been the 1842 Act in the UK (**378**), followed by Austria/Hungary in 1854, France 1874, Germany 1878, the Netherlands 1906, Italy 1907, Norway 1915, Bulgaria 1917, Belgium 1919, Poland 1924, Hungary 1928 and Romania 1928 (Romano and Papastefanaki 2020, pp. 203-4). The lateness of legislation in some places was because it was simply confirmation of long-standing practice, such as laws in 1900 in Spain (Verde, Soto and Gómez 2020; **324**) and Sweden (Henriksson 1996, p. 202).

Outside Europe legislation was passed in Australia in 1901/29, South Africa 1911, Canada 1917/25, Nigeria 1917, China 1923, New Zealand 1915/26, Japan 1928, India 1929, Chile 1931 and Brazil 1932 (Romano and Papastefanaki 2020, pp. 203-4). In India, Japan and Taiwan miners worked in family groups of men, women and children from at least the nineteenth century (Chatterjee 2020, p. 192; **250**), which to some extent protected women from sexual assault. In the early-twentieth century India and Japan were the only industrialised countries where women worked underground in considerable numbers (Khaitan 2020), but at the same time boys as young as nine were working underground in American mines (**426**). In Taiwan legislation in 1964 was driven by the danger of both husband and wife being killed together, leaving children without parents (**344**). Many museums have displays describing the work of women underground (**90**), especially in Japan (**250, 253.2, 273, 277**).

Often there was little employment available for women after they stopped working underground (**146**) and many continued to work on the surface, especially sorting coal from waste material (**244.2, 264, 378.3**; see also image in Henriksson 1996, p. 191, in Sweden). Women often played an important support role during mining disputes; for example, in America (**416, 420, 431**), where children were also sometimes involved (**431**), and the UK (**373, 394**), and with the formation of the Miners Women's Auxiliary in Australia in 1934 (**18, 19**).

### *Slavery and forced labour*

Coal mining is, and always has been, a dirty job involving considerable danger and hard labour. It is not surprising, therefore, that in the past some mine owners have used forced labour. These workers fall into various categories. The most extreme were the seventeenth- and eighteenth-century collier-serfs of Scotland, who were bound to their masters for their lifetime (**387**), and the nineteenth-century slaves in America, who worked under the supervision of free skilled miners (Frantel 2008). This was also the case with convicts, prisoners-of-war and slaves in Roman mines (Travis 2008, p. 110).

In recent centuries coal has played an important part in warfare ([Debrabant] 2016). In areas occupied by enemy forces, it was common for the occupiers to use forced labour. In some cases, prisoners-of-war were transferred to work in the mines, whether or not they were skilled miners. In the Ruhr during WWI German miners who had joined the army (**127**) were replaced by French prisoners-of-war, who were often trained miners and, despite the 1878 legislation, by French women (**150**). During WWII when energy sources, including coal, were at a premium, half of the workforce in the mines of Czechoslovakia were prisoners-of-war (**69**). Russian prisoners-of-war were forced to work in mines in Germany (**115**), Poland and Belgium, only to be replaced in Belgium after the war by German prisoners-of-war (**26**). Other forced labourers worked in the mines in Serbia and Germany (**97, 127, 129, 187**).

Forced labour was also used in nineteenth-century Japanese coal mines, with the death penalty for those who tried to escape (**250**), as well as convicts (**introduction to Japan**). In the twentieth century Japan used forced Chinese and Korean labourers in coal mines in Japan and Japanese-occupied China (**introduction to China**), and during WWII more than 13,000 prisoners-of-war (**259**). Between 1837 and 1878 work in the mines of Tasmania in Australia was punishment for the most recalcitrant British convicts, with those who were already skilled miners working at the coalface (**7**). Other countries also forced prisoners to work underground. Convicts from Hong Kong and India were shipped to Malaysia, although this unskilled labour led to increased accidents and lower levels of productivity (**278**). Similarly prisoners were brought from all the Dutch colonies of Asia to work in Indonesia under very oppressive conditions (**239**). Russian prison-labour camps, known as *gulags*, were established in the far north-west of Russia above the Arctic Circle, where prisoners worked in the coal mines between 1937 and 1943 (**313**).

Asian indentured labour was a system that replaced slave labour in which unskilled workers, known offensively as coolies, signed contracts to work overseas for five or more years without pay, either to repay debt or for compensation at the end of the contract. Indentured labourers from India worked in the Natal coal mines in South Africa from the middle of the nineteenth century (**318**). Finally there are the cases of Asian labourers, such as those in Canada, who were paid only half the rate of their white counterparts (**49**).

### *Religion*

Mining is a dangerous occupation and miners have often sought divine protection. Throughout mainland Europe supplication has been to Saint Barbara (**96**), also known as the Great Martyr Varvara (**306**). Images of Saint Barbara were found at collieries, as well as chapels and churches dedicated to her in mining areas (**26, 77.3, 109**), along with some to the Virgin, named as Our Lady of Miners (**77**). Coal-mining centres were called after the saint (**322**). There is an underground chapel of Saint Barbara in Slovenia (**316**) and claims of similar underground chapels in Bulgaria (**45**). Chapels formed meeting places for miners (**103**) and in some places they gathered for prayers before descending underground (**227**). A chapel was provided for the miners at one French mine as early as 1804 (**91**).

In many places in continental Europe the feast of Saint Barbara is still celebrated each year on 4 December, the date of her supposed martyrdom (**96**). There are church services, parades, preparation of special dishes, shooting salutes and meetings of Barbara associations (**322**). Some mining museums have entire sections devoted to her cult (**131, 299**) and many museums have statues (**95, 109, 226**), paintings (**306**), an altar tablecloth (**91.3**) and a banner (**226**); the Deutsches Bergbau Museum (**108**) has probably the largest collection of such items relating to the saint. Celebration of her martyrdom day was taken to Turkey (**introduction to Turkey**) and the tradition of invocation of Saint Barbara to Brazil (**43**).

In the extreme west of Europe, in Ireland, Saint Barbara was supplanted by the Virgin Mary, with a statue of Our Lady over the entrance to the mine (**240.2**). In eastern Europe invocation in Bulgaria was to Saint Ivan of Rila and an iconostasis of the saint is displayed underground (**45**). In the eastern Orthodox Church the Holy Great Martyr Saint Propocius of Scythopolis is the protector of miners, so in Serbia a church of Saint Procopius was dedicated in 1900 to commemorate miners killed on his name day (**314**). Nevertheless there was veneration of Saint Barbara even in Orthodox areas, such as memorials in Russia (**309**) and Ukraine (**349.3**), perhaps because her supposed relics were kept in Kyiv (**96**).

Protestantism and in particular non-conformism, or Methodism, in Wales (**394**) and parts of northern England, was opposed to the veneration of saints. So there was no tradition of invocation of saints for protection, even in Catholic areas such as Glasgow, despite one claim otherwise in the French press (**355**). Chapels were found underground in Wales (**388**) and during the 1904/1905 Great Revival most pits in Wales seem to have held a prayer meeting before starting work (Thomas 1986, p. 24). Prayers were said underground elsewhere, such as in the north-east (**388**). In disasters where miners survived several days underground before dying, heart-breaking messages were found referring to prayers by individuals or groups of doomed miners (**375, 400**). Protestant traditions were also strong in America, where a preacher even gave a sermon underground (**419**). Churches were built at mining camps, which also served as community centres (**407.3**).

In Japan simple shrines were common in mines, sometimes to the goddess of the mine (**272**), although most were to Ōyamazumi no Kami, the god of mountains, oceans and mines (**272**). It was common for miners to pray at the shrine each day before descending underground (**255**). Replicas or the original shrines are still preserved (**248.3, 255, 272.2**). As in Europe, there were special festivities on 11 and 12 May, the feast day of the Shinto shrine (**272**). In South Korea one museum has a shrine where relatives can pay their respects to dead miners and in the adjacent village there is a traditional shrine with a well used by the community (**320**).

### *Health*

Miners are vulnerable to a number of medical problems ([Debrabant] 2020). Those who survived such incidents as accidents with machinery, explosions and roof falls, often suffered from broken bones, crushing or problems with breathing. Hospitals in areas where coal mining was a major industry usually had lung specialists for pulmonary disease and orthopaedic surgeons skilled in dealing with extensive bodily injuries. Nystagmus, caused by straining to see in poor lighting, was mainly a condition from the nineteenth century, but miners suffered from silicosis and pneumoconiosis well into the twentieth century. Silicosis is caused by breathing in tiny pieces of silica, from rock, causing lung scarring. This is found in all types of mining, and in coal mining usually occurs in miners responsible for driving roadways through stone. Pneumoconiosis is the commonest respiratory illness found in coal miners, caused by breathing in coal dust. As recently as 2018 in Appalachia in America, 21 *per cent* of long-term coal miners were found to have some degree of pneumoconiosis (https://www.ncbi.nlm.nih.gov/pmc/articles/PMC7055360/) and the Mungyeong SeokTan BakMulGwan has a shrine for those who have died from pneumoconiosis (**320**).

There is no cure for silicosis or pneumoconiosis, only treatments to ameliorate the symptoms, such as provision of oxygen. The most important preventative measures are the use of masks (Musa 2011, p. 232) and to try to reduce exposure to stone or coal dust. This was realised as early as the mid-sixteenth century by Agricola (**108**), who stressed the need for ventilation underground. Effective respiratory equipment was being provided in Europe from the later-nineteenth century (**67**), as well as the watering of roadways to dampen down dust (which also reduces the risk of coal-dust explosions), but such provisions for the health of miners were absent in India (Chatterjee 2020). Many museum displays deal especially with pneumoconiosis, which affected miners throughout the world (**96, 241, 276, 320, 323**). In Japan in the Edo period many miners died before the age of 30 and the Tomoko miners' organisation had provision for the proper burial of miners (**276**). In Austria the Bruderlade and in Germany the Knappschaftskassen provided social-security benefits to its members from as early as the Middle Ages (**108, 297**).

What is not usually covered in museums are the general health and injury problems suffered especially by miners, such as the blue scars by which miners always recognise miners from other areas, caused by coal dust being sealed inside cuts (Orwell 1937, p. 32). Although many museums display examples of miners' housing (**4.1-2, 57.2, 77.1-2, 111.1, 118.1-2, 257.1, 260.1-2, 360.3, 389.2, 398.3, 407.2**), these are often sanitised and do not show the often-appalling pre-WWII European dwellings that contributed to the ill health of the occupants (Orwell 1937, pp. 46-68, pls [7-14]).

Provisions to support the health of miners began in France in the early-nineteenth century, when relief funds provided by the mining companies paid for hospitals, clinics and pharmacies for miners and their families (**77**). In Spain, even in the twentieth century, only the Mina San Vicente gave treatment to families (**327**), although the industrialised area of San Blas did have a small pharmacy for both workers and their families (**328**). In Germany it became normal from the early-twentieth century to provide shelter for miners in wet clothes going from underground to the pithead baths (**185**). Some mines had underground treatment rooms (**424**), but most first-aid facilities were on the surface, as displayed by a number of museums (**16, 327, 328.3**). The Musée de la Mine Marcel Maulini is devoted entirely to treatment of the Ronchamp miners' lung diseases between 1947 and 1978 (**92**).

### *Trade unions and strike action*

Coal mining throughout the world involved hard labour and dangerous conditions, as well often as poor living conditions. The result was that miners joined together into trade unions to try to improve their situation. Many museums include general information about miners' organisations, covering the records of mining trade unions and their activities in Africa, America, Australasia, Europe and Japan (**9, 32, 225, 243, 249, 276, 286, 287, 314, 327, 349, 410, 416**), including collections and displays of the banners that are so associated with mining unions (**5, 236, 368.3, 373.1-2, 385, 386, 387, 388, 404**). Also important are the roles of the Japanese Tomoko organisation in training and authorising miners (**249, 276**), and the Austrian Bruderlade and the German Knappschaftskassen in representing the employment interests of their members (**108, 297**); these mutual-support organisations developed fraternal traditions that survive throughout continental Europe, including the hammer-and-chisel insignia (**206**), the black miner's costume (**68.2**) worn in traditional annual processions to church services on Saint Barbara's day (**96**), and the miners' greeting (**110, 210, 226**).

To improve matters, miners' unions frequently resorted to strike action: over safety concerns in Canada (**50**), France (**80**) and America (**426**); for improved working conditions in Canada (**49, 50**), Ireland (**241**), Italy (**245**), Slovenia (**316**), Spain (**332, 335**) and the United States (**423**); together with the demand for higher wages in Canada (**50**), Spain (**335**), Wales (**388, 394**) and Vietnam (**432**). Negotiations without strike action did sometimes result in better conditions, as in Germany (**195**), while occasionally, in Australia (**18**) and Spain (**327**), miners were able to take over production at their mine.

One particular grievance was how mining companies controlled the way that miners lived, by sinking pits in areas distant from any other employment; the provision of housing from which miners' families could be evicted during strikes;

and instead of cash, payment in the form of tokens that had to be redeemed at company-controlled stores, where prices were higher than in non-company stores. This was the case in the nineteenth century in Europe (**23, 77, 382**) and Japan (**250, 258**), but continued into the twentieth century in America and Chile (**introduction to Chile, 406, 417, 426**).

Some strikes were political in nature, such as the Labin miners' strike in Croatia in 1921 in protest against fascist violence (**65**) and passive resistance and strikes by German miners in 1923 against the French occupation of the Ruhr (**127**). Mining unions participated in general strikes during times of economic depression: in Australia during the Great Depression (**19**) and in Belgium during the general strikes of 1936 and 1960/1 (**24**). The British prime minister, Margaret Thatcher, framed the 1984/5 Miners' Strike as being politically motivated (**404**).

Later, as coal failed to compete with oil and gas, there were protests against redundancies and pit closures, such as the sit-ins underground in Australia in 1962 (**18**) and at the Sardinian mines in 1984 and in 2012 (**introduction to Italy, 246**). There were also demonstrations against closures, with banners and placards (**35, 225**). In the UK there were major strikes against pit closures in 1972 and 1974 (**370**), followed by a strike in 1981 when the government backed down (**373**), then culminating in an epic struggle in 1984/5 between the union and the prime minister. This effectively ended in defeat for the union (**368, 382, 388, 394, 403, 404**), as also happened in Japan in 1912 and 1960 (**276**). In Australia, the UK and the United States, women's groups were formed to support the struggle of the unions (**18, 19, 373, 420, 431**).

The authorities often fought back against what they saw as union militancy. As early as 1812 the bishop of Durham used his own troops to break up a strike (**368**). Following the 1906 Courrières disaster, Clemenceau sent in the troops to disperse the demonstrators (**80**), as Churchill did in the UK in the 1910/11 Cambrian Combine Dispute (**394**) and Franco in Spain in 1962 (**335**). In Canada it was believed that the union organiser, Ginger Goodwin, was shot deliberately in 1918 to silence him (**49**). In Poland in 1919 seven miners were killed by border guards (**303**), seven were shot dead in Italy in 1920 and in 1929 troops fired on the miners in Romania (**306**), as German occupiers threatened to do in the Netherlands during WWII (**introduction to the Netherlands**). But it was in America that there was the most violent reaction to union activity: in 1877/78 20 miners belonging to the Mollie Maguires were hanged, or judicially murdered (**426**), and in 1914 the dead, including women and children, appear to have numbered up to 56 (**431**).

## Disasters and their results

The one thing that everyone seems to know about coal mines is that there are disastrous explosions killing large numbers of miners at regular intervals. Apart from the serious loss of life and its disastrous effect on the lives of those left behind, major disasters had two main effects. The first was legislative and the second was on the political and social life of the area or in fact the country.

### *Legislation following a disaster*

It was a sad fact that all too often legislation was enacted only after a disaster, rather than proactively, even though the dangers were often already very obvious. In the UK after the 1838 Huskar Pit disaster, the condolences expressed by Queen Victoria and publicity about the conditions in which small children were working all gave impetus to the passing of the 1842 Mines and Collieries Act forbidding women and girls and boys younger than 10 working underground. Then in 1872 the Mines Regulation Act was passed forbidding the full-time employment of boys under the age of 12 (**378**).

Within seven months of the 1862 Hartley Colliery disaster, legislation was passed requiring pits to have two shafts (**375**). The 1893 Thornhill Colliery disaster stopped the practice of using naked-flame lamps in wet pits and the enquiry after the 1947 disaster at the same colliery recommended that lead rivets in lamps be replaced with magnetic locks and a better relighter be adopted (**400**).

From 1850 British managers had to keep a working plan of their colliery for government inspectors, who had first been appointed in 1843, and in 1872 plans of abandoned workings had to be deposited, as well as requiring managers to have state certification of their training. It was, however, the 38 deaths at the Montagu View Pit in 1925 that led to a central repository for plans being established to avoid miners breaking through into flooded abandoned workings, as had already been demanded after the 75 deaths at Heaton Colliery, 110 years earlier (**384**).

The 1934 Gresford disaster with 266 deaths contributed to the pressure that led eventually to the nationalisation of the collieries in 1946 (**359**). The 1969 Mines and Quarries (Tips) Act to ensure that spoil heaps were safe resulted from the 144 deaths in 1966 at Aberfan and this was further consolidated in the 1999 Quarries Regulations to avoid any instability of spoil heaps (**351**). Finally the 1973 Lofthouse Colliery disaster, in which seven men died, led to the 1979 regulations to guard against inrushes of gas or water (**383**).

In the eighteenth century France was in advance of other countries, with the first mines inspector being appointed in 1781, but after that tended to lag behind Germany and the UK in safety legislation, so practices often continued long after they had been forbidden in those countries. The 1906 Courrières disaster (**80**), the worst in Europe, was almost certainly caused by a methane-gas explosion ([Debrabant] 2006). The result was finally a ban in France on naked-flame lamps underground. Two independent means of escape were also made mandatory, more than 40 years after this had been enacted in the UK. This was followed in 1910 with the opening of the central Liévin mines-rescue station.

After seven miners were killed in Germany in 1875, the authorities insisted on a second shaft to improve ventilation at the Zeche Unser Fritz (**151**), while legislation replacing gasoline safety lamps with electrical safety lamps resulted from the deaths of 349 miners at the Zeche Radbod in 1908 (**146, 198**). In adjacent Austria, the authorities banned the *Firstenulmbau* method after six deaths in 1902. There was probably more proactive initiating of legislation and regulations without the impetus of responding to a disaster in Germany than in other countries.

One major difference was that in the UK controls were done nationally, with legislation enacted by the government, whereas in Germany it was sometimes done at the local level. Thus in the UK the 1887 Coal Mines Regulation Act controlled all aspects of closed flame-safety lamps and the 1911 Coal Mines Regulation Act included that fans should be able to be reversed to pump in fresh air (**350**), while also granting miners an eight-hour day. The first mines-rescue station was set up in 1902 (**80**). In Germany naked-flame lamps had been banned in 1846 in the Ruhr (**80**), but not until 1900 in Dortmund (**199**); electric mining lamps were introduced by regulation in 1909 (**80**). In America it was not until 1952 that the lobbying of John Llewellyn Lewis led to the first Federal Mine Safety Act (**416**).

***Political consequences following a disaster***

The 1906 Courrières disaster led to widespread strikes in the Nord-Pas-de-Calais, which Clemenceau put down using the army. This strong response aided him to become prime minister of France the same year (**80**). The 1908 Zeche Radbod disaster led to the widows being jailed for disorderly conduct for demonstrating for more financial support (**146**). The militia were sent in to put down industrial unrest on Vancouver Island in 1914 after hundreds of miners were killed in explosions (**50**).

After WWII many western European countries were short of labour and brought in men from poorer countries to work in the mines. Sixteen different nationalities worked in the Belgian mines (**28**), until over 150 Italian miners were killed in disasters in 1953 and 1956 (**29, 31**). These ended the movement of Italians to Belgium and resulted in much stricter mining regulations in Europe. Yet in Germany a further 13 Turkish miners were killed in 1988, along with 38 Germans (**121**), which ended mining in Hesse.

***Aftermath of disasters***

There were often formal enquiries after a disaster, leading to legislation, but relatively few prosecutions for breaching legislation. There were relatively negligible fines for negligence after the 1913 Senghenydd explosion (**350**). In a number of cases, the disaster resulted in the closure of the colliery concerned (**179**).

Relief funds were usually set up to assist the widows and children left behind, and royalty, corporations and individuals gave generously (**351, 375**), although the funds often ran out while dependents were still destitute. Items were sold to raise funding, such as crockery (**355.1**), engraved glasses (**356**) or prints showing the scene (**227**). The rescuers were also sometimes rewarded with cash or medals (**375.3, 388.3**) or a monument in cases where they were killed (**355**).

***Memorials***

So numerous were major accidents that it would be impossible to cover all the memorials to the dead in this publication, so only the memorials of a few important disasters are considered. Memorials to the dead of major mine disasters take various forms and were erected at various locations. On occasion the memorial has been placed at the site of the disaster (**285**), at the entrance (**31**), including where the colliery concerned has become a mining museum (**29**), in parks (**121, 227**), in the centre of their community (**80**) or in prominent locations (**355**).

Some monuments were erected in the churchyard or the cemetery, frequently amongst the burials of the dead miners (**137, 146, 355.3, 370.1, 375.1**). In Europe miners were very badly paid, so often the families were so poor that they could not afford gravestones (**355, 400**). Memorials bearing the names of all those killed served to remember all the victims, although that at Méricourt marks the grave just of victims of the 1906 Courrières disaster whose bodies could not be identified (**80.3**). Monuments were usually erected soon after the disaster (**227**), although poor communities continued to remember their dead and, when they could afford it, later erected memorials, which were often smaller and more restrained (**29**). New memorials have been erected in recent years, often 100 or 150 years later, and are often works of art in their own right (**350, 355, 383, 400**).

The form of these monuments foreshadows those erected to commemorate the dead of WWI (Fontaine 2016, p. 211; Saunders 2004). The memorial to the 27 rescuers of the Oaks disaster, erected in 1913, the year before WWI broke out, shows the winged Greek goddess, Pallas Athena, carrying a wounded warrior (**355**). This is comparable to the winged angels sometimes shown on war memorials supporting dead soldiers or bearing their bodies up to heaven.

As well as conventional monuments, some disasters were commemorated by artwork, such as a bronze sphere (**29**) or a bronze bust (**398**). The paintings in remembrance of the 1862 Hartley Colliery disaster (**375.2**) and the 1912 Zeche Lothringen disaster (**187.1**) both show the reaction of the monarch of the time. It became the accepted response for royalty either to visit (**227**) or to send a message of sympathy (**378, 400, 403**). Memorial cards were also printed (**400, 403.3**) and some tragedies were immortalised in song, as after the 1904 Gresford disaster (**359**) and about the 1914 Ludlow Massacre (**431**). The most expensive monument is probably the church erected after the 1893 Senjski Rudnik mine disaster (**314**). At the

Mungyeong SeokTan ParkMuelGwan museum there is a shrine in honour of miners who died of pneumoconiosis (**320**).

A separate class of monuments are those erected in memory of all miners, not just those who were victims of major disasters. The steady holocaust of miners killed every day by rock falls or accidents with machinery is brought home by the provision in some collieries of a mortuary to receive the bodies of miners killed in everyday accidents (**167**). In eastern Saxony there were more than 1,000 deaths from 1914 until mining ended (**110**), while in Canada a database lists over 2,500 fatalities in Nova Scotia between 1838 and 1992 (**54**). In the UK the full tally of mining deaths is shown by a website listing over 23,000 miners (of coal and metal) who can be identified as having been killed in northern England (**369**), where in the 1930s it was estimated that the odds on a miner being killed during his career were 20 to one (Orwell 1937, p. 39). Most such memorials were dedicated to miners in a particular area: Alabama (**406**); central Jutland (**74**); the region of Cornwall (**10**); Durham (**368**); the Emil Mayrisch Mine (**100**); the district of Grey (**285**); Hesse (**121**); Hückelhoven (**131**); Kent (**380.2**); the Kleinzeche Renate Mine (**103**); the Kuznetsk (**309**); Walsenburg (**431**); and miners in the western Donbass killed while at work (**349.3**). The 2016 memorial at the Pozo Sotón commemorates all miners killed in the employment of the Hunosa company since 1967 (**335**), while the memorial at Søby commemorates 51 miners killed at work (**74**). One unusual monument is to over 2,000 pit ponies of the Illawarra coalfield (**18**).

There are a few memorials not relating to underground colliery disasters. Miners played an important role at the front during WWI and they have only recently been honoured (**79.1**). The most poignant memorials are the graves of children: the 26 of the 1838 Huskar disaster (**378**) and the 116 of the 1966 Aberfan catastrophe (**351.1**). Of the three that she sent, the message from Queen Victoria after the Huskar disaster helped the seventh earl of Shaftesbury to get legislation passed to prevent such a loss of young life again. It is fitting that the one mining-related memorial in the very centre of the capital, London, is a fountain in honour of the earl's humanitarian work (**378.4**).

### Operational sites open to the public

Very few tours are offered of operating underground mines (**56, 315**) and most tours of working mines are usually of opencast mines. Use of coal is being phased out in western Europe, but is still a major source of power in much of the world, while exports are a valuable source of income. The largest opencast mine in the world is in Colombia (**61**) and the second largest is in Russia (**308**). Much opencast coal is exported, like that from Australia. The coal produced in American and Canadian opencast operations is mainly for internal consumption (**51, 408, 411-14**). In the Russian Federation, tours of the opencast mine have made Neryungri into a tourist destination (**312**), as also in Germany (**230**). In Czechia (**72**) and Spain (**333**) specialised tour operators offer a variety of tours suited to the age and interests of the visitors, while also giving tours of the restored landscape following opencasting, as in Thailand (**345**) and many sites in Germany. Opencast machinery is extremely large and impressive; where operations have finished, only a few places can offer the space to display some of the machinery used (**106.1-2, 114.1-3, 121.2, 309**). An alternative is to save only part of the machine and explain its operation with models (**15.1-2**). Guided tours are also offered of traditional mining village where tourists can meet the residents (**248, 260, 342**). The most unusual tours offered to tourists are by mine locomotive to view fireflies at night in Taiwan (**344.1**).

### Mining museums

Coal-mining communities have frequently wished to preserve a memory of their earlier work, especially as coal was being displaced as the major form of energy. One of the earliest coal-mining museums in Europe, and probably in the world, is the 1907 Musée de la Mine in Le-Molay Littry (**91**), although the underground of the great copper mine at Falun in Sweden was available for visits from the time of the Napoleonic wars (1803-15).

Coal-mining museums occupy a surprising variety of buildings, but effectively fall into two main groups: those occupying former colliery sites and those in either purpose-built structures or other buildings converted for use by the museum. This section will not consider every example, but will look at just some outstanding examples, with German museum buildings dealt with in the introduction to Germany.

#### *Architecture*

It is said by coal miners that the first shovel-full of coal taken out of a mine is the first shovel-full towards its closure. So generally colliery buildings were not constructed to be permanent and this was used, for example, as an excuse for not providing pithead baths (Orwell 1937, p. 33). At most collieries, the one building on which care was lavished was the winding-engine house (**129.4, 255.1-2, 302.1, 338.2, 365.1, 382.1-2, 386.1, 395.1, 404.1**). This was usually of stone, rather than wood or corrugated iron, was tiled inside and kept spotlessly clean by the winding enginemen, who were the only ones allowed inside.

The modernist headstocks and associated building of Clipstone Colliery were very unusual in being designed by professional architects (**366**) and the winding-engine house of Snowdon Colliery was preserved as a good example of modernist architecture in an industrial context (**396**), although this was the norm in Germany. The 1923 concrete winding unit of the O'gara Mine Number 12 (**425**) is reminiscent of modernist design. The Nederlands Mijnmuseum is housed in an elegant square red-brick building beside the winding-engine house (**280.1**), the 1918 red-brick buildings of the Szyb Wilson were architect-designed (**305**), those of Národní památkový ústav were designed in 1915 by the professor of urbanism (**70**) and the Site du Grand-Hornu is renowned for its superb neoclassical architecture (**37.1-2**). The concrete and red-brick buildings in the neoregional style of Le Centre Denis Papin were saved from demolition by their quality (**82**) and the wooden derrick of the Museo de la Mina de Arnao is considered to be a masterpiece of industrial architecture (**324**). Some buildings have become landmarks of the city where they are located, such as the administrative buildings of the colliery companies now used by the Pernik – Podzemen Minen Muzei (**45**) and the National Museum of Unity (**293**), as well as the headframe of the Schacht Gerdt (**157**). In Japan the Sasebo-shi Sechibaru Tankō Shiryōkan occupies an unusual building constructed in 1912 in the western-European style (**266.1**), while in China the old housing for miners and other industrial workers covers three very different styles (**58**).

A variety of non-colliery buildings have been utilised to house mining museums. The Muzium Chimney is located in a colonial-style building (**278.1**) and the museum in Aldenhoven is housed in a former Capuchin monastery (**100**), the Bergbau- und Heimatmuseum in another monastery (**102**), the Heimat- und Braunkohlemuseum, in a seventeenth-century parsonage (**119**), and the Hessisches Braunkohle Berbaumuseum Borken occupies a timber-framed building (**121.1**). The most impressive are probably the Etnografski muzej Istre, which is housed in a former castle dating back to 983 (**63**); the Narodni muzej Labin in a baroque palace (**65**); the Központi Bányászati in the baroque Esterhazy Palace (**235.1**); the neogothic building of the Museo de la Siderurgia y Minería de Castilla y León (**328.1-2**); and the early-nineteenth-century neoclassical building with classical interiors of the mining school now used for the Sankt-Peterburgskiy Gornyy Universitet (**311.1-2**). The Bidala Audyogik Tatha Praudyogikee Sangrahaalay occupies an impressive and elegant building designed in 1959 (**238.1**).

The 1930 purpose-built monolithic main building of the Deutsches Bergbau Museum is very typical of its era (**108.1**). Approaches to new buildings have varied in recent years. New buildings at the NCMME (**386**) between 1985 and 2010 reflected the changes that were taking place in colliery architecture in the UK over those 25 years, while the main building on the site of the Pozo Norte Mine is based on a coal-preparation plant, with the original headgear of the mine in the centre (**326**).

In other cases striking new buildings have been erected, even on colliery sites. The Cutter at Woodhorn is designed to look like a horizontal cutting machine (**404.2**) and the building housing the Arigna Mining Experience echoes sloping coal seams (**240.1**). The Svalbard Museum occupies a low dark building that stands out against the snow-covered background, but does not obstruct the view (**295.1**). The new building on the Museu de les Mines de Cercs colliery site is faced with a reflective surface (**332.3**), while the building of the Muzej Rudarstva i Željeznice is designed to look like a miner's lamp (**41**).

### *Art*

Many museums have art collections, including audio-visual productions, coins, drawings, engraved glass, engravings, films, furniture, installations, jewellery, paintings, photographs, posters, prints, sculpture, tapestries and decorative items with representations of mines, such as clocks, crockery and furniture. The largest museums have comprehensive collections covering all, or almost all, these facets (**78, 108, 311, 386, 388**). The Nakamura Museum, with ceramics, glasswork, inkstone, paintings and sculpture, is presented as 'the museum of artwork from the period of coal mining' (**273**).

Illustrations in the form of drawings, engravings, paintings and sculptures showing the hard life and work of miners are found in Belgium (**33**), Germany (**98**), Japan (**276**), Slovenia (**316**) and Ukraine (**349.2**). The Centro Italiano della Cultura del Carbone has a comprehensive archive of historic photographs, including images taken in 1950 of women working in the screens (**244.2**), while the Archivio Storico Minerario has nineteenth-century images of armed guards (**243**). The Winding House has a photographic record of Welsh miners and their family life in the 1970s (**403**). Sculptures commemorating miners are to be found in Belgium (**29**), Germany (**113**), Spain (**326**) and the UK (**380.2**), and other sculptures are to be found in Belgium (**33**), Germany (**212**), Japan (**277**) and Russia (**307.2**).

Some museums have collections of paintings or other items made by miners, such as in Hungary (**235**), Poland (**302**), Russia (**307, 313**) and Ukraine (**349**). The works of the Pitmen Painters are now very valuable, and two museums have large collections of their work (**386, 404**). Collieries in Belgium (**25**) and Germany (**130, 160, 171, 186, 188, 215**) have been used for art installations based on mining equipment and sites.

The most famous works of literature about coal mining and miners are those by Takahashi Kiichirō (**276**) and Zola's novel *Germinal* (**33**), which was used as the basis for a film in 1993. Some of the props used were borrowed from the Musée de la Mine d'Harchies (**33**), while displays at the Musée de la Mine de Blanzy (**87**) and the Besucherbergwerk Rischbachstollen (**107**) are based on the descriptions in *Germinal*. The novel *Subterra* by Baldomero Lillo about the lives of the miners of Lota was made into a film in 2003 (**57**).

Particularly valuable are the historical items, including jewellery made from jet on display in China (**60**) and Russia (**311**), and the 1751 *Selmecbánya* table decoration in the form of the surface of a mine (**235.2**). In some cases there are also more esoteric items like decorated stoves (**89, 98, 110, 120, 252.2, 363, 426**) and choral performances, which are a feature of the Cape Breton Miners' Museum (**48**) and the Parque Minero (**333**).

***Fossils***

Coal mines have been an important source of information about extinct animals and plants. The Nova Scotia coalfield is so famous for Carboniferous fossils that a discovery centre is dedicated just to those fossils (**55**). Many museums have displays about the formation of coal in the Carboniferous period. A few have full-size reconstructions of the conditions at that time (**99.4, 226, 241**) or display fossilised remains from that period (**252.1, 387**), while fossilised freshwater mussels from Germany (**131**) show that the coal seams there came from inland water. There are also displays of the range of fossils found during mining (**62, 250**) or a model of the first dinosaur found in Spain (**331**). An internationally important collection of 30 complete igaunodons was found in a Belgian mine (**33**). Some German and Japanese museums have outdoor wooded areas with plant species from which coal developed, such as dawn redwood trees (**120, 263, 277**).

Most important are where new species were identified from specimens found in coal mines. The *Pantelosaurus saxonicus* is known only from parts of six individuals found in the Königin-Carola-Schacht (**99.2**) and the *Keraterpeton galvani* was identified at Jarrow Pit (**241.2**). The only evidence for the presence of pantodonts in Europe comes from Gruve 7 in Svalbard (**295.3**).

**The future**

Underground coal mining has now essentially ceased in western Europe and Japan, but still continues in large measure in Australia, China, eastern Europe, India, Mongolia and Russia. It had been calculated that by 2020 renewable energy would be cheaper than new coal in many of the countries dealt with in this publication: America, Australia, Austria, Bulgaria, Croatia, Czechia, Denmark, France, Germany, Hungary, India, Ireland, Italy, Japan, Malaysia, Netherlands, Poland, Romania, Russia, Slovakia, Slovenia, South Africa, South Korea, Spain, Sweden, Turkey and the UK (Gray and Sundaresan 2020).

A number of museums have accepted that coal mining cannot have a secure long-term future in light of the 2016 Paris Agreement to control global warming, so are including displays on alternative energy. Sometimes this is to bring the story of fuel up to the present day, as in Australia (**1**) and in Denmark (**73**), while the Beckley Exhibition Coal Mine deals with clean-coal technology (**407**) and the Mae Moh Mine Museum stresses the importance of saving energy (**345**). ENERGETICON in Germany opened in 2014 to explain the need to move from fossil fuels and nuclear energy to renewable power (**109**). The Arigna Mining Experience demonstrates new renewable-energy systems (**240**), while the statutes of the Ecomusée du Bois-du-Luc include the objective of raising public awareness of sustainable development, and changed its name to reflect this aim (**23**). At present most tend to be relatively straightforward, concentrating especially on solar energy, wind power and heat-pump technology (**394**), without examining in too much detail drawbacks, such as the number of birds and bats that are killed every year by wind-turbine blades.

# Gazetteer

I worked on preparing this publication between 2012 and 2022, including while I was director of the NCMME (1986 to 2015). This gazetteer was completed in 2022. In a few cases, sites have been included that did not respond to requests for information and may have closed. Without visiting the country concerned, it has not been possible to ascertain whether this is the case.

**Sites covered in the survey**

Seven main types of sites have been included in this publication:

1. Archives dedicated to coal-mining records.
2. Coal-mining museums and heritage sites, whether situated on the sites of former collieries or in modern buildings.
3. Colliery sites that have been converted to other purposes where sufficient significant colliery buildings and/or machinery still survive.
4. Colliery sites that have been given statutory protection to ensure their long-term preservation.
5. Memorials to coal miners and to major coal-mining disasters, where these are prominent features of the landscape or where the disaster led to a significant change in legislation.
6. Opencast-mining sites that permit public access or are clearly visible from outside the site.
7. Sites that deal with important aspects of coal-mining heritage, such as miners' housing.

In general sites have been included only if they are devoted solely to coal mining. In countries where there is no such museum, such as Denmark or Mongolia, museums with important coal-mining collections have been included. So also have museums that have large sections devoted entirely to coal mining, such as the Museum of Science and Industry, Chicago (**421**), and the museum of the Sankt-Peterburgskiy Gornyy Universitet (**311**). Also included are those that have a simulated underground, such as the Technisches Museum (**21.2-3**) and the Tekniska Museet (**339.1**).

It is very likely that there are other coal-mining sites that have not been identified, but every endeavour has been made to produce as comprehensive a list as possible. Every site in the publication was contacted to request information additional to what was given on their website, to obtain photographs for publication and copyright approval to use those images. All assistance with additional information is given in the references after each individual entry. Where available, staff numbers and visitor figures are given either as an average over a number of years or those in 2019, the last full year of operation before figures were affected by closures caused by the covid-19 virus. All those who sent photographs readily agreed to the images being used in the publication free of charge. Where no photographs or acknowledgements are given in a particular entry, it can be assumed that that site did not respond to the request for information and photographs; information about these sites has been taken from their websites, from other local informants and any other information available on the internet. The references after each entry are in strict alphabetical order, not in order of importance. In each case references to websites are those operating in 2021. Other websites that were operating between 2012 and 2020 and were used in preparing the various entries, but are no longer in operation, have not been included in the references. Wherever possible, information on websites has been confirmed with local sources; for example, the information from a website for Saint Ivan (**45**) was confirmed by Dr Milanova.

There is a short oversight of the types of coal and the history of coal mining at the start of the section on each of the 48 countries listed. Wherever possible, those entries have been checked by the relevant expert on the history of coal mining in that country.

**Sources of information**

Information about sites was obtained from the following sources:

1. For international sites from the showcaves website. This gives details of locations, opening hours, contact details, a brief summary of the history of the site and website addresses. It should be consulted for further details for those places included in the showcaves' alphabetical index of these sites under 'mines'.
2. For sites in Europe from the ERIH website. This gives brief details and a photograph of each site, with a link to the website of each place.
3. By asking for information from the representative of the International Council of Museums (ICOM) in each country where ICOM has representation.
4. From google searches for every country in the world.

Once an initial list had been compiled, the information was shared with other museums and with informants from various countries to enable other sites to be identified. Where possible, dates of birth and death are given for people mentioned; details of people whose dates are given can usually be found on the internet.

**Arrangement of gazetteer**

The gazetteer is arranged alphabetically by country. Within each country, sites are listed alphabetically by their official name, with the English translation in brackets for those in non-English-speaking countries. Every site has been given a unique number. For cross-referencing between sites, this number is given in bold in brackets. Some countries have official routes; these have been entered under the name of the route concerned, with individual coal-mining sites on the route given in bold italics and each numbered separately.

Entries about each site are, wherever possible, based on either information available on the internet or sent by somebody personally associated with the site. References to Facebook pages are not included, unless the site concerned does not have a website. References to published works about each site have been included where information has been taken from those publications in addition to what is available on the internet, especially for those that do not have a dedicated website. In a small number of cases information has had to be taken from a publicity leaflet.

**Terminology and transcriptions**

Different terms are used for the same facilities, machinery or mining processes in different countries. To enable comparison of like with like, the terminology used is that employed in the UK; for example, in north America the terms tippler or mine tipple apply to what in the UK would be known as a screens building. Similarly in America pithead baths are called wash houses or wash shanties (**424.1**), while coal-preparation plants are known as breakers, and in much of Europe as washeries. The alternative terms used, for example, in America or Canada, can be found by accessing the websites for those places. Local terms developed in most coal-mining areas; for example, in the Yorkshire coalfield in the UK, the cage was often known as the chair and the goaf as the gob or goff. This is a specialised area which is not generally dealt with in this publication, except where it is relevant to the history (**26**).

Chinese names have been transcribed according to the Hànyǔ Pīnyīn system. The practice adopted with regard to tone-marking diacritics and macrons has been to use names exactly as they appear in the published sources used.

Names in the Japanese section have been transcribed using the modified Hepburn style (http://dept.sophia.ac.jp/monumenta/pdf/MN%20Style%20Sheet.pdf).

Korean names are written in the Hanguel script and have been transcribed using the Revised Romanised system (https://www.gov.uk/government/uploads/system/uploads/attachment_data/file/501589/ROMANIZATION_SYSTEM_FOR_KOREAN-_MOCT_for_ROK.pdf).

Transcription of Vietnamese names can be complicated with regard to tone-marking diacritics and macrons. The practice adopted has been to use names exactly as they appear in the published sources used.

# Australia

Australia has been called 'the lucky country', amongst other things for its natural resources, including coal (Horne 1986, pp. 132, 139, 234-5), which is found in every state. High-quality bituminous coal is found mostly in New South Wales and Queensland, 30 *per cent* of which is used for electricity generation. In the past it was also exported to the other states, which mainly have lower-quality lignite (CMMCP 2016, fig. 2.1). In 2018 Australia produced 510 million tonnes, more than double that in the 1990s; 75 *per cent* of coking and thermal coal was exported (Cunningham, van Uffelen and Chambers 2019). Australia is the largest exporter of coking coal and the second largest of thermal coal (Hui 2021). In October 2020 relations deteriorated significantly between China and Australia, and China instituted an unofficial boycott of Australian coal. The Chinese Government granted approval to power plants to import coal without clearance restrictions, except for coal from Australia. Between January and August 2021 exports to China dropped by 98.6 *per cent* compared with the same period in 2020, from 68.8 million tonnes to 780,000 tonnes. As a result Australian coal became cheaper than similar grades of coal from South Africa, so India doubled its purchases of coking and thermal coal to about 2 million tonnes (*ibid.*).

There is evidence for the use of coal by indigenous peoples (**1, 5, 14**) before the 1788 British settlement, after which coal was first identified by shipwreck survivors at Coalcliff south of Sydney in 1797. The first exports were to South Africa in 1800 (Murray 2011, p. 210) and 150 tonnes to India in 1801. Eventually there were more than 1,900 mines in the three main coalfields of New South Wales (**17**). It was so important that in western Australia an award was offered for the identification of coal (**8**), although Australia did not have a national geological survey office until 1910 (Ranestad and Sharp 2020, p. 5).

Initially mining was carried out by convicts in appalling conditions (Murray 2011, p. 210), with the worst being in Tasmania (**7**). Mining from 66 mines began in the Illawarra and Greta coalfields of eastern New South Wales in 1849 and 1861 respectively, exporting to America, China and Japan (Andrews 2011; Delaney 1998). These mines were worked by free labour, mostly from the UK, who brought a tradition of unionism and militant action to improve the arduous and dangerous working conditions (Murray 2011; Murray 2013). Tonnages in Australian mines reached a peak during the 1920s, then declined during the Depression (1929-39). Demand for coal increased during WWII, when Australia was virtually isolated, but declined again afterwards in competition with oil. The first fully mechanised retreat longwall system (**382**) in Australia was not installed until 1965 at Kamira Colliery (**18**). Many deep mines have now closed, with Austar Colliery, the last underground mine in the Cessnock coalfield, shutting in 2021 after 105 years of operation. Nevertheless there are still 17 underground mines operating in New South Wales and 13 in Queensland. Most coal now comes from massive opencast sites; for example, the Mount Pleasant deep mine worked from 1861 until the 1930s (**2**) and the same eight seams, with reserves of 474 million tonnes, have been worked by opencasting since 2017. Electricity generation from coal accounts for 29 *per cent* of Australia's greenhouse-gas emissions. When its exported coal is taken into account, Australia is one of the three largest polluters in the world, along with China and India.

### 1. Australian Coal Mining Museum

The Australian Coal Mining Museum is located at the Blackwater International Coal Centre in what is currently Australia's largest coal-mining region, the Bowen Basin (**3, 9**) in Queensland, in the north-east of Australia. The indigenous Birra Gubba people in the northern Bowen Basin were using coal (personal communication, Ray Christison ((**5**)), 2021). The museum tells the story of coal mining in the area, covering how coal deposits were originally laid down, the early settlement of the area and the settlers themselves, different methods of coal mining, including opencasting, and the various uses of coal. The story is brought up to the present with displays on the future of coal mining and clean-coal technology. There are miniature models of mining equipment and simulator rides of a coal train, dragline and tugboat. The museum appears to be funded through a large corporate mining company.

The centre also offers guided tours lasting three hours once a week of the adjacent mine, covering the whole process of opencasting, from stripping of the surface through excavation and then processing and transport. The dragline and dump trucks can be seen operating.

*Reference*

http://bicc.com.au/coal-mine-tours-blackwater/

### 2. Balgownie Heritage School Museum (formerly Balgownie Primary School Museum)

Balgownie, in eastern New South Wales, eastern Australia, developed with the opening by Patrick Lahiff (died 1897) and Benjamin Fawcett (1817-90) in 1861 of the Mount Pleasant Coal Mine, 5.6 kilometres north-west of Wollongong. In the first years after opening, the colliery employed 35 labourers and 55 miners who extracted 246 tonnes of coal. The 1890s depression led to reduced production, but there

**Figure 2.1. Exterior of Balgownie Heritage School Museum (© Balgownie Heritage School Museum).**

**Figure 2.2. Model of the working area at a coalface, as it would have been in 1860 (© Balgownie Heritage School Museum).**

was a recovery in 1905. The colliery finally shut completely during the Great Depression (1929-39).

The Balgownie Primary School Museum opened in 1989 to preserve the history of Balgownie, including the coal-mining industry. The three-room building has heritage listing. The museum, which is run by volunteers, is open once a month or by appointment. Displays include a model of the working area at a coalface as it would have been in 1860 and a photographic display of the Mount Pleasant Coal Mine. There is equipment used at the mine, such as a display of blacksmith's tools and a cabinet of safety lamps. These include one used by the mine manager, Thomas Cook, in the 1902 rescue after the disaster at the Mount Kembla Mine (**13**), donated in 2003 by his granddaughter, Mrs Hilary Elliott.

*References*

http://www.wollongongheritageandstories.com/museums/balgownie-heritage-school-museum

personal communication, Elizabeth Woods, museum co-ordinator, 2021

### 3. Blair Athol Coal Mine

Coal was first discovered in 1864 at Blair Athol, 22 kilometres north of Clermont in Queensland in the north-east of Australia. Coal in the Bowen Basin (**1, 9**) comprises 83 *per cent* of Queensland's coal production. The mine has coal reserves amounting to 195 million tonnes of thermal coal, believed to be one of the largest coal reserves in the world. Blair Athol Coal Mine Tours organise trips that leave Clermont twice a week for a 180-minute exploration of the opencast site, guided by local experts.

*References*

https://www.askroz.com.au/blog/blair-athol-coal-mine-mount-victoria/

showcaves

### 4. Bulli Miners' Cottage

The Bulli Miners' Cottage is located on the Princes Highway at Bulli on the eastern coast of New South Wales in eastern Australia. The cottage was constructed by a prominent local identity, Henry Strange Fry (1829-1907). This was in order to rent it to miners working at the nearby Bulli Colliery, where the first major Australian coal-mining disaster occurred in 1887; there were 81 deaths when a lamp ignited methane gas. The cottage is believed to date to the period 1871/4, although some reports suggest it could be as early as 1850/6. The property was purchased in 1874 for £56 by Frederick Barlogio, the first miner to own the cottage.

The single-storey cottage is constructed of rough-hewn pit-sawn timber planks and has been deemed to be 'a very rare substantially intact survivor of a building style common in Bulli in the second half of the nineteenth century'. It is now roofed in traditional-Australian style with corrugated iron, although it appears originally to have had an ironbark-timber shingle roof. There is a traditional

**Figure 4.1. Bulli Miners' Cottage; date not recorded (© Wollongong City Libraries).**

**Figure 4.2. Bulli Miners' Cottage recorded in 1995 (© Wollongong City Libraries).**

Australian verandah at the front of the two-roomed cottage, with a kitchen and laundry at the rear, a separate outside privy and a reconstructed timber-picket fence.

The cottage, which is protected by a permanent conservation order under the Heritage Act of New South Wales, was acquired by Wollongong City Council in 1990 and was occupied by miners or their descendants after that. The cottage was restored in detail and converted to a museum with funding from the New South Wales state government. It was previously run by a management committee with representatives of the community, the council, the mining industry and the mining unions. It is, however, no longer open to the public and the collections have been dispersed, although the exterior is still of interest.

Plans are now underway to adapt the cottage for use by an artist-in-residence and a gallery space. The New South Wales Roads and Transport Department require the improvement of the driveway to accommodate increased traffic. It has been agreed that landscaping of the area around the cottage and the improved surfacing of the driveway should limit the use of concrete and hardstanding as much as possible. Currently costings are being obtained for the design phase and for funding from the city council.

*References*

http://coalfaceexperience.com.au/bulli-colliery-disaster-1887

https://apps.environment.nsw.gov.au/dpcheritageapp/ViewHeritageItemDetails.aspx?ID=2700068

https://www.illawarra-heritage-trail.com.au/miners-cottage/

personal communication, Jenny Mcconchie, local studies team leader, Wollongong City Libraries, 2021

## 5. City of Greater Lithgow State Mining Museum

The City of Greater Lithgow Mining Museum Incorporated was established in 1990 to develop a mining museum at the site of the former Lithgow State Coal Mine, in central New South Wales in eastern Australia, 155 kilometres west of Sydney. The organisation, which is a community-based, not-for-profit organisation operated by volunteers, specialises in presenting the history of the New South Wales western coalfield. This covers an area stretching from the Blue Mountains to the western slopes of the Great Dividing Range.

The society owns the Lithgow State Coal Mine site and a significant collection of mining and industrial artefacts. Before white settlement the indigenous Wiradjuri people were collecting outcropping coal from the Wolgan Valley,

**Figure 5.1. Former State Mine office (© Ray Christison).**

one of the deep canyons on the western side of the sandstone Blue Mountains. The mine operated between 1916 and 1964, with coal being worked by pillar-and-stall, with stalls, or bords, between 5 metres and 8 metres wide and 2.3 metres high, and pillars 44 metres by 44 metres. The underground is flooded so cannot be opened to the public.

An open-air museum has been created by using the former Lithgow State Mine branch-line railway to link the mine site to other important industrial sites, including the former Lithgow blast furnace and its related railway lines. The museum has partnered with local environmental organisations to carry out remediation, bushland regeneration and protection of nearby wilderness sites of World-Heritage status. The museum has also developed a programme of resident artisans who occupy studio spaces on the site.

Children are encouraged to touch and feel, and interactive experiences are provided for visitors, such as a mine-transport ride and blacksmithing demonstrations. Historical resources are available on-line for teachers and there is a display of trade-union miners' banners. The 'fire in the mine' interpretation provides a multimedia experience outlining the dangers of working in the region's collieries. The museum maintains a comprehensive and growing collection of images and also manages an active publication programme, focusing on the history of coal mining in the New South Wales western coalfield.

*References*

http://www.statemine.org.au/

personal communications, Ray Christison, president, 2012, 2013, 2021

showcaves

### 6. Coal Creek Heritage Village

The Coal Creek Heritage Village is located in Korumburra in the south-eastern Australian state of Victoria, to the south-east of Melbourne. It is an outdoor museum portraying the early development of the South Gippsland area (**15, 19**) between 1870 and 1920 and life in a coal-mining town during that period. The village is centred on the original coal-mine workings and the headframe, known as the poppet head. Since being set up in 1974, a total of 63 original buildings from the area have been restored and re-erected at the village, which is set in 16 hectares of bushland. The timber and dairy-farming industries are illustrated as well as coal mining. This is a

**Figure 5.2. Pithead baths and workshops (© Ray Christison).**

working village with volunteer interpreters and traders; for example, a practising blacksmith. The village is run by a not-for-profit organisation under the auspices of the South Gippsland Shire Council.

*Reference*

https://coalcreekcommunityparkandmuseum.com/

### 7. Coal Mines Historic Site

The Port Arthur penal settlement on the Tasman peninsula in Tasmania, in the extreme south-east of Australia, was established in the 1830s for some 1,100 of the most recalcitrant of Australia's convicts. The coal mines were opened three years later as Tasmania's first operational mines and produced most of the coal used in Tasmania. The mines were worked between 1837 and 1878 as a place of punishment for convicts from Port Arthur who had committed serious offences while in the colony. Convicts who were skilled miners worked at the coalface, producing some 3 tonnes of coal a day, each with three convict labourers to move the coal. The coal mines were unprofitable and in 1848 were leased to private operators, until finally closing in 1901. The mines began as adits into a seam 2 metres thick, but the coal was later worked from a main shaft sunk to depth of 100 metres.

The underground is no longer accessible. The ruins still remain of the houses of the superintendent and of the jailers and their families, military barracks for soldiers to guard the prisoners, the two-storey block in which the convicts were housed, the offices and the solitary-punishment cells. The Coal Mines Historic Site is one of 11 sites making up the Australian Convict Sites World Heritage listing.

*Reference*

https://portarthur.org.au/

### 8. Collie Coalfields Museum

Coal was so important to the economy that, after the first white settlement at Swan River in 1829, a free grant of land was offered by the government to the first person to find coal in Western Australia. Coal was discovered in the Collie area in the early 1880s, with the coalfield being developed in the late 1890s. A local coal miner, Stan Cull, set up the Collie Coalfields Museum, which is located in offices of the Roads Board in use between 1900 and 1930.

Coal mining in the area was so successful that a railway to move the coal was constructed in 1895. The Collie Steam Locomotive Museum now depicts how coal was taken from the mines to the railways for use in industry elsewhere. Also available for visitors is a 30-minute tour of the Tourist Coal Mine, which is a simulated underground coal mine, fitted out with horse-drawn skips, a crib cabin and a Jeffrey coal cutter, one of the first machines used in the coalfields (*cf.* **386**).

*Reference*

https://www.collierivervalley.com.au/local-listings/coalfields-museum-historical-research-centre/

**9. Collinsville Coalface Experience**

Collinsville is located in the Bowen Basin region (**1, 3**) of central Queensland, 1,245 kilometres north of Brisbane and 87 kilometres south-west of Bowen on the north-eastern coast of Australia. Coal was discovered in 1866 and large-scale mining operations began in 1919. The town was known at first by its indigenous name of *Moongunya* (place of coal). The Collinsville Coalface Experience is located on the top level of the Collinsville Scottville Workers Club. It is a mining museum dedicated to the seven men who lost their lives in the Collinsville Mine disaster on 13 October 1954, as well as 19 who had been killed underground previously. The mine ceased operations in 1961 and in 1990 the last two working pit ponies in Australia were retired from Collinsville Number 2 Mine. The museum, which is approximately 260 square metres in size, is funded by the Queensland State Government and opened in 2004. The committee is comprised of volunteers from the community. All eight staff involved in the day-to-day running of the museum, care and maintenance, and tours are volunteers, in particular retired miners from Collinsville.

The Coalface Experience begins with a video introducing the everyday life of the coal miners and their families and including interviews with locals to highlight the important role of the trade unions in the development of safety and fair working conditions. The evolution of coal mining in Collinsville is told through the use of interpretive panels, interactive audio-systems and touch-screen displays. There are models of a horse and miners working underground. The final display uses a pepper's ghost film to recreate the 1954 mine disaster and commemorate the men who died. Stage 2 of the experience focuses on the development of safety underground over time, with an interactive touch-screen display illustrating the various dangers found in mining. This includes a film showing how miners today ensure their personal safety underground.

*References*

http://coalfaceexperience.com.au/

personal communications, Jenn Honnery, multimedia designer, 2012, 2021

**Figure 9.1. Model of horse underground (© Jenn Honnery, Coalface Experience).**

## 10. Cornwall Coalminers Heritage Wall

Coal was discovered in 1843 in Cornwall when a farmhand saw his dog digging in an animal hole and scratching out coal. Coal production began in the Mount Nicholas Range in 1886 at Cornwall, 7 kilometres west of Saint Mary's in Tasmania in the extreme south-east of Australia. Coal was important in Tasmania with many industries, including the railways, requiring coal for energy. In 2006 the Cornwall Coalminers Heritage Wall was unveiled in Cornwall Park in the centre of town. It gives details of how early miners worked at hand-getting the coal, together with their pit ponies, and contains the names of some 1,200 men who pioneered the coal industry in the area. A plaque acknowledges those who perished in the process.

*References*

http://www.tassielocal.com.au/towns/stmarys.php

www.thegardensfamily.com/cemeteries/Cornwall/Coalminers.htm

## 11. Gunnedah Rural Museum

The Gunnedah Rural Museum is one of Australia's largest collections of early-agriculture and transport memorabilia, with displays of over 1,600 items in some 14,614 square metres of land. It is located in the inland New England region of New South Wales in eastern Australia. The displays include a working foundry and a collection of mining machinery.

*Reference*

https://mgnsw.org.au/organisations/gunnedah-rural-museum/

## 12. Kandos Museum (formerly Kandos Bicentennial Industrial Museum)

The Kandos Museum is located in central New South Wales in eastern Australia. The museum occupies the former Kandos Methodist church, a unique building constructed to an American plan, brought to the town by the manager of the New South Wales Cement, Lime and Coal Company in 1915. The museum occupies two 1,000-square-metre blocks of land. The Kandos Bicentennial Industrial Museum was established by Rylstone Shire Council, which amalgamated with other councils to become the Mid-Western Regional Council. The council operated the museum and then closed it in July 2013. It reopened as the Kandos Museum in October 2014 and is owned and operated by the Kandos Museum Incorporated Association, which has a seven-member management committee and 29 volunteers.

The cement works at Kandos closed in 2012 after almost a century of production. In its heyday it was the largest cement factory in the southern hemisphere and its superior-quality cement was used in the construction of the Sydney Harbour Bridge, which opened in 1932. Three coal mines supplied the works. Number 1 Colliery closed in 1941; Number 2 Colliery operated between 1940 and 1964, producing 2.36 million tonnes of coal; and Number 3 Colliery operated between 1959 and 2001.

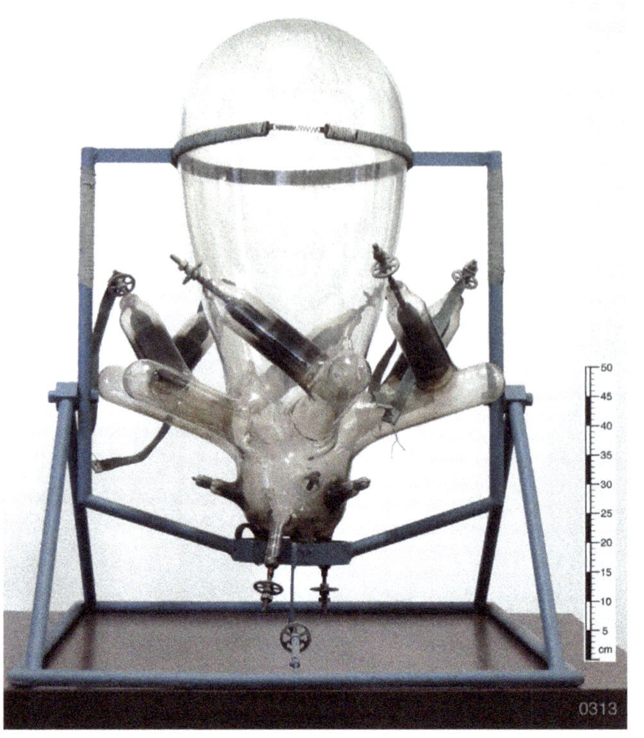


**Figure 12.1. Display of a mercury arc rectifier used in the coal mine (© Kandos Museum).**

Though not primarily a coal-mining museum, the museum focuses on Kandos' industrial heritage and has many objects from these three collieries and other local mines. Amongst items on display are a mercury arc rectifier used in the coal mine and an old cap worn underground by a miner. There is a significant collection of twentieth-century artefacts and displays, including the communications and transport history of the area; 75 *per cent* of the collections are on display. The museum has strong links with the local community and acts as a cultural facility.

*References*

https://www.kandosmuseum.org.au/

personal communications, Leanne Wicks, public programmes officer, 2012, 2013; Fiona MacDonald, secretary, 2021

## 13. Mount Kembla Heritage Centre

Mount Kembla is located in New South Wales on the east coast of Australia, 89 kilometres south of Sydney. Mount Kembla Mining Heritage, a not-for-profit organisation, was originally formed to conduct an annual festival to commemorate the anniversary of the Mount Kembla mine disaster on 31 July 1902. Torches used for lighting caused

**Figure 12.2. Photograph of entrance to Number 1 Colliery (© Kandos Museum).**

an explosion of gas and then of the coal dust thrown up into the air (**80**), which killed 96 men and boys. The dead miners left 33 widows and 120 children. A safety lamp used in the rescue is held by the Balgownie Heritage School Museum (**2**).

This continued to be the worst peacetime disaster in Australia for more than 100 years. The Mount Kembla Heritage Centre was set up to cover the events of the disaster in a display which developed from one that accompanied the festival. In 2020 ownership of the collection and assets were transferred to a new body, Wollongong Heritage Collections Incorporated, and the bulk of the collection is currently in storage. The collection consists of many coal-mining items of all sizes, from a clip to an egg-ended boiler weighing about 20 tonnes. There are also many social and family items from the village, spanning the period from about 1850 to the present day.

*Reference*

https://www.mtkembla.org.au/

### 14. Newcastle Museum

Newcastle is located in New South Wales on the east coast of Australia, 117 kilometres to the north of Sydney. The indigenous Awabakal people of this area cooked using outcrop coal, which is referred to in their legends. They named Lake Macquarie *Ni-kin-ba* (place of coal: http://www.coalandcommunity.com/coal-and-the-aboriginal-people.php).

The Newcastle Museum is a 10,000-square-metre heritage site in the heart of Newcastle's cultural precinct. Newcastle Museum has a team of 10 permanent staff and a number of casual visitor-services staff who carry out all the important functions needed to run the museum, from exhibition planning to building maintenance, education-programme delivery, event management and collections care. Newcastle Museum opened in the Honeysuckle Railway workshops in 2011, with displays dealing with Newcastle's past, present and future. The Moal gallery displays items relating to coal mining in the area, including a model of a dragline bucket used in contemporary opencast mining in the Hunter Valley. In the entrance there is a coal locomotive, the Buck.

*Reference*

http://www.newcastlemuseum.com.au/

### 15. Old Brown Coal Mine Museum

The Old Brown Coal Mine Museum is located in the small country town of Yallourn North in Gippsland (**6, 19**), Victoria, in south-eastern Australia. The Great Morwell Coal Mining Company took up a lease in the Latrobe Valley in 1887. At first the brown-coal mine was worked with shafts and tunnels. It later became an opencast mine,

**Figure 14.1. Newcastle Museum at night with coal locomotive, the Buck, visible (© Newcastle Museum).**

**Figure 14.2. Model of dragline bucket used in contemporary opencast mining in the Hunter Valley (© Newcastle Museum).**

but was found to be unprofitable in competition with black coal and so closed in 1899. The mine was reopened in 1916 to provide fuel for Victoria during a strike in the New South Wales coal mines.

The colliery was renamed Old Brown Coal Mine to distinguish it from the opencast, which was run by the Mines Department of Victoria. The State Electricity Commission of Victoria came into being in 1920 and the first sod was turned for the Yallourn project in 1921. The Brown Coal Mine opencast lay between the Latrobe River and the camp on the hill that would become known as Brown Coal Mine township. A post office with this name was opened in 1917 and the settlement was renamed Yallourn North in 1947.

The Old Brown Coal Mine Museum was opened in 2002 in the former Young Men's Christian Association building, built in 1947 as a recreation centre for the single men's camp in the Brown Coal Mine township. The museum is run by members of the Yallourn North and District Historical Society. The museum holds an extensive photographic collection dealing with the early days of the old town, the opencast mines and the Yallourn power stations. There are also scale models of opencast dredgers used to remove the overburden and the coal beneath it. Number 8 bucket-chain dredger from the Yallourn Open Cut Mine was built in the early 1960s. It stood 27.4 metres high, weighed about 2,165 tonnes, could dig nearly 24.3 metres above and 26.5 metres below its track level and could remove 1,720 tonnes of coal per hour. It was scrapped in 2004, but the driver's cabin was donated to the museum, together with the plant manuals and drawings. The cabin, now on display at the museum, has been repaired, reglazed and painted.

*References*

personal communication, Robyn Rosenboom, vice-president, 2021

www.browncoalminemuseum.websyte.com.au

**16. Richmond Vale Railway Museum**

Richmond Main Colliery in the Hunter Valley region of New South Wales in eastern Australia was purchased from a Melbourne syndicate in 1897 by the J. & A. Brown Company. In 1900 the Richmond Vale Coal Mine Railway Act was passed to allow the company to construct what became known as the Richmond Vale railway. Development of the Richmond Main Colliery began in 1910, with the mine reaching full production in 1918. The

**Figure 15.1. Model of bucket-wheel dredger and bulldozer (© Yallourn North & District Historical Society).**

**Figure 15.2. Driver's cabin from number 8 dredger (© Yallourn North & District Historical Society).**

colliery was developed by John Brown (1852-1930), who controlled everything. He owned his own brickworks, which produced the bricks from which all the buildings at the mine were constructed. The coal was placed onto railway waggons and pulled by steam engines, all owned by John Brown. The trains ran on John Brown's railway to Newcastle, where the coal was loaded onto John Brown's ships for export. One of the reasons for the long time taken to develop the mine was that John Brown wanted it to be a showpiece with everything on a grand scale, including a modern power station, which provided power for operating the colliery. The mine still holds the Australian record of 3,120 tonnes of coal mined by hand and raised to the surface in a single eight-hour shift.

The 3.4-kilometre Richmond Vale railway serviced four collieries and was the last commercially operated in Australia to use steam locomotives. Richmond Main Colliery closed in 1967, with some buildings being demolished soon afterwards. After the closure of the Richmond Main power station in 1976, Cessnock City Council acquired the abandoned colliery, together with 16 hectares surrounding the buildings. In 1979 the Richmond Vale Railway Museum was formed as a volunteer not-for-profit organisation with responsibility for the railway and the mining museum. Also known as the Richmond Vale Preservation Co-operative Society Limited, it has about 150 members, some 20 to 30 of whom take guided tours around the colliery. They also operate the railway, which now runs from the colliery to Pelaw Main Colliery.

The ground floor of the administration building contained the boardroom and offices for the officials and clerks, while the upper floor had accommodation for John Brown when he visited the mine. The building has now been restored and houses a small mining museum, including maps, photographs, safety equipment and tools. The first-aid room in the administration building still contains first-aid posters and medical equipment current at the time the mine was operating. Along the northern wall of this

**Figure 16.1. Richmond Vale Railway Museum, housed in the former administration building (© Tracey Hamilton, Richmond Vale Railway Museum).**

**Figure 16.2. Restored non-air-brakes wooden coal waggons (© Tracey Hamilton, Richmond Vale Railway Museum).**

building are the windows where the miners collected their pay. The signs requesting the miners to 'check money before moving away from the window' are still in place.

The power-station building also remains in place, although internal access is not allowed. The cooling towers and dam that served the power station can be seen beside the administration building. The boiler house was part of the electricity-generation plant that powered the mine site and part of the township of Kurri Kurri. The museum has a large collection of mining equipment, mainly from local collieries. In the grounds there is half of a fly wheel whose bearings were so good that, even with brakes, it took a week to stop spinning. To the west of the power station is the pit-top building of number 2 mine shaft. All the equipment, horses, miners, supplies and equipment went down this shaft and all the coal came up it.

*References*

personal communications, Tracey Hamilton, honorary secretary, 2012, 2013, 2021

www.richmondvalerailwaymuseum.org

### 17. Sir Edgeworth David Memorial Museum

The Sir Edgeworth David Memorial Museum is located in Abermain, in the centre of the South Maitland coalfield of the Hunter Valley, in New South Wales in eastern Australia. It is named after Sir Tannatt William Edgeworth David (1858-1934), Antarctic explorer and professor of geology at Sydney University, who was the first to identify the South Maitland coalfield. The museum began in 1971 and is located in the Abermain School of Arts Building. Whilst it is not specifically a mining museum, but rather a history museum, the museum contains more items relating to the coal-mining heritage than any other museum in the region.

The collections number over 4,000 items, representing the rural and coalfields heritage of the region. The Jim Comerford library, including rare and old books dating from the 1830s, attracts over 5,000 visitors annually. Jim Comerford (1913-2006) was a leader of the miners' union and wrote several books on mining history. The library contains some 400 photographs, 100 postcards and 400 books, with over a million pages covering the history of the region's coal mining, including the history of some 850 regional collieries. The WW I section records the role played by coal miners in the various Australian battalions fighting in Europe (**79**). In 2009 the museum was chosen to research and write a 500-page reference book, detailing all the 1,900 plus coal mines that have ever existed in New South Wales, commencing in the convict days of the early-nineteenth century.

*References*

http://monumentaustralia.org.au/themes/people/humanitarian/display/20756-jim-comerford

https://aumuseums.com/nsw/hunter/sir-edgeworth-david-memorial-museum

https://coalfieldshistory.org/the-museum.php

personal communications, Brian J. Andrews, secretary, Coalfields Heritage Group, 2012; Lexie Matthews, secretary, 2021

**Figure 17.1. Exterior of Sir Edgeworth David Memorial Museum (© Coalfields Heritage Group).**

## 18. Tarrawanna Travelling Road Miner and Pit Pony Tribute

The Miners' Tribute Committee has created a memorial of three structures in Tarrawanna just north of Wollongong, on the eastern coast of New South Wales in the Illawarra coalfield, 82 kilometres south of Sydney. Coal mining in the Illawarra began in 1848 at Mount Keira and the first coal exports to Sydney from the Illawarra left Wollongong harbour in 1849. In the Illawarra there have been more than 60 known sites of coal mining since the mid-nineteenth century. Over 2,000 pit ponies were employed in the Illawarra coalfield, usually with five working and one resting, and even that one would hang around the fans so it could smell the familiar odour of the tunnels (*cf.* **370**).

The monument of an archway and a wall is located on the historic pathway taken by the ponies and the miners on their way to work at Corrimal Colliery. The reserve was previously used as a paddock for the pit ponies and was the original access point to Corrimal Colliery from the horse paddocks. A plaque commemorates the place where the mining ponies used to rest. The interpretative tribute uses text and images to tell the story of the Corrimal Colliery pit ponies and miners to provide a context for the history and significance of their role in creating the coal-mining industry. The tribute shows where the mines were situated. The memorial also commemorates the achievement of the Illawarra in having the first colliery where the workers took over production, at South Clifton in 1972, and the start of the first Miners Women's Auxiliary at Scarborough in 1938 (**19**). There are also more-recent photographs, including an image from 1982 of a sit-in underground by 30 men against redundancies at Kamira Colliery, which lasted 16 days and represented the last spontaneous industrial action in the region.

*References*

https://www.abc.net.au/news/2012-02-22/tribute-to-illawarra-pit-ponies/3845520

www.abc.net.au/local/stories/2012/06/29/3535845.htm

## 19. State Coal Mine

The State Coal Mine at Wonthaggi is the only historic coal-mine experience with a genuine underground now operating in Australasia. Wonthaggi is located in Gippsland (**6, 15**) in Victoria in south-eastern Australia, 130 kilometres south-east of Melbourne. The settlement began as a tented encampment. Between opening in 1909 and closing in 1968, the mine produced some 17 million tonnes of black coal, mainly for the steam trains of the railways of the state of Victoria. The mine was unique in Australian mining history in being a large-scale government-run enterprise, as it was owned and operated by the Victorian State Government. While early technological advances here were considered to be world firsts, by the 1950s mining methods and working conditions were some of the worst

**Figure 19.1. Present underground of Wonthaggi State Coal Mine with overhead haulage running above the tubs, timber roof supports and installed lighting (© Parks Victoria).**

in Australia. The first women's auxiliary of the mining union was established during the five-month Wonthaggi coal strike in 1934.

During 1983 the flooded mine was pumped dry and renovated for the making of the film 'Strikebound', focusing on coal mining in South Gippsland and the importance and influence of the miners' union during the years of the Great Depression (1929-39). The film was the catalyst for the opening of part of the underground to the public in late 1983. The underground component was closed between 2004 and 2009 to convert the whole area into an underground gallery with access for the public. As mining safety standards are now much higher than they were in 1983, a considerable amount of work had to be undertaken to ensure that the underground complied with modern safety requirements. Most of the modern works have been hidden behind timbering to ensure the authenticity of the underground experience. The coal seams can still be seen undisturbed. Over $A4 million were spent on making the whole site safe for visitors, above and below ground.

The Wonthaggi State Coal Mine is operated and funded by Parks Victoria, the Victorian state-government organisation responsible for national parks in the state. The museum has seven sites in total, with the largest, the Eastern Area Mine, occupying about 50 hectares. It employs four members of staff, assisted by some 50 volunteers who give some 26,000 hours a year of support. Every Tuesday night the volunteers work underground retimbering and doing other works to ensure the stability of the mine.

In 2008 the museum attracted some 6,000 visitors and by 2014 this had risen to 27,000. In 2019 there were 56,000 visitors, 10,000 of whom went underground. The tour depicts life underground in a coal mine in the early years of the twentieth century, beginning with a 280-metre walk into the Eastern Area Mine. Working conditions, disasters and accidents, unionism, mining methods and life in a mining town are all covered. Working conditions underground are described to groups escorted by volunteer guides, many of whom are descendants of coal miners. The historic mine buildings are preserved above ground, containing mining machinery and photographic displays. Photographs include historic views of the coal-preparation plant (known as the brace building) and of the underground in operation, such as one of a slusher. One shows a miner with a carbide lamp on his helmet working with a pick alongside timber roof supports; it is captioned COPPER SLAVIN, referring to a miner called John Thomas Slavin and nicknamed Copper, because he had red hair. Also on display is a steam locomotive that used to be powered with Wonthaggi coal. The visitor centre provides information about the mine and mining life, including an audio-visual interpretation about the development of Wonthaggi and a film about the history of the Wonthaggi mines.

*References*

https://www.womenaustralia.info/biogs/AWE0638b.htm

personal communications, Braxton Laine, operations manager, 2014, 2021

showcaves

https://www.parks.vic.gov.au/places-to-see/parks/state-coal-mine-heritage-area/things-to-do/state-coal-mine

**Figure 19.2. Historic photograph of underground of Wonthaggi State Coal Mine showing a slusher bucket either loading onto a conveyor or taking material off the conveyor (© Parks Victoria).**

# Austria

Lignite and bituminous coal are found in small quantities in the east, north-west and south-east of Austria, a total reserve of 333 million tonnes (euracoal). Following its discovery in Fohnsdorf in the district of Murtal in Styria in the south-east of the country in 1670, mining of coal began in 1675, although its high-sulphur content meant that it was used mainly for the extraction of alum from coal ash. Mining began in the Tyrol in 1781, while coal was also discovered in Grünbach am Schneeberg in the east of Austria in 1823, and the mine became an important supplier of coal with the extension of the Schneeberg railway. Austria's Federal Geological Office was set up in 1849 (Ranestad and Sharp 2020, p. 5). The Bruderlade miners' organisation for mutual support developed in Austria during the nineteenth century, out of the German Knappschaftskassen (**108, 297**).

In 1882 work began on a new shaft at Fohnsdorf by the Őesterreichisch-Alpine Montangesellschaft, the largest coal entrepreneur in Austria, and by 1908 there were 2,605 employees. At that time Fohnsdorf was one of the most modern, mechanised and efficient collieries in the Austro-Hungarian empire (Kopp 2015). Opencast mining became important by 1900 and the first longwall mining began in Silesia in 1903. Also in 1903 the authorities in western Styria in southern Austria banned the *Firstenulmbau* method of mining, in which a hole was made into the coalface, only 4 metres wide but 7 metres deep, so that fresh air could not reach into the narrow workings, and between 5 and 15 metres high; the miner then took down the coal from the roof, which was supported only by the side walls. In 1902 there were six deaths and 43 serious accidents, a much higher rate than elsewhere in Austria, resulting from material falling from above, especially with *Firstenulmbau* working (Busson 1906, pp. 117-41).

In 1943 22 east Europeans and eight French prisoners-of-war, as well as 70 Germans, were killed in a methane explosion in Fohnsdorf. When Soviet troops occupied Austria in 1945, the German troops intended to blow up the infrastructure of the Grünbach, which by this stage had reached a depth of 800 metres. The miners managed to prevent this, although it took until 1950 for all the war damage to be repaired. By 1963 Grünbach was producing 95.6 *per cent* of anthracite in Austria. The Grünbach mine closed in 1965, because of increased operational costs, and the Fohnsdorf mine in 1978, because of the low price of coal (see Lukasczyk 2000 for the history of coal mining in twentieth-century Austria).

Coal mining was never a very important industry in Austria, amounting to 1.152 million tonnes in 2003 and there are only a couple of museums dealing with coal mining. Between 1964 and 2007 the proprietors of the Zum Bergmann inn in Grünbach collected mining implements and their daughter now provides mining-related entertainment with an actor once a month (https://www.wieneralpen.at/en/infrastructure/a-station-gruenbach-am-schneeberg; www.bergbaumuseum-gruenbach.at). In 2020 Austria became only the second country in Europe (after Belgium) to phase out coal-fired electricity supplies with the closure of the Mellach district-heating scheme (https://energycentral.com/news/austria-phases-out-coal-closure-verbunds-246-mw-mellach-plant).

### 20. Bergbaumuseum Fohnsdorf (Fohnsdorf Mining Museum)

The Bergbaumuseum Fohnsdorf, which occupies an area of 5,000 square metres in Fohnsdorf, opened in 1983 and belongs to the Museumsverein Fohnsdorf, a not-for-profit association. It attracts some 2,000 visitors *per annum* and is run by a director, with eight honorary members of staff. The site is leased by the museum, which is supported by the community of Fohnsdorf with about €2,500 a year. The museum charges an admission fee and organises concerts and banquets to raise funds.

The Bergbaumuseum Fohnsdorf is the only coal-mining museum in Austria. The museum has an extremely large 3,600-horsepower steam-winding engine, with a trade-union miners' banner displayed above it, and an impressive 46-metre-high winding tower, also shown in an historic photograph on display of Fohnsdorf Colliery in operation. The museum owns a 1952 locomotive engine from Greenwood and Batley, engineering manufacturers in Leeds, UK. This has two axles and weighs 4.5 tonnes. The exhibition, which covers some 1,000 square metres, deals with the working conditions of the miners, the history of mining from 1670 to 1978, the social history and traditions of the miners, machinery, technical equipment, tools, mining lamps from different periods, plans of mines and the emergency service in action. There is a model of what is believed to have once been the deepest brown-coal mine in the world, whose deepest gallery was 1,130 metres below the surface. The display roadway is 160 metres long. Guided tours last between one and two hours, depending on the nature of the group. The museum is open to the public from the beginning of May until the end of October each year.

*References*

https://www.gracesguide.co.uk/Joseph_Emanuel_Fischer

http://www.bergbaumuseum-fohnsdorf.at/

**Figure 20.1. Headgear of the Bergbaumuseum Fohnsdorf (© Museumsverein Fohnsdorf).**

personal communications, Heinz Kopp, director, 2013, 2021; Maria Schäpers, scientific adviser, Deutsches Bergbau Museum, 2021

showcaves

## 21. Technisches Museum Wien (Vienna Museum of Technology)

The Technisches Museum Wien is situated in the eastern part of Austria in the capital, Vienna. The museum is housed in an impressive art-nouveau building and is one of the major museums in Austria, having about 200 staff and some 300,000 visitors each year. It has eight main areas of display, covering approximately 22,000 square metres. The museum deals with heavy industry, transport and energy, with displays outlining the history of coal mining, including the role of British inventors such as Sir Humphry Davy (**355**).

In the early-eighteenth century the imperial gold and silver mines were threatened by excess water underground. The engineer, Joseph Emanuel Fischer von Erlach (1693-1742), was sent to England. He and the English engineer, Isaac Potter (1690-1735), built the first atmospheric steam engine for the monarchy in 1721 at the *Königsberg* (German) or *Újbánya* (Hungarian; now Nová Baňa in modern-day Slovakia) gold mine, near *Schemnitz* (German) or *Selmecbánya* (Hungarian; now Banská Štiavnica), to drain water from the mountain. A second was installed in 1722 in the garden of the Palais Schwarzenberg (see also **235**). This saved the Habsburg state from bankruptcy, while also making the technology available to the coal-mining industry. The Technisches Museum Wien has a model at the scale of 1 to 28.8 of a Newcomen atmospheric steam engine (**360, 371**).

There is a simulated underground of a coal mine in the basement of the museum. This is a replica of a genuine underground for the mining of both black and brown coal. The whole mine covers an area of approximately 300 square metres, built on two floors. The total height is about 5 metres, so the actual height in the galleries varies between 1.8 metres and 2.2 metres. Taking the short cut, the loop road is about 100 metres long, but 140 metres when including the newer parts of the mine with extensions made in the 1950s and 1990s. The mine is based on a coal mine in Karviná, Czechia, at a scale of about 1:300, so 10 metres in the gallery (the average distance between two objects of interest) represent a distance of 3 kilometres in the original mine.

**Figure 20.2. Roadway cutter-loader, alpine miner, with boom-type cutting head for cutting an arched roadway profile; gathering arms collected cuttings onto a central chain and then onto conveyors (© Museumsverein Fohnsdorf).**

**Figure 21.1. Scale model of Newcomen atmospheric steam engine (© Technisches Museum Wien).**

Tours of the underground are undertaken, at least once a day, with a guide to explain the working environment. The tour takes about 45 minutes and there are currently 12 trained guides. An early scene has a model of a miner operating a compressed-air boring machine while using wooden sprags to steady it and push it forward. On display there are a percussion drill, old pneumatic shearer, filling station, electric rotary drill, pneumatic hammer, sloping plane, underground locomotive with provision for the driver to stand not sit, massive electric disc shearer from the 1980s, part of a face-heading machine and a room with miners' tools. A roadways junction includes a direct haulage in a safety roadway. Some of the machinery is still able to be operated and is turned on during a guided tour, to give an impression of the typical noise in a mine.

### *References*

http://www.technischesmuseum.at

personal communications, Barbara Hafok, head of public relations, 2013; Bernhard Schmidt, head of energy and mining, 2013, 2021

**Figure 21.2. Direct haulage in a safety gate with a junction to the right (© Technisches Museum Wien).**

**Figure 21.3. Ranging-drum shearer with extended arm to allow the cutting of a gate profile (© Technisches Museum Wien).**

# Belgium

Coal is found in the north of Belgium in north-east Flanders (including Limburg in Flanders in the east and the Kempen coalfield) and in Wallonia in the south (including the Borinage coalfield and Hainaut in the west). There are still reserves of 4,100 million tonnes of black coal (euracoal). Coal was found used for heating in a Roman hypocaust in Liège (Forti 2011, p. 199). It seems to have been used for something like 900 years in Wallonia, with a dedicated mining company being formed in 1685 (**23**). The Geological Survey of Belgium was established in 1896 (Ranestad and Sharp 2020, p. 5) and by 1900 Belgium was the fourth largest producer in the world. Women (and children) worked underground (Forti 2011, p. 200) until prohibited in 1919 (Romano and Papastefanaki 2020, p. 204).

Belgian mines were exploited during the occupation by the Germans in WWI; the occupiers had intended destroying the mining infrastructure in 1918, but this was mostly avoided by the Belgians. Production dropped steeply during the German occupation of WWII, and the mines suffered considerably from bombing (Forti 2016, pp. 266-7). In 1946 the mines were nationalised (Houp 2011, p. 242), but Belgium was short of colliers, so first used German prisoners-of-war, despite the Geneva Convention (Forti 2011, p. 201). In 1946 Belgium signed an agreement under which 50,000 workers would come from Italy in exchange for supplies of coal to Italy. They would supplement the Belgian workforce by working in the mines during what was known as *la bataille de charbon* (the fight for coal). Men came from Greece, Italy, Morocco, Portugal, Spain and Turkey (Delbroek 2011, p. 223); altogether 16 different nationalities worked in the Belgian mines (**28, 29, 31**). This practice ceased after the 1956 disaster at Le Bois du Cazier (**29**), which resulted in much stricter mining regulations in Europe. The decline in coal usage caused by the advent of oil was accentuated by the European Coal and Steel Community programme of the 1960s and 1970s and the last mine closed in 1992.

In 2012 Blegny-Mine (**22**), Bois-du-Luc (**23**), Le Bois du Cazier (**29**) and the Site du Grand Hornu (**37**) in Wallonia, between Aachen in Germany and the Nord-Pas-de-Calais in France, were recognised by UNESCO as a joint World Heritage Site. Together they represent the various stages of development of coal mining in Wallonia, and have each been developed in different ways, so that they complement one another. One feature in Belgium is that a number of colliery buildings survive in a derelict condition (**24, 26, 31, 36, 39, 40**), some in town centres (**30**). Not all of these are protected as designated monuments and taxes on brownfield sites encourage their owners to demolish them. In 2016 Belgium became the first country in Europe to phase out the use of coal completely (https://www.nsenergybusiness.com/features/coal-phase-out/#).

### 22. Blegny-Mine (Blegny-Mine)

Coal mining in eastern Belgium in the area of Liège was mainly to feed the important iron blast furnaces and for home heating. When it closed in 1980, the Argenteau-Trembleur Mine at Blegny, located between Liège and Maastricht, was the last remaining colliery in the region. In 1980 it was converted into a museum to represent coal exploitation in the area of Liège over a period of eight centuries. The museum is run by an association, which includes the tourism section of the Walloon Government, the province of Liège, the area's two communes and local associations. Approximately 40 *per cent* of the running costs come from the Walloon Government and the province of Liège and the museum raises the other 60 *per cent* from admission charges and trading activities. Visitor numbers are in the region of 140,000 to 150,000 a year. The museum employs the equivalent of about 40 full-time staff. The museum provides an audio-guide in English and German.

The most striking aspect of the site is the winding tower of the former Argenteau-Trembleur Mine and the two winding engines, one of 50 and the other of 660 horsepower. The first was restored first around 1882 and then in 1920, before being changed from steam to electricity in 1949. The second dates to about 1945, with restoration in 2020. Visitors descend underground in an authentic double-deck mine cage together with a guide. The underground tour illustrates coal mining in Liège after WWII. This includes demonstrations of working tools, such as compressed-air drills. The roof of the underground roadways is supported with arched girders. The displays include a centrifugal pump mounted on a tram and a chute loading into a minecar below. In Belgium the coal seams slope very steeply, so that miners had to work lying on their sides. After completing the first leg of the tour at the 30-metre level, visitors descend to the 60-metre level using a steep concrete stairway, which gives them a good view of the sloping coal seams.

Above ground, the museum is located in the 1849 colliery. It covers the history of industrial mining from the hand-working period some 200 years ago up to the automated extraction methods of the late-twentieth century. Displays and films deal with the local production of coal and visits to the screens building and washery are included. Blegny-Mine, the province of Liège and the Walloon Government are considering developing an energy-interpretation centre, to present coal in the global context of energy challenges.

**Figure 22.1. Headgear of Li Trembleu Mine with associated buildings (© Blegny-Mine).**

**Figure 22.2. Tubs about to enter the washery (© Blegny-Mine).**

**Figure 22.3. Underground roadway with arched girders and chute above loading into a minecar below (© Blegny-Mine).**

Adjacent to the site is the spoil heap, made up mainly of schist, on which works of art are displayed. A steel staircase has been provided to access the top of the spoil heap. A road train (a tractor with waggons) transports visitors around the local countryside with explanations provided by a sound system.

*References*

ERIH, *s.n.* Blegny

http://www.blegnymine.be/

personal communications, Jacques Crul, director, 2013; Bruno Guidolin, librarian and archivist, 2021

showcaves

www.sitesminiersmajeurs.be

### 23. Bois-du-Luc, Musée de la Mine et du Développement Durable (Bois-du-Luc, Museum of Mine and Substainable Development; formerly Ecomusée du Bois-du-Luc: Bois-du-Luc Ecomuseum)

Mining was undertaken in the La Louvière region of Wallonia by the Societé des Charbonnages du Bois-du-Luc between 1685 and 1973. The Bois-du-Luc, Musée de la Mine et du Développement Durable, is located at the Saint Emmanuel Pit, sunk in 1846. Saint Emmanuel was a model colliery, with an associated village of houses for the miners, church, community centre, hospital, park, retirement home, schools and shops. This was intended to allow the mining company to control the workforce. The close control of living and working conditions by the company led to strikes and threats of violence in 1893. To protect the mine, it was surrounded by a strong wall with large, thick, steel doors mounted in vertical grooves, known as guillotine doors, flanked by two round towers of medieval appearance.

The last mine in the area, Le Quesnoy at Trivières, closed in 1973. Local people, who were living in the Bois-du-Luc miners' village, lobbied for the site to be preserved. This was because of its outstanding architecture, as the elegant brick buildings on the site were designed in 1907 by the Brussels architect Charles-Emile Janlet (1839-1919). They have elaborate interiors in the *trompe l'œil* style, imitating marble and bronze, and furniture made in the colliery workshops. The Ecomusée du Bois-du-Luc opened in 1983, and was further developed in 1992 with funding from the EU. The buildings have all been conserved, together with the slender steel headgear. In 2016 the name was changed to Bois-du-Luc, Musée de la Mine et du Développement Durable. Since then the statutes of the museum have included the objective of raising public awareness of current and future issues related to sustainable development.

The museum is run by a not-for-profit association, with funding from the Wallonia-Brussels Community and the Walloon Region, the province of Hainaut (which has seconded a member of staff to act as director of the museum) and the city of La Louvière. The museum has 13 staff, including a curator and four guides. The archives have been deposited in the Sauvegarde des Archives Industrielles du Couchant de Mons.

The museum is open to the public throughout the year and attracts about 15,000 visitors annually. The main themes are the living and social conditions of the miners. The Castelain gallery is a reconstruction of the nearby Saint-Joseph Pit, showing typical coal-mining techniques and working conditions underground. The tour of the site, titled 'Between men and machines', explores a typical day in the life of a collier, with visits to the headgear and the director´s office, the colliery workshops with craftsmen at work and the interior of a typical worker's house from the early-twentieth century. The displays show mining equipment, such as lamps and hammers. The collections that are not on display include aspects of the mining company and social history, including clothing and musical instruments belonging to the brass band of the Bois-du-Luc mine. The Societé des Charbonnages du Bois-du-Luc encouraged the band, which still plays at ceremonies and festivals.

**Figure 23.1. Aerial view of the Ecomusée du Bois-du-Luc (© Bois-du-Luc, MMDD).**

**Figure 23.2. The guillotine doors (© Bois-du-Luc, MMDD).**

*References*

ERIH, *s.n.* La Louvière

personal communications, Daisy Vansteene, director, 2012; Karima Haoudy, curator, 2013; Vanessa Bebronne, scientific officer, 2021

showcaves

www.boisdulucmmdd.be

### 24. Charleroi Mines (Charleroi Mines)

Charleroi is located in the province of Hainaut. It is situated at the centre of a major coal-mining area, which had been worked for some 900 years, but really developed in the nineteenth century, giving the region the name of the black country or coal country. Development of the mines of the largest coalfield in Belgium was mainly carried out by the Société Générale, with an average annual production of 907,000 tonnes. Strikes were frequent in the Charleroi mines, including during the general strikes of 1936 and 1960/1. As in so much of Europe, gradually coal came to be replaced by oil and the last mine in the Charleroi coalfield closed in 1984. The remains of the mines in Charleroi still stand, but are derelict.

*Reference*

personal communication, Monique van den Berg, architect, 2013

### 25. C-Mine Expeditie (C-Mine Expedition)

C-Mine Expeditie is located in Genk between Antwerp and Liège in the province of Limburg. It is based on the site of the former Winterslag Mine and has been developed as a site to stimulate creativity. C-Mine Expeditie was opened in 2012 at an approximate cost for the restoration of €4,788,765. Activities at C-Mine Expeditie cover artistic creation and presentation, creative economy, creative recreation and education. The energy building has staff available to orientate visitors and houses the cultural centre, design centre, seminar rooms, visitors' reception and the entrance to the C-Mine expedition. C-Mine Expeditie attracts approximately 40,000 visitors a year and employs the equivalent of six full-time staff. There are also some 185 volunteers providing services each year, although this can increase to 210, depending on special organised events.

The original authentic environment of the surface of the mine has been preserved, so that respect for the industrial past is interwoven with contemporary artistic activity. This is not a museum or a simulated mine, but an interactive walk through the 1.5 kilometres of the old roadways of Winterslag Mine. There is also a new installation, providing access by 360 steps to the top of the 72-metre-high headframe. This is the highest surviving example in Belgium, with a wide view over the surrounding landscape. The design team sought a balance between preserving the authenticity of the spaces and the creation of an additional

**Figure 24.1. Derelict remains of Charleroi Coal Mine (© Monique van den Berg).**

**Figure 25.1. C-Mine at night (© C-Mine Expeditie – city of Genk).**

**Figure 25.2. Stairs to underground at C-Mine (© C-Mine Expeditie – city of Genk).**

experience to offer visitors a challenging and instructive experience. Stories and images are provided in the 'memory tubes', for which writers, photographers, film makers and sound-effects artists drew inspiration from the mining past.

Guided tours are offered for groups and for schools. Visitors are able to explore the site on their own using an iPod tour, 'In the footsteps of Cyril'. This is a game in which students follow a character, Cyril, on a day in the mine, using old photographs, and experience the sights, smells, sounds and spaces of the mine. The 90-minute tour for non-school groups is accompanied by a local guide.

*References*

http://www.c-mine-expeditie.be/

http://www.cosimo.be/onsmijnverleden/schacht-winterslag

personal communication, Hatice Deveci, administrative assistant, 2014

### 26. Eisden Kolenmijn (Eisden Coal Mine)

The Eisden Kolenmijn was sunk in 1906/7 in the municipality of Maasmechelen. As elsewhere in Flanders, the owners had to try to attract miners to this rural area, including workers from eastern Europe. So good housing was constructed, with curving roads, hedges, trees and a church dedicated to Saint Barbara (**96**). There was also a railway link to Genk and vans toured the local villages to transport workers to the mine. During WWII Russian prisoners-of-war were forced by the German occupiers to work in the mine, followed after the war by German prisoners-of-war. In the post-war period workers came from Morocco and Turkey. They were taught Flemish by language teachers brought from the capital, Brussels. Unfortunately they were taught 'classic' Dutch, or Flemish, not the local dialect. So they still had communication difficulties with the local miners, especially with the local technical terms (**29**).

Underground tours of the mine were given until shortly after it closed in 1987. The visitors were taken underground by train and then had to walk and then crawl some distrance through the roadways. After closure, the shafts were filled and the underground roadways flooded, so that the mine is no longer accessible. Some of the derelict mine buildings were saved from demolition and have been restored. Part of the mine site has become an outlet village and in 2013 the entrance to the Hoge Kempen national park was located here. The two large headframes still remain in a derelict state, although with statutory protection (see also **27**).

*References*

http://nl.wikipedia.org/wiki/Steenkoolmijn_van_Eisden

http://noviomagus.tripod.com/mines.htm

### 27. Koolmijn André Dumont Genk Waterschei (André Dumont Coal Mine Genk Waterschei)

The Carbonmines Waterschei was one of seven mines in the Kempen coalfield, in the Waterschei district of Genk. In 1965 there were about 50,000 miners working in the Limburg collieries, but the last mine in the Limburg closed in 1992. Coal was first mined at the Carbonmines Waterschei in 1924 by the Charbonnages André Dumont at depths of up to 1,040 metres. In 1949 the company

**Figure 26.1. Derelict headframes of the former Eisden Kolenmijn (© Monique van den Berg).**

**Figure 27.1. Main building of the mine, converted for special exhibitions (© Blegny-Mine).**

employed 6,834 miners. The mine was connected to the Eisden Kolenmijn (**26**) to ensure sufficient ventilation. Consideration was given to sinking a second shaft, but the 1958 coal crisis led to the abandonment of these plans. The mine closed in 1987.

The miners' association, Mijn-Verleden (My-Past), has converted the former warehouse of the mine into an exhibition area, known as Mijndepot (Mine Depot), giving an overview of the history and work of the Limburg mines. The displays include a simulated underground, a film of work on the surface, a large collection of historical photographs of the mine and a collection of machines, tools and other objects.

The surface of the mine has been converted into the Thor science park, with the main building of 22,000 square metres being used for special exhibitions. The buildings have statutory protection. The site is being developed as a high-quality concept in architecture, environmental quality, ownership and sustainability. In 2012 the main building was the venue for Manifesta 9, the itinerant European Biennial of Contemporary Art. For the first time the biennial had a display of historic paintings of coal mines and miners from Belgium, France and the UK, as well as a small number of historical artefacts.

*References*

https://inventaris.onroerenderfgoed.be/dibe/geheel/122122

https://nl.wikipedia.org/wiki/Steenkoolmijn_van_Waterschei

https://www.youtube.com/watch?v=LYOsBg4pCfY

http://www.mijndepot.be/

### 28. La Chaîne des Terrils (The Chain of Spoil Heaps)

Aiseau-Presles, about 6 kilometres east of Charleroi in the south-west of Belgium, is a small community, which is the headquarters of La Chaîne des Terrils. The organisation was founded in 1986 to conserve, interpret and protect spoil heaps across Wallonia. The chain runs from Bernissart on the French frontier near Valenciennes through the Borinage coalfield, Mons, La Louvière, Charleroi, Sambreville, Flemalle and Liège to Herstal near the borders with Germany and the Netherlands. In the Wasmes (Colfontaine, Hainaut Province) section of La Chaîne des Terrils, it is possible to see, in one view from left to right, spoil heap numbers 6 (Hornu and Wasmes Pits) and 18 (Grand Buisson), and the Marcasse (**31**) spoil heap.

The trail runs past 43 towns and villages, and 340 large spoil heaps from the period of the Industrial Revolution (**379**) or later, as well as numerous smaller tips. The trail incorporates the tracks that miners once followed from their homes to the collieries where they worked and the lines of mineral railways that once carried away the coal. The pit heaps have developed their own microclimates, and many are rich in animal and plant life. The biological significance, landscape qualities and recreational value of the spoil heaps add to their heritage value, which is safeguarded by regulating what activities are permitted on them and regular biological monitoring.

The trail provides evidence of miners of 16 nationalities who have worked in the mines along its route. It incorporates several conserved industrial-heritage sites (**22, 23, 35, 37**). La Chaîne des Terrils is similar to La Chaîne des Terrils (**81**) association in France.

**Figure 28.1. Spoil heap of Marcasse mine seen from the house where Vincent van Gogh lived while working in the Borinage (© National Coal Mining Museum for England).**

**Figure 28.2. Spoil heaps of the Saint Emmanuel Mine, Bois-de-Luc (© Bois-du-Luc, MMDD).**

*References*


ERIH, *s.n.* Aiseau-Presles

http://www.chainedesterrils.eu/


**29. Le Bois du Cazier (The Bois du Cazier)**

The Bois du Cazier museum, at Marcinelle, several kilometres south of Charleroi in Wallonia, occupies the site of the worst disaster in Belgian mining history. In 1956 there were some 30,000 colliers employed in the mines of the Charleroi region. On 8 August that year the mine caught fire, and smoke and gas spreading through the underground galleries resulted in the deaths of 262 men, with only 13 escaping the inferno. Of the dead, 95 were Belgian and 136 Italian (see also **31**), who had come to work on the Belgian coalfields under the 1946 agreement. The disaster resulted from lack of investment and the problems of communication underground without a good grasp of the local language (Forti 2011, pp. 201-2; **26**). The mine itself closed in 1967. The site was given statutory protection in 1990 and opened to the public in 2002. A bronze sphere, 'The belly of the Earth' by Italian artist Antonio Nocera (born 1949), symbolising the dangers of mining and of emigration, was donated in 2006 to mark the fiftieth anniversary of the disaster. There is a wall of remembrance on the site for the victims of the disaster and a ceremony is held annually on its anniversary. The museum belongs to the International Coalition of Sites of Conscience.

**Figure 29.1. Winding-engine building in line with the axis of the winding engine and headgear visible through windows (© National Coal Mining Museum for England).**

**Figure 29.2. One of the spoil heaps of Marcinelle, as viewed through circular window in the former power plant (© National Coal Mining Museum for England).**

The museum is the property of the region of Wallonia and is managed by a not-for-profit organisation. There are 35 employees and 19 volunteers, with the museum attracting some 39,000 visitors each year. The museum is set in 26 hectares of woodland, with three large landscaped spoil heaps, one of which has an open-air theatre, and a memorial walkway with trees from the countries of the 12 nationalities represented amongst the fatalities.

The aim of the museum is to preserve and record the memory of the immigration and of the miners and their working conditions, including maintaining an archive. The site preserves all the main buildings to be expected at a colliery, some of very fine brick construction. There are two steel headframes and winding gear, lamproom, pithead baths, including lockers for the miners' clothes and possessions, and a locomotive shed, together with a number of other administrative buildings and displays of mining machinery. The permanent exhibition of the industry museum includes displays about the 1956 disaster and the industries of the region that used coal, such as iron and steel making, glass manufacture and chemicals, with steam engines, a steel rolling mill and, in the colliery workshops, operational forge hammers brought from the Providence forge at Marchienne-au-Pont. The Glass Museum completes the coal/iron/glass trilogy which brought fame and prosperity to the Charleroi area.

*References*


ERIH, *s.n.* Marcinelle

http://www.leboisducazier.be/

personal communications, Jean-Louis Delaet, director, 2013; Alain Forti, curator, 2021

showcaves

**Figure 29.3. Air intake and winding shaft seen from the Museum of Glass (© National Coal Mining Museum for England).**

### 30. Le Charbonnage du Hasard de Cheratte (The Hasard de Cheratte Coal Mine)

The Hasard de Cheratte Coal Mine is located in Cheratte, part of the town of Visé in Wallonia. The first two shafts were sunk in 1850, to depths of 179 metres and 250 metres. It closed following an accident in 1877, when water broke into the pit and several miners were killed. The mine reopened in 1907, and by 1930 there were 1,500 people working in Cheratte and the mine was producing 907 tonnes every day from a depth of 420 metres. It was the main pit of the coal-mining company, Charbonnages du Hasard. Number 2 shaft was dismantled in 1947 when number 3 shaft was constructed. Number 4 shaft was used for ventilation only. The headquarters of the company in Micheroux were closed in 1974 and the mine itself shut in 1977. The associated town of Cheratte was built by the company in 1925, with 200 six-room houses and a 128-bedroom hostel for unmarried workers. When the mine closed a regional social-housing company bought the company town. The colliery site was purchased by a Flemish developer, who intended to demolish all buildings on the site. In view of strong local opposition, the site was given protection by decrees published in 1978, 1982 and 1992, thereby preventing demolition.

By 2013 the installations of numbers 1 and 3 shafts were in ruins, and the buildings of number 1 shaft were classified as a local-heritage site. The number 2 shaft had been destroyed, as well as several buildings around it. The reinforced-concrete buildings of number 4 shaft were well-preserved. In the early-twenty-first century, this building became well known as an urban ruin. In 2013 the site was sold by the developer and became public property. Conservation of the site is estimated at €2,070,000 and the demolition of unlisted buildings is planned.

*References*

https://en.wikipedia.org/wiki/Coal_mine_of_Hasard_de_Cheratte

http://tchorski.morkitu.org/5/cheratte-01.htm

http://www.forbidden-places.net/urban-exploration-hasard-collieries-cheratte#1

### 31. Marcasse

The site of the former mine of Marcasse is 3 kilometres south of the Site du Grand-Hornu (**37**) in Wallonia. At the end of 1878 the Dutch artist, Vincent van Gogh (1853-90), became a preacher in the Borinage for the small Protestant community at Petit-Wasmes, where he had a six-month training contract. He remained in the Borinage until 1880, where he lived with a worker's family in Cuesmes, near Mons, from which he had a good view of the spoil heap of Marcasse mine (**28.1**).

Figure 30.1. Number 1 shaft of Le Charbonnage du Hasard de Cheratte seen from the town (© Monique van den Berg).

Figure 30.2. Close up of the buildings of number 1 shaft (© Blegny-Mine).

**Figure 31.1. Memorial to the victims of the 1953 disaster (© National Coal Mining Museum for England).**

**Figure 31.2. Heapstead building where headgear was previously located (© National Coal Mining Museum for England).**

While living at Petit-Wasmes (Berckmans [2016], pp. 397-8), he went down the Marcasse mine to a depth of 700 metres, where he spent about six hours in order to experience the working conditions underground of members of his congregation. He wrote a detailed letter about this visit to his younger brother, Theo (1857-91). Other letters to Theo van Gogh described the living conditions of the miners and visits to the surface of various collieries.

Several of the Marcasse buildings still exist today and belong to a private owner, who is trying to restore them with private funding. At the entrance to the site there is a memorial to the 24 men, many of them Italian (see also **29**), who died in a disaster here on 13 January 1953. The engine house and the former heapstead building are still standing, although derelict. The course of the former mineral railway line on the site can also still be seen, together with the spoil heap.

Mons was designated as the Cultural Capital of Europe for 2015, with Vincent van Gogh as one of the key people of the event, so Marcasse was very much in the spotlight. The Van Gogh Museum in Amsterdam in the Netherlands launched Van Gogh Europe to create a Van Gogh route through Europe, with the Borinage included on the circuit.*

*References*

Berckmans [2016], pp. 116-22, 397-403

http://www.vangoghletters.eu, letter 151

https://www.vangoghroute.com/belgium/borinage/mine/

personal communications, Filip Depuydt (**37**), 2012; Monique van den Berg, architect, 2013

## 32. Mijnmuseum Beringen (Beringen Mining Museum; formerly Vlaams Mijnmuseum: Flemish Mining Museum)

The Mijnmuseum Beringen is located in the buildings of the former colliery of Beringen in Limburg. This coalfield was only developed in the early-twentieth century when the Belgian Government decided to promote coal mining in Flanders, a rural agricultural area, to meet the demand from steel producers for high-quality coal. The last colliery in the area closed in 1992. The pit buildings here were constructed about 1901, while a new area of housing was built at the same time for the incoming colliery workers. The estate became part of the museum complex, comprising the housing, pit buildings, spoil heaps and associated infrastructure of the mine.

**Figure 32.1. View of main building, headgear and (foreground) simulated arched girders and underground railway (© Jonathan Vaes).**

* I am grateful to Monsieur and Madame Barberio-Gravis, the owners of Marcasse, for access to the site, and to Filip Depuydt (**37**) for arranging access.

The museum is run by a not-for-profit association. It is funded by a combination of entry fees and subsidies from the city of Beringen and the province of Limburg. There are about 50 volunteers and the equivalent of two full-time staff. The museum has approximately 37,000 visitors a year, who can either follow a designated route with a museum guide or do a self-directed visit using an audio-guide, available in Dutch, English, French, German, Italian and Turkish.

The Mijnmuseum opened to the public in 1986. In 1993 the site was given statutory protection as an historic monument, so most of the former mine buildings, which are still in good condition, are preserved. The site gives a complete view of a typical coal mine of the region and is an example of a large industrial-heritage site. It covers all aspects of coal mining in Flanders, from mining engineering and working in the mine to the social life of the mining communities.

The museum reopened in 2012 after a major facelift. The exhibition covers working underground, the miners' social life in and around the mines and the continuing search for manpower, as well as miners' strikes and pit closures, all of which were part of the Flemish mining heritage. An authentic simulation in the basement of the museum gives visitors an idea what work underground was like.

The huge amount of coal that was mined in the region is emphasised by the size of the site, with four pithead buildings and two headframes. A model of a colliery, complete with sounds and illuminations, explains how the mine operated. Because the pit was sunk relatively late, there were good facilities for the miners, including each man having his own individual cubicle in the pithead baths. The colliers' trail leads from the pithead baths, via the lamproom, wages office and coal-preparation plant, to one of the headgears and then to the neighbouring miners' houses. Finally it is possible to view the whole site from the summit of the spoil heap, which is now grassed over. Adjacent to the museum are the miners' houses and the sociocultural infrastructure built by the colliery.

*References*

ERIH, *s.n.* Beringen

personal communications, Johanna Rediers, communications officer, 2013; Filip Delarbre, curator, 2021

showcaves

www.mijnmuseum.be

### 33. Musée de la Mine d'Harchies (Harchies Mining Museum)

The small Musée de la Mine d'Harchies is located west of Bernissart at Harchies, in Wallonia. It is based on the site of the former Rivage Colliery, which closed in 1968. The Museum of Natural Sciences in Brussels has an internationally important collection of 30 complete skeletons of iguanodons (an ornithopod dinosaur) found underground here in 1878. The mining museum was created in 1992 by Jeannot Duquesnoy, a former mineworker, and opened to the public in 1993. The museum has a large collection of depth indicators, lights, machines, materials, pumps, tipping trucks, tools and ventilators. Large equipment, such as a kibble (**355**), is on the ground floor of the original office building of the mine, together with archives, helmets, lamps, old accounting machines and photographs, paintings and engravings showing the working life of the mine and the hard life of the workers, including works by the sculptor Ernest Dubois (1863-1931).

**Figure 33.1. Steam-powered winch (© Ghislaine Delestrée).**

Several of the museum's exhibits were borrowed by Claude Berri (1934-2009) for use as props in the 1993 movie of the 1885 novel *Germinal* by Émile Zola (1840-1902) about coal miners in the mid-nineteenth century. Also on the ground floor are the former workshops, with mining tools and machinery, including the cage that was used to lower pit ponies down the shaft. There are also specific mining-related exhibitions, such as displays of fossils and minerals, maps, photographs, paintings (including one of a trussed horse being lowered down the shaft) and other mining art. Included is a stained-glass window from the Hensies Pommeroeul power station decorated with the miners' symbol of crossed pick and hammer (**206**). Outside are items such as the flywheel from a François air compressor and a steam-powered winch.

*References*

Berckmans [2016], pp. 275-80.

https://www.visitwapi.be/patrimoine-musees/musee-de-la-mine-memoire-ouvriere

ERIH, *s.n.* Harchies

personal communications, Filip Depuydt (**37**), 2012; Ghislaine Delestrée, owner, 2012

showcaves

Figure 33.2. Kibble used to lower or raise men in the shaft (© Ghislaine Delestrée).

Figure 33.3. Display of stained-glass window from power station (© Ghislaine Delestrée).

### 34. Museum van de Mijnwerkerswoning (Museum of the Miner's House)

The Museum van de Mijnwerkerswoning is situated in the garden city of Eisden-Maasmechelen in the eastern part of Flanders. It was constructed in 1925 as two brick semi-detached houses with a fake-timber facade. The homes in this area were based on the English concept of a garden city, with all houses having gardens, hedges, lanes, *etc.* The aim was to reduce the possible militancy found in the bleak working-class districts of Wallonia. The house was purchased by the municipality of Maasmechelen in 1985 and opened to the public in 1995. In 2004 the archive, building, contents and gardens were jointly designated as a protected monument. It was refurbished with materials collected by residents of the mining estate and has been furnished and decorated as it would have been in the 1930s; for example, the coal stove in the kitchen was the only source of heat in the house (miners received free coal every month); the beds have straw mattresses; and the wall decorations are cheap reproductions of pictures of the Holy Family.

*Reference*

http://www.erfgoedeisden.be/pagina%27s/frameset.htm

### 35. Parc d'Aventures Scientifique (Scientific Adventures Park)

The Parc d'Aventures Scientifique is located on the site of the former Crachet-Picquery Mine, at Frameries, 8 kilometres south-west of Mons in the Borinage. This science-adventure park (generally known as Pass) occupies a 28-hectare site. There are 12,000 square metres of exhibition space, surrounded by gardens. It opened to the public in 2000, after renovation overseen by the architect Jean Nouvel (born 1945) with the aim of maintaining the uniqueness of the site, its history, memories and soul. The intention was to combine futuristic ultra-modern buildings with the historical fabric in order to form a link between the present and the past.

The museum is run by a cooperative company with limited liability and social purpose, which is partly subsidised by the Walloon Region with €3.5 million each year. The museum has a management contract with the Walloon Region to contribute to the dissemination of scientific and technological culture, in order to raise awareness and better understanding of changes in society. Ten of the board of directors are appointed by the Walloon Region, two by the municipality of Frameries

**Figure 35.1. Aerial view of Pass from Mont Ostène (© Parc d'Aventures Scientifiques (Pass)).**

**Figure 35.2. Model of the Crachet-Picquery Mine when it was an operational colliery, with headgear, screening plant, conveyor belt, spoil heap and ancillary buildings (© Parc d'Aventures Scientifiques (Pass)).**

and one by the Invest Mons Borinage. In 2019 there were 139,511 visitors and Pass employed 40 staff (equating to 38.5 full-time equivalents).

Some of the buildings of the colliery were demolished after its closure in 1960. The 1947 headframe of the former colliery has been a listed structure since 1989, together with its associated brick building. Pass is the last mining site in the Borinage whose steel headframe is still standing, 64 metres above the nineteenth-century 418-metre-deep shaft. Also extant are the machine hall and the conveyor for moving coal. The Pass'erelle is a 210-metre-long moving pavement that follows the route taken by the coal. This runs from the shaft to the site of the coal-preparation plant, which no longer exists, up to the belvedere, 17 metres above the ground. Next to the site is the spoil heap, which rises more than 70 metres above the museum.

In addition to temporary exhibitions on artistic and scientific themes, a permanent display on three huge screens shows three separate, but related, films about working conditions, social struggles and the leisure pursuits of miners. One of the photographs in the exhibition shows miners protesting against the closure of the Crachet Colliery with a notice saying, *VOUS NE FERMEREZ PAS CRACHET* (you will not close Crachet). These are complemented by the noises of the cage, the pneumatic drills and the brass band, the clamour of protesters and the cries of the children hurtling down the spoil heap. The exhibitions and walks around the site explain how the colliery was transformed into a science museum.

*References*


ERIH, *s.n.* Mons Framerie

http://www.pass.be/

personal communications, Arnaud De Coster, director of marketing and communications, 2015; Nadine Scoyez, administrative, financial and human resources directorate, 2021


### 36. Sauwartan

The Sauwartan mine, in the Bois de Saint-Ghislain in the Borinage, was sunk in the second half of the nineteenth century. The concrete headframe and associated buildings, such as the platemaking building, were erected in 1928 to replace a wooden structure. All metal elements such as machines or rails were sold as scrap after its closure in 1938, and the gatehouse, locker rooms, offices and workshops were demolished.

The pithead and headframe still survive in a derelict state in woodland on private land. This is the only pithead constructed of concrete still standing in the Borinage. The surviving remains were protected as a designated monument in 1991. The associated spoil heap has become a Natura 2000 natural zone, part of an ecological network of special areas of conservation across the EU for protecting wildlife and habitats. The undergrowth around the pithead is currently being cleared with a view to restoration of the site.

*References*


Berckmans [2016], pp. 77, 229-33

http://fr.wikipedia.org/wiki/Puits_de_Sauwartan

personal communications, Filip Depuydt (**37**), 2013, 2021


### 37. Site du Grand-Hornu (Site of Grand-Hornu)

The Site du Grand-Hornu is a former coal mine renowned for its superb neoclassical architecture, which covers about 10,000 square metres of the 6-hectare site. Exploitation of the coal seams started around 1750 at Le Grand Hornu. In 1814 the Sainte-Eugénie Pit reached a profitable seam, resulting in the rapid expansion of the company. Two years later construction began of 440 houses, of high-quality with hot water and gardens to attract a workforce, along with workshops, offices and a residence for the owner, Henri de Gorge (1774-1832). Attendance at the school provided was compulsory for both boys and girls. In 1830 the site was linked by a horse-powered railway to the canal from Mons to Condè-sur-l'Escaut, at 1.8 kilometres in length believed to be the longest at that time in Europe. By 1832 the mine was producing 118,000 tonnes of coal a year and employing 1,500 workers. The mine closed in 1954, as part of the European Coal and Steel Community programme, and the site was purchased by the province of Hainaut in 1989, although restoration by a private architect had already begun in 1971.

It is used nowadays as a centre for temporary exhibitions of applied art, architecture, contemporary art and design. The two courtyards, still containing remnants of the machinery, are respectively 150 metres by 80 metres and 60 metres by 80 metres. The miners' houses have also been preserved.

The Centre d'Innovation et de Design (Centre of Innovation and Design) is a not-for-profit organisation, established

**Figure 36.1. View of derelict Sauwartan concrete headgear standing in woodland (© National Coal Mining Museum for England).**

Figure 37.1. Back of courtyard (© Ph de Gobert).

Figure 37.2. Workshops (© Le Grand-Hornu).

under the auspices of the province of Hainaut. It is responsible for the buildings and for organising temporary exhibitions. The premises also house the Museum of Contemporary Art, which opened in 2002 with works belonging to the Wallonia-Brussels Region. Together the two institutions employ about 80 administrative, educational, security and technical staff.

*References*

Berckmans [2016], pp. 90-1, 296-307

http://www.cid-grand-hornu.be /

http://www.grand-hornu.eu/

personal communications, Filip Depuydt, guide-animator, 2012, 2021

showcaves

### 38. Site du Levant (Levant Site)

A former colliery complex, the Site du Levant, situated between Cuesmes and Flénu, used to be one of the major centres of coal production in the Borinage. It is now a private property* used for leisure activities. The original workshops still retain their overhead crane and have been converted into an indoor riding school. Another colliery building has been fitted out with a stage for performances. While doing levelling works adjacent to the workshops, the owner discovered a small brick-lined roadway running into the hillside to the bottom of the shaft. The roadway still had surviving rails and four mine tubs, although its purpose is not certain.

The owners, Willy Dreaux and his sister, own 15 spoil heaps in the Borinage, each containing about 20 *per cent* of coal. Shortly after they had acquired them, most of the spoil heaps were given statutory protection, so cannot be washed to recover the coal. The main spoil heap on the Levant site is now used for paragliding. This heap emits smoke because it is on fire, with a temperature of around 700 degrees Celsius at a depth of about 1.5 metres.

*References*

Berckmans [2016], pp. 181-96

personal communication, Filip Depuydt (**37**), 2012

### 39. Steenkoolmijn van Heusden-Zolder (Heusden-Zolder Coal Mine)

Zolder is located in Limburg. In 1901 André Dumont (1809-57) identified the seams of the Kempen coalfield and the Charbonnage d'Helchteren-Attic was established

**Figure 38.1. Main building, including workshop, of the Site du Levant (© National Coal Mining Museum for England).**

* I am grateful to Monsieur Dreaux for showing me around the site and to Filip Depuydt (**37**) for arranging access.

**Figure 38.2. Interior of workshop with rails for overhead crane, now used as an indoor riding school (© National Coal Mining Museum for England).**

**Figure 38.3. Entrance to roadway running into hillside and leading to shaft (© National Coal Mining Museum for England).**

**Figure 39.1. Main building and number 1 headgear on right and water tower on left (© Monique van den Berg).**

in 1907, with its main office in Voort, a hamlet of Zolder, to do test drillings in the region. Development was delayed by WWI, when the Germans occupiers confiscated much of the materials involved in the drilling. Water and soil difficulties meant that freezing the ground was required and it was not until 1930 that the first coal came to the surface at the Steenkoolmijn van Heusden-Zolder. In 1931 781 tonnes of coal were produced and this had increased to 2,868 tonnes by 1939. Peak production was reached in 1955, with more than 1.57 million tonnes of coal.

In 1964 the Société Générale, which owned both collieries, merged Heusden-Zolder with Houthalen (**40**), with the records transferred to the Houthalen administrative building. The combined pits achieved a peak production in 1967 of 2.377 million tonnes with a maximum of 9,379 miners. Total production amounted to 86.5 million tonnes, of which 13 million tonnes came from Houthalen. The mine was the last major productive pit in Belgium when it finally closed in 1992.

Number 1 headframe, with a metal arched roof, was erected in 1924 and the smaller number 2 headframe in 1925. The coal came into the main building, which was constructed under and between the legs of the number 1 headframe. Sloping seams (*cf.* **22**) were worked at depths of 720, 800 and 900 metres. Number 1 shaft, with a depth of 825 metres, was used for ventilation and lowering materials, while number 2 shaft, with a depth of 850 metres, was used for raising and lowering men and raising coal.

Buildings on the site, which have statutory protection, are being conserved and converted to other uses, with the first phase costing €16,980,650. The pithead baths are now occupied by the Centre for Sustainable Building and work is continuing on the second phase with private financing.

*References*

https://inventaris.onroerenderfgoed.be/dibe/geheel/120956

https://nl.wikipedia.org/wiki/Steenkoolmijn_van_Zolder

### 40. Steenkoolmijn van Houthalen (Houthalen Coal Mine)

Houthalen-Helchteren is a municipality located in Limburg. In 1902 drilling by André Dumont (1809-57) identified coal seams at Houthalen. The Steenkoolmijn van Houthalen was one of the seven mines in the Kempen coalfield. In 1927 two wooden headframes were erected, which were replaced between 1937 and 1939 by two iron headgears, both 71 metres high. Like the Steenkoolmijn van Heusden-Zolder (**39**), it was necessary to freeze the ground, and the first coal was not raised until 1939. Each shaft had a cage with six levels, which could carry either 74 miners or two tubs per level, amounting to 11.8 tonnes

**Figure 40.1. Steel headframes with back of administration building in the far distance (© Monique van den Berg).**

of coal on each wind. Peak production was achieved in 1956, with 1,281,000 tonnes being mined that year, but declined after that.

With coal being displaced as a fuel by oil, plus a lack of coal reserves, Houthalen was merged with Heusden-Zolder in 1964. The two mines were linked underground in 1965 and 1978 and most of the workforce moved to Heusden-Zolder. From 1965 coal mined in both mines was brought to the surface at Heusden-Zolder, while the shafts at Houthalen were used for ventilation, supply of materials and retrieval of tailings. The closure had dire effects for the region: cultural, social and sporting organisations struggled with the loss of financial support from the mine and the municipality lost a considerable amount of revenue in taxes; for example, in 1966 it had received 4 million Belgian francs (€99,000) in taxes from the colliery.

The municipality purchased a number of buildings on the site, most of which they demolished in 1994 and 1997. The two headframes, called locally *bellefleurs* (beautiful flowers), have been retained as a landmark. The administration building is an example of brick modernism, with an art-deco interior, and was sold to the Cleantech Campus to form a centre for environmentally friendly high technology. The buildings have statutory protection. In 2012 the municipality centralised services in a new building erected on the former mine site.

*References*

https://nl.wikipedia.org/wiki/Steenkoolmijn_van_Houthalen

https://www.toerismehouthalen-helchteren.be/mijnbeleving

# Bosnia-Herzegovina

Bosnian coal is mainly lignite and sub-bituminous. The largest deposits are in the Tuzla Valley in the north-east, where Kreka Colliery was sunk in 1884 with 35 miners who produced 50 tonnes of coal in the first week. Kreka still has reserves of 1.12 billion tonnes, one of the largest lignite reserves in the world, and Bosnia has 2.264 billion tonnes in total. Only China, Mongolia, North Korea and South Africa have a higher reliance on coal (euracoal).

The predominant method of mining is opencast, but reserves close to the surface have been exhausted and Bosnia is moving towards underground mining. Bosnia-Herzegovina produces approximately 12 million tonnes of coal annually, which is used mainly in the four existing major coal-fired power stations. A fifth power station is planned (see Hobbs 2017 for the history of coal mining in Bosnia-Herzegovina).

### 41. Muzej Rudarstva i Železnice (Museum of Mining and Railways)

The Muzej Rudarstva i Železnice is located in Banovići in the canton of Tuzla in north-eastern Bosnia-Herzegovina. The museum is a project of the Brown Coal Mine Banovići, created in cooperation with the municipality of Banovići and the local Centre for the Protection, Development and Promotion of Tourism, established to preserve the tourist potential and the cultural and historical heritage of Banovići. The setting up of the museum was one of the biggest tourism projects in Bosnia-Herzegovina, carried out in two parts. The first was the museum building, designed to look like a miner's lamp, with two floors of small items in an area of 850 square metres. Large items, such as dump trucks, excavators and locomotives, are displayed outside. There are steam operations for railway enthusiasts on the old colliery railway, and some exhibits relating to mining techniques are stored by the railway.

*References*

https://baum.ba/turizam-u-banovicima-muzej-rudarstva-i-zeljeznice/

http://www.dgt.uns.ac.rs/turizam/arhiva/vol_1802_1.pdf

# Brazil

Coal was first discovered by an English company in the state of Santa Catarina in southern Brazil in the 1820s. The main fields are of sub-bituminous coal, in the south of Brazil on the east coast and in the states from the extreme south northwards of Rio Grande do Sul, Santa Catarina and Parana (CMMCP 2016, fig. 4.1). As it was poor quality, with a high ash and low-carbon content, coal continued to be imported from England and Germany. The coal industry developed during WWI when imports from Europe were cut off. The first coal mine in Latin America opened in Arroio dos Ratos, in the Lower Jacui coalfield, which was so important to Brazil that Rio Grande do Sul came to be known as the cradle of the national coal industry. It was coal from Arroio dos Ratos that fuelled the Usina do Gasometer, the main power-generation plant in Porto Alegre, the capital of Rio Grande do Sul. The plant did not use gas, but was named from the area where it was situated, the Gasometer Volta. Between 1924 and 1956 the railway company and mines of Saint Jerome ran the first thermo-electric plant in Brazil.

Getúlio Dornelles Vargas (1882-1954), president of Brazil (1951-4), insisted on the use of coal for domestic heating and the industry developed further with the oil crisis of 1973, leading to further modernisation. Working conditions and environmental considerations, however, remain poor. The low quality of the coal means that there is no export market. Coal is now used in steel making and for only 5 *per cent* of power generation in Brazil, as renewable energy becomes increasingly important.

*References*


https://en.wikipedia.org/wiki/Coal_mining_in_Brazil

https://www.azomining.com/Article.aspx?ArticleID=51


### 42. Museu Estadual do Carvão (State Coal Museum)

The Museu Estadual do Carvão is located in Arroio dos Ratos. Shortages of coal eventually developed in Arroio dos Ratos and, with the subsequent decline in coal mining, the municipality encouraged agriculture and livestock farming to provide employment for its residents. In 1994 the thermo-electric plant building was restored for the museum, which was established to preserve the history of coal mining in Rio Grande do Sul and the cultural, scientific and technical aspects of the coal-mining community. It opened in 1986 with an area of approximately 11 hectares. The museum is under the management of the secretariat of state of Rio Grande do Sul.

The collections include mining tools and utensils, porcelain parts for electricity and refractory bricks from Europe, as well as books, maps and photographs that record the history of the mines. There are galleries dealing with thermo-electricity. The museum preserves documents about the beginning of coal mining in Brazil. The roadways of the old Number 1 Pit, which opened in 1908, are available for visits. The museum also acts as a cultural centre for the promotion of different aspects of culture and organises activities, such as dance groups, scouts' meetings and theatre workshops.

*References*


http://museucarvao.blogspot.com.br/

https://pt.wikipedia.org/wiki/Museu_do_Carvão


### 43. Mina de Visitaçã Octávio Fontana (Octávio Fontana Mine Visit)

The Mina de Visitaçã Octávio Fontana is located in Criciuma in the state of Santa Catarina. It occupies an area of 15,000 square metres and is the only complete coal mine in Brazil open to visitors. Access to the underground is by cars drawn by a mini-locomotive, which is a replica of one dating to 1922. The route is 300 metres long, with eight stops to tell the story of coal mining in Criciuma, which was known as the coal capital. The underground has examples of stone packing and wooden pit props and the tour takes about 30 minutes. At one stop visitors watch a video and can take photographs of the underground. The museum opened in 2011 and includes a visit to the cave of Saint Barbara (**96**). There were 20,000 visitors in the first six months after opening. The only limitation on visits underground is that children must be aged three or older. The museum offers lunch or dinner underground for groups of between 20 and 50 people. In the museum there is information about coal seams, mining equipment and safety underground.

*References*


https://pt.wikipedia.org/wiki/Mina_de_Visitação_Octávio_Fontana

https://www.youtube.com/watch?v=-0AQBXwh7mM

# Brunei Darussalam

As Brunei Darussalam is rich in oil and natural gas, coal has played little part in its energy mix. Muara Coal Mine, which began production in 1883, was later renamed as Brooketon Colliery. It is located in Surasa in the Brunei-Muara district in the north-east of Brunei. From 1889 it was operated by Rajah Charles Brooke (1829-1917). Annual exports of coal varied between 10,000 and 25,000 tonnes each year. At first it was opencast, until this became more difficult and underground mining was introduced. Brooketon Colliery was connected to the deep-water anchorage at Muara by a now-disused wooden railway line, 2.4 kilometres long. Altogether the mine produced 600 million kilograms between 1888 and its closure in 1924, when it became uneconomic in the world recession. During WWII Brunei was occupied in 1942 by the Japanese, who tried unsuccessfully to reopen the coal mine.

**44. Brooketon Colliery**

The overgrown railway, locomotives and mine entrances still survive on the 62-hectare site, which is protected under the Antiquities and Treasure Trove Act of Brunei Darussalam. It is hoped that it may become an open-air museum run by the Brunei Museums Department.

*References*

en.wikipedia.org/wiki/Economy_of_Brunei

http://en.wikipedia.org/wiki/Brooketon_Colliery

https://bruneiresources.blogspot.com/2006/06/brunei-coal-mine.html

# Bulgaria

Bulgaria has lignite coalfields in the north-west and south-east and anthracite in the west, south-west and north-east (euracoal). Underground resources were owned exclusively by the state, but the Mining Law of 1891 allowed the granting of permits to prospectors, at first for two years and then, from 1906, in perpetuity. Many of the mines continued, however, to be state-owned and operated.

In 2017 16.8 million tonnes of lignite were produced, comprising 48 *per cent* of power generation. With large reserves, coal is still important in providing power in Bulgaria, according to the national strategy for developing the mining industry to 2030. A new energy strategy to 2050 sets out priorities for guaranteeing energy security, including liberalisation of the electricity market in line with EU legislation, implementing an energy-efficiency programme and increasing renewable-energy sources (see euracoal for energy in Bulgaria).

### 45. Pernik – Podzemen Minen Muzei (Pernik Underground Mining Museum)

Pernik is located in western Bulgaria, about 20 kilometres south-west of the capital Sofia. The city, first inhabited at the end of the sixth millennium BCE, is assumed to have been named after the Slavic god of thunder and fire, Perun. Later the patron of mining, Saint Barbara, was also linked with explosions, fire and lightening (**96**). Coal is claimed to have been dug from the tenth century CE and moved by hand from points where it outcropped in the Pernik region. The city really developed with the 1891 Mining Law, although the Pernik mine, in the region of Kulata, was always state-owned. Pernik developed rapidly in the early-twentieth century with the exploitation of the rich coal seams of the region. Large amounts were transported to the nearby capital for household use and for fuel for the railways. The first housing for miners was constructed alongside the river Struma, 1 kilometre east of Pernik. Pernik acquired the status of a town in 1929 and then of a city in 1958, when production of coal reached its peak. In 1949 it was renamed *Dimitrovo*, in honour of the Bulgarian communist leader, Georgi Dimitrov (1882-1949), but regained its original name in 1962. Colloquially it has been known as the town of black gold, referring to the coal. The city is still a centre of industry, including the mining of lignite and the manufacture of heavy mining machinery.

**Figure 45.1. Drift portal which forms entrance to underground with inscription *Starite Rudnitsi*, 'the Old Pits', and tree fossil found during mining operations to the right (© Pernik — Podzemen Minen Muzei).**

The Pernik – Podzemen Minen Muzei occupies two underground roadways, driven in 1891, 50 metres underground and 630 metres in length, with large amounts of brick used for strength and permanent construction. The mine was worked until 1966, when its deposits were exhausted. In the 1980s miners from Pernik visited the museum of rock-salt extraction in Wieliczka, Poland, and the mining museum in Saint Petersburg (**311**), giving them the inspiration for establishing the museum. The mine was then developed into the mining museum, which opened in 1986. The Pernik – Podzemen Minen Muzei has five members of staff and attracts about 5,500 visitors a year. It is funded by the government and since 2014 has been managed by the Regionalen Istoricheski Muzeĭ – Pernik (Regional Historical Museum – Pernik).

Works of art are preserved in the administrative building of the colliery, which is of high architectural quality and a landmark of the city. The ground floor of the mining directorate, which is privately owned, has an exhibition on the geology of the Pernik coalfield and the development of mining history and technology.

The drift portal forms the entrance to the underground museum, with an inscription *Starite Rudnitsi* (the Old Pits) and the date 1891 above the portal, together with a tree fossil found during mining operations. Tours of the museum in groups of between five and 20 people take about 40 minutes, conducted in Bulgarian, although pre-booked tours in English and Russian can be arranged. There are 30 static scenarios along the underground route, dealing with the development of mining in Bulgaria. One of these is an iconostasis (portable icon stand) of Saint Ivan Rilski, or Saint Ivan of Rila (876-946), the patron saint of Bulgaria. He was the first Bulgarian hermit, who was regarded as a saint during his own lifetime. One legend says that Saint Ivan appeared in a dream, pointing to a place and saying, "start digging and there you will find black gold". Later, when the mine was working, a roof collapse buried some miners, but a man dressed in white and carrying a lamp led them to safety. The man who had had the dream recognised the man in white as Saint Ivan, making him the patron saint of miners in Bulgaria. The website https://opoznai.bg/view/podzemen-minen-muzei-pernik claims that there were underground chapels in some mines, but Dr Milanova has no knowledge of this.

Various types of roof supports and lighting are displayed in the underground roadway, including a stretch with traditional timbering, as well as protective equipment, self-rescuers and methods of communication, such as mine telephones. Methods of transporting men and materials are on display, ranging from pit ponies to modern mechanised transport, such as minecars and various makes of locomotive. These include a German Siemens 1925 locomotive, the first to be introduced into the Pernik mines.

*References*

https://opoznai.bg/view/podzemen-minen-muzei-pernik

http://www.museumpernik.com/minen.php

personal communications, Desislava Milanova, public relations, 2016, 2021

**Figure 45.2. Electrically driven direct-rope haulage for moving tubs and materials (© Pernik — Podzemen Minen Muzei).**

# Canada

Canada has anthracite, bituminous, lignite and sub-bituminous coal, about 95 *per cent* of which is found in the west in Alberta and British Columbia and in the south in Saskatchewan (CMMCP 2016, fig. 6.1). The first coal mine, however, opened in eastern Canada, on Cape Breton Island in Nova Scotia in 1672. High-quality coal was discovered in 1849 at Fort Rupert near Vancouver Island on the east coast. The Hudson's Bay Company opened a coal mine there, worked by the Kwakwaka'wakw First Nation people, who went on strike when better-paid miners were brought in from the UK (https://theorca.ca/resident-pod/fort-rupert-and-the-forgotten-coalfields-of-suquash/). The Geological Survey of Canada was established in 1842 (Ranestad and Sharp 2020, p. 5). The western mines developed between 1850 and 1890, after which production increased from 3 million tonnes in 1890 to 17 million tonnes in 1942.

The Drumheller Valley coal is called soft rock in western Canada, meaning the mines are dug into clay and sandstone, rather than the harder geology found further west in the Rocky Mountains. To accommodate the shifting nature of soft-rock geology, timber posts and roof booms were reset or replaced as they shifted or rotted, especially in longer-term structures like conveyors and transport tunnels.

Conditions in Canadian mines were poor, with major safety issues and low pay, resulting in many strikes, which were often violent (**49**). Initially production declined after WWII, but rising energy prices in the early 1970s led to the exploitation of new areas in western Canada, reaching 68 million tonnes in 2013. Safety and working conditions also improved as production moved increasingly to opencasting in the late 1960s, with coal being exported to Japan (see Page and Shapiro 2006 for the history of coal mining in Canada).

### 46. Atlas Coal Mine National Historic Site

The Atlas Coal Mine National Historic Site is located in the Drumheller Valley, Alberta. The coal outcrops in this region, so is clearly visible. A railway was built to the Drumheller Valley and the first coal was shipped out in 1911. A total of 139 coal mines operated in the 40-kilometre-long valley between 1911 and 1979. Gradually, however, coal was displaced by natural gas and oil. The Atlas Coal Mine was the last colliery to operate in the Drumheller Valley.

There were four consecutive Atlas mines in the valley between 1917 and 1984. The current historic site encompasses the underground workings of the Atlas Number 3 Mine (active 1936-56) and the surface workings of the Atlas Number 3 and Atlas Number 4 Mines (active 1956-79); the mines used the same management structures, office buildings, pithead baths and processing structures. The last coal was mined from the Atlas Number 4 Mine in

**Figure 46.1. Run of mine cars and main screening structure (© Atlas Coal Mine National Historic Site).**

1979 and the last load of coal was sold in 1984. The property was reopened to the public in 1987 by the Dinosaur Valley Historical Society, owners of the nearby East Coulee School Museum, which illustrates life in a mining settlement.

The site was designated as an Alberta provincial historic resource in 1989, when the Atlas Coal Mine Historical Society was founded to preserve what is believed to be the most complete historical coal mine in Canada and a complete colliery blacksmiths' workshop. It became a national historic site of Canada in 2002. The site comprises 24 buildings set in 30 hectares and attracts about 20,000 visitors a year. There are six permanent members of staff and 15 seasonal staff. Funding is comprised of 10 *per cent* government grants, 10 *per cent* donations and 80 *per cent* visitor charges. On the surface there are the blacksmiths' workshop, lamproom, pithead baths, an original miner's shack (constructed from dung, mud and straw) and the last wooden screens building in Canada, standing eight storeys high. There are more than 20,000 artefacts in the collections and 8,000 documents in the archives.

Number 3 Mine had a drift entrance located about 61 metres above the valley floor and Number 4 Mine was a shaft mine at prairie level. The surface-processing workings of both mines were located on the valley floor. So the conveyor tunnel allowed the coal to be transported from the upper levels, down through part of the lower valley wall to the gantries connected to the main screening structure, similar to a European screens facility, but with just the bunkering part, and then to the external rail line.

The site offers tours underground and of the screens. The 75-minute tour for groups of up to 20 uses 76 metres of the Atlas Number 3 and Number 4 Mines conveyor roadway driven in 1936. The museum owns a further 152 metres of roadway not open to the public. The low roadways are fully timbered with a now-stationary conveyor belt running their length. The height of the roadways means that the tours use the walkway originally employed for inspecting the conveyor belt.

There are also narrow-gauge rail trips in 2-tonne mine cars hauled by a battery-powered underground Mancha Electric Mule mine locomotive. This was constructed in the 1930s and restored in the early 2000s to a black livery with the name in white along both sides. At the Atlas mines the locomotives were named after the women office staff, so the tradition is continued with this locomotive called Linda after a previous executive director.

*References*

https://atlascoalmine.ab.ca/

personal communications, Jay Russell, executive director, 2016; Sarah Newstead, executive director, 2021

showcaves

### 47. Bellevue Underground Mine

Bellevue Mine, in Crowsnest Pass in the Rocky Mountains in Alberta, opened in 1903. Most of the coal was used in steam engines. The colliery closed in 1961. The Bellevue Underground Mine is run by the not-for-profit Crowsnest Pass Ecomuseum Trust, which was set up to interpret, maintain, preserve and restore features and objects of historical, natural and industrial interest in the Crowsnest Pass and the Bellevue Mine. The site opened in 1989 and operates seasonally from May to September. It attracts

**Figure 47.1. Coal tubs, with trolley between them for transporting wooden pit props (© Bellevue Underground Mine).**

**Figure 47.2. Entrance to the underground tour (© Bellevue Underground Mine).**

between 10,000 and 14,000 visitors a year and has a staff of 12 to run the tour.

The hour-long underground tour covers 300 metres, with a guide who explains coal-mining methods in the late-nineteenth and early-twentieth centuries. Included on the tour are a coal chute, original mine machinery and an area illustrating pillar-and-stall, known in northern America as room-and-pillar working. Pillars are left to support the roof and coal is extracted only from the stalls, or rooms. The pillars were often removed subsequently, allowing the roof to collapse. This is the only genuine underground tour in Alberta. The site focuses primarily on coal-mining artefacts and techniques from 1900 to 1960, including a 68-horsepower locomotive fitted with a Gardner 4LW engine and dating to the late 1940s to 1950s, which was a standard product of the UK Leeds Hunslet engine company. There is also some machinery on display dating to the 1970s and 1980s.

*References*

http://www.bellevuemine.com/

personal communications, Steven Brown, head historical interpreter, 2016; Chris Matthews, chief executive officer, 2021

showcaves

## 48. Cape Breton Miners' Museum

The main growth in mining on Cape Breton Island in Nova Scotia was in the late-nineteenth and early-twentieth centuries. The last mine closed in 2001. Ocean Deeps Colliery, on the coast of the island in Glace Bay near New Waterford, occupies a 6-hectare site. It opened in 1967 as a centennial project (the celebration of the hundredth anniversary of the union of the colonies to form the Canadian Confederation). The museum shows working conditions and life in and around the coal-mining industry of Cape Breton from the 1930s to the 1950s.

The museum is a charity with a small amount of funding from the government, but relies on earned income to cover operating costs. There are 15 members of staff and about 14 volunteers. The museum attracts some 15,000 visitors a year.

On the surface there is mining machinery and four buildings, including displays with historic documents, mining artefacts and explanations about the geological development of the Cape Breton coalfield. Other buildings include a replica company store, with live interpreters, and a miner's home showing life in the 1850s and the early 1900s. There is also the miners' village restaurant, containing early-style furniture, a large brick fireplace, plastered walls and kerosene lamps. The Men of the Deeps, Cape Breton's coal-miner chorus, performs in the museum's auditorium in the summer months. The only qualifications required to be a member of the chorus are to be a miner or former miner and to be able to sing.

The underground, beneath the museum building, shows coal-mining techniques, with guided tours by former coal miners. Visitors are provided with a miner's helmet for the 90-minute tour and a poncho, because of the wet conditions underground. The mine is 600 metres beneath the Atlantic Ocean and has seams a metre thick and

**Figure 48.1. Entrance to Cape Breton Miners' Museum (© Cape Breton Miners' Museum).**

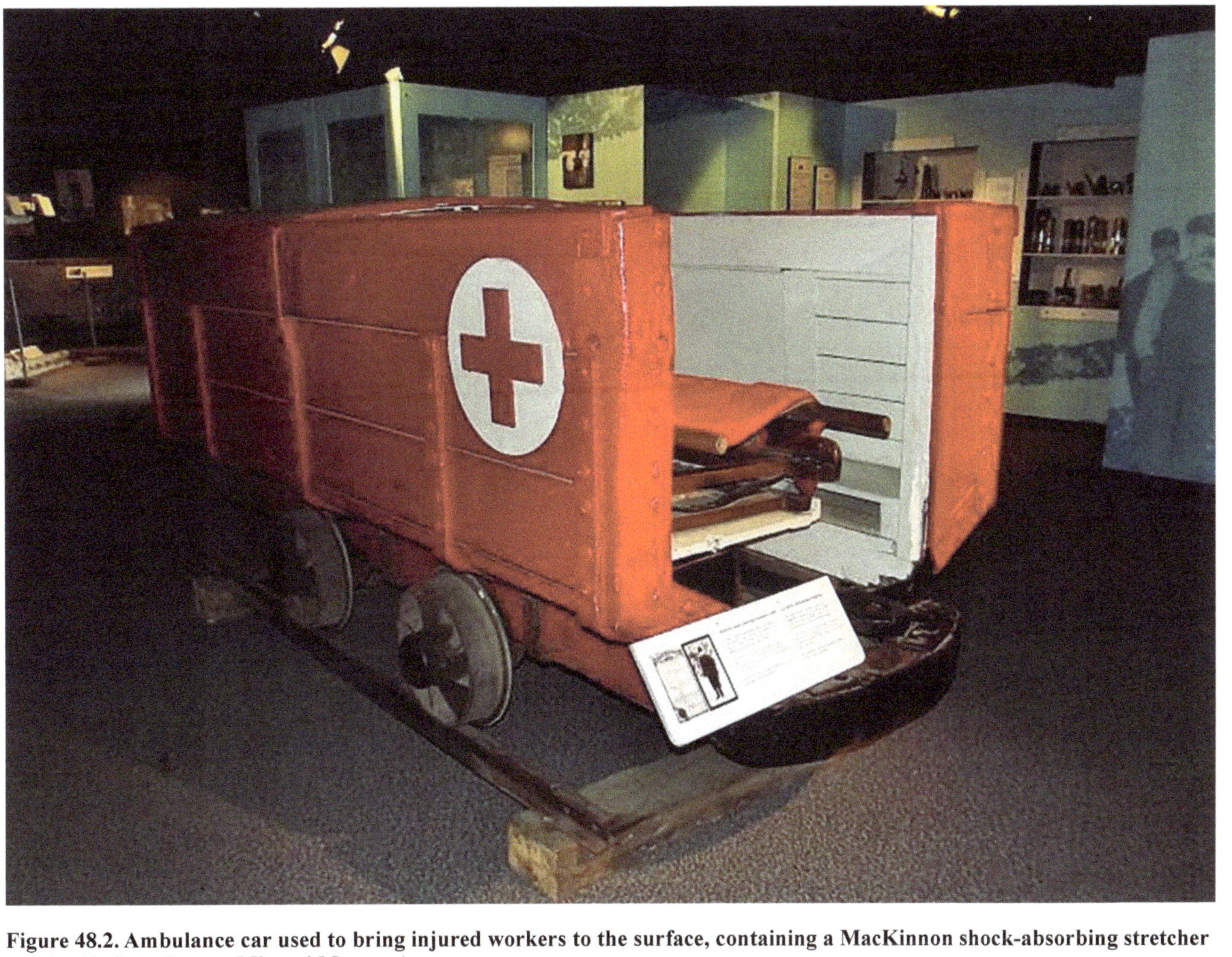

**Figure 48.2. Ambulance car used to bring injured workers to the surface, containing a MacKinnon shock-absorbing stretcher carrier (© Cape Breton Miners' Museum).**

roadways 1.4 metres high. It includes a reconstruction of a 1932 pillar-and-stall mine (**47**).

*References*

http://www.minersmuseum.com/

personal communications, Gail Gillis, administration assistant, 2014; Mary Pat Mombourquette, executive director, 2021

showcaves

**49. Cumberland Museum and Archives**

The Cumberland Museum and Archives is located in Cumberland in the Comox Valley on Vancouver Island. Coal was discovered here in 1852 and the Union Coal Company was founded in 1871, named in honour of the union of the province of British Columbia with Canada in that year. In 1884 the company was purchased by Robert Dunsmuir (1825-89), who founded the town in 1888, which he named Cumberland after the county in north-western England. Prior to the Union Mine, he had already bought the Baynes Sound Mine in 1881, the first to begin production in the area in 1875, and he subsequently purchased the Perseverance Mine in 1888. By 1897 the Union Mine employed 600 men, producing between 700 and 1,000 tonnes of coal a day, and there were 3,000 people living in Cumberland. As the number of mines expanded to 10 in total, they attracted workers from Japan, who lived with their families 2,400 metres to the west of the town, and from China, who lived 800 metres to the west of the town, but without their families. The Japanese and Chinese miners were paid only half what their white counterparts earned and in 1923 government legislation barred them from the more lucrative jobs underground. Many of the Chinese moved back to China during the Great Depression (1929-39) and in 1968 most of their houses were demolished.

In 1942, during WWII, the 31 Japanese families were interned and their possessions sold. From 2002 the village of Cumberland has been developing the 40-hectare Coal Creek historic park, including the Wellington railway trail and the sites of Chinatown and the number 1 Japanese settlement. In 2009 31 flowering cherry trees were planted in memory of the Japanese families.

The Cumberland mines produced a considerable amount of methane gas, leading to explosions, including one in 1901 which killed 105 men; between 1874 and 1966 295 miners died. There were major union disputes from 1912 to 1914 and in 1917 in an attempt to achieve an eight-hour day. These were led by a local labour organiser, Albert 'Ginger' Goodwin (1887-1918). During WWI he and other union members hid in the nearby hills to avoid conscription into the army. They were found about 27 July 1918 by a member of the Dominion police, who shot Goodwin. He claimed that this had been in self-defence, but local people believed Goodwin was killed deliberately to remove a troublesome union organiser. The police force and two magistrates in Victoria recommended, unsuccessfully, that the policeman be committed to stand trial for manslaughter. Goodwin's death provoked a 24-hour general strike in Vancouver in protest. He is buried in the municipal cemetery and every June tribute is paid to him and other fallen workers at the miners' memorial weekend, hosted by the museum. Labour problems and decreasing amounts of available coal led to the closure of the mines, until by 1955 only the Tsable River Mine was still operating, and it finally closed in 1966. The Chinese and Japanese graveyards, both the pitheads and some of former miners' houses still survive. The Cumberland area remains a focus for coal exploration, although future developments are unlikely to be within walking distance of the village centre, as some of the older mines were.

The Cumberland Museum and Archives opened in 1981. It occupies a building of 280 square metres with two floors. There are two permanent members of staff and two seasonal staff, with support from 50 volunteers. The museum attracts about 5,000 visitors a year. It is funded by grants from the federal government, the municipality and the province, together with donations and fundraising initiatives by the museum.

A 15-metre-long replica coal mine is a reconstruction of two working coalfaces in Tsable River Mine. These have equipment and timbering installed by retired coal miners and displays showing how the coal was drilled, blasted and loaded into rail cars. The first reconstruction is of a coalface worked by pillar-and-stall (**47**), with a timber-support system and hand-filled tubs from the first half of the twentieth century. The second reconstruction is of a mine roadway, with a display of typical equipment of the first half of the twentieth century, including flameproof magneto telephone, hand tools and single-shot plunger exploder. There is a figure of a miner with a hand drill and stand for drilling holes for shotfiring explosives to advance the coalface.

A Joy loader is displayed outside the entrance to the museum. Although in the UK Joy loaders were mainly a tool for development work, they had a far wider use in North American-style pillar-and-stall working for mechanised coal clearance, as opposed to hand-filling. This particular machine was used at the Tsable River Mine and was donated to the museum in 1980, after having been found in a pile of bulldozed material. The machine is 8.2 metres long and 1.8 metres wide at the back. It is probably an 8BU variant, which came into use during the 1940s, although earlier variants had existed since the 1910s, and continued in use until the 1960s. The machine does have some features that differ from those shown for the 8BU in contemporary manuals, in particular the valve controls mounted on the right-hand track guard, which may well have been an in-service modification. Unfortunately the machine has lost its tail conveyor boom, but is nevertheless one of the earliest surviving examples.

The museum holds artefacts, books, historical documents, maps, photographs and archival records about the history of Cumberland and the surrounding area, with a focus on labour and coal-mining history. The museum is currently collecting mine-management and technical documents to supplement the already-strong holdings in labour history. In 1876 the Union Coal Company received a quality award for its coal at the Centennial International Exhibition in Philadelphia in America. The award is on display in the museum, which is known locally as the people's museum, because it preserves and tells the stories of the rich, the poor, the powerful, the rebellious, the righteous and the radical. It is also regarded as a labour-history museum because of the information about Ginger Goodwin.

The stories of people who came to Cumberland from China and Japan, as well as from Europe and America, are covered in exhibitions about the community. A community-research room and film-viewing area are provided to enable visitors to explore the past. The Japanese Senjiro Hayashi photographic studio operated in Cumberland between 1913 and 1929 and took numerous portraits of residents of Cumberland who were working in the coal mines or developing businesses to support the industry. The museum holds the entire collection of historic photographs, much of which has been placed on-line.

*References*


https://cumberland.ca/cumberland-cemetery/

https://cumberland.ca/history/

https://cumberland.ca/parks/

https://en.wikipedia.org/wiki/Cumberland,_British_Columbia

personal communications, Michelle Willard, executive director, 2016; Gwyneth Cathyl-Huhn, colliery viewer, Dunsmuir Coal, 2016

www.cumberlandmuseum.ca


### 50. Ladysmith Museum

Ladysmith is located on the east coast of Vancouver Island. The town was founded in 1898 to serve as a tidewater coal port and to house the families of the miners of the Nainamo area. Coal miners in this area went on strike in 1912 because of safety concerns following gas explosions, which had killed hundreds of miners. The militia was sent in to put down the unrest that followed rioting that occurred during what became known as the great strike of 1913/14. Most coal mines on Vancouver Island were closed at one time or another during this period as the miners' union fought, unsuccessfully, to achieve better wages and working conditions.

The Ladysmith Museum and the Archives are housed in separate buildings, but both operated by the Ladysmith and District Historical Society, set up in 1999. The society promotes awareness of the history of Ladysmith and works to preserve its heritage. The museum has displays on mining and community life, as well as logging, and the archives house documents and an extensive collection of coal-mining photographs.

*References*


personal communication, Eric Ricker (**52**), 2015

www.ladysmithhistoricalsociety.ca/


### 51. Michel-Natal-Sparwood Heritage

Sparwood, located near the border with Alberta in the Rocky Mountains of southern British Columbia, was founded in 1897. It was originally a railway point from which timber was shipped to the coast to make wooden spars for ships, from which it took its name. At the same time coal-mining communities developed a few kilometres away at Natal and Michel. The settlement at Michel grew up around a coal mine operated by the Crows Nest Pass Coal Company Limited, the first coal company in the area. By 1966 Natal and Michel had been so badly affected by coal dust that the communities were relocated to Sparwood. Sparwood had been developed in 1939 by the company, which constructed six cottages, then later offered building lots to the miners for $C1 each. In 1968 Kaiser Resources bought the coal mine and offered miners the incentive of a second mortgage, with repayment postponed for up to 10 years.

Michel-Natal-Sparwood Heritage is run by the Michel-Natal-Sparwood Heritage Society. The museum has 59 square metres of displays, plus storage, and a board room upstairs for meetings. The museum is open seasonally four days a week and gets up to 150 visitors a year. It is run by four volunteers each day, with a total of 12 volunteers when projects are underway. The museum runs on fundraising and a monthly grant of $C1,000 dollars from the district of Sparwood, which pays utility bills and the rent of the building.

The society began by collecting oral histories in 1983 and artefacts and other historical items when the society became incorporated in 1988. After moving twice previously, the present museum opened in 1997. The displays include pictures that had not been shown publicly before. There is also an archive of the Crows Nest Pass Coal Company, with pictures of the underground and the machines used.

There are still five opencast mines in the valley, all owned and operated by Teck, the largest diversified mining company in Canada. The Sparwood Chamber of Commerce conducts tours of the nearby opencast, Elkview Mine, in July and August each year on three afternoons a week.

In the middle of Sparwood is what is claimed to be the largest dump truck in the world: a Terex Titan truck, which was used in the opencast operations. Built by General Motors of Canada, it is 20.09 metres long, weighs 350 tonnes and is 17.07 metres high with the box raised.

**Figure 51.1. Terex Titan truck in middle of Sparwood (© Sparwood Chamber of Commerce).**

*References*

https://sparwoodchamber.bc.ca/coal-mine-tours/

https://sparwoodmuseum.ca/

personal communications, Norma McDougall, manager, Sparwood Chamber of Commerce, 2016; Rosalie Fornasier, president, Michel-Natal-Sparwood Heritage Society, 2016

showcaves, *s.n.* Sparwood mine tours

### 52. Morden Mine

Morden Colliery in South Wellington, British Columbia, just south of Nanaimo, opened in 1914 and closed in 1921; it reopened for a short time in 1930 and then closed permanently. Construction began on the site in 1913 of a concrete headframe and screens (**104**). This was the same year that reinforced concrete was also used to a lesser extent for the foundation structures and small buildings at the nearby Reserve Mines 1 and 2 (**53**), now demolished. This development may have been related to the fact that a new cement plant opened on Vancouver Island at the same time that these mine sites were being developed. The Morden headframe and screens cost $C10,000 dollars to build in 1913.

They were acquired for exactly the same amount in 1968 by the government of British Colombia, along with the colliery property and the 6-kilometre railway right of way, which joined the mine site to tidewater at Boat Harbour, where there was a large shipping and coal-preparation plant. In 1972 the 4.23-hectare site was purchased by the provincial government. It was made into a provincial park, along with most of the railway right of way, which is now maintained as a regional trail from Morden to the nearby Nanaimo River. A plan has been developed by local-government bodies to bridge the river where the railway trestles once stood. Once completed, the bridges will fill the last gap in the 6-kilometre trail that extends from Morden Colliery Historic Provincial Park to Hemer Provincial Park in nearby Cedar.

The Friends of Morden Mine is a not-for-profit society founded in 2003, which has had as many as 100 members. Until recently they published the *Morden Mine News* three times a year, dealing with many topics of interest to mining historians and all available on their website. The records of the mine and of the society are held by the Nanaimo Community Archives.

The Friends are working to conserve the site under a partnership agreement with the provincial government. The site comprises the most complete above-ground remains of an historic coal mine on Vancouver Island. The spoil heaps survive, covered by vegetation, and the most important structure is the 23-metre-high concrete headframe and adjoining screens. Having been erected in 1913, this is believed to be the oldest concrete headframe/screens structure in northern America. The Morden Mine headgear is also believed to be one of only two extant concrete headframes in northern America (**425**).

**Figure 52.1. General view of headframe and tippler of Morden Mine (© Friends of the Morden Mine).**

The structure is vulnerable, especially to continuing erosion, falling trees and earthquake tremors. An engineering report prepared in 2014 on the state of the structures and conservation costs indicated that a complete repair programme – including earthquake strengthening – could cost as much as $C2.8 million. Although Morden is a provincial park, the government of British Columbia has not committed to the upkeep and repair of the main structures, which are now over 100 years old. The Friends approached the local authority to request, unsuccessfully, that an application be made for federal infrastructure funding to repair the headframe and screens.

The public can tour the park, but a perimeter fence prevents direct access to the main structures. Interpretive signage and the remains of a boiler house are located on a short walking trail that skirts the structures. Also recently installed on the site is a memorial monument to the miners of the Vancouver Island coal-mining industry. Twice a year the Friends offer tours of Morden Mine and other mine sites in the South Wellington area, as well as custom-tailored tours on other occasions.

**Figure 52.2. Close up of headframe showing encroaching vegetation (© Friends of the Morden Mine).**

*References*

personal communications, Eric Ricker, co-president, Friends of Morden Mine Society, 2012, 2014, 2015, 2021; Simon Vaillant, head of research, Centre d'Archives d'Architecture du XX^e^ siècle, Paris, 2014

showcaves

www.mordenmine.com

## 53. Nanaimo Museum

The first coal from the Nainamo coalfields, on Vancouver Island, was produced in 1852 and found to be ideal for steam production. The deepest shaft on Vancouver Island was sunk 305 metres to the Douglas seam in 1913, at the Reserve Mine (**52**). The coal was worked by miners from China and many parts of Europe, and was shipped to Chile, Vancouver in Canada, and Alaska, Hawaii and San Francisco in America. Over the next 100 years over 50 million tonnes of coal were produced. Coal production began to decrease during WWI and this decrease accelerated with the exploitation of petroleum reserves in the 1920s. Mining generally finished in the mid-1950s, although the Wolf Mountain coal mine continued to work the Wellington seam into the early 1980s.

Nanaimo Museum occupies 650 square metres, of which the coal-mine exhibit is one section. This is a replica coal mine and shaft, constructed with the assistance of scenic artists from the movie industry. There are also geology displays and a recording of a local miner telling his story. The museum employs seven full-time and three part-time staff, augmented by volunteers. It is run by a not-for-profit society governed by a board, which receives support from the city of Nanaimo.

The museum also has a large collection on the coal-mining history of the area, and interprets the number 19 locomotive owned by the city of Nanaimo and on display in a shelter at nearby Piper Park in Nanaimo. This is an example of the type of locomotive that was commonly used on Vancouver Island. Also located in the park is a typical early miner's cottage, built in the 1890s for rental to local miners by Frederick Rowbottom, who had emigrated from Yorkshire, UK, in 1872. Constructed originally with only a living area and a bedroom, later a kitchen and back room were added. It housed up to eight people, although measuring less than 37 square metres. The last owners, the Suttons, donated the house to the Nanaimo Historical Society in 1977, which moved it to its current location in 1980. It has been on the Register of Historic Places of Canada since 2004.

*References*

https://nanaimomuseum.ca/

personal communications, Debbie Trueman, general manager, 2012; Aimee Greenaway, curator, 2021

showcaves

**Figure 53.1. Historic photograph of number 19 locomotive (© Nanaimo Museum).**

**Figure 53.2. Number 19 locomotive arriving on site (© Nanaimo Museum).**

### 54. Nova Scotia Museum of Industry

The Nova Scotia Museum of Industry is the only museum in Canada dedicated to the industrial heritage of an entire province. It has a detailed section on coal mines. The museum opened in 1995 and is funded and operated directly by the province of Nova Scotia. It is approximately 80,000 square feet in area and attracts some 38,000 visitors each year. There are between 10 and 15 members of staff and about 15 volunteers. The museum is dedicated to the story of Nova Scotia work and workers, including the significance of the Nova Scotia coalfield for the economic development of Canada.

Situated 170 kilometres north-east of the provincial capital of Halifax, the museum is located near the site of the first major coal mine in Nova Scotia. The first shaft, known as the Storr Pit, was sunk in 1827. The town of *Albion Mines* (later Stellarton) grew up to service the industry and to accommodate the skilled miners who came from the UK with the General Mining Association of London (**55**). The seam in the Stellarton area comes quite close to the surface and then slopes steeply from south to north. The Association began their workings at the shallower end and subsequent mines, each a little farther north, had to go deeper each time before they reached coal. One of these, the Food Pit, was believed to be the deepest in the world in 1867.

The subsequent development of the coal-mining industry had a major impact on the culture and economy of the north-eastern part of the province. The coal-company towns of Cape Breton Island and mainland Nova Scotia and the once-significant steel industry, created in Sydney, Cape Breton Island, still exercise an influence on developments today.

'Coal and grit' is a permanent exhibition telling the story of mining coal in Pictou County from 1827 up to and including the opencast mines in operation in 2008. It is based on the collections of the former Stellarton Mining Museum of artefacts and photographs donated by many people over a period of 50 years. An extensive collection of photographs of miners is on display, many identified by name. The 1992 Westray disaster, which killed 26 miners at the newest mine in the Pictou coalfield, is highlighted. The exhibition was originally intended to be temporary to mark the opening of the Museum of Industry in 1995. Due, however, to its popularity and the support of the Westray Families Group, it has been developed and expanded to include events associated with the Westray disaster after 1995, and to represent the recent story of opencast mining on the surface.

Linked to the museum's work are three virtual exhibitions developed by the Nova Scotia Archives, using over 1,100 historical photographs, art works, government records, maps, personal documents and published items, supplemented by film clips and sound recordings. The three main themes are the history of mining, life above ground and disasters in the mines, with a searchable database containing over 2,500 names of fatalities recorded between 1838 and 1992. An historical timeline, 1604-1992, provides a chronological framework for the development of the province's mining resources. Items of particular interest or rarity are an 1838 Timothy Hackworth locomotive named Samson, which is believed to be the oldest surviving steam locomotive in Canada, and the Albion locomotive, a Rayne and Burn steam locomotive of about 1854.

*References*

museumofindustry.novascotia.ca

personal communications, Debra McNabb, director, 2014; Andrew Phillips, curator, 2021

### 55. Sydney Mines Heritage Museum

The Sydney Mines Heritage Museum tells the story of coal mining, railways and steel making on Cape Breton Island, which is joined by a causeway to mainland Nova Scotia. The first mining in Sydney Mines took place in 1766 on the north side of Sydney harbour, using iron bars to pry the coal into nearby boats from the exposed coal seams along the cliffs. The only company engaged in the mining of coal in Nova Scotia in the nineteenth century was the General Mining Association of London (**54**), whose agent between 1829 and 1864 was Richard Brown (born 1805) and then from 1864 until 1901 his son, Richard Henry Brown. Richard Brown collected rare geological specimens and fossil remains. His timber and brick-clad house was built in 1828/9 and has statutory protection as a provincial heritage property, in particular for its long association with coal mining in Cape Breton. The early-twentieth century saw a boom in mining and steel making and the doubling of the population in 10 years. Sydney Mines became one of the leading industrial centres of northern America.

The coalfield runs for 34 kilometres along the Cape Breton coast and dips beneath the Atlantic Ocean. Richard Brown was the first to try mining here, with roadways running under the sea. The shales and sandstone of the coalfield are famous for fossils from the Carboniferous period, when the coal was laid down, and Sydney Mines is one of eight important fossil sites in Nova Scotia. These fossils are on display at the Cape Breton Fossil Discovery Centre.

The Sydney Mines Heritage Museum is located in the former Sydney Mines train station. It is operated by the not-for-profit Sydney Mines Heritage Society. The museum is open each year between June and mid-October. The museum has a collection of historical artefacts and archives, photographs, pottery and memorabilia, with descriptions of the artefacts and external historical walks.

*References*

http://sydneyminesheritage.ca/

http://www.historicplaces.ca/en/rep-reg/place-lieu.aspx?id=2988

# Chile

There are three zones in Chile with sub-bituminous coal, all close to the southern coast. Use of coal began in the central zone as early as 1557, with large-scale mining starting in the mid-nineteenth century in response to the needs of steam ships at the port of Talcahuana. The northern zone supplied coal to the cement works, while the southern zone was the major producer in Chile, providing coal for electricity generation.

The coal mines of the city of Lota, on the western coast of southern Chile in the southern zone, 43 kilometres south of Concepción, started production in the early 1800s. It was effectively a company town from the middle of the nineteenth century, being the first industrial town in Chile and the second in South America. The mines were nationalised by Salvador Allende (1908-73) while president (1970-3), but privatised again under Augusto Pinochet (1915-2006), while president (1974-90). Working and living conditions for the miners in the nineteenth century appear to have been like those in Europe: most of the roadways were too low to stand upright and the temperature was always above 23 degrees Celsius. The mine suffered from gas and the miners worked long shifts with poor air quality. The mine does not appear to have been mechanised. The miners were not paid in cash, but with tokens which could be redeemed for basic food and other items only in the shop run by the mine company, owned by the Cousiños family (**57**). This truck system was made illegal much earlier in Europe (**406**). Housing provided by the mine company was also of a very low standard.

In 1997 cheaper Colombian coal came on the market. In view of this, the appalling working conditions and the distance from the shaft making working expensive, the Lota mines were closed, although there are still reserves available (see de Dinechin 2001 for the history of coal mining in Chile). By 2010 there were 900 private mines remaining in Chile, with very low safety standards, and only 12 mines inspectors (Drexler 2011, pp. 244-5).

### 56. El Chiflón del Diablo (The Whistle of the Devil)

Lota is one of the poorest cities in Chile and tried to develop tourism to replace coal mining for employment of local people and to improve its economy. The Chilean Government set up the Corporació de Fomento de la Producción de Chile (Chile Production Development Corporation) to support this initiative. They created the

**Figure 56.1. The entrance to the Chiflón del Diablo Mine (© Fundación ProCultura).**

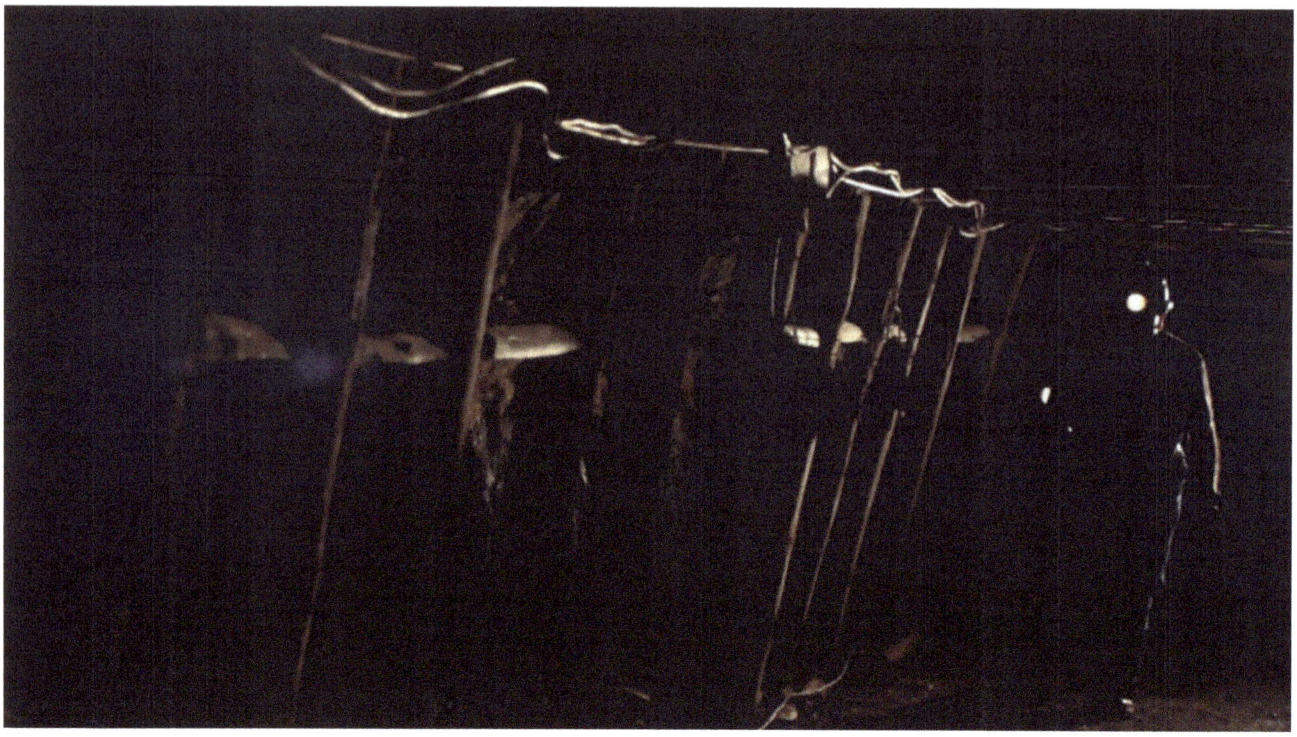

**Figure 56.2. Guide in a roadway showing types of support with services attached to the supports (© Fundación ProCultura).**

Lota Sorprendente (Amazing Lota) tourist circuit of El Museo Histórico de Lota (**57**), El Pueblito Minero (parts of the miners' village), a city park built by the mine company and El Chiflón del Diablo Mine. This has 1,200 metres of roadway going 850 metres out under the Pacific Ocean, with the advantage of being naturally ventilated, thereby reducing running costs. Guides point out different types of roadway supports, such as wooden or prop bars and meshing, with services attached to the supports. The guides are all former miners, so that they can relate first-hand stories and what it was like to work there, as well as the history of the mine.

On 22 February 2010 Chile was hit by an earthquake of 8.8 magnitude just off the coast near Concepción, which triggered a tsunami. Altogether 1,655 people were killed and 2 million were displaced in Chile. Lota was very badly affected, with water and electricity cut off. Roads collapsed and Lota's tourist attractions suffered severe damage.

**Figure 56.3. La Compuerta [roadway] No. 12 with wooden roof supports (© Fundación ProCultura).**

The mine was closed, but a Patrimonial Rescue Programme was instituted in the commercial area of the commune. This enabled the Fundació Chile to reopen the Lota Sorprendente tourist circuit in August 2011. From January 2012 the Fundación ProCultura has been able to open the three main tourist attractions every day of the year: the Isidora Cousiño Park, El Museo Histórico de Lota and El Chiflón del Diablo, including the miners' village. In 2013 El Pueblito Minero and El Chiflón del Diablo attracted 63,598 visitors.

*References*

http://en.wikipedia.org/wiki/Lota,_Chile

https://lotasorprendente.cl/chiflon/

http://unesdoc.unesco.org/images/0019/001921/192194e.pdf

personal communications, Sebastián Lorenzini, consul of Chile, 2014; Andrea Carvajal Lillo, translator and interpreter, Lota Sorprendente-Coorporación Baldomero Lillo, 2015

### 57. El Museo Histórico de Lota (The Historical Coal Museum of Lota; formerly El Museo Histórico del Carbón: The Historical Museum of Coal)

This museum is located in an old house in Lota Alto, on the western coast of southern Chile, 100 metres from the entrance to the Isidora Cousiño Park (**56**). The house was built in 1864 as the administrative building for the

**Figure 57.1. Front of the Museo Histórico del Carbón (© Fundación ProCultura).**

Empresa Nacional del Carbón company and the residential home of first Matías Cousiño (1810-63) and then his son Luis (1835-73), who married his stepsister, Isidora Goyanechea (1836-97). In the late-twentieth century it was subsequently converted into El Museo Histórico del Carbón, which opened in 1998, now called El Museo Histórico de Lota.

The two-storey house is approximately 500 square metres. On the ground floor there was a dining room for the miners, the grocery store (**56**) and a formal dining room. The house contains a collection of items used by the miners in the past and objects that tell the story of the origins of Lota, such as objects used by the indigenous inhabitants of the area and articles dating back to colonisation in 1540. There is also information about the history of the Cousiño family, mainly through a collection of pictures, and about the various economic activities that the Cousiño family developed around mining, such as a collection of ceramics. There are models of the two ships that the Cousiño family lent to the government of Chile for the Pacific War (1879-83), in which Chile defeated Bolivia and Peru. Guides for the tour of the museum dress as Isidora Goyaneche, who managed the coal mines herself from the death of her husband until her own death and who was considered to be one of the wealthiest people in the world. The museum employs four guides and 16 other staff, but no volunteers.

Scenes for the 2003 movie 'Subterra' (Underground) were filmed here. The film deals with the attempt of the miners to establish a trade union in 1897 in the face of dangerous working conditions, low pay, long working hours and debts to the company store. It is presented by Baldomero Lillo (1867-1923), a clerk at the general store, who later became a writer dealing with social matters, including the terrible conditions in which the coal miners worked. His most famous work, *Subterra* (1904), depicted the harsh lives of the miners of Lota and especially of El Chiflón del Diablo with a story about a woman who loses her husband and all three sons in accidents in the mine and commits suicide by jumping down the mine shaft. Visitors can see the store as well as visiting the miners' houses and the laundry rooms.

*References*


https://en.wikipedia.org/wiki/Baldomero_Lillo

https://lotasorprendente.cl/chiflon/

personal communication, Lorena Lillo Chávez, president, Corporación Baldomero Lillo, 2014

Figure 57.2. El Pueblito Minero (© Fundación ProCultura).

Figure 57.3. Ground floor of El Pabellón Minero (© Fundación ProCultura).

# China

Coal deposits are found in all the provinces of China, with five major coalfields (Cao, Wang, Ning, *et al* 2020). These comprise 54 *per cent* bituminous, 29 *per cent* sub-bituminous and 16 *per cent* lignite, the third largest reserve in the world behind America and Russia (Aden, Fridley and Zheng 2008; CMMCP 2020, fig. 20.1). China is still heavily reliant on coal for power.

Mining in China began in the Neolithic period about 7,000 years ago, using both shafts and drifts (Xue 2013, pp. 186-7). In CE 1298 the Venetian merchant, Marco Polo, recorded the widespread use of coal for heating, especially of water in the bathhouses, as it was cheap and plentiful, and there was insufficient firewood for the frequent baths taken by all the numerous population (Cliff 2016, pp. 134-5). In the 1860s a westernising movement developed mining companies, with 15 modern mines opened between 1875 and 1894, but with appalling working conditions, such as miners working for 100 days without a break (Xue 2011, p. 205). One ballad commented, 'all sorts of occupations, mining hardest' (*ibid.*, p. 205). Miners often came from poverty-stricken rural communities; as one ballad said:

> If I had half a bowl of porridge at home,
> I would not go to Mentougou [Colliery](*ibid.*, p. 207).

The mines were usually worked by pillar-and-stall (**47**), followed by longwall mining or opencasting in the 1930s, and then full mechanisation between the 1950s and 1970s (*ibid.*, pp. 203-4; Xue 2013, pp. 188-9).

Between 1904 and 1945 the Japanese gradually occupied China, taking over hundreds of coal mines, from which they extracted hundreds of millions of tonnes. Their method was to extract as much as possible without regard to careful working, causing considerable damage to the geological structures of the mines. The mines were worked by forced Chinese labour, which it has been estimated caused the deaths of millions of Chinese workers (Xue 2011, p. 209; Xue 2016, p. 217). The worst mining disaster in the world was the methane and coal-dust explosion in Benxihu Colliery on 26 April 1942, in which 1,549 miners died. Most were suffocated by carbon-monoxide poisoning when the Japanese operators cut off the ventilation to stop the underground fires. A miners' ballad says:

> Japanese devils are jackals and wolves.
> They oppress workers like the King of Hell (Xue 2011, p. 208).

Many small mines continue to be operated by local communities and these are especially dangerous. When the author visited the Zhōngguó Méitàn Bówùguǎn (**60**) in 2011, she was told by the Deputy Governor of Shanxi Province that there were 32,000 coal mines in the province, a quarter of the Chinese total, and that the aim was to close down 31,000 of these, leaving 1,000 very large collieries (see also Xue 2013, p. 186). Originally the communist government stopped information being disseminated about disasters, but since 1991 it has released information (Escudier 2006, p. 308), although there are still many deaths, with 73 in 2020 (https://en.wikipedia.org/wiki/List_of_coal_mining_accidents_in_China).

As it has little anthracite, China continues to import large quantities of coal from overseas, especially from Mongolia (299.7 million tonnes of coal in 2019), while imports between January and April 2020 increased by 26.9 *per cent*. China has, however, reduced the amount of electricity generated from coal from more than 80 *per cent* in 2017 to just under 60 *per cent* in 2021 (Hui 2021). China continues to be one of the three largest polluters in the world, along with India and Australia, although in a speech to the United Nations general assembly in 2020 the president, Xi Jinping (born 1953), said that China's emissions would peak before 2030 and that the country would reach net-zero carbon by 2060.

It is understood that there are about 60 coal-mining museums in China, but it has been possible to find information about only three of those.

### 58. Jiā Yáng Guójiā Kuàngyè Gōngyuán Bówùguǎn (Jiā Yáng National Mining Park Museum)

The Jiā Yáng Guójiā Kuàngyè Gōngyuán Bówùguǎn is located in the Jiā Yáng National Mining Park in Qianwei County. This is about 180 kilometres south of Chengdu, in the south of Sichuan province in south-western China, in the old mining area of Bajiaogou. The Jiā Yáng coal mine was founded in 1938 as a Sino-British joint venture. After the outbreak of the war against the Japanese (1931-45), the company moved from the north of China into Sichuan to work the coal seams under the mountains of the area. The Jiā Yáng Colliery worked thin coal seams using techniques which became famous throughout the whole country. Many different kinds of mining machinery have been preserved. Huangcun was the number 1 shaft of the Jiā Yáng Colliery, which stopped mining in 1988.

Jiā Yáng National Mining Park was among the second group of national-level mining parks approved by the Ministry of Land and Resources in 2010. The park is located in the city of Leshan. It is connected to the town of Shixi by the Minjang River in the west and with the Huangcun shaft at Bajiaogou in the east, an area of

38 square kilometers. The 20-kilometre-long narrow-gauge Jiā Yáng railway links three key scenic spots in the park: the terminus of the train, Huangcun Pit and the Old Industrial Town of Bajiaogou. The 762-millimetre-gauge railway was originally constructed to move coal from three local collieries to the Minjang River at Xishi to supply the Chongqing iron and steelworks. The coal from the one colliery that is still operating is all used to generate electricity in the power station at the end of the line. The railway still serves the local mining communities along its route as well as being a scenic line using steam trains. It runs through the mining park, which is also a living-history museum. There is a small railway museum at Mifengyan, which attracts 100,000 people every year. The Old Industrial Town of Bajiaogou has buildings representing British-type village construction, Soviet-style buildings and the architecture of south-western China between the 1930s and 1960s. Because of its long history and rich cultural deposits, it has been designated as the Chinese and Foreign Architecture Garden.

The Jiā Yáng Guójiā Kuàngyè Gōngyuán Bówùguǎn has three sections: exhibitions, a technology experience and a simulated underground. The exhibitions cover 980 square metres on the ground floor of a block of Soviet-style offices, which has protected status as being of historical interest. There are 15 galleries, including the history of mining, geology and the life of the miner, using models, multimedia, texts, videos and more than 600 images. The museum also displays a British-type lathe, the machinery for coal processing in the early liberation period of the 1940s and a model of machinery for working thinner coal seams.

The 600 metres of roadways making up the underground were constructed by Huangcun Mine and opened in 2008. The underground is at a depth of over 50 metres, with a coalface 80 centimetres thick. The technology-experience hall is under construction. It is located in the main square of the mining park, covering an area of 960 square metres. It is planned to use some modern high-technology equipment, such as a four-dimensional movie theatre, cartoons, a digital sand table to illustrate the formation of coal and information about coal-mining technology, mine disasters and popular scientific education.

*References*

https://themangoroad.com/the-jiayang-national-mining-park-walking-the-line/

personal communications, Li Xihai, director, and Hu Gaowei, deputy director, Zhōngguó Méitàn Bówùguǎn, 2015

### 59. Kāi Lu Guójiā Kuàngshān Bówùguǎn (Kailuan National Mine Museum)

The Kāi Lu Guójiā Kuàngshān Bówùguǎn is located in Tangshan in the north-east of Hebei Province on the coast of the north-eastern part of the North China Plain, north of the Yellow River. Kailuan Mine was sunk in 1878, claimed to be the first modern coal mine constructed in China and known as the 'Chinese first good colliery'. The mine was crucial to the economic development of Tangshan and neighbouring Qinhuangdao. The coal industry encouraged the development of railways to transport the coal and so Tangshan became a centre of locomotive manufacture. Kailuan Mine is still a major colliery, producing more than 20 million tonnes of coal a year.

Building of the museum, covering 8,000 square metres, began at the end of 2007, construction was completed in 2008 and the museum opened to public in October 2009. A number of industrial sites are preserved around the museum; for example, the century-old arched stone tunnel and the starting point of the 1881 railway, which ran 9.7 kilometres from Tangshan to Xugezhuang and was the first standard-gauge railway built by the Chinese themselves. There are exhibition galleries covering 4,000 square metres and dealing with the 1878 mine, the era of electrical power from 1906 and the winding engine in China.

The museum aims to demonstrate the theme of modern industrial development and the cultural characteristics of heavy-duty coal mining. Major themes covered in the 'black river' mining gallery are geology and the formation of coal, the story of ancient coal mining, the early Industrial Revolution (**379**) in China, the history of Kailuan Mine and the cultural collections of several generations of Kailuan miners in the city.

The museum's collections number more than 20,000 items, including technological items from Kailuan Mine. In China items of national importance are designated, and the museum possesses 47 pieces of national-level cultural items, 72 secondary-cultural items and 326 cultural pieces of tertiary status. Primary-cultural items include the stockbook of the Kaiping Mining Bureau from 1881, which is the earliest stock book preserved in China; the track of the 1881 Tangxu railway; and 1,661 management account books written on sheepskin vellum, which record conditions in Kailuan Mine between 1901 and 1952.

Visits are available with an English-speaking guide to the underground, where mining techniques from different periods are illustrated, before visitors ride out of the underground on a colliery train. The guides also deal with the magnitude 7.6 earthquake of 28 July 1976, which struck just before 4 a.m. and destroyed 85 *per cent* of the buildings in Tangshan and killed at least 242,000 people. At the time there were about 10,000 miners working underground, all of whom apparently survived. The roadways remained standing, but without electrical power there was no ventilation and no winding engines to power the cages to bring up the miners, some of whom took a fortnight to reach the surface. There was considerable flooding resulting from cracking of the linings in the shafts where they passed through water-bearing strata and no power for the pumps to keep the underground dry. Some electrical power was restored after three days, but dewatering and restoration on the surface and underground meant that production did not recover fully for about 18 months.

The next phase of the museum is planned to demonstrate the early industrial history of China, the westernisation movement, the revolutionary struggle (1946-9) and the development of the Communist Party and the labour movement.

*References*

https://en.wikipedia.org/wiki/1976_Tangshan_earthquake

https://www.tripadvisor.com/Attraction_Review-g659916-d8862610-Reviews-Kailuan_National_Mine_Park-Tangshan_Hebei.html#REVIEWS

http://www.railway-fasteners.com/news/history-of-chinese-railway-development.html

personal communications, Li Xihai, director, and Hu Gaowei, deputy director, Zhōngguó Méitàn Bówùguǎn, 2015

### 60. Zhōngguó Méitàn Bówùguǎn (Coal Museum of China)

The Zhōngguó Méitàn Bówùguǎn is the national coal-mining museum of China. It is located in the centre of Taiyuan, the capital of Shanxi province in north-eastern China. The museum opened in 1989 and in 2009 issued a solid-silver (50 grams) medal showing the building on the obverse and the inscription *zhongguo meitan bowuguan chengli ershi zhounian* (the twentieth anniversary of the founding of the Coal Museum of China), all as one word without any tone-marking diacritics and macrons, on the reverse around the miners' symbol of the crossed hammer and pick (**206**). The museum occupies the ground floor of an impressive building. This also houses one of the leading coal-mining companies of China, which funds the operation of the museum. The building occupies a site of 11 hectares, with 90,000 square metres of displays, and also houses the Modern Science and Technology Institute and the China Coal Science and Cultural Exchange Centre. The museum has a formal relationship with the Coal Information Institute of China of the University of Mining and Technology of China. The museum is open from 8 a.m. until 6 p.m. throughout the year.

The museum is made up of two wings, one side of which deals with coal and the other with copies of frescoes from Buddhist institutions, mainly those on the sacred Mount Wutai, a World Heritage Site since 2009. There are now about 240 convents, monasteries and temples out of the approximately 360 there before the depredations of the Cultural Revolution (1966-76). Many historical artefacts were damaged and many unique frescoes were destroyed. The Zhōngguó Méitàn Bówùguǎn developed a new photographic technique which enabled the exact recording of all the remaining frescoes, some up to 2,000 years old, without the distortion normally found with large-scale photographic recording.

The mining galleries cover geology and how coal is created, including an enormous fossilised tree trunk and roots, exploration for coal, and the history of coal mining and of coal-mining technology. The museum displays a number of historical items made from jet, a form of lignite produced from wood changed under extreme compression. It has been used in Asia and Europe for jewellery (**329**) and other valuable items, especially in the Roman and Victorian periods in Europe.

There are guided tours of a simulated coal mine, demonstrating the development of coal mining in China. A four-dimensional cinema uses stereoscopic scenes and special effects to recreate thunder, fogs and heavy snow.

*References*

https://www.chinahighlights.com/taiyuan/attraction/coal-museum-of-china.htm

personal knowledge of the author, a visiting fellow of the Zhōngguó Méitàn Bówùguǎn

**Figure 60.1. Obverse of silver medal showing building housing the Zhōngguó Méitàn Bówùguǎn on the ground floor (© Margaret L. Faull).**

**Figure 60.2. Reverse of medal with crossed hammer and pick symbol (© Margaret L. Faull).**

# Colombia

Coal is widespread throughout Colombia, 94 *per cent* of which is high-quality anthracite (CMMCP 2016, figs 8.1, 8.2). Colombia has less than 2 *per cent* of the world's resources, but in 2020 was the twelfth largest producer of coal. In South America Colombia has the largest reserves of coal, which is its second largest export. The major coalfield is that of La Guagira, 100 kilometres south of Rohacha on the northern tip of Colombia, with reserves of 4.54 billion tonnes of coal. These are worked by the El Cerrejon and La Loma mines.

Copper, gold and haematite have been mined since pre-Columbian times (pre-1492), but coal mining only really developed with the discovery and exploitation of coal in the 1980s. Output increased from 3.6 million tonnes in 1981 to 59.5 million tonnes in 2006. A new mining code was produced in 2001, after which the government divested itself of its shares in El Cerrejón and concentrated on its role as a regulator. A 2013 report by the American Department of Labor stated that Colombian mines employed under-age children in underground mines (https://en.wikipedia.org/wiki/Mineral_industry_of_Colombia#Coal). These are especially dangerous, with 73 killed in 2010 in the San Fernando Mine, which had no methane detectors (Drexler 2011, p. 244), and in 2020 gas explosions at the Veracruz Colliery killed 11 and a further six at the Montgomery Colliery (https://www.mining-technology.com/mining-safety/coal-mine-explosion-kills-11-miners-in-colombia#).

Colombia is currently experiencing a major crisis in its coal mining. The industry has been affected by strikes, criticism of its standards of environmental and human rights, especially of adjacent indigenous communities, a sharp drop in prices and the effects on labour of the covid-19 pandemic, which caused a drop of 7 *per cent* in production in the first quarter of 2020. Sales to America and Europe are declining with the move to clean energy. In 2015 47 *per cent* of Columbian coal was exported to Europe, which had dropped to 9 *per cent* in 2020. The high cost of freight means that China is unlikely to become an alternative major market for Colombian coal (https://dialogochino.net/en/extractive-industries/38033-colombia-coal-industry-pins-hopes-of-salvation-on-china/).

### 61. El Cerrejón (The Cerrejón)

El Cerrejón opencast mine is located near Hato Nuevo on the northern tip of Colombia in the department of La Guagira, 100 kilometres south of Rohacha. Mining operations began at El Cerrejón in 1980, with the first coal being produced in 1986. This is the largest opencast coal mine in the world, 300 metres deep and measuring 7,200 metres by 1,900 metres. It is also the largest coal-mining operation in Colombia, with 32.28 million tonnes exported in 2012 and reserves estimated to be sufficient to last until 2034.

Pre-booked tours of the mine, lasting two hours, are available, including a bus tour of the camp where the workers live and a video introduction at the visitor centre at the entrance to the mine. Visitor numbers are about 19,000 a year. Tours can also be taken to see the coal being loaded onto ships at Puerto Bolívar, 150 kilometres away, after being transported there on a standard-gauge railway. The Museo desQbre guajira (**62**) is being developed to complement the tour of the opencast mine.

*References*


https://www.cerrejon.com/index.php/visit-cerrejon/

http://www.cerrejon.com/site

showcaves


### 62. Museo desQbre guajira (Museum desQbre guajira)

The Museo desQbre guajira is being developed to complement the tour of El Cerrejón (**61**). The Fundación Cerrejón para el Progreso de la Guajira (Cerrejon Foundation for the Progress of La Guajira) is a not-for-profit organisation with the objective of contributing to the development of programmes for the general benefit of communities within the La Guajira Department. The Fundación proposed the creation of an interactive museum to the La Guajira Department and its municipalities. Technical support is provided by Maloka, a not-for-profit institution established in 1997 working under the Science and Technology Law. Maloka is responsible for the design, production and installation of the contents of the museum, and is also responsible for its operation. For the palaeontology activities, the museum receives support from the Smithsonian Tropical Research Institute (part of the Smithsonian Museum in Washington, America), which has been collecting fossils found in the mining operations at El Cerrejón.

It is anticipated that the museum will have 22,800 visitors each year. It has 18 staff and 10 volunteer guides. The museum has three interactive rooms that offer different experiences for visitors, including audio-visuals, dioramas and models, experimentation modules, fossils, immersion scenarios, interactive modules, multimedia applications, scenography and interpretation by guides. The first room deals with the biodiversity, geology and palaeontology of Guajira. The second room covers Wayuu Cosmogony, with the final room exploring the cultural heritage of Guajira, especially the Wayuu ethnic group. The museum also has administrative offices, cafeteria, handicraft store, medical room, storage room, temporary-exhibition room, ticket office and a plaza in reception with enough space to organise cultural events.

*References*

personal communication, Carlos Franco, manager, Social Standards and International Engagement Department, 2014

www.desqbreguajira.org

# Croatia

Anthracite occurs in the eastern part of the Istrian peninsula in western Croatia and in the county of Međimurje in the extreme north. Istrian coal, generally known as Raša coal, has a high-sulphur content, of up to 14 *per cent*. It was worked from the eighteenth century and the four coalfields eventually had seven mines, including Labin (**63, 65**).

Outcropping coal seams were discovered in the Međimurje region after WWI, leading to the development of coal mines there. Mursko Središće, the northernmost city in Croatia on the border between Croatia, Hungary and Slovenia, came to be known as the city of miners. Mining continued here until 1972. Mining of Raša coal ended in 1999, when it stopped being used in the coal-fired Plomin power station, which now relies on imported coal (Medunic, Radenović, Bairamović and Švec 2016). In 2018 Plomin used 595,000 tonnes (euracoal).

### 63. Etnografski muzej Istre (Istria Ethnographic Museum)

The permanent display of the Etnografski muzej Istre, in Pazin in the geographical centre of the Istrian peninsula, deals partly with coal mining in Istria. The museum was founded by the county of Istria and is also partly funded by the Croatian Ministry of Culture. It is located in a former medieval castle dating back to 983. The museum opened in 1962, occupying an area of some 1,367 square metres. There are 11 employees and one volunteer, and the museum attracts some 14,000 visitors each year.

At present the museum holds about 7,500 objects relating to agriculture, everyday items, handicrafts and textiles from the late-nineteenth and early-twentieth centuries. Many of these are on display, including reconstructed carpenters' and blacksmiths' workshops. The permanent exhibition represents ways of life in Istria through imaginary characters from different periods and of different age, gender, social status, *etc.* They include a farmer from Žminj, a winemaker from Momjan, a fisherman from Fažana, a weaver from Križanštine and a mid-twentieth-century miner from Labin. The displays illustrate the transition from agricultural employment to industrialisation and mining with the development of socialism in Istria. They show something of the life of the mine, including the bird-cage lockers in which the miners' clothes were stored.

*References*

personal communication, Ivona Orlic, curator, 2013

www.emi.hr

### 64. Memorijalni Centar u Rudarstvu (Mining Memorial Centre)

The only building in Mursko Središće surviving from the mining period is the coal-preparation plant. This has been converted into the Memorijalni Centar u Rudarstvu to preserve, promote and protect the mining heritage of the region. It is part of a tourist complex opened in 2006, known as Cimper from the local term for the supports of the mine shaft. The tourist complex became insolvent, but the Memorijalni Centar u Rudarstvu, whose aim is to show both the green landscape of the region and its industrial heritage, remains open.

The centre displays hoppers into which the coal was sorted, mining tools and old photographs. There is a multimedia area where stories told by miners are projected. A miner's home is also preserved, with photographs of the life of miners in Mursko Središće.

**Figure 63.1. Life at the mine in Raša showing miners' lamps and suspended basket for storing miners' underground clothing (© Etnografski Muzej Istre).**

*References*

https://croatia.hr/en-GB/eden/cimper-tourist-complex

https://hvm.mdc.hr/spomen-dom-rudarstva.865MSR/hr/info

**65. Narodni muzej Labin (Labin People's Museum)**

The Narodni muzej Labin was established in 1960 as a local-history museum. It is located in the centre of the city in a former seventeenth-century baroque palace. New permanent displays were created in 2008, including showing four centuries of mining in the town of Labin and the surrounding municipality of Kršan. There are texts available in Croatian, English, German and Italian. The Labin miners' strike in March and April 1921 in protest against fascist violence is included in the history of the establishment of the short-lived Labin (Croatian), or Albona (Italian) Republic in 1921. There are pictures of Podlabina from 1942, when it was built as a new coal-mining town, with the brutalist buildings typical also of Italian fascist architecture.

In 1961/4 the museum created the simulated underground of a mine in the basement, with the assistance of the Raša coal mine in Istria. There are approximately 150 metres of reconstructed roadways with objects once used in the mines, such as conveyor belts, different types of roof support, tools, underground railway, waggons and a variety of types of equipment.

*References*

http://hvm.mdc.hr/narodni-muzej-labin,510:LAB/hr/info/

personal communication, Vedran Kos, collections manager, 2014

**66. Tehnički muzej Nikola Tesla (Nikola Tesla Technical Museum; formerly Tehnički muzej Zagreb: Zagreb Technical Museum)**

The Tehnički muzej Zagreb was founded in 1954 in the capital Zagreb, situated in the north-west of Croatia. In 2015 it was renamed as the Tehnički muzej Nikola Tesla, in memory of the Croatian electrical engineer (1856-1943) who filed some 112 patents during his lifetime. The museum attracted 141,000 visitors in 2018.

In 1961 the museum began construction of a replica mine in the basement to show visitors the atmosphere and environment of life and work in an underground coal mine. The mine was renovated in 1995 and renamed as the Jama Barbara Pit, named after Saint Barbara (**96**). The roadway is about 300 metres long and was designed with support from the Faculty of Mining, Geology and Petroleum Engineering of Zagreb University. Some of this is based on the former mines of the Zagorje region, north of Zagreb, and the Raša region. It demonstrates methods of transport, of mine ventilation and of roof support, including wooden pit props and bricks, concrete and stone.

*References*

https://en.wikipedia.org/wiki/Technical_Museum,_Zagreb

https://tmnt.hr/hr-hr/posjet

# Czechia (formerly Czech Republic)

Anthracite is found mainly in northern Moravia in eastern Czechia, with the most important being the Ostrava-Karviná deposits in the Upper Silesian coalfield. Lignite is found mainly in the Northern Bohemian coalfield in north-western Czechia. Coal is the only significant energy resource in Czechia, with reserves estimated at 705 million tonnes, 95 *per cent* of which is lignite, most now produced from opencast mines (euracoal; **72**).

Ostrava is the third city of Czechia and the largest in northern Moravia in the eastern part of the country. This region has the earliest known use of outcropping coal in the world (**70**). The Žacléř region of north-eastern Bohemia belonged to the Křešov monastery in Poland and the first record of mining in this region is found in a document of 1570 in which Abbot Kaspar authorised the exploration and extraction of coal by Jakob Rabe from the village of Opawa in Silesia (now in Poland). In the early-seventeenth century Abbot Tobias recorded that around Žacléř there was more emphasis on the mining of coal than of iron ore. The industry was developed by a large number of small business owners to meet the demand for coal from manufacturing and households because of the lack of firewood.

Czechia is rich in metal ores and coal became important for processing the ores as a replacement was necessary as timber became scarcer. So the Austrian Government supported prospecting for coal in the eighteenth century, with the first mining in Ostrava being carried out by Count Václav Kořenský of Těrešov in 1753. The Kladno part of the Rakovník-Slaný coalfield district in the central Bohemian region was worked extensively from the eighteenth century, with hundreds of small mines operating. Industrial-scale mining started in the 1830s with the sinking of deep mines, which were equipped with steam-winding engines for raising the coal. Regular mining, but by opencasting, also began in the nineteenth century in the Northern Bohemian coalfield, which has amongst the richest European deposits of lignite.

The Ostrava-Karviná region had record outputs in WWI, when the district was militarised; there was an increase in the number of women working in the mines, and 11-hour shifts and Sunday working were introduced. The military administration collapsed with a general strike in 1918. Between 1918 and 1939 coal was exported to south-eastern Europe. In 1939 Germany took over the mines for their high-quality coking coal to fuel its metallurgical industry (essential for producing planes, tanks, *etc.*, for the war effort). There was considerable damage to the mines during the advance of the Allied front in 1945. The Sudetenland was restored to Czechoslovakia after WWII and its coal mines were nationalised, along with those elsewhere in the country. The six Czech mining companies were amalgamated to form a state-owned company, the Ostravsko-karvinské doly (OKD), which owned 32 mines and 10 power plants. Peak production for the whole country of 24.3 million tonnes of coal was recorded in 1980. State control of the company ended in 1991, when it was replaced by a new joint-stock company, also known as the OKD (for the history of coal mining in Czechia see euracoal and http://www.okd.cz/en/coal-mining/the-history-of-coal-mining).

### 67. Hornické muzeum OKD Landek (OKD Mining Museum Landek)

Hornické muzeum OKD Landek is also known as the Ostrava Mining Museum. It is located in the suburb of Petrkovice in Ostrava, which was a major coal-mining area (**70**). Near the museum, on the south-eastern side of the hill, is an exposed cross-section of outcrops of a number of coal seams. The museum is located on the site of the former Anselm Mine in Landek Park, an area of 35 hectares with recreational activities and a restaurant offering traditional mining dishes. Landek was declared a national natural monument in 1992. The museum opened in 1993 on the feast of Saint Barbara (**96**).

Visitors are first kitted out in the original changing rooms, where the chains from which the miners suspended their clothes after changing into their mining overalls still survive. They then go underground with a former miner as guide to see the coal seams, together with mining equipment and machinery. There are 250 metres of roadways and visitors are taken along these in mine cars pulled by an authentic 1960s mine locomotive. The guides demonstrate hand-getting coal and cutting coal using mechanised ploughs. Sound effects are used to recreate the atmosphere underground of a working mine.

On the surface the compressor house is open to the public and there is an outdoor display of large-scale heavy mining equipment, including drills, loaders, mine locomotives and shearers. Smaller items of mining equipment are displayed in the house originally provided for the manager of the mine. A display about mine transport is located in the former joiners' workshop, which had previously been a carriage house and stables. The museum also has an exhibition of respiratory equipment from 1883 to the present day. Former miners explain how this equipment is used and about rescue work underground. The miners also demonstrate the traditional initiation of apprentice miners of 'jumping the leather' (**316**).

*References*

showcaves

www.landekpark.cz

### 68. Hornický Skanzen Mayrau ve Vinařicích (Mayrau Mining Museum in Vinařice)

The Hornický Skanzen Mayrau in the village of Vinařice, in the Kladno region, 30 kilometres west of Prague, is located on the site of the Důl Mayrau (Mayrau Colliery). Sinking of the Důl Mayrau began in 1874 by the společnost Pražská železářská (Prague Iron Company). The company had a long history going back to the 1700s; it subsequently traded under the Tatra brand for much of the twentieth century and is now part of the Siemens group.

In 1877 the shaft reached a depth of 515 metres. The main seam of high-quality anthracite underground was up to 12 metres thick, while a second seam of poor-quality coal had a thickness of up to 1 metre. The mine was named after Kajetan Mayer von Mayrau (1811-83), a German industrialist who was chair of the company's board of directors. A number of technical innovations were introduced for the first time at the Důl Mayrau. A tachograph to record speed and distance was first used on the railways in 1844; a tachograph for use in mine shafts was trialled at the Důl Mayrau and was then patented in 1899 by Jan Karlik (1840-1915). In 1905 a two-cylinder steam-winding engine, works number 1861, produced by the Ringhoffer Company of Smichov in Prague, was purchased at the Paris World Exhibition. This was at the same time as a MAG Ruston Praha *Koepe* winding system (**127**) was installed. In 1925 the main mines-rescue station with its training facilities was moved here. An electric-winding engine by Škoda Plzeň was installed in 1932 and a new compressor in 1936. During WWII the mines came under German control and then were systematically destroyed by the retreating German forces. The mines were nationalised after the war and a new boiler room, fan building and pithead baths were constructed.

The mine closed in 1997, having extracted a total of 32.3 million tonnes of coal while operating, and the shafts were filled in 1998. The Hornický Skanzen Mayrau ve Vinařicích is now the only mining site in the Rakovník-Slaný coalfield open to the public. The museum, which is a branch of the Sládečkovo vlastivědné muzeum v

**Figure 68.1. Exterior of the Hornický Skanzen Mayrau ve Vinařicích (© Sládečkovo vlastivědné muzeum v Kladně, příspěvková organizace).**

**Figure 68.2. Ringhoffer two-cylinder steam-winding engine, works number 1861, dating to 1905 (© Sládečkovo vlastivědné muzeum v Kladně, příspěvková organizace).**

Kladně (Sládeček History Museum in Kladno), opened to the public in 1994. The museum occupies a site of 15.8 hectares and there are eight members of staff. Funding is provided by the příspěvková organizace Středočeského kraje (Central-Region Allowance Organisation). The museum attracts about 11,000 visitors a year.

The aim of the museum is to display the pit exactly as it was on the last working day, as if the last miners have only just left the site. So cracks in the walls have not been filled and the banisters have been left unpainted. Visitors follow the route of the miner, with former miners as guides, through the admission to the guild (**316**), changing rooms, pithead baths, lamproom and shaft building. A replica underground roadway extends 119 metres into Homola Hill, with mining machinery from the early-twentieth century and a demonstration of mechanised longwall working.

The headgear, the compressor and the three winding systems are preserved. The duplex-cylinder steam-winding engine is kept in operating condition and can be demonstrated in motion by the guides. The mining guides wear traditional dress, showing the customs, fraternity and traditions of Czech, or more extensively central-European, coal mining. The ceremonial pick and dress, including the hat, mirror that in Germany, such as the crossed pick-and-hammer badges (**206**). In the mines-rescue station there are displays of maps and plans of the mine, and of fossils and other geological items. The museum covers the history of mining in the Kladno region from the opening of the mine up to the present day. Colliery blacksmiths worked on site to sharpen tools, particularly picks; to repair equipment and machinery, such as replating worn tubs; making fittings for underground, such as pipe brackets and fishplates for steel arches; and horseshoes for the pit ponies. Several times a year there are working demonstrations by blacksmiths in the mine‘s workshops.

A nature trail has been created through the spoil heaps of the Důl Mayrau, showing the flora, fauna, geology and methods of reclamation of the post-industrial landscape. The museum also organises excursions to old mine workings in the Vrapice area.

*References*


ERIH, *s.n.* Vinarice

http://mayrau.omk.cz/

https://en.wikipedia.org/wiki/Mining_Museum_P%C5%99%C3%ADbram

http://www.muzeum-pribram.cz/

personal communication, Tomáš Voldráb, curator of collections, 2016

**Figure 68.3. Ringhoffer winding engine with two guides in traditional ceremonial costumes (© Sládečkovo vlastivědné muzeum v Kladně, příspěvková organizace).**

### 69. Hornický Skanzen Žacléř (Žacléř Mining Museum)

The Hornický Skanzen Žacléř is located on the site of the former coal mine, the Důl Jan Šverma (Jan Šverma Mine), in the town of Žacléř in the Hradec Králové region. The largest mining company was owned by Baron Adolf Silbersteins and managed by his nephew. By the middle of the nineteenth century they employed about 100 miners producing 9,840 tonnes of coal a year and by the 1870s these mines were producing 73,800 tonnes a year. The industry went into crisis about 1875, with many colliery owners becoming bankrupt, and the Silbersteins mines passed into the ownership of the Žacléřská ZBAS Company, which continued to operate them until 1945. After the annexation of the Sudetenland from Czechoslovakia by Germany in 1938, the mines came under the control of the Sudetenländische Bergbau AG. By 1944 production stood at 295,200 tonnes and half of the workforce were prisoners-of-war. In the area of Žacléř the six deep mines and three adits, most constructed in the first half of the nineteenth century, were amalgamated after 1945 into one mine, the Důl Jan Šverma. It was named after Jan Šverma (1901-44), a communist activist, Czechoslovak journalist and resistance fighter against the Nazi-backed Slovak state. He died of exhaustion in the Low Tatras Mountains while leading an insurrection of Slovak communists.

Between 1954 and 1965 production from those mines still operating increased from 200,000 tonnes a year to 290,000 tonnes. Cost of production per tonne gradually increased, leading to the closure of most of the mines. The oldest of the shafts, the Jáma Františka, sunk in 1814, closed in 2003. The Důl Julie, originally the deepest shaft with a depth of 865 metres, a diameter of 6.5 metres and two counter-balanced cages, was sunk in 1853. Until it closed in 2005, materials were brought to the shaft top by narrow-gauge surface railway. The most modem and deepest shaft, 955 metres deep, the Důl Jan Šverma, was sunk in 1958 in place of the Eliška and Marie shafts and took over the coal winding from the Eliška, Julie and Marie shafts. Unlike its predecessor, Důl Jan Šverma was equipped with a modern tub circuit using automatically controlled creepers and retarders. This sent full mine cars to the coal-preparation plant and returned empty cars to the shaft for refilling underground. This is similar to the system devised in 1945 in the UK (**383**). In 1992 the government of the Czech republic decided to cease deep mining due to the high costs. The mine gradually closed level by level, with the final level shutting in 2004; over time some 26

**Figure 69.1. Distant view from the west of the spoil heap and headgear of the Hornický Skanzen Žacléř (© Důl Jan Šverma; photographer Stanislav Krčmář).**

million tonnes of coal had been extracted. It was the only mine in Czechia to be filled completely underground with material produced from large-volume waste (mainly ash from power plants and iron production) in order to avoid subsidence on the surface and the venting of methane gas. Opencast mining continued from 1998 to 2008, but produced only about 330,000 tonnes.

The Důl Jan Šverma has been preserved as a monument to deep coal mining in eastern Bohemia. Amongst the surface buildings of the Důl Jan and Důl Julie, the headgears, heapstead and winding-engine rooms have been given statutory protection as cultural monuments of Czechia, along with all the machinery and a bridge connecting the two shafts. These were made accessible to the public in 2012, including the Důl Jan Šverma headgear with panoramic views, from a height of 52 metres, of the colliery site and over the surrounding area, encompassing the eastern Krkonoše Mountains with Sněžka and the massif of Vraní hory with Královecký Špičák. There is also a panoramic view from the spoil heap of the colliery workshops, power-station chimney with boiler house and power station behind, fed by overhead conveyor with coal from the rail bunker, railway waggons loading into a bunker, with the top of the Důl Julie headgear visible behind it, and the 1950s Důl Jan Šverma headgear surrounded by the coal-preparation plant.

The museum opened in 2012, with its aims including maintaining mining sites in the Lower Silesian coalfield. There are two full-time and two part-time members of staff and the museum attracts about 3,000 visitors a year. It has the status of a public-service company and occupies a site of 6 hectares. Financial support is provided by the GEMEC-Union, a joint-stock company, which founded the non-profit-making company, the Důl Jan Šverma o.p.s., which maintains and operates the Hornický Skanzen Žacléř.

Some of the buildings of the Důl Jan Šverma are currently used by the GEMEC-Union to deal with water problems from the closed mines (**143**). The Hornický Skanzen Žacléř occupies the headframes, shaft buildings and winding-engine houses. The Důl Jan Šverma winding-engine house contains the late-1950s electric-winding engine that wound cages until 2004. It also has an exhibition about the history of mining at this pit and about opencast mining in Žacléř. The ventilation duct of the Julie shaft is also

**Figure 69.2. View of the site of the Důl Jan mine from the east taken from the spoil heap (© Důl Jan Šverma; photographer Stanislav Krčmář).**

available to be visited. The main colliery ventilation fan is an axial-flow fan, in which the airflow passed straight though the fan; part of the ductwork has been removed so that visitors can see the fan blades inside.

Restoration of the coal-preparation plant, including the machinery, was finished in 2018. In 2020 a reconstruction was completed in the museum grounds of an underground roadway of the Žacléř coal seam, which houses a display of roadheading equipment, including a coal tub for waste removal, compressed-air drills for shotfiring and an Eimco rocker-shovel. There is an exhibition of old mining maps and photographs of the last phases of the mine. Some unique examples of palaeontology found during opencast-mining operations are displayed in the winding-engine house. Mining machinery is displayed outside.

*References*


http://www.djs-ops.cz/

http://www.zdarbuh.cz/reviry/vud/dul-jan-sverma-v-zacleri/

personal communications, Karel Novotný, director, 2016, 2021


### 70. Národní památkový ústav, Česká republika (National Heritage Institute, Czech Republic)

The Národní památkový ústav, Česká republika, is located in Ostrava. In the nineteenth and twentieth centuries it was the centre of a heavy industrial region, similar to that in Silesia to the north. The area has the earliest recorded use of coal, with burnt coal being found in fireplaces dating back to the late Palaeolithic (Old Stone Age), considered to have ended here about 8,000 BCE. Despite claims that it may have been from an outcrop ignited by a forest fire, this appears to be deliberate use of coal (Travis 2008, p. 9). This coal, and that used later, came from areas where the seams outcropped on the surface. There were many underground coal mines in Ostrava between 1760 and the closure of the collieries in the late 1990s. The region had a mainly German population and was originally known as *Schlesien*. It was renamed when it was divided between Poland and Czechoslovakia at the end of WWII, although many of the mining terms and names of the coal mines are still German.

Důl Michal (Michal Mine), in the suburb of Michálkovice, has been a Czech national cultural-heritage site since 1995 and opened to the public in 2000, displaying items from the Ostrava and Karvina mines. There are six permanent members of staff at the Národní kulturní památka Důl Michal (National Cultural Monument Michal Mine) and five guides who work annually from April until October, with two vocational trainees from the university each year. The museum occupies some 4.5 hectares of land and there are about 12,000 visitors a year. Důl Michal is a state-funded institution belonging to the Kroměříž Regional Department of the National Heritage Institute, the largest organisation of the Ministry of Culture.

The colliery dates back to 1843, when two mine shafts were sunk as part of the mining programme of the Austrian state. In 1850 one of these shafts was named Michal, after the court councillor, Michal Laier, who had recently died, and this name then became attached to the entire colliery. The surface underwent several changes, but is today effectively as it was at the time of the final reconstruction of 1913/15. All surviving buildings date from 1915, designed by Frantisek Fiala (1895-1957), professor of urbanism at the Czech Technical University in Prague. They have been conserved as they were in the early-twentieth century. The first electrical mining machinery, including compressors and rotating converters for changing alternating current to direct current, was installed in 1912. The original machinery continued to work until the closure of the mine in 1993. The areas with machinery and equipment have been left intact intentionally, for authenticity and to give the impression that work there has just ended. The overriding principle in the presentation of the buildings has been the concept of the last working day. The main aims are to show visitors a realistic picture of coal mining in Ostrava and to provide a venue for science and research.

The 90-minute guided tour of the site encompasses all the above-ground working areas that a miner would have encountered during the course of his shift. The most imposing structure is the classic European-style latticework headgear (**204**) with steeply inclined raker legs and winding wheels, with a roof above the wheels. This is situated above the now-disused shaft, 960 metres deep. A high-level walkway gave access from the shaft to the pithead baths, situated to the south of the headgear. A second high-level walkway was part of the tub circuit leading from the shaft to the washery. The changing rooms display a European-style system for storing clothes, with a series of baskets that could be raised and lowered, containing personal belongings or workwear. Each miner would have his own key for the appropriate suspension cable on the locking rack that runs across the right-hand side. The winding-engine house contains the 1912 electric-winding engine, compressors and rotary electrical converters that powered the winding engine and associated equipment. Of particular interest is the driveshaft coming off the winding-drum axle. It runs around the rear of the drum via a series of gearboxes mounted on short columns for supplying winding revolutions to the winding indicator and control gear.

The museum also holds temporary exhibitions, often of work by foreign artists. The restoration of the newest building on the site was completed in 2011. The administration and reception building has been converted into a visitor centre, with a newly built reception area, entrance exhibition and cafe. Extensive guided trails through the city's mining landscapes are also available.

*References*

ERIH, *s.n.* Ostrava

http://www.dul-michal.cz/

Michal Colliery leaflet

personal communication, Alexandr Zaspal, mine manager, 2016

www.npu.cz

**71. Podkrušnohorské technické muzeum (Podkrušnohorské Technical Museum)**

The Podkrušnohorské technické muzeum is located on the site of the former village of Kopisty near Most in the Usti region, north-western Czechia. The colliery was opened during the period of the Austro-Hungarian Empire (1867-1917) and mining has been almost continuous here for 120 years. Kopisty was demolished between 1974 and 1979 to allow the area to be mined for lignite. The mine closed in the early 1990s. The Podkrušnohorské technické muzeum was established in 2003 with support from the economic and social council of Most. The aims were the restoration of landscapes affected by mining activities and to rescue and preserve industrial and technical monuments, mainly from the fields of underground and surface mining, and the treatment and use of coal. The museum occupies the 5-hectare site of the former Dolu Julius III (Julius Number 3 Mine) and opened to the public in 2008. There are 10 members of staff at the museum, which is funded by grants from the city of Most, the Department of Labour and the resources of the coal companies. The museum has the status of a public-service company and attracts about 9,000 visitors a year.

The museum concentrates on the history of the mining and processing of coal in the central part of the North Bohemian coalfield. Part of the museum surface has been declared a cultural monument, so the mine buildings have been preserved to demonstrate the industrial architecture and technical equipment used in lignite mining in the 1880s. The surviving buildings are the headgear, heapstead, chimney, original boiler room and winding-engine room. The double-drum duplex steam-winding engine dates to 1891, when the shaft it served was sunk; it is the second oldest in existence in the state and is a protected technical monument in its own right. A number of modifications have been made to the winding engine over the years. As the boiler installation is no longer in existence, it remains static at present, but there are hopes of installing a hidden electric motor in the future to enable it to run. There are also two-storey residences and contemporary buildings from the workers' estate.

There are displays of deep-mining and opencast techniques, including mines rescue, a model of the historical appearance of the mine and interactive games on mining and the skills involved. A scraper unit used in opencast mining for exposing coal seams is displayed outside the museum buildings. Scrapers like this were originally developed for removing overburden on heavy construction projects, such as road or runway construction. The first practical scrapers came into use in the 1920s and were towed by tractors, whereas modern versions, such as this example, are self-powered. In cooperation with various appropriate companies and the regional museum in Most, it is intended to expand the museum's displays to include exhibitions about the other industries that provided employment in the area, especially those of chemicals, energy, forestry, glass and textiles.

It is not possible at present to go underground, although the downcast shaft survives, so that option still remains and could be developed in the future. In 2007 a simulated underground tour was opened in one of the surface buildings, with conveyor belts and underground mining machinery. Tours are offered every hour and last about two hours with trained guides, who demonstrate the machinery and conveyor belts in operation. A replica section of timber-supported roadway contains a display of mines-rescue equipment, including representative examples of rescue breathing apparatus manufactured by companies such as Dräger, mobile emergency-communications equipment and hydraulic roof props.

The most important item in the museum is a narrow-gauge locomotive, number 853, which came from a coal-preparation plant and is on external display. This type 900 BS 200, a 0-4-0 side-tank steam-shunting engine produced by Skoda, was used underground until the 1970s, with a high point in production during the 1940s. The vast opencast lignite mines were one of the main users for moving coal and equipment on their narrow-gauge internal-rail networks. Originally more than 860 examples were produced, but it is believed that there are now only three left in existence.

A second locomotive is displayed with a section of monorail. These monorail systems have been produced in Czechia for manriding and materials transport by the Ferrite Company in floor-mounted or roof-suspended versions, as is the case with this example. This type of locomotive dates to the 1970s and continued in use through to the 1990s; in fact modern examples are still in production. The roadway reconstruction also depicts a typical conveyor installation for coal clearance.

*References*

https://cs.wikipedia.org/wiki/Podkrušnohorské_technické_muzeum

http://www.mulitvinov.cz/podkrusnohorske-technicke-muzeum-o-p-s/os-42094

http://www.ptm.cz

personal communication, Zbyněk Jakš, museum director, 2016

## 72. Uhelné Safari (Coal Safari)

Uhelné Safari operates out of the area around the city of Most, in the Usti region, in north-western Czechia. The safaris are organised pre-booked trips to see working opencast lignite mines operated by the Severní energetická (Northern Energy) and Vršanská uhelná (Vršanská Coal) incorporated companies, as well as reclaimed areas after opencasting. The tours are run between May and September, all only in the Czech language. There are five trips of differing lengths, some using supplied off-road transport and some using participants' own vehicles, with three more routes planned. The enterprise was launched in 2009, and there have been more than 26,000 participants since then. The safaris employ six former staff of the mining companies and, as of 2016, the project has been breaking even.

The Severní energetická company manages the Československá Armáda opencast site and the Komořany coal-preparation plant. The company was the founder of the main mines-rescue station in Most. It is part of the Severní energetická group, which operates the Chvaletice power station. The Československá Armáda opencast produces highly calorific coal, especially suitable for combined heat-and-power generation. Lignite provides heating to approximately a million Czech households. Extraction at the Československá Armáda site is constrained by a 1991 resolution of the government on limits on coal mining. Some 9,000 tonnes of Czechia's most-calorific

**Figure 72.1. Vršany opencast mine (© Uhelné Safari).**

**Figure 72.2. RK5000 overburden excavator working in the the Československá Armáda opencast site (© Uhelné Safari).**

lignite exceed these limits, so extraction at this site will be discontinued in 2024.

Vršanská uhelná extracts lignite at the Vršany site in the central part of the North Bohemian coalfield, whose reserves stretch beyond 2050 within the existing government mining limits. In 2013 Vršanská uhelná signed a 50-year agreement to supply coal to the Počerady power station. The Czech Mining Act obliges mining companies to restore the landscape to its previous state. Some 140 million trees have been planted and over 7,000 hectares reclaimed in the last 60 years. This is mostly agricultural land with lakes filling the voids left by the opencasting.

Route 1 visits the KU800 and RK5000 overburden excavators, the KU300 mining excavator and the indoor observation deck of the Československá Armáda opencast site, as well as the spoil heap of overburden soil being saved for reclamation work. One area has already been reclaimed and water has filled the void. Route 2 visits the KU800 overburden excavator, the KU300 mining excavator, the Vršany viewing terrace and the bridge over the lake. Route 3 visits the Vršany viewing terrace, the Vrbenský waterworks, the bridge and various reclamation works.

Route 4 is designed for families with children from six years of age. This tour encompasses the view over the quarry and the RK5000 overburden excavator, which is about eight storeys high and is believed to be the last remaining example in the world. It then visits the Podkrušnohorské technické muzeum (**71**). Route 5 is only for people over 18 years of age and physically fit, as it involves climbing the main stair around the bucket-chain drive on the RK5000 excavator to the driver's cab on the roof and then descending into its interior.

Three new routes have been designed to operate from 2021. The first includes visiting the highest point of the inner spoil heap of the Československá Armáda, not currently open to the public, and then the Hedvika and Marcela Lakes, created as part of reclamation and named after the deep mines. The second route concentrates on the large number of birds in the wetlands and the údolí Matoušův průplav (Matoušův Canal Valley), currently being landscaped, where there will eventually be a lake of over 260 hectares. Finally there will be a tour to view the reclamation work by helicopter.

*References*

personal communications, Ruzena Martínková, project coordinator, 2016; Iveta Janošová, communications officer, 2021

www.uhelnesafari.cz

# Denmark

In the eighteenth century a large population and major military expenditure including building timber ships for the navy led to the depletion of Danish forests, so an alternative source of power was required. The development of steam-powered ships, railways and machinery meant that coal became an essential fuel, which generally had to be imported. Some 70 extensive searches were made for coal in Denmark between 1571 and 1939, mostly on Bornholm, a small island in the Baltic Sea. Some mines resulted but usually closed quickly with major financial losses, as the Bornholm coal was poor quality, produced a lot of ash and was difficult and expensive to work, being in thin seams separated by sand. Bornholm coal was used mainly for domestic heating. The Geological Survey of Denmark was established in 1888 tasked, among other things, with locating coal deposits.

Denmark was occupied by German troops between 1940 and 1945 and so was cut off from coal imports and had to rely especially on coal from Søby in northern Denmark in the centre of Jutland. Jutland used to be known as the Danish Ruhr (the mining area of Germany). With high levels of unemployment, in 1940 the government ordained that coal production in northern Jutland should be done manually in opencast pits, whose walls were prone to collapse. More than 100 people died before the law was repealed in 1954. Denmark now has North Sea oil and gas, becoming self-sufficient in power in 1999 and an exporter of energy in 2004. It is expected that by 2030 Denmark will produce all electricity from renewable sources (euracoal). Coal already contributes only a negligible amount to electricity generation, more than 80 *per cent* of which is now produced from renewables (see Ranestad and Sharp 2020 for the history of coal mining in Denmark).

### 73. Energimuseet (Danish Museum of Energy; formerly Elmuseet: Museum of Electricity)

Energimuseet is situated at Bjerringbro, close to Viborg on the Jutland peninsula. It is located at the largest hydro-electric power plant in Denmark, which is still operating and producing electricity. Elmuseet opened to the public in 1984 and changed its name to Energimuseet in 2011, although the organisation's preferred English name is Danish Museum of Energy. There are 11 members of staff and the museum attracts about 40,000 visitors a year. It is an independent state-authorised museum under the Danish Law of Museums, funded jointly by the Danish

**Figure 73.1. Aerial photograph of the power plant, museum and artificial lake (© Den Danske Energimuseet).**

Government, the local community of Viborg and the energy company, Ernerginet.dk.

The museum has 3,000 square metres of exhibitions, including the production, distribution and consumption of electricity in everyday life and society. The exhibitions deal with the conventional production of electricity using coal and oil, as well as of renewable energy, such as solar and wind power. There is an open-air section presenting the role of electricity in Danish life in the 1920s, 1930s and 1960s. The collections comprise more than 10,000 objects related to the production, distribution and consumption of energy. The collection also contains films and pictures, together with interviews with engineers who have operated the boilers and have worked at power plants and on the transport and storing of coal in Denmark.

*References*

personal communications, Jytte Thorndahl, curator and vice-director, 2013, 2021

www.energimuseet.dk

**74. Søby Brunkulsmuseum (Søby Lignite Museum)**

The Søby Brunkulsmuseum in the small community of Søby is a small private museum that deals with the extraction of lignite. The museum opened in 1977 and deals with the mining of lignite in the period between 1940 and 1970, using old photographs, objects and tools from that period. It portrays the working and daily lives of the 4,000 to 5,000 people who lived in Søby. The displays include three local houses dating back to 1940 and a grocery store of 1941, with coupons for the period of rationing in WWII and the post-war period (1940-53). There were three weighbridges for weighing trucks and their coal contents. A 1946 Libra weighbridge is fully functional and displayed at the museum with a 1959 Opel Blitz truck.

Tours of the area, lasting two hours, are offered with guides who have experience of the period of mining. The tours finish at the Memorial Park, containing 57 memorials to miners killed during mining. This was moved to Søby in 2008 from its previous position north of the village of Arnborg, a region that has now been landscaped and shows few traces of its mining history.

In 2010 the Søby Lignite Private Foundation received a Carlsberg Memorial Grant of 20,000 Danish kronor. This is to cover the cost of the purchase of information technology to set up an archive of the period of lignite mining between 1939 and 1970.

*Reference*

http://www.brunkulsmuseum.dk/index.htm

# Europe

Coal is found in most of the countries of Europe. The major mining resources are in Belgium, Czechia, France, Germany, Hungary, the Netherlands, Poland, Russia, Spain, the UK and Ukraine, with lesser amounts in Austria, Bosnia-Herzogovina, Bulgaria, Croatia, Denmark, the Faroe Islands, Ireland, Italy, Norway, Romania, Serbia, the Slovak Republic, Slovenia, Sweden and Switzerland. There are museums dealing with coal mining in all these countries. The UK and Germany were the first to mechanise their mines. It was often cheaper for other countries with less coal, poor transport or difficult mining conditions to import rather than to mine their own coal (**247**).

The European Coal and Steel Community, founded in 1952, pooled the coal and steel resources of Belgium, France, Germany, Italy, Luxembourg and the Netherlands, thereby creating a market that made coal uneconomic. It successor body, the EU, to which most of the major coal producers of western Europe belonged, decreed that state aid was unacceptable. So state aid was gradually reduced until it reached zero, although little consideration was given to the social costs involved. Coal came under economic pressure with competition from oil and cheaply mined coal from countries such as Colombia (**61**); although this effectively meant transferring the problems to countries with worse working conditions and worse environmental, health and safety legislation. Creating clean-coal burning requires removing or sequestrating carbon dioxide, nitrogen oxide and sulphur dioxide. The EU preference has been to develop renewable energy, such as wind turbines or photovoltaic panels, of which Germany has been the leader. Between 2005 and 2010 wind-turbine power in Italy increased from 1,000 to 5,000 megawatts (see Cherchi 2011 and Dias, Kanellopoulos, Medarac, *et al* 2019 for the decline of coal mining and consequences in EU countries).

### 75. European Route of Industrial Heritage

As stated by the website of the European Route of Industrial Heritage (ERIH), north-western Europe was the cradle of the Industrial Revolution (**379**). The thematic routes relate to the industrial history of Europe and highlight potential links between very different industrial monuments all over Europe. The sites chosen as anchor points are the major places on each route, relating to energy; housing and architecture; industrial landscapes; industry and war; iron and steel; manufacturing; mining; paper; salt; service and leisure industries; textiles; transport and communications; and water. All mining sites that are anchor points have their own entry in this publication, with reference to the name under which they appear on the ERIH website.

In 1987 the Council of Europe launched its Cultural Routes programme, demonstrating how the heritage of the different countries and cultures of Europe contributes to a shared and living cultural heritage by means of a journey through space and time. The Cultural Routes are grass-roots networks promoting the principles of the Council of Europe: cultural democracy, cultural diversity, human rights, mutual understanding and exchanges across boundaries. They act as channels for intercultural dialogue and promote a better knowledge and understanding of European history. In 2016 the Council received a delegate of ERIH for the first information meeting about the possibility of ERIH being certified as a Cultural Route of the Council of Europe. ERIH is not yet, however, listed as a formal candidate for accreditation as a Cultural Route and it is not known when that may occur.

In 2002 the national coal-mining museum of France (**78**) started holding regular international conferences on coal-mining topics, whose proceedings were subsequently published ([Paris] 2002; [Debrabant] 2006, 2009, 2011, 2013, 2016, 2020). In 2003 staff from the museum visited their English counterpart (**386**) with a view to joint working. After informal meetings of the major European coal-mining museums at the conferences in France, the Network of European Coal Mining Museums was formally established in 2013. There is one member each from Belgium (**29**), England (**386**), France (**78**), Germany (**108**), Italy (**244**), Poland (**297**) and Spain (**327**), who meet once a year, rotating venues. A representative of the Council of Europe attended a meeting held in Poland and proposed that the Network should create a European Route of Coal-Mining Heritage. With support from the Council, the Network submitted a bid to the EU for funding to establish this route. The bid was unsuccessful and, without finance, the Network agreed that it was not possible to proceed with this project.

*References*

http://culture-routes.net/cultural-routes

http://www.erih.net/european-theme-routes/mining.html

# Faroe Islands

The Faroe Islands are an archipelago, located approximately halfway between Norway and Iceland and about 320 kilometres north-north-west of mainland Scotland. The Faroe Islands have been a self-governing autonomous country within Denmark since 1948. Lignite mining started in the Faroe Islands around 1770 at Fámjin, Hvalba and Trongisvágur. Attempts at commercial mining were made on the island of Suðuroy, the only one where coal is found, but could not compete with cheaper coal from the UK. It continued to be an important source of energy for the domestic market of the remote islands until the end of WWII.

### 76. Hvalba

Hvalba, located on the east coast, is one of the northernmost villages in Suðuroy, which is the southernmost island in the Faroes. Hvalba is the only village in the Faroe Islands which still has active coal mining. A few men continue to work in one coal mine in Hvalba. Guided hiking tours to the coal mines of Hvalba are arranged by Visit Suðuroy, the tourist information offices on Suðuroy.

*References*


http://en.wikipedia.org/wiki/Hvalba

http://www.visitsuduroy.fo/uk/index.asp?pID={03A2DA00-3578-43C8-B2E1-FE556781F3A5}

personal communication, Sólfríð Bech, manager, Visit Suðuroy, 2021

**Figure 76.1. Tour guide on the way out of an Hvalba mine (© Visit Suðuroy).**

**Figure 76.2. Sieve, coal weight and miners' shed (© Visit Suðuroy).**

# France

France has coalfields in Commentry in central France; the Loire in central south-eastern France; the Vosges and Jura coalfield in eastern France, where there are four distinct coal deposits; the Normandy region in the north-west; north-east of Marseille in the south; and the Midi-Pyrenees region, also in the south. The Nord-Pas-de-Calais, in the extreme north, is the most important coalfield in France, which in the twentieth century provided two-thirds of national coal output ([Debrabant and Dumont] 2014, p. 10). The Saar-Warndt coalfield in north-eastern France, straddling the French-German border, was worked from the beginning of the nineteenth century, but only passed into French ownership between 1918 and 1933 and again between 1945 and 1957. France currently has reserves of 425 million tonnes of anthracite and 300 million tonnes of lignite (euracoal).

Diodorus Siculus, writing between 60 and 30 BCE, based on the account of the travels of Posidonius about 90 BCE (Tierney 1960, p. 203), is quoted as having described the Gauls of France as having:

> ... fireplaces heaped with coals, and on them are cauldrons and spits holding whole pieces of meat (Oldfather 1939, V, 28).

The term *πυρός* (Tierney 1960, p. 227), in the genitive singular, is translated by Oldfather as 'coals'. However, *πυρ* means 'fire' in both ancient and modern Greek; the word for both 'charcoal' and 'coal' is *ανθραξ*. So Diodorus does not provide documented evidence for the use of coal in Iron Age France. Pliny the Elder (CE 23-79), however, describing the use of coal for smelting (see **introduction to Italy**), states (Rackham 1952, XXXIV, 96) that this was *in Gallia maxime sentitur*, 'especially the case in Gaul', providing evidence of the use of coal in France after *Gallia* became a Roman province in 27 BCE.

Mining began in the Tarn region of southern France in the twelfth century. The use of coal from outcrops or shallow workings in the Cévennes area of the Gard Département of the Languedoc-Roussillon region in southern France began at the initiative of the abbot of the Benedictine abbey of Cendras in the 1230s. It was generally used by lime burners and by blacksmiths in their forges. Coal mining started in the Auvergne region (**85**) of central-southern France in the fourteenth century. The first mention of coal mining in the Provence-Alpes-Côte d'Azur region (**93**) of southern France dates to 1443 and in the Nièvre region (**84**) of central France from the fifteenth century on, but wet conditions underground were always a problem. Coal mining in the Nord-Pas-de-Calais (**77-8**, **80-2**) really began in the 1720s and eventually encompassed approximately one-twelfth of the region, with the deepest shafts, at Lens, reaching 1,200 metres. The first coal seam was identified in Littry (**91**) in Normandy, in 1741, leading to two centuries of industrial activity in the region. The coal of the Moselle Département of Lorraine (**95**) in north-eastern France only began to be exploited in 1856; this is a continuation of the seams found in the Ruhr of western Germany. Artisan mining stopped in the nineteenth century with the development of large mining concerns to power the Industrial Revolution (**379**).

Girls under the age of 14 and boys were employed underground until employment under the age of 10 was forbidden in 1813 (Dumont 2011, pp. 192, 195). Women were extremely rare underground and finally prohibited in 1874 (*ibid.*, p. 195). France was more advanced than other European coal-mining countries in its provision of health facilities, housing, pensions and alternative employment for disabled miners from as early as the later-eighteenth century (*ibid.*, p. 193; see also [Debrabant] 2020, pp. 182-8, 197-9, 218-20). France was also an early adopter of new techniques, such as the installation of a Newcomen engine as early as 1750 (**91**); the first inspector of mines was appointed in 1781, whereas in England it was not until 1843 (**375**); and a school of mines was set up in 1783 (Preece 2007, p. 163). But in the later-nineteenth century and early-twentieth century France lagged behind the UK and Germany in its levels of mechanisation and safety standards (**80**).

France was occupied by Germany during WWI and WWII, and there was extensive fighting and destruction on French soil (**79**), with the Nord-Pas-de-Calais coalfield cut in two by the front ([Debrabant and Dumont] 2014, p. 58). The production from these mines was directed towards the German war effort and in WWII many French miners were transferred to work the mines in the German Ruhr, while Russian, Serbian and Ukrainian prisoners-of-war were brought to work the French mines. In 1941 100,000 miners went on strike, resulting in 244 being sent to the Sachsenhausen concentration camp, 122 of whom perished there (*ibid.*, pp. 62-81, 91-173). As the Germans retreated they destroyed the infrastructure of the mines (*ibid.*, pp. 10-11, 49; Fontaine 2016, p. 210; Guiollard 2013, p. 176); altogether 103 collieries were demolished and their undergrounds flooded, 16,000 miners' cottages and 800 kilometres of railway track were destroyed and Lens was completely flattened (([Debrabant and Dumont] 2014, p. 170). Large-scale restoration led to the introduction of innovative and improved standards of machinery (Guiollard 2013). In 1946 all mines were nationalised under the Charbonnages de France (*ibid.*, pp. 177-8) and the 21 Cévennes mines were modernised, with 23,000 workers employed in 1947. They were not, however, able to compete with cheaper imported coal, so by 1974 there were only four mines left. The last mines of the Nord-Pas-de-Calais closed in 1990 and all Cévennes mining had ceased by the beginning of 2001. Closure of the La Houve mine in the Lorraine in 2004 ended anthracite mining in France and the Charbonnages de France ceased activity in 2008 (euracoal). The coalfields of France all have museums dedicated to coal mining, the Musée de la Mine, Le-Molay Littry (**91**) having opened as early as 1907 (see Besson, Malolepszy, Gigault, *et al* 2019 and [Debrabant and Dumont] 2014 for the history of coal mining in France).

### 77. Bassin Minier du Nord-Pas-de-Calais inscrit au Patrimoine Mondial de l'UNESCO (UNESCO Nord-Pas-de-Calais Coalfield World Heritage Site)

In 2012 UNESCO designated the 120,000-hectare coalfield of the Nord-Pas-de-Calais in northern France as a World Heritage Site. This is the French part of the north-western European coalfield, running for some 120 kilometres through the Nord-Pas-de-Calais and Pas-de-Calais Départements. The Nord-Pas-de-Calais is one of five landscapes in France where the 'combined work of man and nature' has been accepted as worthy of World Heritage Site status. The designation recognised the importance of the continuity and homogeneity of the cultural coal landscape covering the period from the late-eighteenth century to the end of the twentieth century. This relates especially to the working and living conditions of the coal miners, the solidarity to which the conditions gave rise and urban planning creating model workers' housing. The World Heritage Site encompasses 109 individual elements, grouped into 13 local sections corresponding to the former mining companies, and includes a total of 353 sites making up the mining landscape. The region as a whole illustrates a significant period in the history of industrial Europe.

Included in the World Heritage Site are the following aspects:

1. Collieries, including all surface facilities, the associated shafts and winding infrastructure dating back to the early 1850s, the main period of industrial development of the coalfield. All the major periods of mining and construction-technology development are represented in the preserved sites. Of especial importance is the Centre Historique Minier (**78**), which presents the full history of coal mining not only in the area, but also in the whole of France.
2. Headframes of metal or concrete, which are a major feature of the mining landscape.
3. Spoil heaps, including the two heaps of the 11/19 Lens Colliery that spread over 90 hectares and exceed 140 metres in height.
4. The infrastructure for transporting coal, either by rail or water. The canals, railways and rivers have helped

**Figure 77.1. Miners' housing at the Cité Notre Dame at Waziers (© Centre Historique Minier).**

**Figure 77.2. Miners' housing at Valenciennes (© Centre Historique Minier).**

**Figure 77.3. Church of Saint Barbara at Bruay (© Centre Historique Minier).**

shape the landscape, linking up the various elements of the coalfield.
5. Railway stations, which form central hubs in the mining towns.
6. The ponds of water resulting from mining subsidence on the surface, mainly dating to the first half of the twentieth century.
7. Housing estates and social housing forming blocks of workers' dwellings in the mining countryside, usually with repetitive brick facades, following a symmetrical alignment along straight paved streets. These represent both the paternalism of the nineteenth-century coal owners and their desire to control the miners (**382**). Some 600 have been identified as representative, based on their degree of authenticity. There is a variety of types, ranging from the traditional to suburban cities and garden cities with rental properties.
8. Schools, vocational-training centres, housewife schools for girls, *etc.*, were built by the mining companies in the nineteenth century as part of the social provisions for the workers. These also provided specific training of boys for the mine. In the twentieth century the Catholic Church provided educational resources.
9. Places of worship, such as the Church of Saint Barbara (**96**) at Bruay and the Church of Our Lady of Miners at Waziers, were built by many companies to promote religious practice and good moral conduct and to supervise the social practices of miners and their families. They also aimed to curb the rise of socialist and communist ideas. Religious buildings of high architectural merit were constructed, often being designed by renowned architects.
10. The medical institutions are amongst the most important legacies of mining. In the early-nineteenth century mining companies created relief funds for the social protection of workers. These funds paid for many clinics, hospitals, maternity hospitals, pharmacies, *etc.*, for miners and their families.
11. From the end of the nineteenth century halls and sports facilities were constructed to enable the development of cultural activities (such as musical groups) and sporting activities (such as gymnastics and football) for the miners and their families.
12. Monuments and places of commemoration representing the dangers of the working life in the mine (**80**).
13. Buildings relating to the socio-economic life of the mine, such as the offices of the coal companies and the premises of unions and workers' cooperatives.
14. The residential accommodation for the employers and senior executives was generally located in the immediate vicinity of the mine, such as houses for the mine manager and his engineers. Owners and top executives of the companies had homes in enclosed parks, a short distance from the collieries and the urban centres. These buildings also encompassed management functions, such as boardrooms, lounges, prestigious offices, *etc.*, to reflect the strength of the mining companies.
15. Civic buildings, such as town halls, built in the urban centres.

*References*

http://whc.unesco.org/en/list/1360

http://www.bassinminier-patrimoinemondial.org/

**78. Centre Historique Minier (Mining History Centre)**

The Centre Historique Minier is located at the former Delloye Colliery in Lewarde, 8 kilometres to the east of Douai, in the Nord-Pas-de-Calais. The centre functions as the national coal-mining museum for France. Delloye Colliery began extraction in 1931, reaching its peak in 1963 and closing in 1971. The two headframes of the colliery are preserved, as well as the two electrical winding engines, both of 430 horsepower. The site was preserved because of its representative character and its location at the centre of the coalfield and near the motorway network, and was classified as an historical monument in 2009. The museum is one of five major sites in the Nord-Pas-de-Calais coalfield World Heritage Site (**77**).

The museum was originally managed by a not-for-profit association. In 2016 the status of the museum changed to that of an *établissement public de coopération culturelle* (public institution for cultural cooperation). It is now managed by a board with representation from the French state, the regional council, two community groups, the mayor of the city of Lewarde, two representatives of the employees and eight specialist individuals. In 2019 there were 157,784 visitors. More than 4 million people have visited since the centre opened to the public in 1984.

The museum has 8,000 square metres of buildings and structures on an 8-hectare site. It has a full range of colliery surface buildings, although some have been converted to other uses. A surface-railway system has been reconstructed with several locomotives and coal waggons. There is a one-hour guided tour of a simulated underground created on the surface next to the number 2 shaft, with 450 metres of galleries filled with authentic mining machinery. The tour depicts the evolution of mining. Audio-guides are available in Dutch, English, French, German, Italian, Polish and Spanish.

A new reception block includes an auditorium, café, information centre, meeting rooms, offices, shop and temporary-exhibition rooms. The external architecture echoes the form of the original colliery buildings, but with modern contemporary glazing. The permanent exhibitions display the coal industry of the Nord-Pas-de-Calais from the eighteenth century to the closure of the last mine. They cover the discovery and exploitation of coal, the evolution of working conditions and transport, development of industry, rise in power of the mining companies, emergence of labour movements, and economic and social history.

The Centre de Culture Scientifique de l'Énergie (Centre of Scientific Culture and Energy) at the museum aims to put the history of coal mining in the context of the broader history of energy in France and the rest of the

**Figure 78.1. Entrance to the Centre Historique Minier, showing one headgear; original building to the left and new glass reception building to the right (© Centre Historique Minier).**

**Figure 78.2. Diesel locomotive with waggons that are filled by rapid-loading bunkers to form liner or rapid-discharge trains (© Centre Historique Minier).**

world. The museum has a scientific council set up in 1999 to support the scientific team. It consists of former mining engineers and mine specialists in all academic disciplines (ethnology, geology, history, law, linguistics, sociology and technology). In 2021 a new exhibition, dealing with a variety of energy sources and using audio-recordings, films and maquettes, was added to the permanent displays.

The collections comprise about 12,000 objects relating to coal mining and to life in the community of the mining area from the late-nineteenth century up to the 1990s. Particularly well represented are the history of mining techniques, the daily life of the miners and geology, represented by about 1,600 coal, fossil and rock samples. There are everyday household objects and also items relating to the major waves of immigration. Leisure items include holiday postcards, the practice of pigeon racing and breeding, and musical scores and instruments, emphasising the importance of music.

The art collections (drawings, film, paintings, posters, sculptures, tapestries and watercolours) and decorative items with representations of mines (clocks, coins, crockery, furniture and medals) comprise some 15,000 objects relating to the history of the Nord-Pas-de-Calais coalfield. One outstanding example is the artwork, 'La remontée au jour des revenants de Courrières' (**80.4**). The photographic collection illustrates both the technical aspects of mining and coal-processing industries, and the mining towns and their social activities from the late-nineteenth century up to the present day. Since 1996 the centre has held copies of all audio-visual productions from the coalfield.

The archives include 7,000 books, 500 films, 2,500 linear metres of paper documents and 300,000 photographs. In 1984 the collieries of the Nord-Pas-de-Calais coalfield received authorisation from the Direction des Archives de France (Administration of French Archives) to entrust the collection, management and development of coal-mining archives to the museum. From 1985 to 1992 the museum received documents produced by the old mining companies as the pits closed. They cover the entire period of coal mining in the Nord-Pas-de-Calais coalfield (1720-1990). The holdings also include nearly 300 French and foreign journals, books on legal issues, historical and regional magazines and historical works on mining companies, technical publications, and trade-union and business journals.

*References*

ERIH, *s.n.* Lewarde

http://www.chm-lewarde.com/en/index.html

personal communications, Virginie Malolepszy, director, Archives and Resource Centre, 2012, 2021

showcaves

### 79. Givenchy-lès-la-Bassée Memorial

In France during WWI the Allied forces, and the German troops opposing them, came virtually to a standstill by the end of December 1914, with trenches facing each other from which both sides would launch attacks on the others. A further method of attack employed by both sides was to attempt to tunnel underneath the enemy's lines and also to try to intercept tunnels being dug by the enemy, much in the manner of medieval castle warfare. Two techniques were used by the Allied side. The clay-kicking method was adopted early in 1915 and was first used at Givenchy-lès-la-Bassée in the Pas-de-Calais Département of the Hauts-de-France region, 31 kilometres south-west of Lille. This method had been developed in England for driving tunnels through clay soils at a relatively shallow depth; for example, for sewers or railway tunnels. Clay kickers already had experience of working in clay, which was something that coal miners did not possess, although they quickly picked up the method. Soon tunnelling companies were comprised of a mixture of clay kickers, conventional miners and men with an aptitude for working underground. If the geology was clay, then clay kicking would be used, whereas coal-mining techniques would be used in other types of geology.

The UK Royal Engineers formed 21 companies of tunnellers, alongside three Australian (**17**), one New Zealand and three Canadian companies (Preece 2016, p. 251). As well as offensive and defensive tunnels, they also constructed underground shelters, such as hospitals, military headquarters and signals centres, which below ground were safe from shelling.

Around Givenchy-lès-la-Bassée the two opposing sides reached a virtual impasse for almost the whole four years of the war, during which an increasingly complicated and extensive system of tunnels was driven. In June 1916 a German group of miners blew up a large depth charge, resulting in five British miners being trapped underground in what was known as the Shaftesbury gallery. Three were rescued, but the fourth was too badly injured to get out through the hole driven by the rescuers, and the fifth, Sapper William Hackett (1873-1916), refused to leave him, saying, "I am a tunneller and must look after the others first". Before the two could be rescued, the hole closed and the tunnel collapsed. Sapper Hackett had worked in both the Nottinghamshire and Yorkshire coalfields and had been rejected three times for service in the army as being too old. He was an experienced miner who was well aware of the danger of what he was doing and who had also always had a horror of being buried alive. The two men are still entombed there and Sapper Hackett became the only tunneller during the war to be awarded the Victoria Cross, the highest British honour for bravery in the face of the enemy.

In 2010 a memorial to Sapper Hackett and to all the Australian, British, Canadian and New Zealand tunnellers was unveiled at Givenchy-lès-la-Bassée. The memorial

**Figure 79.1. Memorial to WWI tunnellers with yellow-glass T aligned on the location of the Shaftesbury shaft and explanatory panel to the right (© Tunnellers' Memorial; www.tunnellersmemorial.com).**

was designed by Peter Barton (born 1955). It has a central panel of sea-green slate, 120 centimetres high and 80 centimetres wide, exactly the same as the dimensions of the Shaftesbury gallery. A frame of black slate represents the dimensions of the timber framework that supported the tunnel underground. The circular base is engraved with a compass rose to show the orientation of the memorial and has the same dimensions as the Shaftesbury shaft which led to the Shaftesbury gallery. The shoulder flashes of tunnellers were a simple T, which has been reproduced in yellow glass on the memorial, above the inscription. The T is aligned so that the view through the central axis is to the location of the Shaftesbury shaft and the resting place of the two tunnellers.

*References*

https://en.wikipedia.org/wiki/William_Hackett_(VC)

http://www.tunnellersmemorial.com

personal communications, Jeremy Banning, military historian, 2017, 2021

### 80. La Catastrophe des Mines de Courrières (The Courrières Mines Disaster)

The disaster at the three conjoined pits of the Compagnie des Mines de Courrières was at the time the worst that had ever occurred in a coal mine anywhere in the world and still remains the worst in European coal-mining history. Courrières is located 11 kilometres north-east of Lens in the extreme north of France in the region of Hauts-de-France in the Pas-de-Calais Département. The roadways of the three pits extended over an extensive area and were in a complex arrangement on many different levels.

At 6.30 a.m. on 10 March 1906 the mine was rocked by a massive explosion. Discussions at a conference in 2006 at the Centre Historique Minier (**78**), held to commemorate the hundredth anniversary of the disaster, agreed that the most likely explanation for the catastrophe was a coal-dust explosion, possibly following a methane-gas explosion ([Debrabant] 2006). The scientists Michael Faraday (1791-1867) and Sir Charles Lyell (1797-1875) prepared a report for an inquest into the 204 deaths in the 1844 disaster at Hanwell Colliery in County Durham, UK. This was the first time that it had been suggested that coal dust in the air could cause an explosion. In the UK there were a number of major disasters between the 1870s and the 1890s, including at Astley Pit in 1874, Altofts Colliery in 1886, Elemore Colliery in 1887, Mossfield Colliery in 1889 and Thornhill Colliery (**400**) in 1893, some of them caused by coal dust. As a result a royal commission on explosions from coal dust in mines was set up with the remit:

> To inquire into the effect of coal dust in the cause of explosions in mines and whether there were any practicable means to prevent or mitigate any dangers caused by coal dust.

The commission published its report in 1894. The suspicion that coal dust in the air was a cause of explosions was confirmed in experiments in 1908 on the surface at Altofts Colliery in Yorkshire by Sir William Edward Garforth (1843-1921), who had been inspired to look into this by the Astley and Altofts explosions. Photographs of his experiments were published in Rice 1910, pp. 80-1.

Coal-dust explosions were usually triggered by an initial explosion of methane gas, throwing the dust up into the air where, mixed with oxygen, it exploded. It had long been known that similar flour-dust explosions could occur in bakeries. A flour-dust explosion is the most likely explanation for the Great Fire of London (2-5 September 1666), which is recorded as having started in a bakery and destroyed 80 *per cent* of the houses in London. The sinking of the liner Lusitania in 1915, which resulted in America coming into WWI, is also likely to have been an explosion of coal dust thrown into the air in the bunkers when the coal-fired liner was hit by a single German missile.

Coal-dust explosions were usually prevented by stone-dusting, but safety standards were not high in the mines of the Nord-Pas-de-Calais coalfield, including at Courrières. The explosion threw the cage of number 3 shaft up into the headgear, a cage raised at number 2 shaft contained only dead or unconscious miners and the explosion rushed along the 110 kilometres of the complex of roadways. About 600 miners managed to reach the surface after the explosion, although many were seriously burned or suffering from the effects of mine gases.

At the time there were already highly trained mines-rescue services in Germany and the UK, but at Courrières firemen had to be brought from Paris to carry out the rescue. There were immediately offers of help from Germany. These were initially rejected, because of ill feeling against Germany, as it was only 35 years since the Franco-Prussian War (1870-1). By the time that German rescuers arrived on 12 March, they found only dead bodies underground, although one last survivor was found on 4 April. One group of 13 miners did manage to emerge 20 days after the explosion by working their way along roadways in the dark. They survived because they had with them a pit pony, which they ate (*cf.* **384**). The two who led the party were awarded the highest French honour, the Légion d'Honneur (Legion of Honour), and the others, three of whom were younger than 18 years old, received the Médaille d'Or du Courage (Golden Medal of Courage).

In all 1,099 miners died. The only worse disaster since then has been that in 1942 in China, in which 1,549 miners died. The dead at Courrières included 404 from Méricourt, 304 from Sallaumines, 114 from Billy-Montigny and 104 from Noyelles-sous-Lens. There are monuments to the dead at all these places. That at Sallaumines was paid for by public subscription and is inscribed 'by the town to its children, victims of the catastrophe', and that at Noyelles-sous-Lens states that the bodies of 104 victims of the catastrophe of the mines of Courrières are buried beneath the mound, and it is dedicated to the victims of work. The monument known as Le Silo at Méricourt marks the communal grave of 272 men whose bodies were so badly damaged that they could not be identified.

The disaster had several long-term effects beyond the immediate tragedy. The catastrophe was the first to be extensively covered in the press, including the india-ink, pencil and gouache artwork, 'La remontée au jour des revenants de Courrières' (the return to the surface of the ghosts of Courrières), by Frantisek Kupka (1871-1957), which was published in *L'Illustration* in 1906 and is now held at the Centre Historique Minier **(78).** As the reporters from Paris and Lille were unfamiliar with coalfield terminology, they did not know how to spell *terri*, the French term for a spoil heap. They asked the locals whether it was spelled like *fusil* (shotgun), which the poorly educated local people inaccurately confirmed, so the reporters added the final *l*, thus giving the modern form, *terril*.

Another more serious result was that local people saw how much better equipped the trained rescuers from Germany were and realised how poorly administered and equipped the Nord-Pas-de-Calais collieries had been. The first funeral was held on 13 March, sparking fury amongst the community and leading to extensive strikes by workers throughout the coalfield. Georges Benjamin Clemenceau (1841-1929) visited the coalfield twice in his role as minister of the interior. On 1 May he called in the army to disperse the strikers, as he also did with the textile workers' strike in Montpellier and the wine growers' strike in Languedoc-Roussillon in 1907. His strong response aided his political career, becoming prime minister of France between 1906 and 1909 and from 1917 to 1920 and taking a leading and hardline role in the negotiations with Germany at the 1919 Paris peace conference following WWI.

The worst mining catastrophe ever in Europe marked a turning point in terms of safety underground in France. The disaster led to a ban in France on naked-flame lamps underground. In the UK locked lamps had been required under certain circumstances from 1872 and regulations were introduced in 1887 on the construction, examination and use of closed flame-safety lamps. In the Ruhr in Germany the first police mine-air order to ban naked-flame lamps was made in 1846 and renewed in 1863, with electric mining lamps introduced by regulation in 1909. Two independent means of escape also became mandatory in France, which effectively meant two separate shafts. This had been the case in the UK since 1862 following the Hartley Colliery disaster (**375**). In the Ruhr in Germany there was an order to establish two independent shafts, one upcast and the other downcast, in the 1880s.

In 1910 the central Liévin mines-rescue station was established to serve the mines of several different companies. During WWI the original building was destroyed, along with most of the town, and was rebuilt in

U. S. GEOLOGICAL SURVEY BULLETIN 425 PLATE VI

*A.* ALTOFTS EXPERIMENTAL GALLERY AFTER EXPLOSION OF DUST OCCUPYING 450 LINEAR FEET OF GALLERY, AUGUST 11, 1908.

The vibration of the explosion was felt at a distance of 7 miles. (From Colliery Guardian.)

*B.* EXPERIMENT 58 AT ALTOFTS GALLERY, SHOWING STONE DUST AND SMOKE ISSUING FROM GALLERY AS THE RESULT OF AN EXPLOSION OF THE 275-FOOT COAL-DUST ZONE.

The flame penetrated only 54 feet into a stone-dust zone of 100 feet, therefore not igniting another zone of coal dust at the outer end of the gallery. (From Colliery Guardian.)

*C.* EXPERIMENT 99 AT ALTOFTS GALLERY, SHOWING FLAME AND SMOKE ISSUING TO A DISTANCE OF 160 FEET FROM EXPLOSION OF 375 FEET OF COAL-DUST ZONE.

Flame shows white in the view. (From Colliery Guardian.)

**Figure 80.1. Photographs of coal-dust experiments in 1908 at Altofts Colliery (out of copyright; image courtesy of National Coal Mining Museum for England).**

**Figure 80.2. Arrival of the German mines-rescue team on 12 March 1906 (© Centre Historique Minier).**

**Figure 80.3. Le Silo monument at Méricourt, marking the communal grave of 272 miners who died at Courrières and whose bodies could not be identified (© Centre Historique Minier).**

**Figure 80.4. 'La remontée au jour des revenants de Courrières', by Frantisek Kupka, published in *L'Illustration* 1906 (out of copyright; image courtesy of Centre Historique Minier).**

1920. The Centre Historique Minier (**78**) has the archives, photographs and some of the objects from Liévin. The first mines-rescue station in the UK had been set up at Tankersley in Yorkshire in 1902, but in Germany the first central mines-rescue station was not in operation until July 1910.

*References*

https://en.wikipedia.org/wiki/Courri%C3%A8res_mine_disaster

http://www.british-history.ac.uk/office-holders/vol10/pp27-42#h3-0005

http://www.dmm.org.uk/pitwork/html/altofts.htm

personal communications, Virginie Malolepszy (**78**), 2006, 2017; Margarete Merz, personal assistant to the director, Deutsches Bergbau Museum, 2017

### 81. La Chaîne des Terrils (The Chain of Spoil Heaps)

Loos-en-Gohelle in the Nord-Pas-de-Calais, 5 kilometres north-west of Lens, is the base for La Chaîne des Terrils. This is an organisation founded in 1986 that conserves, interprets and protects spoil heaps across the Nord-Pas-de-Calais.

The battle of Loos in WWI was part of the Artois-Loos offensive by the French and British troops in 1915, during which the town was completely destroyed. After the war Loos became a centre of coal mining, until the pits shut in the 1980s. The chain is based at the 11/19 Lens Colliery, which closed in 1986 and has a distinctive steel headframe. Around Loos there are five enormous spoil heaps, two of which, at 146 metres, are believed to be the highest in Europe. They are made of red shale from which protrude mining equipment, such as rails, steel rope and the remains of sheet-metal waggons. The tips support many species of birds and a range of plant life, including fruit trees sprung from apple and pear cores and fig seeds from the meals of earlier miners. The pools in the valleys support various reptiles. Tours can be arranged to see the bird and reptile life and the mining heritage. The tips are used for sports such as mountain biking, orienteering, paragliding and, in winter, skiing and luge. La Chaîne des Terrils is similar to the La Chaîne des Terrils association in Belgium (**28**).

*References*

ERIH, *s.n.* Loos-en-Gohelle

http://www.chainedesterrils.eu/

http://www.hfinster.de/StahlArt2/archive-FranceNorth-en.html

### 82. Le Centre Denis Papin: Centre de la Mine et du Chemin de Fer (The Denis Papin Centre: Mine and Railway Centre)

Le Centre Denis Papin commemorates the industrial past of the Nord-Pas-de-Calais. The Centre is named after Denis Papin (1647-1713), the French inventor of the steam

Figure 81.1. View of spoil heaps, headframes and buildings at Oignies (© Centre Historique Minier).

Figure 81.2. Spoil heap at Loos-en-Gohelle (© Centre Historique Minier).

engine. It is located at the site of the Number 2 Mine of Oignies (**81.2**), situated between Lille and Douai. The last coal mine in the region closed in 1990. The 475-metre-deep shaft of Number 2 Mine closed in 1977. This had been accessed by a cage with three levels, capable of holding 90 miners.

It had been intended that the impressive concrete and red-brick buildings, in the neoregional style, were to be demolished. They were, however, saved by being classified as historical monuments in 1994 under pressure from the Accusto Seci Association of Former Miners. In 2003 the municipality of Hénin-Carvin purchased the colliery from the coal-mining company in order to undertake the preservation and conversion of the site.

There had been two adjacent pits, which shared the ventilation. All the colliery's buildings – headgear, lamproom, offices, pithead baths, winding-engine house and workshops – survive intact, including the 3,800-horsepower steam-winding engine, which was built in Mulhouse in eastern France in 1932. The displays include exhibits about how coal was mined and the lives of the miners. Nearby can be seen the steel headgears of Numbers 9 and 9A Mines, which are protected monuments, and the colliery spoil heaps 110, 116 and 117 in a wooded landscape.

*References*

ERIH, *s.n.* Oignies

https://www.petitfute.com/v27218-oignies-62590/c1173-visites-points-d-interet/c958-musee/1581463-centre-de-la-mine-et-du-chemin-de-fer.html

http://www.cmcf-oignies.com/welcome.htm

museum leaflet

### 83. Mine Témoin d'Alès (Alès Simulated Mine)

Alès is located in the Cévennes and the Mine Témoin d'Alès is located at the former training centre of the Cévennes collieries in southern France. The museum was developed at the initiative of Raymond Aubaret, a trainer of technicians for the Cévennes coal mines. He used the centre's students to install materials and machines that he had collected from the Alès coalfield. Groups descend in a simulated cage into the former training galleries with a retired miner as guide. Apart from 20 steps, the 650 metres of mine roadways are horizontal, with a constant temperature of 12 to 16 degrees Celsius. The displays show the five major stages in the development of mining technology between 1880 and 1960 and deal with extraction, roof supports, safety and exit from the mine.

*References*

https://www.ales.fr/sortir-bouger/musees/mine-temoin/

showcaves

### 84. Musée de la Mine à La Machine (La Machine Mining Museum)

La Machine is a small town in the Nièvre Département in central France. The local geology prevented a drainage channel being used to remove groundwater. The hamlet took its name from the machine, or horse-powered winding apparatus, installed in 1689 initially to lower and raise men and coal and later to pump water from the mine. Coal production finished here in 1974.

The mining museum opened in 1983 in the former mining-administrative headquarters, on the initiative of former miners who wanted to create an educational, but also entertaining, centre. The museum deals with the working and social lives of the miners and their families. Also displayed are the manager's office and collections of fossils, minerals and miners' lamps. The headgear above the Glénons shaft has been preserved and a mine roadway reconstructed on the surface to illustrate mining technology and working methods. Former miners act as guides. A visit usually takes about 90 minutes.

*References*

http://www.mairie-la-machine.fr/le-musee-de-la-mine

personal communication, Virginie Malolepszy (**78**), 2022

showcaves

### 85. Musée de la Mine (Mining Museum), Brassac-les-Mines

Brassac-les-Mines is a French commune in the Puy-de-Dôme Département of the Auvergne region of central France. In 1886 *les-Mines* (the Mines) was added to the name of Brassac to distinguish it from other French towns of the same name. The mine closed in 1978 and the mining museum, created by former miners, opened in 1990.

The museum at Bayard is housed in the former power plant and winding-engine house of the mine. It depicts the history of coal mining in the Brassac coalfield since the fourteenth century, including both mining techniques and the daily life of the miners. The original colliery buildings, including the headgear and the pit machinery have been preserved, and there is a 160-metre-long simulated underground roadway.

*References*

http://mineur.simplesite.com/

http://www.brassaclesmines.fr/culture-/-patrimoine/musee-de-la-mine

showcaves

### 86. Musée de la Mine (Mining Museum), Cagnac-les-Mines

Cagnac-les-Mines is situated in the Midi-Pyrenees region of the Tarn, about 6 kilometres from Albi. The mines of

Albi were sunk in 1890. The last underground mine closed in 1987 and the Musée de la Mine opened in 1989 to preserve the history of 700 years of mining in the Tarn. The Tarn Département took over ownership of the museum in 2007.

On the surface there is a permanent exhibition on the social and economic history of mining and the role of Jean Jaurès, the leading French socialist of the early-twentieth century, who was born in the Tarn in 1859 and assassinated in 1914. Temporary exhibitions are organised on the themes of mining and the industrial heritage of the region. The entrance is marked by a huge miner's safety lamp.

The headgear has been preserved along with the cage. This leads to a 90-minute guided tour of a simulated underground with 350 metres along reconstructed roadways. These have special sound and light effects, and machinery and tools to show the technology of mining and the working conditions of coal miners.

*References*

http://musees-departementaux.tarn.fr/index.php?id=mine-presentation

showcaves

### 87. Musée de la Mine de Blanzy (Blanzy Mining Museum; formerly the Musée de l'Association la Mine et les Hommes: Museum of the Mine and Men Association)

Blanzy is a commune in the Saône-et-Loire Département of Burgundy in east-central France. There has been coal mining in this region for about 200 years. The Musée de la Mine de Blanzy is located on the site of the Puits Saint-Claude (Saint-Claude Pit), whose headframe is the only one still standing in this coalfield. The Musée de l'Association la Mine et les Hommes was founded by volunteers in 1975 to preserve the buildings and the machinery on the site and the history of mining and the mining communities of the coalfield.

Visitors follow the route that would have been taken by the original miners. First the lamproom and headgear, then the winding-engine house, which contains a steam-winding engine produced in Châlon-sur-Saône by G. Pinet in 1885. It worked at the number 1 Puits Sainte Elisabeth between 1886 and 1920, then between 1921 and 1982 at the number 2 Puits Barrat, before being moved to the Musée de la Mine de Blanzy in 1988. The engine has a duplex-cylinder layout and in the twentieth century was converted to being driven by compressed air instead of steam. It has relatively short cylinders, but it would have been possible to operate the engine, when first constructed with cylinders of that size, with about 300 to 500 horsepower. As it is displayed now, it has exceptionally low gearing and a quite rudimentary main steam valve, which would not have given very fine control when operating at speed. It seems likely that some original parts may not have been available when the multi-part engine was converted and reassembled and that they were replaced with the only parts available, although inappropriate if the machine were actually to operate as originally designed. It does, however, give a good idea of this sort of early winding engine with a type of transmission not used later.

**Figure 87.1. Shaft side and cage of the Puits Saint-Claude (© Eberhard Lantz, Bamberg).**

The shaft side and cage show how the miners were transported underground (*cf.* **280.2**). The reconstructed roadways depict mining ranging from about the time of the 1885 novel, *Germinal* (**33**), up to the 1990s. Paid guides and volunteer former miners explain about communications, the methods of mining, safety, transport and ventilation underground.

*References*

https://www.musee-mine-blanzy.fr/

personal communication, Eberhard Lanzt, photographer, 2016

### 88. Musée de la Mine de Bruay La Buissière (Bruay La Buissière Mining Museum)

The Musée de la Mine de Bruay La Buissière is located in the Pas-de-Calais Département in the Nord-Pas-de-Calais. Former coal miners took eight years creating the museum, which opened in 1989. The museum traces the history of the mines of Bruay from 1855 until their closure in 1979. It includes methods of coal mining, such as working in

**Figure 87.2. Duplex-cylinder steam-winding engine produced in Châlon-sur-Saône by G. Pinet in 1885; now operated by compressed air (© Eberhard Lantz, Bamberg).**

a 40-centimetre seam, the technological development of mining tools and measures taken to prevent disasters, such as dust suppression, methane explosions and roof falls.

As well as 400 metres of mining galleries, through which visitors are escorted by former miners, there are two 40-millimetre films: the first a 25-minute film on coal mining in the Nord-Pas-de-Calais and the other about the underground visit in 1959 of General de Gaulle (1890-1970; prime minister, then president, of France, 1958-1969) at the number 6 shaft at Bruay. Audio-guides are available in English, French, German and Polish. One item of particular note is a block of coal weighing 3.5 tonnes, which was extracted intact in one piece from the number 2 pit of Maries les Mines in 1920.

*Reference*

http://www.apphim.fr/articles.php?lng=fr&pg=680

### 89. Musée de la Mine (Mining Museum), Épinac

Épinac is situated in the Saone-et-Loire Département of Burgundy in east-central France. The museum is located in the town hall of Épinac and deals with the industrial past of the region, especially coal mining, glass making and the railway. The fourth railway line to be constructed in France opened in 1836, to take coal from the mines of Épinac to the Canal de Bourgogne. Maximum production was reached in 1917 with 292,545 tonnes of coal. In 1891 the town became known as *Épinac-les-Mines* (Épinac-the-Mines), but reverted to being simply Épinac in 1966. The last pit was sunk in 1920 and the last mine closed in 1966, despite there still being ample coal reserves underground.

The displays include mining equipment, stoves and photographs of miners working underground and descending in the cage and by rope. Guided tours are offered, including the operation of some working machinery.

*References*

http://epinac.chez.com/histoire.html

showcaves

## 90. Musée de la Mine Jacques Déramaux (Jacques Déramaux Mining Museum)

The Musée de la Mine Jacques Déramaux is located at Auchel in the Pas-de-Calais Département in the Nord-Pas-de-Calais. L'Association, Mémoire de la Mine (Association for the Memory of the Mine) was founded by 12 former coal miners to convert the former mining school into a museum. Volunteers invested some 15,000 hours of work over an 18-month period starting in June 1986, with financial and material support from the municipality of Auchel. The museum attracts about 6,000 visitors a year. There are 11 volunteers, who are responsible for the events, guided tours, maintenance, organisation, reception and secretariat. The museum is self-funded and the municipality of Auchel provides financial support by paying for utilities such as electricity, internet and telephone.

The museum occupies an area of 3,000 square metres. There are 250 metres of simulated underground roadways, with 90-minute tours led by former miners explaining mining methods between 1734 and 1990, including using films of mining activities. The roadways have a typical timber roof-support system using packing walls behind the props. The tour covers the different methods of transport underground, ranging from a model of a pit pony in the late-nineteenth century pulling waggons of coal to an armoured face conveyor discharging its contents onto a tailgate conveyor for onward movement to the shaft bottom. Also on display is a 20-tonne double-ended ranging-drum shearer, 12 metres long, made by Anderson Boyes of Motherwell in Scotland, UK, and used in working longwall faces in coal seams averaging 2 metres in thickness. A troughed belt conveyor is suspended from the roadway roof supports by chains rather than using floor-mounted stands. Safety notices say, 'Warning. The conveyor may start at any time. Do not work on the conveyor without disconnecting from the [electrical] box' and 'Do not clean the rollers while [the conveyor] is operating'.

All mining machines on display are in full working order. Every 4 December, the feast day of Saint Barbara (**96**), the simulated underground is lit and visitors are offered a snack of the miners' meal of black coffee, bread and lard (used on the bread because butter would go rancid in the hot conditions underground).

The museum covers geology, fossils, the stratigraphy of the coal seams of this coalfield, the 300-year history of the mine, the social history of the coal miners and a reconstructed room from a typical house of a miner in about 1930. There is an exhibition on women in the mine and also a model of a mine, made by a former miner and donated to the museum by his widow after his death. Tours are also offered in English, French and Polish of the nearby spoil heap of the Ferfay Number 2 Mine at Saint Pierre les Auchel, with explanations of the origin of the spoil heap and its flora and fauna.

**Figure 90.1. Typical timber roof-support system using pack walls behind the props (© l'Association, Mémoire de la Mine).**

**Figure 90.2. Armoured face conveyor discharging its contents onto a tailgate conveyor for onward movement to the shaft bottom, steel telescopic roof supports and steel roof girders (© l'Association, Mémoire de la Mine).**

*References*


https://www.tourisme-bethune-bruay.fr/a-voir-a-faire/culture-patrimoine/le-musee-de-la-mine-jacques-deramaux-628615

museum leaflet

personal communications, Daniel Vos, president, Association, Mémoire de la Mine, 2016; Serge Hoyez, president, 2021


## 91. Musée de la Mine (Mining Museum), Le-Molay Littry

Le-Molay Littry is located in the Calvados Département of Normandy. The museum is on the site of the Frandemiche Colliery, which began production in 1760 and worked until 1864, at a depth of 97 metres. In 1840 the mine had about 950 workers, but shut in 1880 because of competition from coal produced in England and northern France.

The first steam engine used at the Littry coal mine was a Newcomen steam pump **(360, 371)**, which was in operation in an extraction shaft between 1750 and 1755. In 1804 the building housing the pump was converted into a chapel for the miners. It was demolished about 1900, and no traces remain of either the machine or its building.

The Musée de la Mine opened in 1907 as the first museum in France dedicated to coal mining. It has the designation of a *Musée de France* (Museum of France), which makes its collections inalienable, but the site itself does not have statutory protection, despite its historical importance. When the museum was renovated in 1997, archaeological excavations carried out in its courtyard exposed the foundations of the heapstead with the top of the mine shaft and the foundations of two steam boilers, one dated by an inscription to April 1846. These boilers would have had high chimneys to exhaust the fumes they produced, a method developed in the early-nineteenth century.

Visits usually take about 90 minutes, with audio-devices, display panels and a video in English and French. The museum, which is operated by the municipality, attracts about 5,000 visitors a year and there is one member of staff. Two major themes are covered in the 700 square metres of displays: the history of mining in Littry in the eighteenth and nineteenth centuries and the evolution of mining techniques up to about 1960.

**Figure 91.1. Exterior of museum building and courtyard with wall foundations uncovered during excavations in 1997 in foreground and 1846 chimney to right (© Musée de la Mine, Le Molay-Littry).**

The displays of the museum were renewed at the beginning of 2021. The visit starts in the courtyard with the interpretation of the archaeological excavations of Frandemiche Colliery to introduce the public to the surface facilities of a mine. The permanent exhibition occupies the main building of the museum. A video on the history of the mine is shown at the start, then the work of the mine and the lives of the miners are explained using audio-testimonials, machines, models, paintings, tools and samples of coal and fossils. Interactives, such as moulding fossils, a quiz on the polluting effects of burning coal and puzzles, are used throughout the displays.

The most important item on display is a steam engine used, not for pumping, but to raise coal from the bottom of the pit. This is a condensing steam engine of 4 horsepower, based on the principle of the double-acting engine patented in 1783 by James Watt (1736-1819), operating by direct steam pressure and not on the atmospheric principle like a Newcomen engine **(371)**. The Compagnie des Mines de Littry bought three engines from the Périer brothers, Joseph (1786-1868) and Casimir (1777-1832). Without further research it is not possible to know whether this is the engine supplied by the Périer brothers' Chaillot (Paris) workshop to the company in 1800, the one installed at Frandemiche Colliery in 1802 or the last machine delivered to the company in 1811, although it seems most likely to be the 1802 machine. When the company closed in 1880, the engine was purchased by a private collector, François Sosthène Lefrançois (1833-1905), director of the Portuguese railways. His widow donated it to the museum 25 years later to form the centrepiece when the museum opened in 1907. This engine has undergone modifications since its construction. Much of the metalwork, such as the cylinder and pipework, is original, however, and it is believed to be the oldest surviving stationary steam engine in France (the oldest steam-powered machine is the 1771 Fardier de Cugnot, the first motor vehicle powered by an engine, which is displayed in Paris). This engine has not yet been the subject of a detailed technological study, with disassembly and examination of the interior of the cylinder, the piston, metallurgical study of its main iron and cast-iron parts, *etc.* This lack of study, combined with its age and rarity, means that it is unlikely that the engine

**Figure 91.2. Steam engine, probably dating to 1802, used for winding rather than for pumping; note the braking mechanism just visible on the inside of the large flywheel (© Musée de la Mine, Le Molay-Littry).**

will ever be restored to operate. The options in the future would appear to be to make an animated digital model or, if funding were available, to construct a full-size operating engine.

At the end of the displays, there is a model of the number 5 shaft of the Bruay-en-Artois Hauts Mine in Artois in the Pas-de-Calais at a scale of 1 to 10. This model was displayed at the palace of mines and metallurgy at the Paris world fair in 1900. The model has three monumental buildings side by side: the engine room with the boilers, the air compressors and ventilation engines; the integrated winding-engine house and heapstead, with the headgear, steam-winding engine, lamproom and pithead baths; and the screens building.

A 70-metre-long reconstructed mine gallery shows the evolution of mining in Littry Mine in the eighteenth and nineteenth centuries and the working conditions and technology in the underground mines of northern France up to about 1960. Ponies could not be used underground to haul the coal because of the extremely low roof in Littry Mine. In 1834 the first underground narrow-gauge railway was installed at the mine, with more than 4 kilometres of track in place by 1836. The museum has two of the first wooden mine tubs fitted with cast-iron wheels.

**Figure 91.3. Altar table canvas of 1804, showing Saint Barbara, above a group of miners and a mine building, and below a wooden barrel for measuring coal (© Musée de la Mine, Le Molay-Littry).**

Having started collecting at such an early date, the museum has a number of unique or important objects. An altar table canvas of 1803, which may have been made following a disaster, shows Saint Barbara **(96)**. She is depicted in a black cloud (possibly the result of an explosion), above a group of miners, a mine building and tools on the ground, including a wooden barrel for measuring coal. The museum has such a measure with three iron hoops and two strong iron handles, which could hold about 50 litres of coal. The museum displays a two-volume manuscript dating from 1796/1800, giving the history of Littry Mine and the techniques and applied economics of the Littry Mines Company. This was compiled by Louis-Etienne-François Héricart Thury (1776-1854) for presentation at a mining conference held on 28 May 1800. He subsequently became a member of the board of directors of the Littry Mines Company and then, in 1809, was appointed inspector-general of quarries of Paris.

Lighting underground at Littry Mine was done by candles, until safety lamps were adopted by the company in 1857. The candle was mounted in an iron candleholder, which had a sharp point to enable it to be stuck into the timbering. Also in the collections are eighteenth-century tokens with the inscription *MINE DE LITTRY* (Littry Mine) on one side and on the other a crossed shovel and pick symbolising the mine, together with a number–10, 12, 15 or 18. The numbers show the sum in *sols* (the currency in use before the 1789 French Revolution; 1 *sol* = €0.22) that a miner had earned.

Three wooden wheels, which probably date from the late-eighteenth century, are among the very few such objects preserved in a European mining museum. Before the introduction of steam engines, these were placed at the bottom of the pit to turn the rope to lift coal tubs from the horizontal to the vertical. The winch was secured to a gantry above the shaft and turned by horsepower.

*References*

https://journals.openedition.org/insitu/12091

https://www.ville-molay-littry.fr/tourisme/musees/le-musee-de-la-mine/

personal communications, Philippe Bernouis, curator, 2016, 2017, 2021, 2022

showcaves

### 92. Musée de la Mine Marcel Maulini (Marcel Maulini Mining Museum)

Ronchamp is located in the Haute-Saône Département of the region of Franche-Comté in eastern France, between the Vosges and the Jura Mountains. Mining began here in the middle of the eighteenth century and the Ronchamp mines were visited in 1767 by Antoine Lavoisier (1743-94), the father of modern chemistry; his diary of observations was published in 1777 with information about the beginning of mining from coal outcrops. At the height of its activity,

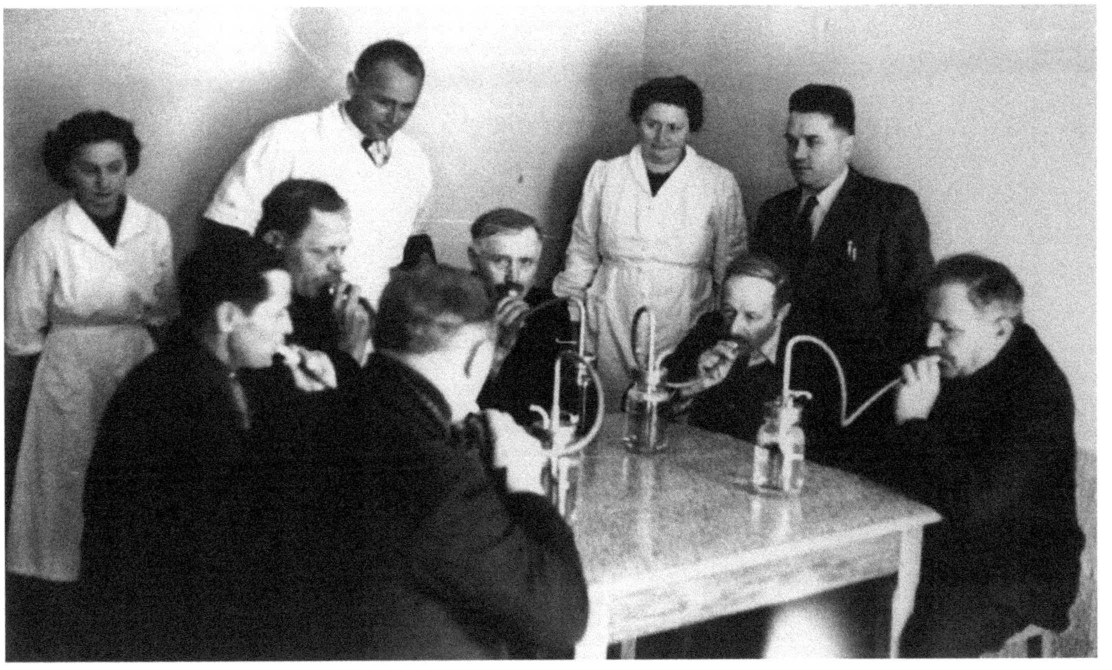

**Figure 92.1. Image from the collections of treatment of miners with silicosis (© Collection Musée de la Mine Marcel Maulini).**

the coalfield employed 1,500 people, who produced about 200,000 tonnes of coal each year. From about 1900 the profitability of the coalfield began to decline and the last six pits in the region, and the associated power station, closed in 1958.

The museum opened in 1976, based on the collections of Dr Marcel Maulini (1913-83), who had been the doctor of the Ronchamp mines between 1946 and 1978 and after whom the museum is named. In 1991 it was acquired by the municipality, which manages and funds the museum. The deed of gift of the collections, however, contained covenants which made its further development impossible. In 2009 these clauses were lifted by deed, so that it is now possible for the museum to evolve further.

The museum traces two centuries of mining activity in the coalfield. The displays cover work in the mine, the development of the mining company, social life, social legislation and the technology and tools used in Ronchamp. There are two members of staff with two volunteers, and the museum attracts about 3,000 visitors a year.

The museum is housed in a three-storey building, occupying 350 square metres. On the ground floor there is currently a large collection, most of which was donated by miners, of original photographs and various types of tools and artefacts associated with mining and the daily life of miners, including a major collection of mine lamps. There are also displays with plans and maps about the discovery and exploitation of coal in the area. The first floor exhibits artistic works related to the mine, including lace and embroidery, paintings and tapestries. There are displays about the large Polish community who worked in the mine after WWII, documents on silicosis and images showing the development of silicosis in the miners' lungs (in which Dr Maulini was especially interested) and treatment of patients.

Amongst the photographs in the collections are ones of the facilities established in Ronchamp in 1947 to treat silicosis and of the Sainte Marie Mine, which opened in 1866 and whose headgear was strengthened in 1924. The site was cleared after the mine closed in 1958 and it was intended to demolish the concrete headgear. Opposition from the local community enabled Dr Maulini to save the headgear in 1972, which was given statutory protection as an historical monument in 2001. Temporary exhibitions are held on the second floor, which also houses the archives of the Ronchamp collieries.

The museum is currently in the course of a major development programme, which is scheduled to be completed by 2024. It is intended to redisplay the permanent exhibition with the aim of bringing in new audiences by developing the cultural, educational and tourist offer and reconnecting with the spirit of Dr Maulini's initial project. There will be building work to provide spaces for temporary exhibitions, a reception area for group visits and premises for administration, conservation and documentation.

*References*

https://en.wikipedia.org/wiki/Sainte_Marie_Coal_Mine

https://fr.wikipedia.org/wiki/Houill%C3%A8res_de_Ronchamp

https://www.mineronchamp.fr/

http://www.abamm.org/

personal communications, Liliane Frahier, heritage assistant, 2016; Stéphanie Millot, cultural animator, 2021

showcaves

**Figure 92.2. Image from the collections of the concrete headgear of the Sainte Marie Mine (© Collection Musée de la Mine Marcel Maulini).**

### 93. Musée de la Mine, Puits Hély d'Oissel (Mining Museum, Hély d'Oissel Pit)

The Musée de la Mine is located on the site of the former Puits Hély d'Oissel, just west of Gréasque, a commune north-east of Marseille in the Provence-Alpes-Côte d'Azur region. It is situated between Aix, Aubagne and Marseille. The mines of the area were profitable, having won the prize for the highest work productivity in France three years in a row during the 1950s. The last mine in the region shut in 2003.

Mining on this site started in 1919 at a depth of 450 metres, one of 51 shafts sunk in the region. The mine closed in 1962. The museum was created by an association named, in the Provençal language, la Carbouniero de Prouvènço (the Coal Mining of Provence). This was set up by the mayor and the president of the regional council, and opened to the public in 2000. It is run by an association of 14 towns and villages of Provence. The museum is funded

**Figure 93.1. Latticework headgear and heapstead (© Musée de la Mine, Puits Hély d'Oissel).**

**Figure 93.2. Electrically driven winding engine (© Musée de la Mine, Puits Hély d'Oissel).**

by those towns and villages, some individual supporters and subsidies from the local and regional administrations. The work of the museum is overseen by a scientific committee. There are three employees and between 30 and 40 volunteers. The museum attracts about 6,000 visitors each year.

The entire site was designated as an historical monument in 1989. Some nearby miners' houses were also classified as historical monuments. Some nineteenth-century mines are preserved in the forest nearby, but have not been protected. The steel-lattice headgear (**204**) stands 25 metres high and was first erected in 1917, although it had to be replaced after it collapsed in 1922. Below the headgear the brick-built engine room houses the museum's mining-historical division, established in 2000, which collects documentation about underground resources, such as rocks and fossils. Guided tours are available in the mornings and early afternoons, with audio-guides available the rest of the time. The tours start from the models room and then the engine room, which contains the electrically driven winding engine. Its cylindro-conical drum is shaped with conical ends to help even out loads at the start of a wind. The cage at the bottom of the shaft had its rope on the smallest diameter, in effect the same as first gear on a car; as the rope worked up to larger diameters, the cage would travel further for each revolution of the drum until on the parallel section of the drum. One of the surviving cages of the mines was removed from the shaft and is displayed adjacent to the heapstead.

A short section of a recreated underground roadway in a closed gallery on the surface shows different methods of roof support, including traditional wooden pit props and steel mechanical props used during the first half of the twentieth century. Displayed here at present are the three main forms of personal light for individual miners: the flame-safety oil, or Davy, lamp (**355**); the carbide, or acetylene-gas, lamp; and the battery-powered cap lamp. There is a permanent display of the lamps of different periods and from various countries in a new space near the simulated underground.

The exhibition building has three rooms on the ground floor and two on the upper floor, totalling 536 square metres. The displays illustrate the history of the mining of lignite in this region, ranging from working coal on the surface where it outcropped, to mines such as the Morandat shaft, 1,350 metres deep, which was the last shaft worked in Provence. They cover the operation of mines, safety and monitoring systems, and problems encountered in closing down the mines. There are displays of large-scale underground mining machinery, including free-steered vehicles, roadheading machinery and a pneumatic wheeled personnel carrier, which would have been used in areas of the mine not equipped with rail tracks. A typical longwall plough-worked coalface (which underground could have been anything up to 150 to 200 metres in length) has been reconstructed on the surface. As well as the plough, its associated armoured face conveyor, powered roof supports, or chocks, and tailgate conveyor are also shown in shortened form. Work is underway on a smoke system to give an impression of the dust found underground.

Guides are available to explain how the machinery worked. There is also a 40-horsepower diesel mining locomotive, weighing 7 tonnes. A geology room displays local fossils and describes local geology, different coals throughout the world and the paleoenvironment of the local countryside, 85 million years ago. Active collection of fossils is continuing.

*References*

https://museeminegreasque.fr

https://museedelaminegreasque.business.site

http://www.ville-greasque.fr/village/puits-hely-doissel

personal communications, Lucile Decombe, manager, 2016, 2021

showcaves

### 94. Musée de la Mine (Mining Museum), Saint Étienne

The area around Saint Etienne was for several centuries one of the principal sources of energy in France. Coal was the basis of the textile and engineering industries of Saint Étienne in the Massif Central, 50 kilometres south-west of Lyon in the Rhône-Alpes region of eastern-central France. The area had been one of the main suppliers of coal in France, but the last mine ceased production in 1984. The Musée de la Mine is located at Couriot Colliery, which worked between 1913 and 1972 and was one of the first in France to have surface machinery operated by electric power. The mine produced about 3,000 tonnes of coal a day and employed 1,500 miners.

The museum opened in 1991 to commemorate mining in the Saint Étienne coalfield. The tall steel headframe has been preserved and there is a 350-metre-long reconstructed roadway on the surface. The beginning of the display has the changing rooms for the miners, together with the pithead baths. Visitors ride along the roadways on an underground-type railway, alongside the conveyor belts installed in most collieries in the twentieth century and the machines used for cutting coal. The winding engine and the generating station that supplied it with power are preserved. There is also an audio-visual programme on the history of mining in the region.

*References*

ERIH, *s.n.* Saint-Etienne

http://www.musee-mine.saint-etienne.fr

### 95. Musée les Mineurs Wendel (Wendel Miners' Museum; formerly Musée du Carreau Wendel: Carreau Wendel Museum)

The Musée les Mineurs Wendel is located in Petite-Rosselle, in the Moselle Département of Lorraine in north-eastern France. The museum is close to the border with Germany and less than 5 kilometres from the Erlebnisbergwerk Velsen (**112**). Wendel Colliery was sunk in 1868 and closed in 1986, although the shafts remained in use for other mines in the area: number 1 shaft closed in 1989; number 2 in 1992; and number 3 in 2001. The museum deals with mining in Lorraine, which was the location of the very last French coal mine, La Houve, which closed in 2004.

In 1985 the Scientific Culture Centre was created and opened the first mining museum in Folschviller, called the Centre de Culture Minière (Centre of Mining Culture). This closed in 1988 and the collections were moved to Carreau Wendel, but the not-for-profit organisation of volunteers did not have the finance available for a proper museum. In 1998 a *syndicat mixte* (joint venture) was formed of various public authorities, which opened the totally renovated museum in 2006 as the Musée du Carreau Wendel. A new drift roadway was inaugurated in 2012 by the president of the Lorraine region, when the museum was renamed as the Musée les Mineurs Wendel.

In the basement there is a reconstructed mine, reached by a simulated cage travelling for two minutes at 12 metres per second. Visitors are escorted by a former miner, who shares stories about working life and the various working techniques. The whole tour is accessible to people with reduced mobility. At the entrance to the underground is a statue of Saint Barbara (**96**). The display covers the development of mining through the various dip slopes, from slightly sloping roadways to ones at an angle of between 30 and 45 degrees. As far as possible, the mining technique used was to bring down the coal from the roof straight onto the conveyor belts. The equipment on display includes an AM100 heading machine, a G210 electric loader, a 2000 shearer, roof supports and conveyor belts. Lighting equipment, pumps, skips, transformers and ventilation systems are also on display. The humidity, smell and temperature of a mine are recreated for authenticity. The trip ends with a film of a miner's working day underground. Visitors are given polarised-light glasses so that they can see in three dimensions.

Part of the site is designated as an historical monument, and includes all the surface buildings needed for the operation of a mine. At the entrance is the 1889 headgear from number 2 shaft, which is the oldest steel headframe extant in Lorraine. The site includes a coal-preparation plant, lamproom, pithead baths, power building, stores building and three workshops. The functions of the various buildings are explained using photographs from the 1950s and 10 terminals, with commentary in English, French and German.

The museum's collections are comprised of almost 5,000 objects and machines, donated by collieries in Lorraine. They cover engineering, geology, underground and surface transport and the everyday life of the miners, and include material from related industries, such as the chemical industry, coking plants and power stations. There are also training materials and safety equipment, including safety posters, stretchers and rescue equipment. Some compressed-air, diesel, electric and monorail locomotives are on permanent display and others are used in temporary exhibitions. The museum is still actively collecting and also has a documentation centre, with books and documents on the history of coal mining in Lorraine, mining techniques, safety, training, *etc.*

*References*

ERIH, *s.n.* Petite-Rosselle

showcaves

www.musee-les-mineurs.fr

# Germany

In Germany the main lignite coalfields are the Lusatian in the east; the Central German area; the Helmstadt area to its north; and the Rhineland area in the west. There are anthracite coalfields in Ibbenbüren in the north-west and the Saar-Warndt in the south-west, which extended across the border into France, which took ownership of the coalfield between 1918 and 1933 and again between 1945 and 1957. The most important coalfield in Germany is the Ruhr in the west, north of the Rhineland area. Germany has 4 billion tonnes of lignite reserves (eurocoal). There are more preserved coal-mining sites in Germany (https://www.bergbau-sammlungen.de/en) than any other country in the world. Germany was one of the leading countries in Europe in technology and mining techniques from the first recorded mining in the early-twelfth century (**97, 99**) until about 1500. Then in the nineteenth and twentieth centuries, although behind the UK in amounts mined, Germany was extremely innovative in technology (**104, 115, 127, 141, 154, 171, 177, 204, 212, 227**). This was then often adopted in adjacent countries, such as Belgium. Boys younger than 16 and women and girls never worked in German coal mines and miners in general had better conditions than other German social groups (Slotta 2011, p. 188).

In 1871 the various princely states of Germany were unified, with the Prussian king, Wilhelm I (1797-1888), becoming emperor of the whole country. This led to an outpouring of nationalism and the development of new institutions, such as the Royal Prussian State Geological Institute, established in 1873 (Ranestad and Sharp 2020, p. 5). With its rich mineral reserves, the Ruhr experienced an industrial boom following unification. In the 1890s more than 70 new mines were developed in the coalfield and between 1880 and 1913 steel production increased from 2 million tonnes a year to 17 million tonnes. With patriotic fervour at its height, a number of mines were named after royalty, such as the Zeche Kaiserstuhl (**137**) or the Bergwerk Auguste Victoria (**140**); after government ministers, such as the Zeche Graf Bismarck (**178**); or after heroic figures from Germanic history, such as the Zeche Radbod (**198**). Other mines were named after their owners, members of the aristocracy or staff members. Partly as a matter of nationalistic pride and partly in competition with each other, the large mining companies strove to outdo each other in the magnificence of their surface architecture (**129, 228, 233**) and the high technical quality of their underground machinery. They employed leading firms of architects, who designed the elaborately decorated facades of the surface buildings to be in harmony with each other and with equally magnificent interior decoration, such as elaborate wrought ironwork, stained-glass windows and tiled floors. As a result, unlike elsewhere in the world where surface buildings tended to be regarded as relatively impermanent and strictly industrial in nature, many of the Ruhr coal mines have buildings that merit preservation in their own right and so can be converted for other uses much more easily. This was not just in the later-nineteenth century and early-twentieth century (**109. 120, 132, 154, 159, 162, 164, 167, 178, 185, 187, 202, 203, 205, 209, 219**), but had been prefigured in 1850 (**213**) and extended into the 1920s (**99, 144, 180, 182, 195, 222**), then into the Nazi era (**113, 212, 233**) and even into the 1950s (**157, 173, 221**). Also architect-designed are the high rectangular-cube winding-engine towers of the Zeche Königsborn (**144**) and Zeche Sachsen (**205**), which are believed to be the original models for modern oil rigs.

During WWI there were serious labour and material shortages, which caused major changes in labour relations, production methods and the workforce (Przigoda 2016). Coal production increased during the later 1930s as Germany prepared for war. At first during WWII there was a shortage of coal, but this was resolved by the annexation of coal-rich Upper Silesia in Poland and the use of forced labour (Farrenkopf 2016, p. 217). This use of forced labour and prisoners-of-war in WWI and WWII is fully acknowledged in museums and publications in Germany (see, for example, **129**), unlike in Japan (see **introductions to Indonesia, Japan and South Korea**).

At the end of WWII the authorities tried to destroy mining infrastructure (**109**). For a period following both WWI and WWII mines in the Ruhr and the Saar came under French control, but returned to Germany following plebiscites of the population in 1935 and 1957. In 1925 a great rationalisation programme was carried out of the whole Ruhr coalfield, leading to the closure of many mines, and many surviving mines were badly damaged during WWII (**175, 186, 210, 212, 217, 225, 232**). A second major round of closures occurred during the 1960s, the first closure of a coal mine after WWII being the Schachtanlage Klosterstollen Mine in 1957 (**124**), and the decommissioning of collieries then continued.

In 1949 Germany was divided into two separate countries, which were only reunited in 1990. During that period the Deutsche Demokratische Republik (German Democratic Republic), commonly known as East Germany, pursued a policy of large-scale opencast mining, including the use of enormous machines (**101, 106, 114, 119, 121, 230**), the removal of entire villages to enable the expansion of mines (**111, 123, 229, 230,** *cf.* **71, 245, 412**) and in one case changing the course of a river (**229**), and the creation of large lakes (**101, 106, 110, 111, 114, 119, 121, 123, 229**).

Although the last two deep mines closed in 2018 (euracoal), Germany is still the largest user of coal in the EU and continues to produce large amounts of carbon dioxide. There are three massive opencast lignite mines in North Rhine-Westphalia, for which some 50 villages have been destroyed by the RWE company. Heavily polluting lignite from opencast mines such as the Tagebau Welzow-Süd (**230**) provides 28 *per cent* of the fuel used by German power stations. In January 2020 the German Government committed to phasing out the use of coal by 2038, and in April 2021 the Federal Constitutional Court forced the government to change its commitment to greenhouse-gas neutrality from 2050 to 2045.

### 96. Barbarastollen (Barbara Gallery)

The Barbarastollen in Cologne in the state of North Rhine-Westphalia is a simulated mine, created in 1932 by the painter and graphic designer, Kurt Holl (1938-2015), from Essen. It was named after Saint Barbara, the patron saint of miners. Saint Barbara was isolated from the world by her father, Dioscorus, who imprisoned her in a tower. She became a Christian despite her isolation. Statues often show her holding a miniature tower or standing next to her tower. Her father had a bath-house constructed for her. It was planned to have two windows in the bath-house, but Saint Barbara had three windows inserted, to represent the Christian Trinity (Father, Son and Holy Ghost). She refused to marry a pagan husband, so she suffered many tortures to try to force her to change her mind. But each night the prison in which she was held was bathed in heavenly light and her wounds were healed. She is also sometimes shown with the chains with which she was bound. Eventually her father beheaded her. In punishment a bolt of lightning struck her father dead. This link to violent explosions made her the patron saint of anyone who worked with explosives or cannons, such as artillerymen, forge workers, gunsmiths, tunnellers or miners. These events supposedly took place on 4 December sometime in the third century CE in Baalbeck in Lebanon or Nicomedia in Turkey, but there is doubt about her historical existence. In the twelfth century her supposed relics were moved from Constantinople (now Istanbul in Turkey) to Kyiv in Ukraine.

During the fourteenth-century plague, she was invoked as the patron saint of the dying, while in the German Rhineland cherry branches were put in warm water on her feast day, 4 December. If flowers then appeared at Christmas, this meant a good harvest and, for marriageable girls, the number of flowers showed the number of suitors to be expected, their appearance and property.

The tunnel was designed for the students of the Wirtschafts- und Sozialwissenschaftlichen Fakultät (Faculty of Economic and Social Sciences) of the Handelshochschule Köln (Graduate School of Cologne), which later became part of the University of Cologne. It was intended to teach the students coal-mining techniques and to provide a visual object and inspiration for artwork. The 40-metre-long tunnel then became part of the Museum für Handel und Industrie (Museum of Commerce and Industry), but was abandoned and forgotten during WWII.

The replica was rediscovered by accident in the early 1980s and Professor Claus Piekarski, Director of the Institute and Polyclinic for Occupational Medicine, Social Medicine and Social Hygiene, organised the restoration of the roadway in 1984. The tunnel reopened in the early 1990s in the ownership of the Institut fur Arbeitsmedizin, Umweltmedizin und Präventionsforschung (Institute for Occupational, Environmental and Preventive Medicine). Tours are taken by professors of the institute, who include material on typical miners' diseases, such as pneumoconiosis and silicosis. The mine still shows conditions in 1930s German mines, similar to those that still exist in many less-technologically advanced countries today.

Access for groups of up to 30 students from the auditorium of the university to the basement-room tunnel is by a fully operational mine cage. The coal used came from the mines near Aachen and the machinery and tools from mines all over Germany. The walls show layers of sedimentary rocks and sloping coal seams. The roof is supported with metal arched girders and the walls have wooden planking, in the German and Polish style. There are rails in the floor for a mine train with some mine cars. The equipment includes compressors, a conveyor belt and drilling machines. The temperature is about 25 degrees Celsius.

Because of its location and nature, the simulated mine comes under German mining legislation. So there are fire doors, a protected telephone system and a working ventilation system, all used to avoid the danger of methane explosions. Similarly objects forbidden underground in working mines, such as tinfoil, are prohibited here also.

*References*

http://www.bahnen-im-rheinland.de/wb/colund.htm

http://www.bergbaumuseum-grube-anna2.de/index.php/276-2/

http://www.philtrat.de/articles/509/

http://www.philtrat.de/articles/1806/

showcaves

### 97. Bergbaudenkmal Adolf (Adolf Mining Monument)

The Bergbaudenkmal Adolf is located in Herzogenrath in the district of Aachen in the state of North Rhine-Westphalia, adjacent to the border with the Netherlands. The settlement is divided artificially between the German town of Herzogenrath and the Dutch town of Kerkrade, which formed the western half of Herzogenrath before the independence of the Netherlands in 1648. Aachen (known in French as Aix-la-Chapelle) is considered to be the oldest documented coal-mining area in continental Europe, because the mining is mentioned as early as 1113 (but *cf.* **45**).

Once rail links were opened between Cologne and Aachen in 1841 and Mönchengladbach and Aachen in 1853, it became possible to move the coal to markets that were further afield, including to transport on the river Rhine from Cologne and Neuss. The result was that gradually control of most of the production passed to the large mining companies, including the Eschweiler Bergwerksverein. Sinking of the Grube Adolf (Adolf Mine) began in 1889. It was named after Baron Adolf von Steffens (1817-98), chair between 1871 and 1898 of the board of the Eschweiler Bergwerksverein, which also owned the Grube Anna (**109**). There were considerable water problems with the sinking and it was not until 1908 that the seam of high-quality bituminous coal was reached. In 1909 problems at the Grube Anna affected the ventilation of the Grube Adolf. The first winding engine, which is a double-action twin-piston steam engine, was installed in 1910 and production finally began in 1913. New estates were also provided in Merkstein for the workforce, based on the model of the garden city of the Krupp factory in Essen.

In 1923 a separate ventilation shaft was sunk. The fan house and the ventilation-shaft/winding-engine house, in which the 1913 winding engine was installed, were built in 1922/3 to designs by the Cologne firm of Erberich and Scheeben. The mine had its own sidings at Merkstein railway station and its own branch line to the coking plant at Grube Anna. In 1929 the Grube Adolf was the first mine owned by the company to be converted to electricity underground and in the 1930s a modern coal-preparation plant was constructed. In 1933 the winding-engine house was extended and positioned above the radial fan, which still survives. In 1939 the central region of the elongated brick hall was extended with a gable and ridge perpendicular to the original building. The still-surviving axial fans were added north of the building in 1954/5.

The mine was damaged by bombing in WWII, during which some 850 forced labourers were killed in the Aachen district. Production was renewed after the war, so that peak production was achieved in 1959, with a total of about 939,700 tonnes of coal. In the early 1970s the mine employed 2,900 men, reaching a production of 3.4 tonnes per miner in 1971. In 1972 the mine was joined underground to become part of the 860-metre level of the Grube Anna, and the Grube Adolf site became superfluous. The Grube Adolf shaft was retained until 1990 to provide underground ventilation.

Most of the buildings were demolished, apart from the winding-engine house and the fan house. The spoil heap next to the mine also survives. The not-for-profit organisation, Bergbaudenkmal Adolf Verein (Adolf Mining Monument Association), was constituted in 1993 to preserve the remainder of the site, as well as surviving traces of mining in the city of Herzogenrath and the Wurm Valley. Other aims of the foundation are to revive local mining traditions, such as the role of Saint Barbara (**96**), and to form links with other mining associations in the Belgian and Dutch regions of Limburg and Wallonia.

Repairs to the building began in 1999. Funding provided by the state allowed the second section to be converted into a visitor centre, which opened in 2007. The 1913 steam-winding engine has been restored by members of the foundation. It is demonstrated to visitors, powered by compressed air, limiting it to 1,000 horsepower, as it had been between 1977 and 1990. There are also displays about the life and work of the miners and about the uses of coal. The adjacent fan house has been converted into a ballroom, meeting room and training room.

*References*

http://bergbaudenkmal-grube-adolf.de/

https://de.wikipedia.org/wiki/Grube_Adolf

museum leaflets

### 98. Bergbaumuseum Ibbenbüren (Ibbenbüren Mining Museum)

The Bergbaumuseum Ibbenbüren is located in the district of Steinfurt in North Rhine-Westphalia. Coal mining here goes back to at least the sixteenth century. The museum is housed in the turbine hall of the former power station on the premises of the RAG Anthracite Ibbenbüren GmbH Company, which operated the last northernmost coal mine in Germany. The company employed almost 2,200 people, producing about 1.9 million tonnes of high-quality anthracite a year from a depth of 1,600 metres, amongst the deepest in Europe.

The displays are of items dating mainly from the eighteenth to the twentieth century, including mining tools and machinery. There are models of ploughs from the early mining period up to the modern day. Face and roof supports are on display, again from early examples with wooden timbering up to modern arched girders and covered roof supports. Visitors are able to use mining hammers and other implements. Many of the machines are fully operational and able to be set in motion for visitors. The most impressive is a traversing steam capstan for raising and lowering pitwork in the shaft, mainly very heavy pumps and the associated pipework. Having the capstan on a traversing platform meant the raised pitwork could be moved safely away from the shaft top and then lowered to the ground. This is unlike a traditional fixed installation, where pitwork would have to be manhandled to immediately beside the shaft, with the consequent risk of it accidentally falling back down the shaft. Being very low geared, it was not a fast method of raising and lowering equipment. Its very high torque, however, allowed it to deal with extremely heavy weights and with a high degree of precision; for example, when trying to line up bolt holes when joining pipework together. It was built by the Gutehoffnungshütte Company at Sterkrade (**207**) and was installed as part of an improved

**Figure 98.1. Traversing steam capstan installed in 1894 for raising and lowering pitwork in the shaft (© Eberhard Lantz, Bamberg).**

pumping regime after the colliery suffered an inrush of water in 1894.

The museum has a range of coal-fired stoves, many of them elaborately decorated. Included in the displays are relevant minerals, together with fossils from the mine and from the geologically diverse region between the rivers Ems and Weser. Examples of paintings of mining activities from the museum's extensive art collection are also on display.

*References*


museum leaflet

personal communication, Eberhard Lanzt, photographer, 2016

showcaves


### 99. Bergbaumuseum Oelsnitz (Oelsnitz Mining Museum)

The Bergbaumuseum Oelsnitz is located in the east German state of Saxony. Mining of anthracite coal began here in the twelfth century, making this one of the oldest coalfields in Europe. The museum is located on the site of the Kaiserin-Augusta-Schacht Mine (Empress-Augusta-Shaft Mine), named after Augusta von Sachsen-Weimar-Eisenach (1811-90), wife of Kaiser Wilhelm I (1797-1888) and first empress of Germany. In 1946, under communist rule, it was renamed as the Karl Liebknecht Schacht (Karl Liebknecht shaft), after the German socialist (1871-1919). The colliery closed in 1971 and the museum opened to the public in 1986 after 10 years preparation. Additional space was added in 2010, giving a total exhibition area of 6,000 square metres and an outdoor display area of 1,000 square metres for large machines. Since 2019 the museum has been undergoing a major redisplay, scheduled for reopening in autumn 2023.

The museum, one of the largest mining museums in Germany (*cf.* **108**), is dedicated to coal mining in Saxony. It has 14 employees and 20 freelance workers, and is run by the Erzgebirgskreises, an administrative body of the Erzgebirge Ore Mountains of Saxony. The red-brick colliery building, the 50.75-metre-high winding tower, and the steam-winding engine were designated as cultural heritage in 1976. The buildings include the gatekeeper's house, generator house, heapstead (incorporating the administration building) with pit bank and cages (similar to **387.4**), lamproom, office building, pay office, pithead baths (called the Revierstuben and Steiger baths) and a stair tower with a railway crossing.

**Figure 99.1. Exterior of museum, showing friction-winder tower and steel headgear (© Bergbaumuseum Oelsnitz).**

The permanent displays deal with the geology of coal and the history of coal mining in Saxony. The economic, technical, technological, social and societal infrastructure are covered, together with the development of the region from an agricultural to an industrial landscape. The displays include a full-scale reproduction of a forest from the Carboniferous and Lower Permian eras (*cf.* **226, 241**), modelled on fossil finds from the Döhlen and Zwickau-Oelsnitzer coalfields. This includes a replica *Pantelosaurus saxonicus* (Saxon complete lizard), an early mammalian precursor, known only from parts of six individuals. These were all collected in 1901 at the Königin-Carola-Schacht, 90 kilometres away. Agricola (**108**) described how a coal seam in Planitz near Zwickau burned underground from 1476. The exhibition shows how from 1837 this geothermal energy was used to grow tropical fruit, such as pineapples, until it was extinguished in 1860. Also covered is the innovative work on mine lighting by the Friemann and Wolf Company, founded in Zwickau in 1884; an offshoot of this company is the Wolf Safety Lamp Company in Sheffield in England (Jackson 2012, p. 1).

The main building is a good example of the art-deco architectural style of the 1920s. The imposing generator house contains a Leonard converter from 1923 with an 18-tonne flywheel; during the period of active coal production, the incoming three-phase alternating-current voltage was converted into direct current to drive the winding engine. The 1923 1,500-horsepower twin steam winder in the winding-engine house is the largest in Saxony that is still operational.

Guides take groups of visitors on tours of the simulated 400-metre-long roadways three times a day, starting from the winding tower erected in 1922/3. Visitors gather in the ticket office, which is the former gatehouse of the colliery. The tour crosses the tracks of the railway by a bridge built in 1939, which leads to the rotunda, which was the administrative building and housed the miners'

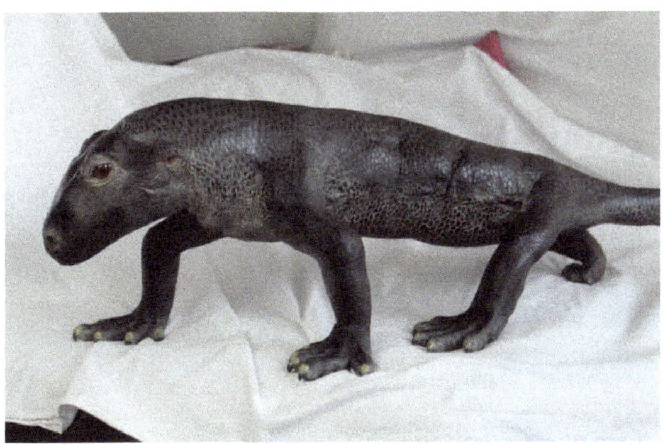

**Figure 99.2. Model of *Pantelosaurus saxonicus*, a pelycosaur formerly classified as a reptile but now considered early land vertebrates that resemble reptiles (© Bergbaumuseum Oelsnitz).**

Figure 99.3. Pit bank and cages, with waggons waiting to go underground (© Bergbaumuseum Oelsnitz).

Figure 99.4. Reproduction of a Carboniferous forest (© Bergbaumuseum Oelsnitz).

canteen. The technical development of coal-mining machinery is demonstrated using machines in action. Some of the types of machines developed in Saxony were transferred to the Ruhr mines of western Germany. The simulated mine contains machinery and tools such as an air pick and cutting machine, with an underground locomotive pulling tubs full of coal.

Large mining machines are on display in the grounds of the museum, including a class 52 steam locomotive. A narrow-gauge suburban railway ran between the cities of Oelsnitz/Erzgebirge and Hohenstein-Ernstthal from 1913 until 1960 to provide transport for the miners. The exhibition displays a car from this system and the locations on the coalfield that were connected by the surface railway.

Further afield the Friends of the Bergbaumuseum Oelsnitz are developing a 240-kilometre Route of Saxon Coal Mining. The trail will run first between Oelsnitz and Zwickau, encompassing industrial architecture, mining routes and spoil heaps. It will then continue to Chemnitz, Freiberg, Olbernhau, Freital (**227**) and Dresden. The route will follow the traditional course by which coal was moved from the coalfield to the main economic and industrial centres of Chemnitz, Dresden and Zwickau.

*References*

ERIH, *s.n.* Oelsnitz

https://nat.museum-digital.de/index.php?t=objekt&oges=596108

https://tu-freiberg.de/en/presse/examination-of-saurian-plate-with-six-skeletons-from-the-permian-period

https://www.mindat.org/loc-16044.html

http://www.bergbaumuseum-oelsnitz.de/

personal communications, Heino Neuber, manager of collections, 2012, 2021

showcaves

### 100. Bergbaumuseum und Bergmannshaus des Bergmännischen Traditionsvereins für Stein- und Braunkohle (Mining Museum and Miner's House of the Association for the Tradition of Coal and Brown-Coal Mining)

The Bergbaumuseum und Bergmannshaus des Bergmännischen Traditionsvereins für Stein- und Braunkohle is located in Aldenhoven in Düren in the very west of the state of North Rhine-Westphalia in the west German Aachen coalfield. In the 1950s the Emil Mayrisch Mine (named after a Luxembourg industrialist, 1862-1928) was worked by miners from all over Germany and from around the world. Eventually the Emil Mayrisch and Anna Mines were amalgamated to work as one, known as the Verbundbergwerk Emil Mayrisch-Anna (Emil Mayrisch-Anna Combined Mine). Klaus Hemmerle (1929-94), bishop of the diocese of Aachen from 1975 to 1994, went underground at the Verbundbergwerk Emil Mayrisch-Anna to get an idea of what was involved in the work of the miners.

The mine closed in 1992, the last in the region to do so. A meeting of 29 former miners the same year arranged to register an association to preserve the traditions of the mine. They collected many items from the pit, such as lighting fixtures, loading machines, locomotives, mine cars, mining lamps, pulleys and trams, and items from the nearby Inden opencast mine. Their first museum of 130 square metres, with a courtyard of 60 square metres, opened in Inden in 1995. The same year fundraising enabled them to erect a monument to the miners. Meanwhile the association grew to have 300 members.

The municipality worked with the association and identified a former Capuchin monastery as a more suitable building to house the museum. A grant of 100,000 Deutschmarks (€51,130) from the municipality and 16,000 hours of work by the association enabled the museum to open to the public 18 months later, in 1997. In 2010 the museum was partly renovated and the exhibition space was expanded to a total area of 1,400 square metres, including an exhibition on the history of lignite mining and power generation from brown coal. The various tools used by miners are displayed in a simulated roadway and there is an extensive minerals collection. The museum is deliberately focused on individual experiences, with a programme recording oral histories from about 20 participants and the collecting of old photographs. Amongst items preserved in the museum are an overhead monorail and a complete underground train, with cars specially produced for the visit of Bishop Hemmerle.

*References*

http://bergmaennischer-traditionsVEREIN.de/

https://de.wikipedia.org/wiki/Museum_Bergmannshaus

museum leaflet

### 101. Bergbau-Technik-Park (Mining-Technology-Park)

The Bergbau-Technik-Park is located in Neuseenland in Leipzig, Saxony, in south-eastern Germany. The mines all closed in the early 1990s as a result of changes to energy policies after the 1990 reunification of Germany. Virtually the whole mining infrastructure has disappeared, leaving lakes in the voids created by opencasting. The Bergbau-Technik-Park was established in 2002 to preserve something of the history of the industry. It opened to the public in 2012 on 5.4 hectares of the site of the former Espenhain opencast mine. The park was set up with the support of the local Espenhain, Grosspösna and Markham municipalities, the city of Leipzig and the state of Saxony, as well as the LMBV and MIBRAG companies. There are six members of staff and the park attracts 20,000 visitors a year. It is financed by sponsors and income from entrance tickets.

**Figure 101.1. Entrance to the Bergbau-Technik-Park (© Bergbau-Technik-Park e.V).**

**Figure 101.2. Aerial view of the park; in the background the Leipzig Lake, filling the void of the former Espenhain opencast mine (© Bergbau-Technik-Park e.V).**

The park aims to illustrate a complete cycle in the operation of a lignite opencast mine. This begins with the stripping of the surface soil and overburden to expose the coal and follows through the removal of the coal, its transport off site and the eventual restoration of the landscape. Leipzig Lake owes its origins to centuries of this process of extraction and refining of lignite.

Guided tours of 60 to 90 minutes explain the various enormous machines used on opencast sites. The core of the exhibits are two machines from the Espenhain opencast mine: a 1985 bucket-wheel excavator, number 1547, weighing 1,300 tonnes, and a 1986 conveyor-belt stacker, number 1115, of about 2,400 tonnes. Both of these were manufactured by TAGRAF Magdeburg and are both fully operational. Other exhibits include a coal-tailings train, a conveyor-belt system and two cabs from the overburden-conveyor bridge, which was dismantled in 2001, from the Zwenkau opencast mine.

Detailed information on technical issues is provided in the adjacent exhibition areas, which originally housed the equipment to generate the electricity needed on the site. The displays also deal with the question of the loss of workers' housing and the disappearance of whole villages when the area on which they were located was needed for opencasting. There is a large collection of tools, workers' clothing and details of opencast-mining systems and subsequent reclamation schemes.

*References*

ERIH, *s.n.* Grosspösna

https://www.ruhrgebiet-industriekultur.de/muttental.html

http://www.bergbau-technik-park.de/

personal communication, Thorsten Hinz, head of park, 2016

### 102. Bergbau- und Heimatmuseum Erbendorf (Erbendorf Mining and Local History Museum)

The Bergbau- und Heimatmuseum is located in Erbendorf in the Upper Palatinate region of Bavaria in south-eastern Germany. Coal mining is first mentioned in this area in the thirteenth century, but all the coal mines had closed by 1927. The museum was opened in 1995 in an old monastery by the Initiative des Heimatpflege-gevereins Erbdendorf e.V (Erbendorf Heritage Village Association).

The Erbendorf region was extraordinarily rich in minerals, with some 60 different types all being worked, including coal, gems, gold, lead, silver and soapstone. The museum deals with the geology of the area and the mining of these minerals. It also covers various local crafts that are no longer practised, such as blacksmithing to produce nails, glass working, tanning and textiles. There are documents, pictures and tools on display related to mining.

*References*

http://fichtelgebirge.de/die-region/staedte-und-gemeinden/erbendorf/kultur/museen-ausstellungen/

showcaves

### 103. Bergbauwanderweg Muttental (Muttental Mining Heritage Trail)

The Bergbauwanderweg Muttental is a mining-heritage trail. It is located in the Witten Valley in the Ruhr and includes the Westfaelisches Industriemuseum (Westphalian Industrial Museum). It is thought likely that the presence of coal in the Ruhr was first identified in the Muttental, where there are the remains of many shafts and small mines. In the Middle Ages (CE 500-1500) peasants acquired coal from small holes where the seams outcropped. In the eighteenth century mining came under state regulation intended, amongst other things, to preserve the landscape by enacting rules for refilling the voids left by mining. The coal-mining industry developed further in the nineteenth century, with the introduction of specialised machinery, but it became unprofitable in competition with the large collieries of the Ruhr. Some mines were reopened during periods of shortage in WWI, WWII and the immediate post-WWII period, but there is now no mining in the Muttental.

The Bergbauwanderweg Muttental was created in 1972 for cyclists, hikers and horse riders and runs for 9 kilometres along the Mutten Valley, following the course of the river Muttenbach. There are over 30 stopping points, many related to coal mining, some with original buildings and some partially reconstructed. The trail begins near the Zeche Nachtigall (**128**), close to a narrow exposed coal seam. On the trail is a *Bethaus,* literally a 'prayer house', *i.e.* a chapel. This was constructed in 1830 and served as a meeting place for the miners of the Muttental coalfield. The building now has statutory protection as the last of its kind in the Ruhr and is now in the ownership of the Witten local authority. It can be booked for meetings and celebrations and can seat up to 25 people.

Near this is the Zechenhaus Herberholz (Herberholz Mine House), with a display of mining tools, such as drilling machines and mining vehicles. Other mining sites with visible remains are the Gruben- und Feldbahnmuseum Zeche Theresia (**116**), the spoil heap of the Vereinigte Hermann (Hermann Combined) coal mine and the tracks of the railway running from the tunnel entrance of the Zeche Jupiter (Jupiter Colliery), now closed by a wrought-iron grille.

The Kleinzeche Renate (Renate Small Colliery) only operated between 1950 and 1954. Its concrete enclosure has been preserved, and the winding-engine house and conveyor system have been reconstructed. Nearby is a memorial stone to miners who died underground. Other mining sites on the route are the Zeche Stettin (Stettin Colliery), the Göpelschacht Moses (Moses winch shaft) of the Zeche

Ankunft und Anclam (Arrival and Anclam Colliery), the winding-engine house above the Schacht Margarethe (Margaret shaft), which was reconstructed after a fire in 2010, and the Zeche Martha (Martha Colliery), whose administrative buildings are now in private ownership.

*References*

https://www.ruhrgebiet-industriekultur.de/muttental.html

personal communication, Cindy Kramer, academic volunteer, 2017

### 104. Bergwerk Camphausen (Camphausen Mine)

Numbers 1 and 2 shafts of the Bergwerk Camphausen in the Saarland, in the west of the country, were sunk in 1871, just after the unification of Germany. The colliery was named after the Prussian minister of the treasury, Otto von Camphausen (1812-96). The construction was undertaken by the Prussian state mining authority, because in the Saar coalfield the mines were state-owned. Construction of Pit Number 4 began in 1908 and of the concrete headgear in 1910. The headgear, which stands 40.7 metres high, was erected in 1911/12 by the Saarbrücken branch of the Carl Brandt Company. The design of the headgear was undertaken by the Austrian civil engineer, Professor Karl Walter Mautner (1881-1949), who is considered one of the pioneers of the use of concrete in Germany. The colliery changed to electrical machinery in 1936-7. The substation to supply electricity for the new engines installed in the headgear was constructed in about 1935. The headgear was restored in 1978/82. As one of the earliest examples of the use of concrete for a headgear (see **52** for northern America), if not the first, it was included on the first inventory of Saarland industrial monuments compiled in 1977. The headgear was classified as a protected monument in 1991 and ceased to be used by the mining company in 1994. It would appear to be the oldest concrete headgear still surviving anywhere in the world (Slotta and Kierdorf 2016). The apparently earliest example in the UK, erected at Bentley Colliery in Yorkshire in 1911/12, was demolished in the mid-1990s.

A number of other components of the site were given statutory protection in 2016 as an historisches Wahrzeichen der Ingenieurbaukunst in Deutschland (historic landmark of civil engineering in Germany). These comprise part of the winding-engine house, built in 1880 to serve Numbers 1 and 2 Pits (although none of the winding engines themselves survive), the 1935 electrical substation and the 1962 steel headgear of Number 2 Pit. Other listed monuments outside the curtilage of the colliery are

**Figure 104.1. Traditional steel headgear to left and concrete headgear for friction winder to right (© Büro für Industriearchäologie).**

the dormitory building constructed in 1924/5 for the mineworkers and the local inspection building of the state mining authority, built in 1890/1. A small mineworkers' settlement, situated next to the inspection building, also survives, although not listed. So also does an extensive building, which included the large pithead baths of the colliery. This has been separated from the site and is now used by a private company, so is not open to the public. In a recent study the Saarland Ministry of Economy assessed the Bergwerk Camphausen as one of the four most important coal-mining sites in the Saar.

*References*

https://de.wikipedia.org/wiki/Grube_Camphausen

http://www.saarland.de/dokumente/thema_energie/bergbaudenkmal_dokumentation.pdf

https://wahrzeichen.ingenieurbaukunst.de/

personal communications, Axel Böcker, head, Industrial Heritage Section, Saarland Heritage Service, 2015, 2021

### 105. Bergwerk Ensdorf (Ensdorf Mine)

Ensdorf is a municipality on the right bank of the river Saar in the district of Saarlouis in Saarland, in the west of the country. In 1807 the École Pratique des Mines (School of Mines) was founded under the direction of Jean-Baptiste Guillot-Duhamel (1767-1847) on the orders of the Emperor Napoleon (1769-1821) of France. The École had the remit of investigating the Saarland coal deposits, the preparation of mining concessions and the training of miners. After the battle of Waterloo in 1815, the region of the Saar passed to Prussia, together with the Saarland collieries. Prussia had little interest in a school of mines in the far west of Germany, so the school was closed. Although only in existence for eight years, the École had provided an impetus to the development of mining in the region.

Work began on the Bergwerk Ensdorf in 1833 and it opened in 1842, with 2,350 metres of roadways. In 1836 the mine surveyor, Friedrich August Prediger, used a mine theodolite here, the first time in a European mine. In 1861 the first coal train ran from the nearby station at Griesborn to the Ensdorf coal-fired power station. The shafts were gradually deepened over the years. Administration of the Saarland passed to the League of Nations following WWI, with the coal mines operated by the French. The Saar shaft, sunk in Ensdorf in 1913, was renamed the Duhamel Schacht after Jean-Baptiste Guillot-Duhamel. By 1925 impressive day-care facilities had been opened, which still exist today.

**Figure 105.1. Duplex-cylinder steam-winding Koepe engine dating to 1918 (© Eberhard Lantz, Bamberg).**

**Figure 105.2. Duplex-cylinder steam-winding Koepe engine dating to 1936 (© Eberhard Lantz, Bamberg).**

By 1972 the output of the Bergwerk Ensdorf was 6,256 kilograms per miner each shift and it was one of the most productive mines in Europe. In 1976 the Bergwerk Ensdorf is believed to have been the first pit in Europe to be fully equipped with a shield above the miners for protection from roof falls. In 1977 it achieved an underground production of 12.334 tonnes per manshift, the first deep mine in Europe to exceed 12 tonnes in a manshift. Work on sinking the northern shaft began in 1981, beginning production in 1987, and on the southern shaft in 1984, starting production in 1986. The Bergwerk Saar was formed in 2004 when the Bergwerk Ensdorf merged with the Warndt/Luisenthal Joint Mine (**112**), which then closed in 2006. In 2007 the Bergwerk Saar produced 3.5 million tonnes of coal. Following an earthquake in 2008, the RAG Deutsche Steinkohle Company decided to close the mine, the last surviving pit in Saarland. The Bergwerk Saar finally ceased production in 2012.

Two steam-winding *Koepe* engines (**127**) have been given statutory protection as historical monuments. A duplex-cylinder steam-winding engine, built by Dingler`sche Maschinenfabrik AG in Zweibrücken in 1918, was installed above the east shaft. It was not decommissioned until 2012, the last such winding engine to operate in Saarland. A second duplex-cylinder steam-winding engine built by Dingler`sche Maschinenfabrik AG in Zweibrücken in 1936 was installed above the west shaft. It was decommissioned in 2000. The two engines have recently been renovated on their original site and are now open to the public.

*References*

http://www.duhamel-park.de/projekt/geschichte/

museum leaflet

personal communications, Eberhard Lantz, photographer, 2016, 2017, 2021

### 106. Besucherbergwerk F60 (F60 Visitor Mine)

The Besucherbergwerk F60 displays what is claimed to be one of the largest surviving surface gantry conveyors ever constructed. It is situated in the countryside near Lichterfeld in Lausatia on the eastern border of Germany. The Besucherbergwerk Förderbrücke F60 (F60 Visitor Mine conveyor bridge) contains about 11,000 tonnes of steel, is 502 metres long, 204 metres wide and almost 80 metres high. As its orientation is horizontal, it is known as 'the Lusatian horizontal Eiffel Tower' (the 1889 French Eiffel Tower in Paris is 324 metres high). From 1958

**Figure 106.1. F60 with Lake Bergheid beyond (© Besucherbergwerk F60).**

**Figure 106.2. F60 lit up at night for a performance (© Besucherbergwerk F60).**

engineers of East Germany developed a series of standard conveyor gantries, each in turn larger than the previous one. Five were constructed of the F60 type, of which this F60 was the end of the series and so the largest.

This Besucherbergwerk F60 example could move up to 29,000 cubic metres of earth per hour. The machine itself comprised: the energy and transformer station, where the incoming 30 kV were transformed into 27,000 kW to power the two excavators (1.2 kWh were used for the removal of the spoil); the main chassis, with two standard-gauge tracks and two additional tracks for the cable car; the main conveyor belts with nine different belts running at speeds of up to 10 metres per second (36 kilometres per hour); and the tipplers. The bridge itself had a maximum speed of 13 metres per minute (780 metres per hour) and a working speed of 9 metres per minute (540 metres per hour). The workshop and the facilities for the workforce, such as the canteen, are 11 metres above the ground. They have a length of 29 metres and a width of 12 metres.

This F60 was built on site between 1989 and 1991 by VEB TAKRAF Lauchhammer (today TAKRAF GmbH Lauchhammer), but only operated from May 1991 until June 1992. During this time it moved some 27 million cubic metres of overburden. When East Germany was reunited with West Germany in 1990, the Lichterfeld F60 was no longer required, because the coal from the Klettwitz-Nord opencast mine could no longer be processed. Four other F60 are still operating in the opencast mines of Lusatia at Jänschwalde, Nochten, Reichwalde and Tagebau Welzow-Süd (**230**) to provide a reliable supply for modern power stations that are still coal-fired.

It was originally intended to blow up this F60. The municipality of Lichterfeld, however, undertook a major development project in 2000, including converting the Klettwitz-Nord opencast mine into a recreational landscape, with the F60 as its centre point. Lake Bergheid was created from one of the opencast voids next to the colliery, where musical galas, opera performances, rock concerts and sporting events are held. The workshop now has information about the site and the history of lignite mining in the Niederlausitz (Lower Lausatia). In 2019 it was employing 20 staff, 12 of whom were permanent, and attracted 66,238 visitors. Visitors can undertake a 1.4-kilometre guided tour, starting from the workshop and climbing to the viewing platform, 80 metres high, to see over the former opencast site. Tours are also offered in the evenings, when the F60 is illuminated. In the summer season the F60 forms a backdrop for concerts, festivals, firework shows and other events. The site has been recognised as significant on two routes (**75, 111**).

*References*

ERIH, *s.n.* Lichterfeld

http://www.f60.de/

personal communication, André Speri, managing director, 2021

**107. Besucherbergwerk Rischbachstollen (Rischbachstollen Visitor Mine)**

The Besucherbergwerk Rischbachstollen is located in Sanct Ingbert in Saarland, in the west of the country, close to France and Luxembourg. The first written references to coal mining in Sanct Ingbert date to 1615. In 1771 the local count started the Grube Sanct Ingbert (Saint Ingbert Pit). The mine was taken over by the state of Bavaria in 1816, which introduced water pumps operated by a steam engine and lifting gear. Number 1 shaft was sunk in 1844, but moving the coal to Sanct Ingbert involved transporting it through the hills on a steep unmade road. The solution was to build a roadway, the Sulzbach, underground from the mine to Sanct Ingbert. The coal was brought along the roadway to the surface at the Risch Bach in the town. A coal-preparation plant was built here in 1897, then a headgear, offices, pithead baths and workshops in 1903. From 1852 the coal was moved underground by horses and carts until the introduction of electric locomotives in 1907. In 1910 compressors, rotary drills and shearers were introduced. The management of mines came under the control of the French Mines Domaniales de la Sarre (State Mines of the Saar) in 1920 as part of the reparations after WWI, but the collieries were sold in 1925 to the German firm of Maschinenfabrik Kaiser. In 1935 the Risch Bach mine came into the possession first of the German state and then in 1951, after WWII, the Jägersfreude Company.

The main operation in the Risch Bach was closed in 1957 and the Grube Sanct Ingbert in 1959. Renovation of the mine started in 1990, undertaken by the Besucherbergwerk Rischbachstollen e.V., a not-for-profit group of former coal miners. The show mine opened to the public in 1994. It is, however, only open for pre-booked tours, which are available in English, French, German and Russian. More than 1 kilometre of roadway, known as the Rischbachstollen (Risch Bach roadway), is now available to visitors.

Tours start on the surface from the former 1849 mine-entrance building, which also contains a café for refreshments after visiting the mine. After an introduction to the geography and geology of the region and the history of the colliery, visitors enter the mine through a sandstone portal. The first displays are laid out as the mine would have been during the 1880s, when the Grube Sanct Ingbert was the most productive mine in Bavaria. Working the coal with hammer and pick and moving it by pit ponies are shown, based on the accounts of French coal mining in the 1885 novel *Germinal* (**33**). Further displays show the technical development of the mine, including drilling techniques, equipment, mining signals and tools, up to the time of its closure. Visitors are equipped with mining protective clothing, including an *Arschleder* (a leather patch on the seat of the trousers); the need for this is shown by accessing the Johann seam, which is only 80 centimetres thick and very steep, so miners would sit on the *Arschleder* to slide down the incline. As well as the entrance building, the lamproom and the pithead baths are preserved on the surface.

The Besucherbergwerk Rischbachstollen is part of the Saarland mining trail to Schnappach. The steep narrow road illustrates the reason for constructing the underground roadway to transport coal from the mine. The trail is 4 kilometres long, with 14 stopping points, including the former administrative building of the mining company, the miners' hospital and several traditional miners' homes.

*References*

ERIH, *s.n.* St. Ingbert

http://www.rischbachstollen.de/

**108. Deutsches Bergbau Museum (German Mining Museum; formerly the Bergbaumuseum: Mining Museum)**

The Deutsches Bergbau Museum is situated in Bochum, which by the 1950s was the largest mining city in the Ruhr and, it is believed, the whole European continent. It is housed in an extensive purpose-built museum, designed by the industrial architect Fritz Schupp (1896-1974), which opened in 1930 as the Bergbaumuseum. It was renamed as the Deutsches Bergbau-Museum Bochum in 1976 to reflect its significance as the largest specialist mining museum in Germany. It is now generally known simply as the Deutsches Bergbau Museum. The museum covers mining in general and is not restricted just to coal. It is funded mainly by the city of Bochum, the government, the North Rhine-Westfalian State and admission charges. The museum has 17 volunteers, who are former miners, and employs about 120 staff. It attracts some 380,000 visitors a year which, together with its extensive collections, makes it probably the largest mining museum in the world.

The museum occupies an area of some 17,000 square metres, spread over three large separate buildings. Towering over the museum is a 60-metre-high headgear constructed in the German A-frame style (*cf.* **231, 233**). This was originally erected in 1944 at the main pit of the Germania Colliery in Dortmund and was transported to Bochum in 1972. The simulated underground tour at the museum is at a depth of between 17 and 25 metres, with a total length of 2,500 metres, 1,200 metres of which are open to the public.

On display in the museum are models of shaft sinking and shaft construction; large pieces of mining machinery and shield supports; and an extensive and very valuable collection of art and decorative-art objects, including silver statuettes. The mining archives contain 6 kilometres of shelving for documents. There are approximately 40,000 books and 40,000 periodicals in the library, and 200,000 photographs as well as 100,000 items in the museum's collections. The museum possesses five original copies (dating between 1556 and 1657) of the seminal work on mining methods, *De Re Metallica* (On the nature of metals) by Georg Bauer, under his name in the latinised form of Georgius Agricola (1556), in English George Farmer.

In 2010 the museum held an exhibition to commemorate 750 years since the founding of the Knappschaftskassen organisation (**297**), which had made Germany a world leader in social security for miners long before Bismarck (**178**) enacted his 1881 social legislation. The earliest record in Germany dates to 28 December 1260, providing support to disabled (ore) miners and to the dependents of deceased miners, making this the earliest social-security system in the world. Originally membership was voluntary, but from 1889 contributions from both employers and every person employed in a coal mine became mandatory.

The museum has relatively generous funding and does not have to maintain historic colliery buildings. Neither does it have the expense of maintaining a genuine underground, with all the costs associated with meeting mining legislation. So the museum has been able to develop an extensive research programme and has become a renowned research establishment for mining history.

The museum is divided into two major sections: the research division, which provides the basis for many of the museum's projects and uses the museum to present its findings; and the museum division, whose facilities and displays build on the work of the research division. The research division is divided into departments: archaeometallurgy; materials science; mining archaeology; and mining history, alongside the research laboratory and the Montanhistorisches Dokumentationszentrum (the Mining-History Documentation Centre), or Montan.dok. The Montan.dik is also the central service facility for the conservation, preservation and restoration of all the collections held by the museum. The personnel structure of this division is, however, project-based, so that staff from different research fields work together as interdisciplinary teams as appropriate on each project.

The museum has extensive laboratories, containing the most up-to-date equipment in the field, and undertakes research projects throughout the world. It has collaboration agreements with a number of universities in Germany and also Istanbul (Turkey), and with non-university institutions, including organisations in Chile, Iran, Oman and Turkey. So extensive is its research work that it has published a detailed report on the nature of the research that it can offer.

During 2016 the museum closed to enable the displays to be renewed completely, together with the construction of a new building of some 10,000 square metres for research and storage. Throughout this period only the underground tour and the headgear were kept open, along with one exhibition of about 800 square metres of highlights of the museum. In 2018 the last deep coal mine in Germany closed and the first stage of the €75 million project opened

**Figure 108.1. Entrance to the Deutsches Bergbau Museum (© Deutsches Bergbau Museum).**

**Figure 108.2. Simulated underground tour showing an outboard double-ended ranging-drum shearer cutting a simulated coalface (© Deutsches Bergbau Museum; photographer Karlheinz Jardner).**

**Figure 108.3. Photographic reproduction of an historic vignette (1899) of Georgius Agricola (© Deutsches Bergbau Museum).**

the same year with an exhibition on a century of German hard-coal mining. The entire project is scheduled to be completed in 2026.

*References*

ERIH, *s.n.* Bochum German Mining Museum

https://de-m-wikipedia-org.translate.goog/wiki/Knappschaft?_x_tr_sl=de&_x_tr_tl=en&_x_tr_hl=en&_x_tr_pto=sc

https://www.bergbaumuseum.de/

personal communications, Michael Ganzelewski, head of mining collections, Montan.dok, 2014; Stefan Brüggerhoff, director, 2015; Michael Farrenkopf, head of Montan.dok, 2021

showcaves

### 109. ENERGETICON (formerly Bergbaumuseum Wurmrevier: Wurmrevier Mining Museum; Bergbaumuseum Grube Anna: Anna Pit Mining Museum)

ENERGETICON has in the past been known both as the Bergbaumuseum Wurmrevier and as the Bergbaumuseum Grube Anna. It is located in Alsdorf in North Rhine-Westphalia on the site of the Grube Anna II, which for many years was the largest mine in the Aachen coalfield. Sinking of the Grube Anna began in 1850 and the first coal was produced in 1854. A steam engine had to be installed because of the high water make in the mine. In 1871 the mine was connected to the railway from Rheinische Eisenbahn to Stolberg and in 1875 to the Aachen industrial railway. A washery was constructed in 1878. In 1865 work started on a second production site, the Grube Anna II, whose shaft was sunk 225 metres deep in 1904. It was named the Eduardschacht (Edward shaft) after the chair, Eduard von Oppenheim (1831-1909). The first electric-winding engine was installed at the Eduardschacht in 1907, and a second electric-winding engine was installed in 1911, when a second coking plant was added to the colliery. The two coking plants were close together, but operated separately. They had a total of 402 furnaces, making it one of the biggest coking plants in Europe.

In 1930 297 miners died in an underground disaster, some killed by a firedamp explosion at the 360-metre level, but most by the subsequent fire. The headframe was torn from its concrete foundations and crashed into the administrative buildings and pithead baths, demolishing them. Further slight damage was caused to the boiler house by grenades dropped during WWII, leading to an interruption to the electicity supply and so to a cessation in the pumping. The result was that eventually the 610-metre level was under 15 metres of water, although mining continued, but with just the minimum number of workers. In 1944 the Nazi regime ordered all industrial plant to be destroyed so that it could not pass into the hands of the Allies. But the mine operators managed to evade this order, apart from dismantling two insignificant turbines after the mine director had been imprisoned temporarily to force him to comply. The same year the mine was occupied by American troops and in 1951 the two mines, Anna I and Anna II, were merged into one combined colliery, making it the largest pit in Wurmrevier. With the decline in the use of coal, Grube Anna was amalgamated with the Emil Mayrisch (**100**) in 1979. The two were linked underground, with the coal being washed on the Emil Mayrisch site. Demolition of the coal-preparation plant and the loading area of the colliery started in 1985. The composite colliery closed in 1992, after which some of the buildings were destroyed.

In 1986 the not-for-profit Verein Bergbaumuseum Wurmrevier e.V. (Wurmrevier Mining Museum Association) was founded to preserve the history and culture of coal mining in the area, and now has more than 700 members. The state of North Rhine-Westphalia took over ownership of the site in 1995. After restoration some of the colliery buildings became the Bergbaumuseum Grube Anna, situated in the Anna Park. Some of the site was used for commercial, industrial and residential buildings. The society had hoped to run a museum of the history of the social and technological history of coal mining in the region, but the provincial government indicated that it would not provide funding for another mining museum in North Rhine-Westphalia. The society does, however, still produce books twice a year, along with a periodical with scientific papers.

**Figure 109.1. Exterior of ENERGETICON in front of the Bergehalde Noppenberg spoil heap (© ENERGETICON).**

**Figure 109.2. Steam locomotive, Anna 8, manufactured in 1938 (© ENERGETICON).**

Under EuRegionale 2008, organised by the Aachen region, Belgium and the Netherlands, three of the buildings on the site were restored, forming a complex of about 30,000 square metres. ENERGETICON opened in 2014 in the winding-engine house on the north-western edge of the former operational area of the Grube Anna. It was founded to explain the need for a transition from fossil fuel and nuclear energy to renewable energy and describes itself as an energy-experience museum. By the end of the year the other buildings and outdoor area had been converted to present an interactive exhibition on energy. Since 2020 operation of the Grube Anna mining-information centre has been carried out by ENERGETICON. In 2019 there were 41,000 visitors and 10 staff.

The winding-engine house was built in 1905 in the neo-romanesque style, with two electric-winding engines: one a drum machine driven by direct current (with a three-phase power plant to convert the incoming alternating power to direct current) and the other a powered three-phase machine. Since 2008 its three halls have served as a multifunction venue for business, culture and science.

The pithead baths were also built in 1905. An extension dating to 1930 was destroyed in the explosion and had to be rebuilt in 1931. The building now houses displays about the social and labour history of mining and the use of solar-powered energy. There are views of the spoil heap and the park from the large windows on the first floor.

The former blacksmiths' workshop was built about 1910. It was also used from the late 1940s into the 1970s as a training centre, with a small teaching mine in the basement in the form of a replica of a standard colliery roadway. The arches use a very heavy U-cross-section (in fact almost V-cross-section) rolled steel and the sheeting is also far more robust than usual. The Verein Bergbaumuseum Wurmrevier has equipped the replica roadway with mining machinery, such as a section of a longwall face, dating from the late 1940s to the early 1960s, an armoured flexible conveyor on the face and a repositional metal prop-and-bar roof-support system. ENERGETICON uses this building for its catering, reception and shop and an exhibition on fossil fuels and solar energy, leading on to the display in the basement.

Displayed outside are underground machinery and equipment collected by the Verein Bergbaumuseum Wurmrevier and too large to be exhibited inside. Also on display outside are modern sculptures and railway vehicles used in mining in various collieries of the Aachen coalfield. The two exhibition routes run through the grounds, which cover in total about 30,000 square metres. The most impressive exhibit is the locomotive, Anna 8, a 0-8-0 well-tank steam locomotive manufactured by Henschel of Kassel, Germany, in 1938 but to a 1920s design. This type of locomotive would have been used primarily for shunting in the colliery sidings and also short-trip working. Anna 8 was one of the longest-serving locomotives in the Aachen area and was still working in 1989. An overhead monorail locomotive is also displayed, manufactured in Alsdorf in 1958, which would have been used to transport men and materials. The company SMT Scharf, which is based in the area, still manufactures this type of locomotive.

The Bergehalde Noppenberg, one of the three spoil heaps of the Grube Anna, stands 110 metres high behind the museum. It is in private ownership, but can still be climbed either independently or on one of the guided tours offered once a month between April and October. Depending on weather conditions, it is also possible to climb the 30 metres to the top of the headframe once a month between April and October for a view over Alsdorf and the Aachen coalfield area. Guides explain the history of the former mining operations en route to the headgear. The museum is also a starting point for the Karbonrouten (coal routes) of hiking trails, which visit various mining sites in the Alsdorf region.

Saint Barbara (**96**) has been especially popular in this area because of an influx of miners from Upper Silesia and East Prussia in the nineteenth century. It was common practice to put a statue of Saint Barbara underground. A metal figure of the saint was hung in Grube Anna at a height of 1.3 metres on the front of a brick-masonry bridge pier where it could be seen by miners both on entering and leaving the mine. The statue was moved in 1986 to the offices of the mining company. In 2006 it passed into the care of the Verein Bergbaumuseum Wurmrevier. In 2015 a chapel dedicated to Saint Barbara was erected on the site to house the statue.

*References*


museum booklet

https://de.wikipedia.org/wiki/Grube_Anna

https://industriemuseen-emr.de/de_DE/grube-anna-ii-alsdorf

http://www.rheinische-industriekultur.de/objekte/Alsdorf/anna/anna.html

personal communications, Harald Richter, director, 2016; Thomas König, director, 2021

www.bergbaumuseum-grube-anna2.de

www.energeticon.de


### 110. Energiefabrik Knappenrode (Knappenrode Energy Factory; formerly Lausitzer Bergbaumuseum Knappenrode: Knappenrode Lausation Mining Museum)

The Energiefabrik Knappenrode is located in Hoyerswerda, eastern Saxony. The Energiefabrik is one of four sites of the Sächsisches Industriemuseum (Saxon Industrial Museum). The museum is housed in a listed building, which was formerly a factory producing briquettes from lignite. The Briquettefabrik Werminghoff, named for its original owner Joseph Weminghoff (1848-1914) and later known as the Briquettefabrik Knappenrode, opened in 1918 with seven plate dryers and 10 steam presses. To provide electricity and steam for the dryers and presses, the factory included its own boiler house and a power plant, with three steam turbines. The oldest, a Brown Bovery, was constructed in the nineteenth century; the second is an AEG dating to 1943; while the youngest engine dates to the 1950s. The workers lived next to the factory gates in a company housing estate, with a company store, guest house, railway station and school. The factory reopened in 1948, now called VEB (*i.e.* a publicly owned company) Braunkohlekraftwerk (lignite-coal power plant) Glückauf (good luck: **210**) Knappenrode, formed from *Knappe* (miner) and *roden* (cutting). In 1965 the plant turned out more than 1.5 million tonnes of briquettes, but ceased production in February 1993. The museum opened in 1994, occupying all the red-brick buildings of the former factory and covering the different elements of the history of coal mining in East Saxony. It employs 16 staff, with six volunteers.

The tour takes in former methods of briquette production and the historical power station, which is also used for the performance of concerts in summer. Having lost all of its machinery in reparations to the Soviet Union following WWII, the machines now on display were collected from other factories. The site does, however, show a uniquely preserved technique and has, therefore, been listed as part of the national heritage.

The pithead baths still retain the traditional storage method for miners' clothes with boots and overalls hanging from the ceiling. Below these is a display of minerals found underground during digging for coal, such as agates, amber, fossilised wood, jasper, petrified wood and rock crystals. The old goods station contains a collection of historic railway technology of the Lausation mining area, including diesel, electric and steam locomotives and mine trains. There are also a tender turned into a snow plough and machinery for large-scale opencast mining, such as the Atomic-Lok from the Berzdorf opencast mine. The site also offers a 2.7-kilometre round trip on a trolley-bus with a 900-millimetre gauge.

The mines-rescue exhibition details the story of accidents and rescue operations carried out by specially trained fire crews and underground-rescue teams. Since 1914 there have been more than 1,000 deaths in the mines of this area. The displays also include some 20 vehicles. The Markscheide (mine-surveying) exhibition deals with the work of the mine surveyor, who delineated and recorded the underground areas to be worked. It was essential that there be an accurate record of all areas that had been worked, especially after a mine had been closed (*cf.* **384**). The displays include historic mining maps, measures, measuring instruments and surveying techniques.

An exhibition about living beside the factory gates explains how the company was responsible for all aspects of the lives of the workforce. The company provided education, healthcare and workmen's clubs, and the pithead baths were used by the families on Saturdays. Heating is very important in the northern European climate; the museum has a collection made by a former Berlin chimney-sweep after he retired in 1988, consisting of more than 3,000 stoves and ovens from 10 different countries, about 700 of which are on display. They range from the status-symbol art-nouveau stoves through to the coal-fired kitchen stoves used in the housing estate.

In 1966 a headgear from the Nochten opencast-mining area was moved to stand outside the buildings on the 25-hectare site. The headgear stood at Nochten for 35 years above a shaft 100 metres deep to access an underground drainage network of 40 kilometres, constructed to dewater the lignite opencast workings (**143**). The headgear stands 17 metres high and weighs 30 tonnes. Inside the pithead building are displays dealing with the technology used for drainage in underground-lignite and opencast mining. Also in the outdoor area are exhibits on the subject of lignite and the environment, including different methods of reclamation of land affected by mining and the development of the post-mining landscape. Related to this is the lignite and environment exhibition housed in the former cold-storage house of the briquette factory. This display looks at the mining and refining of lignite, including the work of the landscape architect, Otto Rindt (1906-94), in converting the local mining landscape into a recreational area.

Nearby a lake has been created in the void caused by opencast mining. This now forms the Grey Heron Lake, so named because it is the largest heronry in Germany. The site has been recognised as significant on two routes (**75, 111**).

*References*

ERIH, *s.n.* Hoyerswerda

https://de.wikipedia.org/wiki/Energiefabrik_Knappenrode

http://www.saechsisches-industriemuseum.de

showcaves

### 111. ENERGIE-Route der Lausitzer Industriekultur (ENERGY-Route of Lusatian Industrial Heritage)

The ENERGIE-Route Lausitz Industriekultur is one of 19 regional routes of the European Route of Industrial Heritage (**75**). Lusatia is a region rich in lignite, with some 200 million tonnes being mined each year in the 1990s and about 60 million tonnes nowadays. Until 1990 its landscape used to be marked by briquette factories, coking works, power stations, textile factories and 80,000 hectares of voids created by former opencast mines. From the early 1960s more than 130 villages and districts and 27,000 homes were destroyed to make way for opencast mining. Their inhabitants were moved into one of the new concrete housing estates, such as Hoyerswerda, which between the 1960s and the 1990s increased in size from 7,000 to 77,000 inhabitants. Deep mining reduced in Lusatia with the reunification of Germany in 1990, and only five opencast mines still operate now, together with three modem coal-fired power plants.

The ENERGIE-Route was developed as a project of the International Building Exhibition (IBA) Fürst-Pückler-Land (Prince Pückler Country), 2000-2010 (**123**), and opened in 2007. It was named after Prince Hermann Ludwig Heinrich von Pückler-Muskau (1785-1871), who was an accomplished artist of landscape gardens. The IBA developed an economic and ecological programme through a process of regional discussion. The aim was to transform the landscape of Lusatia to improve living conditions for the local people and to attract tourists. The route is organised as an interest group. In the beginning it was financed by the IBA and membership fees from the 11 sites of energy production along the route. Then, for a period of five years, it was financed by a private sponsor and membership fees. Since 2016 it has been financed by the Ministry of Economy of the federated state of Brandenburg and membership fees. The route attracts some 152,000 visitors a year.

The route deals with 150 years of industrial activity and the challenges and benefits of energy production, as well as the industrial origins of the post-mining landscape. The 11 sites show how coal has been used to produce energy and how the miners lived and worked. All the sites are accessible for tourists. At some sites guided tours are offered all year round, while at other sites tours are available only on specific dates or by prior arrangement. Some sites are still operational; for example, the Schwarze Pumpe, one of the most modern lignite-fired power plants in the world. Others have been converted to alternative uses, such as the Cottbus diesel power plant, which is now a museum. The flooding of the voids of former opencast mines has created more than 20 new lakes, which provide water sports and other recreational activities, and are expected to form Europe's largest artificial chain of lakes, the Lusatian Lakeland.

The route includes four places that are described in more detail elsewhere in this publication (**106, 110, 123, 230**). In addition to the staff employed at these sites, the other sites on the route between them employ 23 staff, together with 49 volunteers.

*References*

http://www.energie-route-lausitz.de/

personal communication, Antje Boshold, project coordinator, 2016

**Figure 111.1. Housing for the miners at the adjacent lignite mine built by the Ilse Bergbau Company in the first planned garden city in Germany, Gartenstadt Marga (© Peter Radke).**

### 112. Erlebnisbergwerk Velsen (Velsen Mine Experience; formerly the Alte Grube Velsen: Velsen Old Mine)

The Erlebnisbergwerk Velsen is located at Saarbrűcken, the capital of Saarland, in the west of the country. The museum is less than 5 kilometres from the French Musée les Mineurs Wendel (**95**). Velsen Colliery was sunk in 1899, named after a Prussian mining official called Gustav von Velsen (1847-1923). The local housing estates reflected the strict hierarchical nature of Prussian society, with a rigid demarcation between dwellings for directors, white-collar workers and miners.

Early galleries driven into the hillside beside the underground mine served as an air-raid shelter for the workforce during WWII, after which it was expanded to include an above-ground accurate training gallery for new recruits. This is now open to the public. Grube Velsen closed in 1965, although one shaft remained open to provide ventilation and access to the underground for the new integrated Warndt Colliery (**105**). The colliery closed in 2005 and the training of new miners ended in 2011, just before the closure in 2012 of the Bergwerk Ensdorf (**105**).

The Erlebnisbergwerk Velsen opened to the public in 2012. It is run by the Verein Erlebnisbergwerk Velsen e.V (Erlebnisbergwerk Velsen Association), which is a not-for-profit organisation and has more than 400 members, 30 of whom actively carry out maintenance. There are 20 volunteer guides and one volunteer organiser, but no paid staff. It occupies an area of some 26,000 square metres. Tours are offered in Dutch, English, French, German and Italian.

On the surface the entrance is through the 1907 gatehouse, which has a hipped roof. This is next to the former stables for the horses, which became redundant when the colliery was joined to the railways in 1907. Nearby is the 1916 steel headgear. The 1936 2,400-horsepower engine is still in operation, as is the 1916 2,200-horsepower Maschinenfabrik Dingler steam-powered twin-winding engine, located above the original number 2 Gustav shaft and operated with compressed air. In 1936 this was converted from conventional to *Koepe* winding (**127**). This is the oldest steam-winding engine left in Saarland and has been conserved by volunteers, who also maintain it.

The aim of the Erlebnisbergwerk Velsen is to demonstrate modern operational coal mining, rather than to illustrate historical techniques. Visitors start by riding the mine train, before walking along the coalface. The whole mine is level, so is suitable for visitors with reduced mobility. On display in the roadways are a coal cutter, drills, loading machines, roof supports and shearers. Conveyor belts and an electric locomotive show how coal, materials and men

**Figure 112.1. Exterior of the Erlebnisbergwerk Velsen with 1916 steel headgear in centre (© Verein Erlebnisbergwerk Velsen e.V).**

Figure 112.2. Side-loader shovel (© Verein Erlebnisbergwerk Velsen e.V).

Figure 112.3. Model of a miner riding on a conveyor belt and about to get off the belt at an alighting station (© Verein Erlebnisbergwerk Velsen e.V).

were moved in the pit. The Erlebnisbergwerk Velsen has the only working *Koepeförderanlage* (*Koepe* conveyor system) remaining in Germany, which is operated with compressed air. Visitors are able to see how miners used to travel on the manriding conveyor belt.

Compressed-air, hydraulic and pneumatic systems illustrate how machines were powered safely underground without risking explosions. A blind shaft shows the signalling systems for shaft riding. All the machinery underground is in full working order and run for the visitors. So the 800 metres of roadway still come under mining legislation, overseen by the mines inspectorate.

*References*

ERIH, *s.n.* Saarbrücken

https://artsandculture.google.com/streetview/erlebnisbergwerk-velsen-saarbrücken/yAHljMS7GQAoNw

https://www.youtube.com/watch?v=GXmOkT772pc

http://www.erlebnisbergwerkvelsen.de/

http://www.youtube.com/watch?feature=player_embedded&v=PGH49oXVovY

personal communications, Volker Etgen, chair, 2012; Carsten Grammes, vice-chair, 2021

showcaves

### 113. Erlebnisort Reden (Reden Experience)

The Grube Reden, named after the Prussian minister of state, Friedrich Wilhelm Graf von Reden (1752-1815), was located in the Neunkirchen region of the Saarland in the west of the country. The first underground roadway was opened in the Klinkerthal near Landsweiler in 1846. In 1847 sinking began of a vertical shaft and construction of surface buildings began in 1849. On the surface the widely visible conveyor system became the landmark of the coal industry. The railway line from Saarbriicken to Neunkirchen opened in 1851. The line ran right next to the colliery and led to its rapid extension around number 2 and 3 shafts. This included the construction of a coal-preparation plant, smithy and workshops. By WWI there were a total of five shafts in operation.

In 1935 the Nazi Reichsminister (government minister) Hermann Göring (1893-1946) visited the colliery, which was then expanded; for example, with a new compressor house. The new administrative building, which still exists, is a large, geometric building, with two right-angled wings. The west wing has a monumental, two-storey entrance area, supported by two grey, angular pillars. Inside the portal is decorated with a black-grey mosaic of a geometric meandering ribbon. The upper floor is set back, which also emphasises the geometric character of the building. Offices for the management staff were located along the longitudinal walls. A comparison with the buildings at the iron-ore mine Haverlahwiese near Salzgitter, also expanded in 1937 by Hermann Göring, suggests that the new structures at the Grube Reden were designed by the architect Herbert Rimpl (1902-78). The Itzenplitz site was decommissioned in 1940 when the construction work was complete, so that the number 5 shaft of Grube Reden became the central production site. In 1942 a new coking plant was erected and the colliery became the most modern in the Saar coalfield.

After WWII a new conveyor system was constructed for number 5 shaft, with increasing mechanisation between 1947 and 1954. In 1958 radical restructuring began of the Saar mining industry, leading in 1960 to the partial decommissioning of the pit, the workforce reduced from about 6,000 to about 3,000 men and output cut from 7.23 million tonnes to 1.65 million tonnes. Productivity improved, however, from 1,657 tonnes to 3,932 tonnes per manshift. After various restructurings, in 1991 the colliery was amalgamated with the Bergwerk Camphausen (**104**) and the Grube Göttelborn to form the Verbundbergwerk Ost (East Combined Mine). Production was finally discontinued in 1997, in the face of pressure from global competition, and the main buildings of the site were given statutory protection.

Also in 1997 the mining company RAG Deutsche Steinkohle AG developed a decommissioning and renovation concept for the abandoned areas. The project for structural change of the site began in 2001 under the leadership of the Landesgesellschaft Industriekultur Saar GmbH Company. The aim was to retain the unique nature and the topography of the site, while developing it into a centre for leisure, renewable energy and tourism and a regional and supra-regional attraction.

At the entrance stands the 1937 bronze sculpture, the Saarbergmann (Saar miner), by Fritz Koelle (1895-1953). The site has been reorganised into an arrangement of three squares, the central one bounded by the old mine buildings. The pit spoil heap, which stands 90 metres high, has been transformed into a dry-rock garden for endangered amphibians and insects, with 4 kilometres of hiking and skating paths. A central water channel runs down to a green water garden at the lowest point of the site, on the former railyard. Minewater at 30 degrees Celsius is pumped from 800 metres underground to heat the buildings (*cf.* **386**), before flowing through the water garden and into a nearby stream.

The administration building, lamproom and pithead baths now house the administrative offices, collections, exhibitions and libraries of the Centre for Biodocumentation, the Saarland Institute for Landscape Studies, the Saarland Mountain Administration, the Saarland Regional Memorial Office and the Neunkirchen Tourism and Cultural Centre. There is a permanent exhibition on the geology of the region. The state antiquarian collection of the Landesdenkmalamtes (State Monument Office) in Landsweiler organisation is displayed in a small museum of prehistory and early

Figure 113.1. Winding-engine house and headgear of the Zeche Reden (© Eberhard Lantz, Bamberg).

Figure 113.2. Duplex-cylinder Koepe steam-winding engine constructed in 1941 (© Eberhard Lantz, Bamberg).

history. Mining history is dealt with in the Das Erbe (the heritage) exhibition.

There are two winding engines housed in the 1916/17 winding-engine house of number 5 shaft. A duplex-cylinder *Koepe* steam-winding engine (**127**), built in 1941 by the Prinz Rudolph foundry in Dülmen (see also **153, 200, 206, 231**), North Rhine-Westphalia, was decommissioned in 1983. At 8,600 horsepower, this was the most powerful steam-winding engine in Germany, and possibly in Europe or indeed the world. A *Koepe* electric-winding engine, built by the same company in 1983, is still operational.

*References*

https://www.behance.net/gallery/1788612/Garden-Reden-Saarland-Germany

http://www.alamy.com/stock-photo-water-garden-reden-mine-schiffweiler-germany-103470460.html

http://www.erlebnisort-reden.de/

personal communications, Eberhard Lantz, photographer, 2017, 2021

**114. Ferropolis (Town of Iron)**

Ferropolis is located close to Gräfenhainichen, once the centre of the lignite industry of central Germany. It is situated on the site of the former opencast mine, Golpa-Nord, which began operations in 1964 and closed in 1991. The 1,000 hectares of the void created by the mine were flooded in 2000 to form Lake Gremmin, but with a 40-hectare peninsula occupied by Ferropolis. The idea of developing a monument to the industry of the area came from the Bauhaus School of Architecture in Dessau. The main shareholder in the project is the municipality of Gräfenhainichen, with funding coming from the Federal Republic of Germany, the Federal State of Saxony Anhalt and the EU. In 2019 the site had seven employees, with volunteer guides at weekends, who explain the history and the future of the project, as well as technical aspects of the former mining processes and equipment.

The old 30-kilovolt-electricity power station was converted into a museum of the history of mining in the region. In the centre of the site is an arena for international festivals and large-scale open-air concerts, at which internationally well-known groups, such as the American heavy-metal band, Metallica, have performed to up to 25,000 spectators. As the original pit railway is still linked to the mainline, visitors can travel by train direct to the arena, past displays of colliery locomotives and waggons. The five enormous decommissioned and fully restored opencast machines, weighing a combined total of 7,000 tonnes, are the most striking feature of the site. Visitors are able to climb the machinery for a view of the surrounding landscape. Each machine has its own name and history.

Big Wheel, on the eastern edge of the island, was constructed in 1984 by TAKRAF Lauchhammer. It weighs 1,670 tonnes, stands 31 metres tall and has a length of 74.5 metres. The Junior, or Mosquito, was constructed by VEB

**Figure 114.1. Aerial view of Ferropolis on a peninsula in Lake Gremmin, showing the layout of the five opencast machines on the site; from left to right, Medusa, Mad Max, Mosquito, Gemini and Big Wheel (© Ferropolis).**

Figure 114.2. From left to right, Mosquito, Big Wheel and Gemini arranged around the arena (© Ferropolis).

Figure 114.3. Big Wheel (© Ferropolis).

Conveyors Köthen in 1958 for mining under the surface. It is located in the middle of the island and stands 30 metres high, weighs 1,796 tonnes and measures 125 metres in length, with a 60-metre boom. The most striking feature is the feed wheel, which is 8.4 metres high.

Gemini is the heaviest and most powerful of the five excavators on site. It is made up of a primary section and a discharge device. It was constructed in 1962 by VEB Conveyors Köthen. Gemini weighs 1,134 tonnes, is 27.6 metres high and 79.2 metres in length. This is the only machine on site that can be visited independently, as well as on the guided tour.

Mad Max is an excavator equipped with 40 giant buckets, with which it was possible to move 1,920 cubic metres of earth per hour. It was constructed in 1959 by VEB Conveyors Köthen. Mad Max weighs 1,089 tonnes, stands 36 metres high and has a length of 102 metres, including a boom 61 metres long. The machine stands on the edge of the arena facing the lake.

Medusa was constructed in 1941 by Maschinenfabrik Buckau/Magdeburg R. Wolf. It is a 360-degree-swivel repair crane and jib, weighing 718 tonnes, with a height of 27.2 metres and a length of 67.1 metres. Medusa is the oldest, and the smallest, of the machines on site and is still fully operational.

*References*

ERIH, *s.n.* Gräfenhainichen

http://www.ferropolis.de/

personal communications, Isa Feller, tourism manager, 2013; Tom Freiheit, apprentice, 2021

### 115. Göttelborn Grube (Göttelborn Pit)

The imperial Prussian mining authority began to mine coal at Göttelborn Grube in 1887, some 12 kilometres north of Saarbrücken, in the state of Saarland in the west of Germany. Between 1888 and 1912 the German state mining authority erected 23 homes, a school and a café-kitchen. From 1918 until 1935 the pit was under French control and in 1921 the French mining authority built 10 single-storey semi-detached houses.

During WWII the pit was worked partly by forced labour, most of whom were Russian prisoners-of-war. The largest number of employees, 4,309, was reached in 1951, producing 1.8 million tonnes of coal. Göttelborn was always at the forefront of technical developments and in 1972 was one of the top pits in the country, with some 2,662 employees and 100 kilometres of roadways underground. In 1990 400 million Deutschmarks (€204.5 million) were invested in number 4 shaft, which was around 1,200 metres deep with an 87-metre-high white headgear. By 1995, however, Göttelborn was part of the general decline in coal mining, partly as the result of a reduction in coal subsidies. By 2000, when it closed, it was the last coal mine in eastern Saarland.

The town of Quierschied is responsible for the site and the Industrie Kultur Saar (Saar Industrial Culture) company was set up to develop it, including an industrial-heritage workshop and a circular trail around the spoil heap to a viewing point at the top. It is also possible to visit the upper deck of the headgear of number 4 shaft. The 1926 headgear of number 3 shaft, which is 13.45 metres high, was declared a national monument, and the rest of the pit buildings, the workers' hospital and the miners' housing have also been listed. The housing reflects the social hierarchy of the colliery, with free-standing houses for the management and semi-detached houses for the miners, each with a garden, shed and yard.

*References*

de.wikipedia/wiki/Grube_Gottelborn

ERIH, *s.n.* Quierschied-Göttelborn

http://www.memotransfront.uni-saarland.de/grube_goettelborn.shtml

### 116. Gruben- und Feldbahnmuseum Zeche Theresia (Theresia Colliery and Field Railway Museum)

The Gruben- und Feldbahnmuseum Zeche Theresia is a technical museum dedicated to the demonstration and preservation of mine trains. It is located at the site of the former Zeche Theresia in Witten-Bommernont on the southern shore of the river Ruhr in the south-east of the Ruhr. Work on the museum began in 1989 and it opened to the public in 2002. It is run by a not-for-profit charity, the Arbeitsgemeinschaft Muttenthalbahn e.V. (Muttenthal Railway Working Community Association), which supports its work with donations, membership fees and charges for trips on the railway. The Muttenthalbahn in the name refers to the nineteenth-century field railway built to move coal in this area. A *Feldbahn* (field railway), was not a permanent narrow-gauge railway, but a narrow-gauge private light railway, often of a temporary nature, constructed for transporting specific items, such as coal, industrial products or wartime materials such as munitions. A new track has been constructed to accommodate the tourist traffic. It runs from the Nachtigallstrasse car park through the museum premises to the nearby Zeche Nachtigall (**128**). Together with the Zeche Nachtigall, the museum is part of the Bergbauwanderweg Muttental (**103**).

The exhibitions show the history of coal mining in the area, with especial reference to mine trains and field railways. The Geiltingen coal seam, which runs underneath the rock on which Steinhausen Castle stands, has been mined here since the eighteenth century. In 1800 the mineral rights here were granted to the barons Elverfeldt. The Zeche Theresia was one of the early coal mines in the area south of the Ruhr. The colliery began production in 1828 above the water level of the river Ruhr. The Zeche Catharina was sunk in 1850 and by 1870 had reached a depth of 1,000 metres. The Zeche Theresia was joined with the neighbouring Zeche Nachtigall in 1854. The Zeche Catharina closed in 1877, however, and coal production ended in the Theresia coalfield with the closure of the Zeche Nachtigall in 1892. The buildings of the Zeche Theresia were listed and are now used to house the museum.

The exhibition includes 90 locomotives and about 200 railway carriages and waggons from the Ruhrgebiet mining area and the peat mines of nearby Lower Saxony. There are locomotives powered by compressed air (pneumatic locomotives), diesel, electricity and steam, together with numerous waggons from the field railway. There are also some English waggons, which were used during WWI behind the Belgian frontier. A number of carriages have been renovated to carry passengers on the railway.

*References*

https://de.wikipedia.org/wiki/Gruben-_und_Feldbahnmuseum_Zeche_Theresia

https://en.wikipedia.org/wiki/Feldbahn

www.muttenthalbahn.org/

showcaves

### 117. Grube Feggendorfer Stollen (Feggendorfer Pit Roadway)

The Grube Feggendorfer Stollen is located in the Deister Hills in Lower Saxony, north-west Germany, about 25 kilometres south-west of Hanover. The mine was first opened in 1832, mainly to provide domestic coal and fuel for smiths in the local Lauenau region. It closed down in the 1870s, but reopened in 1917 to supply coal during WWI and remained operational until 1947. Local people were permitted to produce small quantities of coal for personal use from 1947 until 1952, when many of the buildings and the mining equipment were demolished.

A local not-for-profit foundation, the Heimat- und Museumsverein Lauenau und Umgebung e.V. (Local History and Museum Association for Lauenau and Surrounding Region), has been working at the mine since 2003. The members reopened the tunnel and made it secure, having removed some 1,000 tonnes of rock. A year later, in 2004, they received a licence from the Landesbergamt Clausthal-Zellerfeld (the authority governing mines) to operate a tourist mine. The founding members included the local mayor, the district forester, the Lauenau Heritage Association and former mineworkers. The foundation has a chair, vice-chair, treasurer, secretary, deputy manager and 17 volunteers, some of whom are former mineworkers. Funding came from the People's Bank, the People's Bank Foundation, the EU and Lauenau District.

The mine reopened in 2005 as an example of a small coal mine of the Deister region, without electric lighting or improvements to the floors of the roadways in order to maintain authenticity. Since then it has been managed and operated by the foundation, offering underground tours of about 400 metres for groups of up to 10 people. The tours take about an hour on level ground in a temperature of approximately 9 degrees Celsius. For people who are physically fit, a second tour takes visitors up a steeply inclined drift to the surface, with a 50-metre stretch where it is necessary to walk in a bent position. This 500-metre tour is for groups limited to eight people.

On the surface there are the mine building, lamproom, workshop and various surface installations, such as tracks, a timber yard and a stores building. The aim has been to restore these to as close to how they were at the closure of the pit in 1947.

*References*

https://feggendorfer-stolln.de/aktuelles/neuigkeiten

showcaves

### 118. Haus des Bergmanns (House of the Miner)

The Haus des Bergmanns is situated 8 kilometres north-west of Moers, in Kamp-Lintfort in the Wesel district of North Rhine-Westphalia. The museum displays a typical house for the families of two miners. It is located inside two miners' cottages, which were part of an estate built in 1907 to house workers at the nearby Friedrich Heinrich Colliery (**174**). The estate was built as a modern garden city, but it was east of the mine, from which the winds tended to blow dirt for most of the year. The housing estate for the higher-ranking staff was located upwind to the west of the mine. The housing is an excellent example of the type of housing provided by colliery companies from the late-nineteenth century until the mid-twentieth century, with high rectangular windows with arched window heads, gables and plaster work.

The museum opened in 2006 and is managed by 25 unpaid volunteers. The city of Kamp-Lintfort lets the building to the Fördergemeinschaft für Bergmannstradition (Association for Mining Traditions), without any charges for rent or additional costs, such as supplies of energy and water. The equipment and the installations are the property of the Fördergemeinschaft.

The first house is decorated with the crossed hammer-and-pick mining symbol (**206**). The exhibition in this house has been restored to its original condition around 1910 and shows different kinds of rooms with typical furnishings. The kitchen has all the dishes, equipment and furniture of the period, including a traditional German cross-stitch embroidered work hanging on the wall with the slogan *Frohes Erwachen* (awake happily). In the master bedroom there is a picture on the wall showing Saint John the Baptist blessing the Virgin Mary, together with Christ as an infant, and in the children's bedroom there are dolls and picture books. Outside the house there are a pit tub

Figure 118.1. Exterior of the Haus des Bergmanns, with pit tub to the left and coal plough to the right (© Fördergemeinschaft für Bergmannstradition linker Niederrhein eV).

Figure 118.2. Dining room-lounge with dresser containing the best china owned by the household, Catholic crucifix and treadle sewing machine (© Fördergemeinschaft für Bergmannstradition linker Niederrhein eV).

and a coal plough of the type used extensively on German longwall coalfaces from the 1940s before the introduction of shearers. This type of plough continued to be used into the twenty-first century, with the addition of modern computer controls and power systems.

The second house has displays about the history and technology of coal mining, including a collection showing the stages of development of miners' lamps. There are documents, historical photographs, mining models and information about coal mining in the region. The cellar houses a replica of the underground of a local mine, with typical machinery and tools. Outside there is a typical garden.

The museum has a collection of films and mining lamps, and general mining pictures and photographs of the Friedrich-Heinrich Colliery (later called the West Colliery) and Rossenray Colliery. These date between 1906 and 1955, with an emphasis on the development of the city of Kamp-Lintfort around the mines.

*References*

http://www.bergmannstradition.de

personal communication, Susanne Rous, head, Kamp-Lintfort Museums Service, 2021

showcaves

### 119. Heimat- und Braunkohlemuseum Steinberg am See (Steinberg am See Local History and Lignite Museum)

Steinberg am See is a municipality in the district of Schwandorf in Bavaria, south-eastern Germany. Lignite mining was carried out in Bavaria from 1904 until 1982, when the coal reserves were exhausted. Voids left by opencast mining filled with water, creating six lakes encompassing some 650 hectares. As a result the name of the municipality was changed from Steinberg to Steinberg am See (Steinberg on the Water). The land reclamation and the relocation of the town of Alt-Wackersdorf during the mining are the topic of the nearby Wackersdorf Museum, with a mining nature trail running to the Heimat- und Braunkohlemuseum Steinberg am See.

The Heimat-und Braunkohlemuseum, which opened in 1994, is situated in the seventeenth-century parsonage of Steinberg am See. It is operated by the not-for-profit Heimatkundlicher Arbeitskreis Steinberg-Wackersdorf e.V. (Steinberg-Wackersdorf Local History Working Group Association). The museum covers lignite mining in the region around Steinberg and documents the history of the Bayerische Braunkohlen Industrie AG. Lignite mining is explained in detail in 20 panels, together with displays of fossils, mine lamps, minerals, models of dredges and documents such as certificates. Further panels detail the work life of the miners and craftsmen, the workshops and rail transport. The operation of an opencast coal mine is explained. The museum owns a considerable number of tools and equipment. Large machines are on display outside, such as drilling machines and the bucket of a huge opencast dredge.

*References*

https://braunkohlemuseum-steinberg.de/

showcaves

### 120. Herrmannschacht Brikettfabrik (Herrmannschacht Briquette Factory)

The Herrmannschacht Brikettfabrik is situated in Zeitz in the Burgenlandkreis district of Saxony-Anhalt in north-eastern Germany. Presses and plate dryers for producing lignite briquettes made in the period 1883 to 1895 continued operating here until the end of 1959. A single 12-horsepower steam engine, which was used to drive the transmission, was subsequently replaced by an electric motor. All the remaining equipment is, however, still in its original condition. This means that the Herrmannschacht Brikettfabrik is the oldest remaining first-generation briquette factory in the world. The industrial monument is generally acknowledged to be the last evidence of lignite processing in central Germany.

The Herrmannschacht Brikettfabrik was named after Richard Herrmann, then the manager of a sugar factory in Zeitz, who laid the foundations for the briquette factory in the later-nineteenth century. He was looking for cheap fuel and decided to buy up coal reserves in the surrounding area. In 1866 he sank a shaft called Neue Sorge (new concern) and at first processed the lignite in a wet-pressing factory, before building the briquette factory in 1889. The factory closed in 1959. By 1961 the site had already been listed on the grounds of its extraordinarily fine condition. With the exception of the boiler house, of which only a section is still standing, all the key buildings involved in briquette production have remained in their basic condition. In 1994 the not-for-profit Verein Mitteldeutscher Umwelt- und Technikpark e.v. (Central German Environmental and Technology Park Association) was set up to protect important industrial monuments, such as this, and the site became part of the European Route of Industrial Heritage (**75**) in 2009.

Three different perspectives are offered to visitors, giving an overall picture of lignite from its formation through processing to its final end uses. A nature trail leads through a wooded area with plant species from which the lignite developed millions of years ago. These range from forests in swamps through to mixed deciduous woodland and gave the landscape its characteristic moors, beneath which the lignite evolved.

The factory illustrates the various stages of briquette production. The kiln museum, which is housed in the tower of the factory, shows the use of briquettes in private households with a variety of stoves and ovens. There is also an exhibition on the Zemag Zeitz engineering company, which constructed conveyors, cranes and other heavy machinery, including for use in mining. A mini-preserved railway features models from the industrial history of the Zeitz-Weissenfelser lignite coalfield. Every September the museum participates in the Days of Industrial Culture programme, which highlights industrial museums and sites in the region.

*References*

ERIH, *s.n.* Zeitz

http://www.mut-zeitz.com/

### 121. Hessisches Braunkohle Bergbaumuseum Borken (Borken Brown-Coal Mining Museum of Hesse)

The Hessisches Braunkohle Bergbaumuseum Borken is located in Borken in Hesse in the western part of central Germany. The lignite coalfields of region around Meissen has been mined since 1578. The museum had a foundation endowment of €1 million from the municipalities of Borken and Hesse when it was set up. It has municipal sponsorship and the Hessian Ministry for Science and Art has given the museum significant support in recent years with grants under the Municipal Financial Equalisation Programme. The museum is part of a cross-regional cooperation with the Cultural Summer Nordhessen, the Educational Council of Kassel Museums, the Geopark Grenz Welten, the Hessian Museum Association and the Network of Industrial Culture of Northern Hesse.

**Figure 121.1. Exterior of the Hessisches Braunkohle Bergbaumuseum Borken in timber-framed building, Am Amtsgericht (© Hessisches Braunkohle Bergbaumuseum Borken).**

**Figure 121.2. Bucket-chain excavator on display in the Kohle und Energie theme park (© Hessisches Braunkohle Bergbaumuseum Borken).**

The museum was founded in 1992. It is housed in a timber-framed building called Am Amtsgericht (the district court), which is the oldest building in the town centre and occupies 400 square metres. The museum has four employees and 25 freelance staff in visitor services. It attracts between 8,000 and 10,000 visitors each year.

The museum shows the history of the Hessian coalfield, the role of brown coal in industrialisation and the ways of life of lignite miners. An overview map shows the former colliery sites. A reconstructed gallery, 100 metres long, displays the methods used for working lignite underground. These range from the early-twentieth century until the industry came to an end in the 1980s, when extensive use was being made of mechanical cutters and hydraulic props. Included in the equipment are compressed-air drills, conveyor belts, mines-rescue and safety equipment, mining lamps, mine-surveyors' geodetic devices and tunnelling machinery.

A mineral collection, the Mineralogische Sammlung M. Schrammel, is open to the public once a week in a separate building. The museum has its own archive of films, historical documents, photographs, postcards and publications, together with construction diaries and draft drawings for the Borken power station.

Large bucket-chain and bucket-wheel excavators used in lignite opencast workings, power-plant grinders, several mine locomotives, turbines and underground coal-mining equipment, such as shearers and tunnelling machines, are displayed in the adjacent Kohle und Energie (coal and energy) theme park. The park extends over 3.5 hectares and was opened in 2003. There is a special experimental area for children aged between six and 12 to learn about energy and the environment. The park demonstrates the production of electricity by a cyclone in a boiler in which coal dust was burned in a circular vortex at the power station in Borken. The cyclone heated water to over 1,400 degrees Celsius as it flowed through pipes inside. The water evaporated into steam under high pressure, which then powered the turbines. The exhibit no longer produces steam, but demonstrates the principles of the use of lignite to produce energy and illustrates the high expenditure involved.

The park has a lookout tower, from which the changes in social structures in the coalfield are apparent, such as the Islamic cultural centre built for the Turkish miners. The park includes a memorial to all the Hessian miners who died while at work, and especially to the 38 German and 13 Turkish miners killed at the Stozenburg Mine in 1988. This coal-dust explosion (**80**) brought coal mining to an end in the region. The memorial was inaugurated the year before the museum opened.

Nearby the Borkener See 140-hectare lake occupies the void left by a former opencast mine, now filled with water. Although the lake water is nutrient poor, the shores provide a habitat for animals, migratory birds and rare plants. A conservation-information centre uses interactive media, plants and live animals to show how land that has been opencast, or that has suffered from mining subsidence, can be reclaimed for agriculture, forestry and water recreation.

The museum is a focal point on the Braunkohl-Rundweg (Brown-Coal Trail), a 35-kilometre route encompassing 20 sites of historical interest in the coalfield. These include the nearby former power station built in 1922-3, powered by lignite and, at its peak, supplied 600,000 people with electricity. In the 1970s new, larger, more-modern power stations were built that produced cheaper electricity to higher environmental standards. So Borken power station closed in 1991, but is now a protected monument.

*References*


ERIH, *s.n.* Borken

http://www.braunkohle-bergbaumuseum.de/

personal communication, Ingo Sielaff, director, 2016

showcaves


### 122. Hüttenstollen Osterwald (Osterwald Worked Gallery)

The Hüttenstollen Osterwald is located on the north-western edge of the district of Grafschaft Bentheim in Lower Saxony in north-western Germany. The first coal mining on the Osterwald (eastern wood) ridge was undertaken in 1585 by Duke Julius of Brunswick-Lüneburg (1528-89). The Hüttenstollen Osterwald was sunk between 1842 and 1845 in the eastern area of the Osterwald coalfield, making it the oldest coal mine in the former kingdom of Hanover (**127**). The mine was close to the glassworks, which had been fired by coal since 1701. There were 300 miners working the five seams, which varied in thickness between 35 and 70 centimetres. The seams were so thin that they had to be worked by miners lying on their sides, with only the main roadways being high enough to allow transport of the coal by underground locomotives. As seams of fireclay are frequently interleaved between the coal seams, as here, the brick industry also developed alongside. The bricks were fired using the coal. Mining, which at that time was in the Gustav seam, ceased in 1926, but the mine reopened in 1948. Coal mining ceased permanently in 1953 and fireclay mining in 1954.

The not-for-profit association, the Verein zur Förderung des Bergmannswesens Osterwald e. V. (Osterwald Association for the Promotion of Mining) was founded in 1980 to preserve the history of the mine. The underground tour opened to the public the same month. Volunteers developed the mine tour over the next 20 years. The roadway was originally driven with a slope to drain minewater and to move the tubs filled with coal more easily by gravity. This means that the mine is naturally drained, without the need for pumps to keep it dry. Illumination has been provided in the mine and 90-minute tours on foot are led by former miners along 450 metres of roadway. The temperature underground is a constant 8 degrees Celsius and there is relatively high humidity. The tour concentrates on the working conditions of miners in the nineteenth century and the history of the coal industry, together with the geology.

Near the mouth of the adit there is a small museum, which opened to the public in 1985. This has galleries dealing with coal processing and mining; energy; geology and earth history; glass production; and quarries. It demonstrates the tools used by the miners, as well as the history of the village of Osterwald. Outside there is a display of mining machinery. In 2016 the museum was accorded accredited-museum status, confirming its professional standards in matters such as conservation and presentation of its collection.

*References*

http://www.der-huettenstollen.de/

showcaves

### 123. IBA Terrassen – Besucherzentrum Lausitzer Seenland (IBA Terraces – Visitor Centre Lusatian Lakeland)

Grossräschen is a town in Lower Lusatia, in the north-eastern German state of Brandenburg. Mining of lignite began here in the late-nineteenth century, when many miners migrated to the region from Bohemia in the west of the Czech Republic, Poland and Silesia (at that time part of Germany). The brick industry also developed here (**122**). Following the 1990 reunification of Germany, many industries closed in the early 1990s and the brick kilns were destroyed as redundant.

For about 100 years coal was obtained from opencast mines, with whole villages being demolished to allow the coal underneath to be extracted. One of the largest opencast sites was Meuro Mine. In 2007 the mine ceased production and the void was allowed to begin filling with water. The creation of Lake Grossräschen in 2018 has led to the development of the tourist industry here with hotels, a lakeside winery and a marina. The lake will be connected to an adjacent, opencast void, which began filling in 2005 and will take until 2023 to fill, to become Lake Sedlitz, the region's largest man-made lake.

Grossräschen is often referred to as IBA-City, because IBA Fürst-Pückler-Land (**111**) had its headquarters in the city between 2000 and 2010. The visitor centre of the IBA, known as the IBA Terrassen (IBA Terraces), was built on the edge of the lake in 2004. The design by the Frankfurt-am-Main architect, Ferdinand Heide (born 1962), won the Brandenburg Architecture Prize in 2005. There are three terraces, a total of 270 metres in length, which are connected by building cubes that also house such facilities as a café and conference rooms. The site employs an average of 1.5 staff with one volunteer. The IBA was devoted to finding imaginative ways of transforming the landscape, such as innovative architecture for the new Seenland (Lakeland). This included floating houses on the flooded opencast mines, and finding new uses for disused industrial buildings, like the sites on the ENERGIE-Route der Lausitzer Industriekultur (**111**).

**Figure 123.1. Terrace of the visitor centre of the IBA Terrassen overlooking Lake Grossräschener (© Cornelia Wobar).**

Guided tours, with names such as 'journey to Mars' and 'jeep safari', are offered through the areas of agricultural cultivation of disused industrial sites and the extensive opencast-mining landscape, where there were some 20 lignite opencast mines. All of these are now being flooded to form the largest group of artificial lakes in Europe, known as the Lausitzer Seenland (Lusatian Lakeland). The IBA visitor centre was used to present their various projects to the public. The IBA Terrassen are now used as the visitor centre for the Lausitzer Seenland and house an exhibition on the concept and history of international building exhibitions in Germany. They are classified as a highlight of the ENERGIE-Route der Lausitzer Industriekultur.

*References*

ERIH, *s.n.* Grossräschen

de.wikipedia.org/wiki/Grossräschen

https://www.lausitzerseenland.de

http://www.iba-terrassen.de

### 124. Klosterstollen Barsinghausen (Barsinghausen Monastery Roadway)

The Klosterstollen Barsinghausen is located in the town of Barsinghausen in the region of Hanover in Lower Saxony in north-western Germany. Coal mining is first mentioned in this region in 1639. The Gemeindebergwerk im Bullerbachtal Mine (Municipal Mine in Bullerbachtal) opened in 1847. The Klosterstollen Barsinghausen was a roadway in this mine, driven in 1856, although it did not reach the coal seam until 1869. The roadway took its name from the Barsinghausen Monastery, first recorded in 1193 and still occupied by religious today. Mining in this region was always costly because the seams are only about 1 metre thick and pumping was required to clear the large amounts of water coming out of the seams. So coal mining ceased at Klosterstollen in 1921. The roadway continued in use as a ventilation tunnel until the Schachtanlage Klosterstollen (Monastery Roadway Mine) closed in 1957.

There were problems with minewater in the region in the 1970s, so that the Klosterstollen had to be drained in the 1980s, during which it was realised that it would be possible to reopen the roadway. Renovation of the site began in 1986 and it opened to the public in 1999. The mining museum is located in the former pithead baths, erected in 1898. The exhibition deals with the history of coal mining on the Deister limestone ridge. A lift takes visitors up the headgear of number 2 shaft, which was reconstructed in 2007.

Visitors travel by mine train about 1,400 metres into the tunnel to see the 70-centimetre-thick coal seam and the associated mining machinery. Guides demonstrate how coal is mined, such as the use of explosives and pneumatic drills. For those in good health there is an adventure tour on which visitors walk the whole way and, in the course of a three-hour tour, see more of the mine than is accessible by the mine train.

Starting from the colliery five surface coal-mining routes have been laid out, the longest of 5 kilometres. At each stopping point there are detailed information panels. The trails pass the former colliery spoil heap, the monastery fishpond and the Kohlenweg, the mountain road by which the coal used to be transported.

*References*

https://www.klosterstollen.de/

showcaves

### 125. Kokerei Hansa (Hansa Coking Plant)

The Kokerei Hansa is located in the Huckarde area of western Dortmund, North Rhine-Westphalia. It is in the care of the SIG (**228**), whose office has been in the former administrative building of the coking plant since 1997. The plant was established in 1928 to convert coal produced by the collieries of the surrounding area. It was the last of the 17 coking plants built during the 1920s rationalisation boom. Each coke oven was emptied every 10 minutes and then reloaded with coal. The coke was supplied to the Dortmund Union iron and steelworks. Materials removed during the process were used by the chemical industry, while some of the waste gas from the ovens was supplied to the chemical works and smelters in Dortmund and the rest was stored in the compressor houses to be fed into the public gas network. The site is important because it is representative of the fully planned integration of various elements of the heavy industry of the Ruhrgebiet. The plant was decommissioned in 1992 and its main production areas were listed in 1998.

The route to the plant is called the Nature and Technology Adventure Trail. It goes past the coal tower, which stands 40 metres high and gives a panoramic view over the area. Much of the surrounding area has been reclaimed by nature, with trees such as birch, buddleia and lilac. Former workers take guided tours through the plant to see the coking ovens and the machine hall with its five unique gas compressors.

*References*

https://creativeindustrial.wordpress.com/2013/06/02/kokerei-hansa/

https://www.komoot.com/highlight/95865

### 126. LWL-Industriemuseum: Westfälisches Landesmuseum für Industriekultur (LWL-Industrial Museum: Westphalian State Museum of Industrial Heritage)

The LWL-Industriemuseum is made up of eight separate sites and museums located around the region of

Westphalia in north-western Germany. The LWL of the name stands for Landschaftsverband Westfalen-Lippe (Regional Association of Westphalia-Lippe), which is a municipal association for the 8.2 million people in the region, employing more than 16,000 staff. The nine cities and 18 districts in Westphalia-Lippe make up the official members of the LWL. They oversee and finance the LWL, whose statutory duties are defined by an association of 116 members from the Westphalian municipalities. The Westphalian Industrial Museum was established in 1979 and became the LWL-Industriemuseum in 2007. The LWL-Industriemuseum is funded by the LWL. After extensive restoration work the Zeche Zollern became the headquarters of the LWL-Industriemuseum in 1999. The LWL-Industriemuseum has museums of brickworks, glass, textiles, and iron and steel, as well as a boat lift and three colliery sites (**127, 128, 129**).

*References*

https://www.lwl-industriemuseum.de/en/

personal communications, Norbert Tempel, head of engineering and conservation, 2017; Britta Handke, public relations, 2021

### 127. *Zeche Hannover (Hanover Colliery)*

The Zeche Hannover is located on the site of the colliery in Bochum in the Ruhr coalfield of North Rhine-Westphalia. The name comes from the former independent kingdom of Hanover which, from 1714 until 1837, was ruled jointly with the UK by the elector of Hanover, who was also king of the UK. It is now part of the federal state of Lower Saxony. Coal was first identified here in 1847, and the two headframes, with the winding-engine house between them, were constructed in 1857. The tower headframe is perhaps the outstanding feature of the site. Its name comes from a defensive Russian fort on a hill called Malakoff, captured by the French during the Crimean War (1853-6), which became a synonym for a monumental construction. This was when the first such towers were beginning to be built at German collieries, although Malakoff tower was not formally adopted as a technical term until 1928.

The first coal came to the surface in 1858, but the colliery did not become profitable until it reached a depth of 162 metres in 1870. The industrialist Alfred Krupp (1812-87) took over in 1872 and the first housing estates for the workforce were built. In 1877 the mine manager, Carl Friedrich Koepe (1835-1922), devised the winding system that still bears his name. This is also sometimes known as friction winding, with the rope going directly around the winding drum, rather than using the traditional headgear with two wheels and a separate winding engine (**99.1, 104.1, 105.1-2**).

Production dropped during WWI, when many of the miners were conscripted into the army. Production dropped further during the French occupation of the Ruhr in 1923, during which the German miners engaged in passive resistance and strikes. In 1931 transport underground was converted from rope haulage to diesel locomotives. The Malakoff tower above number 2 shaft was demolished in 1939, the same year as 20 men were killed in an explosion underground. During WWII the mine was worked by up to 2,000 forced labourers, until the colliery was effectively put out of action by Allied airstrikes in 1944. The mine was reconstructed between 1945 and 1953.

By 1959 production stood at 1,971 kilograms per man *per annum*, reaching a peak of production of 1.8 million tonnes a year. In 1966 number 2 shaft was rebuilt but, with the developing crisis in the coal industry, the colliery was shut in 1973, the last mine in Bochum to close. The 1873 winding tower was demolished in 1976, despite public pressure. Originally it was intended to demolish all the structures, but the fan house, the Malakoff tower over number 1 shaft, the 1857 winding-engine house and some other buildings were preserved as a museum, which opened to the public in 1995. The 6-hectare museum attracts about 35,600 visitors a year and has six members of staff and eight volunteers.

The ventilation building is used as a cafeteria. The 1893 steam-winding engine has been restored, operated with an electric motor, but set up to demonstrate the principle of the *Koepe* system of winding. It is believed to be the oldest remaining steam-winding engine *in situ* in the Ruhr coalfield. Also preserved are three privately built 1890s houses of the Rübenkamp miners' settlement, an area of 2,527 square metres. Each of these would have contained four or five apartments and would have had appropriate gardens. These houses show the living conditions of the workers and their families between the 1860s and the 1950s. They have been a registered monument since 1997 and are now owned by the LWL.

The museum focuses on the cultural, industrial and social history of the Ruhr and especially the contribution of immigration and industrialisation. From the beginning in the 1850s major industrial plants like the Zeche Hannover could only meet the huge demand for workers by systematically advertising and recruiting workers from throughout Germany and abroad, such as from Italy and Poland and, in more recent years, from Russia. These immigrants brought their own individual cultural practices to the Ruhr, traces of which still remain today. The displays describe the general migration trends, as well as using local examples and oral-history recordings. In 2013 the Documentation Centre for the History and Culture of Poles in Germany was established at the Zeche Hannover.

The LWL-Industriemuseum shows exhibitions during the summer. In 2001 the Regional Association of Westphalia-Lippe opened a facility in the museum grounds to enable teams of children to learn how a mine operates. This is known as the Zeche Knirps, which can be translated as something like the 'kids' colliery' (German *Knirps* and

**Figure 127.1. Malakoff tower at the Zeche Hannover (© LWL-Industriemuseum).**

**Figure 127.2. Steam-winding engine dating to 1893 (© LWL-Industriemuseum).**

English 'kids' are slang terms for 'children'). It has its own miniature Malakoff tower, galleries, headgear and a waggon run. Gravel is mined instead of coal and is brought to the surface using the *Koepe*-winding system. Every two years a festival is held at the museum where unique vehicles from several European countries are operated on site.

### *References*


ERIH, *s.n.* Bochum Hannover Colliery

https://de.wikipedia.org/wiki/Malakow-Turm

https://en.wikipedia.org/wiki/Hoist_%28mining%29

https://zeche-hannover.lwl.org/en/

http://www.foerdergerueste.de/geruestarten.htm

personal communications, Norbert Tempel (**126**), 2017; Britta Handke (**126**), 2021

**Figure 127.3. Exterior of ventilation building, now used as a cafeteria (© LWL-Industriemuseum).**

**Figure 127.4. The Am Rübenkamp workers' housing estate (© LWL-Industriemuseum).**

### 128. *Zeche Nachtigall (Nightingale Colliery)*

The Zeche Nachtigall is located in Witten on the south bank of the river Ruhr in the Ruhr Valley in the western part of North Rhine-Westphalia. The Zeche Nachtigall is first mentioned in 1714 and was the first coal mine in this region to have a vertical shaft, as opposed to a horizontal adit. The shafts were named after Roman gods and heroes. The first shaft, Neptune (god of the sea), was sunk in 1832, but it soon became clear that it was too small and that large volumes of water were a major challenge. In 1839 the second shaft, Hercules (a divine hero), was sunk 450 metres deep, enabled by the new steam engines capable of pumping water out of the underground. The colliery went on to become one of the first deep mines in the Ruhr coalfield and one of the largest. The mine closed in 1892 and was replaced by a steam-powered brickworks, which survived until 1963. A large vaulted ring kiln on the site contains an exhibition on the family life and work of the brick makers, who produced some 11 million bricks every year.

The museum opened in 2003 and employs 15 members of staff. It attracts about 35,000 visitors a year. The museum deals with the history of the beginning of coal mining in the Ruhr coalfield, one of the main themes being the hard work and the difficult lives of the miners and their families. Guided tours for groups of up to 20 visitors take place every hour along a nineteenth-century 100-metre-long roadway. This is based around the 1839 Hercules shaft and is equipped with compressed-air pneumatic drills and jigger picks. The temperature is a constant 12 degrees Celsius. Most of the galleries in the visitor mine are part of the small twentieth-century pit. Weddings can be conducted in the visitor mine, with the reception held in the former workshops.

The displays include the Zeche Eimerweise, a replica of a small pit from which coal was hauled in buckets. This shows the primitive way coal was mined by local people immediately after WWII, when there was there was a pressing need for coal because many of the mines had been destroyed during the fighting. Between 1945 and 1976 there were more than 1,000 small mines in operation between Dortmund and Essen, some of them in the area of Witten.

The 1887 steam-winding engine is one of the oldest steam engines in the region, although now worked by electricity. An audio-visual journey in the winding-engine house deals with three centuries of industrialisation in the Ruhr coalfield. There is a background of contemporary music, as well as the song of the nightingale, after which

**Figure 128.1. Zeche Eimerweise, replica of small type of pit from which coal was hauled in buckets immediately after WWII (© LWL-Industriemuseum).**

**Figure 128.2. Quay and life-size replica of a sailing barge illustrating how coal was moved on the river Ruhr (© LWL-Industriemuseum).**

the colliery is named (**163**). Also preserved is the walled quay for unloading coal for the brickworks from after the colliery had closed. A life-size replica, 35 metres long, of an original sailing barge illustrates how coal was moved on the river Ruhr. This was built in 2000 for training unemployed young people. It is named the Ludwig Henz after Ludwig Benjamin Henz (1798-1860), a Hattinger hydraulic engineer. In 1840 he developed waterborne transport on the river Ruhr, which had been fully navigable since 1780 (**169**). For younger visitors to the museum, there is the opportunity to experience mining activities, such as pulling a mine tub on rails while wearing a mining helmet. The tub contains a toolset of all the mining tools and equipment needed by the miner in his daily work

A field railway (**116**) takes visitors from the museum on a circular route of 1.6 kilometres to the old Zeche Theresia (**116**). It is also possible for visitors to walk a hiking trail to see the chapel for the miners (**103**).

*References*


ERIH, *s.n.* Witten

https://zeche-nachtigall.lwl.org/en/

personal communications, Cindy Kramer, academic volunteer, 2017; Britta Handke (**126**), 2021

showcaves


### 129. *Zeche Zollern (Zollern Colliery)*

The Zeche Zollern is located in the Bövinghausen suburb of Dortmund in the centre of North Rhine-Westphalia. The museum employs 25 members of staff with five volunteers. It occupies an area of 140,000 square metres and attracts about 90,000 visitors a year. The first shaft of the Zollern Colliery was sunk in 1873, named after the Prussian royal family of Hohenzollern, who became emperors of Germany after unification of the country in 1871. A new shaft for ventilation had to be sunk in 1897. This, however, accessed further reserves of coal and so became another coal-drawing shaft, rather than the intended ventilation shaft.

Zeche Zollern II/IV (number IV shaft was sunk as a ventilation shaft) was built between 1898 and 1904 by the Gelsenkirchen Mining Company, at that time the largest mining company in the Ruhr coalfield. It was designed by professional architects as a model plant and with the most modern machinery. The winding-engine house contained the first electrically driven main-shaft winding engines in Europe. Other electric winders were already in existence, but not of the scale that was required for a modern colliery at a time when steam was still the main source of energy. Zollern was the first deep colliery to use electric winders capable of dealing with the speeds and heavy loads required.

Unfortunately production did not match expectations and in 1929 the company decided to concentrate their efforts at the Zeche Germania. This meant that the Zeche Zollern was not modernised and its buildings survived as originally constructed. The mine closed in 1963. Number II shaft was filled with debris in 1969 and demolition was planned of the winding-engine house, which was derelict. A number of local individuals campaigned to save the site and the owners were informed at the end of 1969 that demolition was forbidden. The winding-engine house became the first industrial building in Germany to be listed as being of national importance. In 1999,

Figure 129.1. Close-up of pay-office building showing red-brick historism style, with ornate gables and facades with white rendering (© LWL-Industriemuseum).

Figure 129.3. Interior of administration building showing ornate tiling and ironwork (© LWL-Industriemuseum).

Figure 129.2. Exterior of administration building of the Zeche Zollern, flanked by the two headframes (© LWL-Industriemuseum).

Figure 129.4. Exterior of winding-engine house with art-nouveau entrance (© LWL-Industriemuseum).

Figure 129.5. Interior of winding-engine house with stained-glass window and 1902 electrical reciprocating compressor (© LWL-Industriemuseum).

after restoration, the Zeche Zollern II/IV site became the museum for the cultural and social history of mining in the Ruhr, before it became the central site of the LWL-Industriemuseum. In order to retain its original character the Landwehr garden city, built to house the miners and their families, was included in the renovations. Nearby is the colliery spoil heap in the middle of the Deipenbecker Wood nature-protection area.

The Zeche Zollern is a rival to the Zeche Zollverein XII (**233**) for being the most beautiful colliery in Germany, if not in the world. In 1905 the site was described as being 'more like a feudal castle courtyard than an industrial site, and the machine hall, whose size and beauty surpasses that of most staterooms'. The surface buildings were designed by the architect, Paul Knobbe (1867-1956), in a style that was part red-brick gothic and part art-nouveau, with ornate gables and facades with white rendering. The winding-engine house, however, was constructed with steel trusses based on the new style of architecture displayed at the 1902 Dusseldorf exhibition. It was designed by the Berlin architect, Bruno Möhring (1863-1929), with an art-nouveau entrance and stained-glass windows. Two of the 1902 electric-winding engines survive, together with a 1920 electrical reciprocating engine to provide compressed air for use underground. Other historical machines include mercury-arc valves, transformers and further compressors. Six of these were restored to working condition between 2003 and 2007.

Renovation of the winding-engine house did not begin until 2007. It is important as a structure in its own right, but is also used for a variety of purposes, such as for temporary exhibitions and special events. The administration building, with its magnificent entrance hall and staircase, has a permanent exhibition about the history of the Zollern, including the use of forced labour in WWII. In the pithead baths there is information about the education and cultural history of the 1950s, while the wages/office building has exhibitions about the risk of explosion, accidents and mines-rescue services; couples can hold their wedding ceremony in the wages building. The original shaft house was demolished in 1959 and the headgear in 1969. In 1988 the shaft building and 36-metre-high headframe of the Wilhelmine Victoria I/IV Mine in Gelsenkirchen were re-erected at the Zeche Zollern. The shaft house deals with the various uses of coal and the individuals who have worked on the site, estimated at a total of 30,000 over the life of the colliery.

In 2021 the former training mine of Westerholt Colliery (**217**) was transferred to the museum to form a simulated underground, 40 metres long. On display are fully functional hydraulic shields to support the vaulted roof. Visitors can crawl under the supports while a light projection moves along the simulated coalface like a shearer. Original sounds were recorded at a working mine and are played back using modern technology to demonstrate the noises during coal cutting and also the collapse of the roof as the face moves backwards during retreat mining (**382**).

The museum has an important collection of original colliery-railway rolling stock, including steam locomotives. Some of these are in full working order and are used to transport visitors around the site during museum events. On special occasions the electric-winding engine and some of the other machines inside the winding-engine house are operated.

The archives hold material about the history of coal mining in the area, and especially about Zeche Zollern II/IV itself. These include, for example, documents about the origins of the mine, equipment and the mineworkers' clothing. Interviews conducted with various Zollern mineworkers during the 1970s and 1980s give a deeper understanding of the daily work in the mine and the lives and experiences of the miners themselves. All this material is used in the exhibitions on the site.

*References*

ERIH, *s.n.* Dortmund

http://www.lwl.org/industriemuseum/standorte/zeche-zollern

personal communications, Anne Kugler-Mühlhofer, director, 2017; Britta Handke (**126**), 2021

### 130. Maximilian Park (Maximilian Park)

Maximilian Park is situated in Hamm on the north-eastern edge of the Ruhr coalfield in North Rhine-Westphalia. Sinking of two shafts for the Maximilian Colliery by the ironworks company, Maximilianshütte AG from Sulzbach-Rosenberg, started in 1902, but water kept breaking through into the shafts. The first coal was brought to the surface in 1907, but it was not until 1912 that productive mining began. There were about 1,260 miners producing 100,000 tonnes of coal a year. In 1914 the pit flooded again and the colliery had to be evacuated. Two more, unsuccessful, attempts were made to reopen the mine, but it was finally abandoned in 1943. The buildings were not demolished, but were left to nature and became completely overgrown. In 1984 the colliery site was used for the first national garden show to be held in North Rhine-Westphalia, becoming one of the first colliery sites in Germany to be reclaimed. The following year, the Landesgartenschau Hamm (Hamm State garden show) was renamed Maximilian Park.

Traces of the buildings of the colliery are still visible. A very large, glass elephant's head has been added to the former washery, with a lift installed in the elephant's trunk to provide access to a viewing platform, 35 metres above the ground. The washery and the former workshop stand next to the red-brick electricity-generating plant with its large windows, which has been converted into an area for art exhibitions and other events such as concerts. Nature has been allowed to colonise the spoil heaps. There is a display of historic locomotives and waggons in a railway museum on the site. The park attracts 350,000 visitors a year.

*References*

ERIH, *s.n.* Hamm

https://www.maximilianpark.de/en/attraktion/der-park/

**131. Museum der Mineralien (Museum of Minerals)**

The Museum der Mineralien is located in the centre of Hückelhoven in the very west of North Rhine-Westphalia. The not-for-profit museum is run by volunteers and concentrates especially on minerals, although it does have some mining artefacts. There are minerals from around the world, but the focus is on minerals of the former Sophia-Jacoba Mine and on plant and tree-fern fossils. It is pointed out that the presence of fossils of freshwater mussels shows that the local coal seams were not caused by flooding by sea water, but by inland lakes and rivers.

There are simulations of typical conditions in a coal mine, using exhibits from the Sophia-Jacoba Mine, which was sunk in 1914 and closed in 1997. Between 1960 and 1997 it was consdered to be the most modern coal mine in Europe. There were six shafts, of which only number 3 shaft and its headgear are preserved. The simulations include a model of a miner using a drill, safety equipment and a reconstruction of a colliery workshop where mining tools were repaired. The system of bells for signalling is demonstrated. The museum also deals with coal-mining traditions, such as the veneration of Saint Barbara (**96**). In 2000 the local priest dedicated an altar at the entrance to the museum, with a triptych showing Saint Barbara in the central panel and a plaque above representing the symbolic union of the Church and the miners. To the right is a stand, decorated at the front with a hammer and pick (**206**), on which a memorial book lists the names of miners killed at work, with the date and cause of death.

Other activities of the museum include holding a minerals-exchange fair, which attracts collectors from a wide region, and organising a tour each year to other mining regions, such as the Eiffel, the Hunsrück or the Sauerland. The museum offers other museums permanent loans of some of its collections, such as the minerals or the fossils from the Sophia-Jacoba Mine or the collection of mining lamps. Admission to the museum is free.

*References*

https://www.mindat.org/loc-225231.html

http://www.museum-hueckelhoven.de/mh.htm

museum guidebook

**132. Museum Industriekultur Osnabrück (Osnabrück Museum of Industrial Culture)**

The Museum Industriekultur Osnabrück is situated on the north-western outskirts of the city of Osnabrück in Lower Saxony in eastern Germany. It is located in a country park and uses the surface buildings of the former Zeche Piesberg (Piesberg Colliery). This mine in one of the three regions of the Ibbenbürener anthracite coalfield was named from the Piesberg Mountain adjacent to Osnabrück. In the eighteenth century Osnabrück was a farming town. Mining began on this site in 1870, but eventually was stopped by water problems underground. The buildings occupy an area of 1,800 square metres. They are a fine example of the nineteenth-century *Rundbogenstil* (round-arch style), created in the nineteenth century as a German national building style in imitation of the romanesque style.

The museum was opened by the municipality in 1994 and became a not-for-profit limited-liability company in 1998. The majority shareholder is the Sparkasse Osnabrück (Osnabrück Savings Bank), which pays for the materials and personnel costs of the museum. The museum attracts between 42,000 and 50,000 visitors a year and there are nine members of staff.

The museum deals with the role that coal has played in the industrial and business development of the Osnabrück region during the nineteenth and twentieth centuries. The displays include material on how coal was formed. A three-dimensional model of the Piesberg Colliery illustrates how coal seams are layered underground and how the roadways branched out to enable the seams to be worked.

The main museum building occupies the Haseschacht (hare shaft) building. This is named from the hare tunnel (**163**), the adit at a depth of 9 metres, which is 280 metres long. The roadway is reached by visitors using a glass-sided lift. No machinery is displayed in the roadway. Although unlit, sufficient low levels of light reach inside the tunnel for lichens and moss to grow on the walls. Other colliery buildings that can be visited include the explosives store, pithead baths, stables for the pit ponies and the railway station, which used to serve the mine and which is now the headquarters of the Friends of the Industriekultur Osnabrück Museum. The main building also houses two steam-winding engines used in the colliery, dating to 1849 and 1916. These can be set in motion, with one engine driving the line shafting.

*References*

ERIH, *s.n.* Osnabrück

http://www.industriekultur-museumos.de/

personal communication, Rolf Spilker, director, 2016

**133. Nordstern Park (Nordstern Park)**

The Nordstern Park in Gelsenkirchen is located in North Rhine-Westphalia. It occupies the site of a disused coal mine, which was sunk in 1857 and given the name Zeche Nordstern (North Star Colliery) in 1865, because it was the most northerly colliery in the Ruhr. Most of the existing red-brick buildings were erected in 1930. Coal from the mine was stored in a bunker, then moved by conveyor belt to the coal-preparation plant. There the high-quality Nordstern coal was mixed with low-grade imported coal to give a uniform calorific value. The coal was then moved 4.7 kilometres by conveyor belt to the power plant for processing, mainly into coke. The finished product was stored in bunkers. The cooling tower lowered the temperature for the water used in the four compressors, which produced compressed air to provide power underground for tools, such as jigger picks. The water flowed in piping in the compressors, before being returned to the cooling tower for processing and then reuse. In 1955 4,400 miners were producing 1.2 million tonnes of coal each year. The colliery closed in 1993. The buildings were modernised and landscaped gardens created for a national garden show. This formed the 100-hectare park, which attracts 275,000 visitors a year. The coal bunker, coal-preparation plant and covered above-ground conveyor belt are all preserved as industrial monuments in the park.

As part of the garden show, the apprentices on site constructed a simulated mining roadway, 63 metres long, on the surface. Tours of the gallery are guided by former miners, who belong to the Friends of Nordstern. The modern red steel arched bridge over the Rhine-Herne Canal and the classical pit headgear are features of the site. The pithead baths, wages office and storage building now contain the offices of businesses, engineering companies and housing societies, with some 1,500 jobs created on the site. Panoramic views over the surrounding area are available from the number 2 Nordstern tower, which is a listed building, and its above-ground covered walkway, which is 83 metres long. The original pumping station continues to pump water from underground. Nordstern Park participates in the special event, ExtraSchicht. This is held in the Ruhr every year when those restored industrial plants which are not normally accessible are opened to the public for one night of music and other performances.

*References*

ERIH, *s.n.* Gelsenkirchen

https://www.nordsternpark.de/Startseite.html

personal communication, Marielies Lange, head of marketing, internet and social media, Stadt Gelsenkirchen (city of Gelsenkirchen), 2016

### 134. Rabensteiner Stollen (Rabensteiner Roadway)

The Rabensteiner Stollen is located just to the east of Netzkater in the Harz coalfield of Thuringia in eastern-central Germany. Coal mining first started in this region with this mine, which was worked from 1737 to 1770, 1831 to 1836, 1848 to 1880, 1921 to 1924 and 1946 to 1949, the last two periods because of fuel shortages following WWI and WWII. It was never very profitable, especially in competition with coal from the Upper Silesian and Ruhr coalfields. In 1981 the mine was opened as a visitor experience and the not-for-profit Förderverein Rabenesteiner Stollen e.V. (Friends of Rabensteiner Stollen) was founded in 1990 to manage the mine.

Visitors are taken by underground railway for a distance of 500 metres. They are then led on foot by experienced miners to the eighteenth-century coalface, where various types of mining machinery are demonstrated. The temperature underground is a constant 8 degrees Celsius. Mining machinery is exhibited in an outdoor area, together with some mining and narrow-gauge locomotives. Also in the grounds of the museum is a reconstructed miner's hut, which can be booked for visitors to stay in overnight. Steam trains of the Harzquerbahn (Harz railway) stop at the site.

*References*

http://www.ausflugsziele-harz.de/ausflugsziele-sehenswertes/hoehle-be

http://www.rabensteiner-stollen.de/

showcaves

### 135. Route der Industriekultur (Industrial Heritage Route)

The Route der Industriekultur in North Rhine-Westphalia consists of the most important places in the 150 years of industrial history of the Ruhr coalfield, such as blast furnaces, colliery headframes and gasometers, with a total of 25 highlight anchor points, 16 viewing points of panoramas over industrial landscape and 13 settlements from different periods. The route is 400 kilometres long by road and has 700 kilometres of cycle paths. Information about the Route der Industriekultur is available at the Ruhr visitor centre (**225**). The Portal of Industrial Heritage is a series of information stations presenting details of the anchor points along the route, in particular the history of coal and steel. The route of industrial sites in Hokkaido (**268**) was inspired by the Route der Industriekultur.

There are 12 theme routes (TR) that have sites related to coal mining:

Theme Route 2: industrial-cultural landscape of Zollverein.
Theme Route 3: Duisburg; industrial culture on the Rhine.
Theme Route 4: Oberhausen; industry makes city.
Theme Route 6: Dortmund; triad of coal, steel and beer.
Theme Route 7: industrial culture on the Lippe.
Theme Route 8: Erzbahn-Emscherbruch.
Theme Route 10: brine, coal and steam.
Theme Route 11: early industrialisation.
Theme Route 12: past and present of the Ruhr.
Theme Route 16: route of Westphalian mining.
Theme Route 17: Rhenish mining route.
Theme Route 29: Bochum, industrial culture in the heart of the district.

There are 28 special individual trails, including 12 major industrial monuments that are managed by the SIG (**228**).

There are eight anchor points related to coal mining on the routes that are described in more detail elsewhere in this publication (**108, 125, 128, 129, 130, 133, 231, 233**; see also **225**).

*References*

entries under the names of the individual sites in wikipedia

https://www.route-industriekultur.ruhr/

museum guidebooks and leaflets

The entries below deal in alphabetical order with those points on the various routes of the Route der Industriekultur that do not have separate entries in this publication. The theme route on which each site first occurs is given in brackets after the title. Only the first route is given for those places that are listed on more than one route.

### 136. *Bergarbeiter-Wohnmuseum (Museum of Miners' Lives)(TR 7)*

Two apartments are displayed to show the living conditions of miners in the 1930s in Lünen-Brambauer in the Alten Kolonie (Old Colony) of the former Zeche Minister Achenbach (**188**). They are decorated and furnished as they would have been in about 1935, with a traditional outhouse and rabbit hutch in the garden. The Alten Kolonie was built from 1900 immediately next to the colliery to house the workforce. The architects, the Schulze brothers, Dietrich (1864-1938) and Karl (1876-1929), used the same design for the ground floor of all the houses, but employed a variety of designs on the upper parts, so that no one house is like another.

The idea for the museum came during renovation, when craftsmen found the family books and lease book of the first tenant, Jacob Sears, under the rafters of the house. He had moved into the house in 1906 with his family. They lived in three rooms, with a garden, outhouse and pump to draw water. Now the courtyard is set up for catering for visitors and couples can be married in the parlour.

### 137. *Bergbaugedenkstätten auf dem Ostfriedhof (mining memorials in the Ostfriedhof)(TR 16)*

In 1893 61 miners were killed in an explosion of methane gas at Schacht 1 (number 1 shaft) of the Zeche Kaiserstuhl (Seat of the Kaiser Colliery), 48 of whom were buried in the Ostfriedhof (Eastern Cemetery). The names and dates of birth of the dead miners are mounted on two slim circular columns located behind a red sandstone monument modelled on classical architecture. There is an inscription on the front below the mining symbol of crossed hammer and pick (**206**), with a pediment above. Nearby a stone stele in front of their graves commemorates 19 miners who were killed in a methane explosion at Schacht II (number 2 shaft) of the same colliery in 1897. Nearby is a memorial stone to C. W. Tölcke (1817-93), who was a leading figure of the Allgemeiner Deutscher Arbeiter-Verein (General German Workers' Association), founded in 1863 to campaign for improvements in the working conditions of industrial workers.

### 138. *Bergwerk Auguste Victoria, Schacht 1/2 (Augusta Victoria Mine, shaft 1/2)(TR 7)*

The Bergwerk Auguste Victoria was sunk in 1899, named after Augusta Victoria of Schleswig-Holstein (1858-1921), the first wife of Kaiser Wilhelm II (1859-1941). There were two shafts and two headframes, dating to 1905 and 1907. The colliery was closely linked to the chemical works, which needed coal to generate steam from its many boilers to drive the machinery. Later the coal was used to produce coke and by-products, such as ammonia, benzene, sulphuric acid and tar. The Bergwerk Auguste Victoria was also one of the few mines in the Ruhr where ore was mined (**165**).

The two pit headframes still survive, together with their two winding engines, although no longer driven by steam but by compressed air. In front of the head office of the mine is a bronze statue of Augusta Victoria, who was the last empress of Germany.

### 139. *Bergwerk Auguste Victoria, Schacht 3/7 (Augusta Victoria Mine, shaft 3/7)(TR 7)*

At the beginning of the 1920s two new shafts, numbers 3 and 4, were planned for the northern section of the Bergwerk Auguste Victoria (**138**). Number 3 shaft collapsed shortly before it was commissioned. Number 7 shaft was sunk in 1957, when the central conveyor system was expanded, and came into production in 1960. Although further shafts were sunk, the coal continued to be brought up the central shafts, numbers 3/7, to the coal-preparation plant. In 2011 the mine produced 2.7 million tonnes of coal and had an underground-roadway network, 99 kilometres long at a depth of up to 1,320 metres. The mine closed in 2015. Today the lattice-girder headgear (**204**) survives above number 3 shaft and an enclosed headgear above number 7 shaft.

A working wharf for the mine was located north of numbers 3/7 shafts from the early 1930s. Larger amounts of coal and coke were required with increased production by the chemical plant. So six barges were purchased, each able to carry about 1,100 tonnes of coal and coke from the colliery to Ludwigshafen.

### 140. *Bergwerk Auguste Victoria, Schacht 8 (Augusta Victoria Mine, shaft 8)(TR 7)*

Numbers 8 and 9 shafts of the Bergwerk Auguste Victoria (**138**) are both in Haltern-Lippramsdorf. Number 9 shaft was for ventilation and concealed in the woods. Number 8 shaft was sunk in 1963 to ventilate the north-eastern part of the underground area and became a landmark on the north bank of the river Lippe. In 1972 the miners broke through between the roadways served by numbers 3/7 shafts and those served by number 8 shaft. By 1980 shaft 8 had reached its final depth of 1,300 metres, with 1,000 miners employed in this part of the mine.

### 141. *Bergwerk Ost, Schächte Heinrich-Robert (East Mine, Heinrich-Robert shaft)(TR 7)*

In order to provide a cheap and reliable supply of coal and coke for the ironworks, the Schacht Heinrich was sunk in Bergkamen in 1899 and the Schacht Robert in 1901 (see **143** for the names of the shafts) and the Bergwerk Ost came into production in 1904. The mine expanded repeatedly throughout its history, and adjacent mines were connected or taken over. Schacht Robert became the main shaft for raising the coal in the 1960s. The *Hammerkopfturm* headframe (**147**) rises above the surface buildings, some of which date back to the early years of the colliery.

In 1998 the formerly independent mines of Heinrich Robert and Haus Aden were amalgamated to form the Bergwerk Ost, with the Schächte Heinrich and Robert as the main operational shafts of the mine (see also **158**). Production finished in 2010.

### 142. *Deutschland-Bergbauwanderweg (German mining trail)(TR 11)*

In the eighteenth and nineteenth centuries, the region of Sprockhövel experienced an economic boom, with the development of a number of different industries, including coal mining. The 1960s crisis led to mine closures, completed in 1969 (**163**). Numerous traces of former mining activity still survive in the countryside and in 1994 a trail was laid out linking 30 mining features in the landscape, such as shafts and miners' houses. A map is available with directions and there are 11 information panels along the route.

### 143. *Fördergerüst Zeche Heinrich (Heinrich Colliery headgear)(TR 12)*

There were mines in this area from the eighteenth century, with the coal being moved along the adjacent river Ruhr. The first shaft of the Zeche Heinrich was sunk in 1847, with the coal being transported on the newly built Prince William railway. By 1861 the mine had 360 workers. In 1900 the mines were sold to a company owned by the de Wendel brothers, Henri (1844-1906) and Robert (1847-1903). The Schacht Heinrich, given the German form of Henri, was sunk in 1901 and the Schacht Robert in 1903. Peak production was achieved in 1966 with 966,000 tonnes of anthracite coal and more than 3,000 employees. At that time the mine was one of the largest in the Ruhr coalfield. In 1969, however, the colliery closed, the surface buildings were demolished and houses were built on the site.

Number 3 shaft, and that of the former neighbouring Zeche Theodor (Theodore Colliery), were retained to control the underground minewater. Even though the mines are no longer working, in much of the Ruhr the underground workings still produce water which, if not pumped to the surface and treated, would contaminate the ground water and also cause flooding on the surface (*cf.* **214, 221, 386**). The headgear of the Zeche Heinrich on the bank of the river Ruhr is a prominent landmark.

### 144. *Förderturm Bönen-Ostpol (Bönen-Ostpol headgear)(TR 10)*

The winding tower of the Zeche Königsborn (**186**) number 4 shaft was built in 1924/9. It was not constructed as the usual *Hammerkopfturm* (**147**), but in the form of a high rectangular cube, 66.83 metres high. The architect was Alfred Fischer (1881-1950), who from 1911 was head of the School of Applied Arts in Essen until the Nazis had him dismissed in 1933. The mine continued in operation until 1981. The winding tower has been preserved for its importance in the history of the development of 1920s industrial architecture, as it is believed to be the original model for modern oil rigs (*cf.* **205**). The interior stairs give access to a viewing point over the eastern Ruhr region at a height of 55 metres above the ground. Access has been organised since 2003 by a community foundation.

### 145. *Früher Bergbau in Fröndenberg (early mining in Fröndenberg)(TR 10)*

Coal seams were easily worked in the Fröndenberg area, where they outcrop, with evidence of mining from 1577 onwards. Between 1820 and 1848 tunnels were driven for the Zeche Zum Wilden Mann (Mine of the Wild Man). The acute shortage of coal after WWII meant that the 61-metre-deep Zeche Zum Wilden Mann was reopened in 1951 and only ceased production in 1953 after a miner was killed underground. Similarly the small Haggenberg Mine produced 856 tonnes of coal with 10 employees between 1951 and 1953. Notices have been erected on the sites of the early mines on the Route der Industriekultur.

### 146. *Gedenkstätte Zeche Radbod (Radbod Colliery memorial)(TR 7)*

Only 36 bodies were recovered from the 349 killed in the Zeche Radbod explosion (**198**), to be buried in two mass graves in the old cemetery at Hovel. A monument to the dead was erected in the cemetery in Bochum. A relief committee in Münster managed donations to a fund for the families, who had effectively been left destitute. After disasters everywhere, donations did not usually alleviate the suffering caused by the loss of the breadwinners, especially as little or no work was available for women: the 1857 Lundhill disaster in the UK left 90 widows, over half of whom had remarried by 1860 to acquire a provider (Elliott 2006, p. 75). Some widows of the Zeche Radbod disaster complained that they should have a larger pension than that for children and demanded a direct payment. They demonstrated in black mourning dress, but their demands were rejected and they were sentenced to five days in jail for disorderly conduct.

### 147. *Hammerkopfturm Zeche Erin 3 (hammerhead headgear Erin Colliery 3)(TR 16)*

The Zeche Erin (**172**) in Castrop-Rauxel was founded by an Irishman, William Thomas Mulvany (1806-85), who had left Ireland in the 1850s. He had already established collieries elsewhere in Westphalia called Hibernia (the Latin name for Ireland) and Shamrock (a young sprig of clover, which is used as a symbol of Ireland). In honour of his homeland he named this colliery Erin, a poetic variant of the Gaelic name for Ireland, *Éire*. The mine produced its first coal in 1867, but there were several methane explosions underground and the company went bankrupt in 1877. In 1882 the Essen mining entrepreneur, Friedrich Grillo (1825-88; see also **171, 190, 191**), re-established the mine and a third shaft was sunk in 1889. The risk of methane explosions was considerably reduced with greater investment. In 1902 a house was built for the mine official near the shaft. The high headframe, not built until 1918/21, is known as a *Hammerkopfturm*, an enclosed tower winder with the winding engine on the top. The overhang of the winding-equipment room over each side of the winding tower gives the appearance of an upturned hammer. The Erin *Hammerkopfturm* is one of only 10 constructed in the Ruhr during the 1920s. There was no winding-engine house, because friction winding was used (**127**). Number 3 shaft was used only for manriding between 1937 and the closure of the mine in 1983, as the coal came up number 7 shaft (**172**).

The shafts of the mine were filled, but the headgear was restored in 1993 with state funds. It is one of the few remaining *Hammerkopfturm* headframes in the Ruhr and the oldest in Westphalia. The city of Castrop-Rauxel surrounded the headgear with a 'Celtic tree circle', in honour of the Irish founder of the mine. This was because it was believed that the early Celts used a tree circle as a calendar, with each tree representing a particular astrological aspect. Unfortunately this was an invention in 1948 of the British writer, Robert Graves (1895-1985), in his book *The White Goddess*.

### 148. *Kleinzeche Egbert (Egbert Small Colliery)(TR 11)*

There had been numerous small mines in the area of Herbede from the mid-eighteenth century. The Kleinzeche Egbert was opened in 1962, despite the fact that the programme of closure of many mines in the Ruhr coalfield had already begun. The mine worked the 1.25-metre-thick Kreftenscheer seam at a depth of 135 metres with only six men. They raised 2 tonnes of coal each day, similar to the average production per manshift nowadays. Minewater flowed through a sough into the nearby river Ruhr, so that elaborate drainage or pumping arrangements were not required to keep the underground dry. The mine closed in 1976, the last of the small mines of the Ruhr to do so. The surface buildings were restored between 1979 and 1982, including the wooden headgear, compressor house, surface-rail system and equipment such as coal crushers, fans and pumps.

### 149. *Kulturzentrum Zeche Carl (Carl Colliery Cultural Centre)(TR 17)*

The Zeche Carl was an enterprise of the Cölner Bergwerksvereins (Cologne Mining Association), the first company to be established in the Ruhr mining industry. Founded in 1847, it was set up by the Camphausen and Schaafhausen banks of Cologne, which also invested in the Cologne to Minden railway. Altenessen had a station on the main railway from the very beginning in 1847, and the Zeche Carl was connected to this by a branch line, enabling the mine to move its coal efficiently to its markets.

The three-storey brick Malakoff tower (**127**) at the Zeche Carl has symmetrical lateral-storey wings and dates to 1857, making it amongst the oldest such towers preserved in the Ruhr. Steam-winding engines were installed in 1857 and 1863; in 1863 both were used for raising coal and manriding, although it had originally been planned to use only one for this. Also surviving are the 1900 winding-engine house, a boiler plant with a chimney and a 1920s workshop. The administrative and pithead-baths building dates back originally to 1890, and was extended in 1910 and 1925. The surface structures are listed buildings and are now used as a cultural centre.

### 150. *Kulturzentrum Schacht III, Zeche Werne (Shaft III Cultural Centre, Werne Colliery)(TR 7)*

Number III shaft of the Zeche Werne (**216**) was sunk in 1915, intended to be a free-standing sub-unit with its own boiler house, coal-preparation plant, garages, pithead baths and power station. Many of the miners were conscripted for service in the army at the outbreak of WWI, so they were replaced by women and French prisoners-of-war. The French prisoners were often trained miners and were paid for their work, while the women undertook equally hard work, but received only half the pay. The mine was closed between 1930 and 1946 as a result of the Great Depression (1929-39). Zeche Werne number III shaft finally closed in 1960, although was used until 2000 for ventilation and dewatering of the neighbouring mine. The headframe was demolished in the 1980s and the former pithead baths were converted into a cultural centre, the Kulturzentrum Schacht III (see also **216**).

### 151. *Künstlerzeche Unser Fritz (Our Fritz Art Colliery) (TR 8)*

The Zeche Unser Fritz was named after the crown prince of Prussia (1831-88) who, as emperor, Friedrich III, reigned for only 99 days before dying of throat cancer. He was popular for his liberal attitudes and was, therefore, affectionately called *Unser Fritz* (our Fritz), Fritz being a nickname for Friedrich. In 1875 seven miners were killed in a firedamp explosion at shaft number 1. To improve the ventilation, the mining authorities insisted on the provision of a second shaft, which was sunk in 1881, 700 metres north-east of number 1 shaft. These two shafts enabled the mine to produce more than 200,000 tonnes of coal a year. From 1907 number 3 shaft became the main shaft, with number 2 used for ventilation only. The colliery closed for a period in the 1920s, but reopened in 1936. It finally ceased production in the 1960s; the shafts were filled in 1993 and the headframes demolished. In 2007 restoration began of the winding-engine house of shaft number 3, which reopened in 2009 as an art-exhibition and events hall. Künsterleche Unser Fritz is now a cultural centre of national reputation.

### 152. *Kunstwald Zeche Teutoburgia (Teutoburg Colliery Art Forest)(TR 16)*

The Zeche Teutoburgia in the north-east of Herne, near the outskirts of Castrop-Rauxel, began operation in 1911. The mine was named from the Teutoburg Forest, about 140 kilometres north-east of Castrop-Rauxel. Here, at Kalkriese in CE 9, the Germans, led by Arminius (18 BCE-CE 21), the later German name Hermann, annihilated three legions of Roman soldiers, led by Publius Qunctilius Varus (46 BCE-CE 9). The Romans lost between 15,000 and 20,000 men and never again attempted to conquer any territory in Germany east of the river Rhine.

The mining zone available under the contract was too small which, combined with geological faults, prevented the colliery being profitable. The mine was, therefore, decommissioned as an independent unit in 1925 and was integrated underground with the neighbouring Zeche Erin (**147, 172**) in 1929. The winding-engine house and the 1908 headgear have been preserved in an 'art forest' amongst modern sculptures. The Friends of Teutoburgia arrange workshops and exhibitions of modern art in the former winding-engine house. The outline of the walls of the demolished colliery buildings are laid out on the ground.

The estate for the miners was built in 1909, based on the English garden city, with gardens for the houses, allotments and large green open spaces. The central avenue was aligned on the entrance to the coal mine. Eventually the estate covered 21.4 hectares with 459 homes of a number of different designs. Although relatively untouched by WWII, by 1989 the houses required renovation. The houses were all restored to their original condition, so that the Teutoburgia estate is now the best preserved and the most historically and culturally important settlement surviving in the Ruhr.

### 153. *Maschinenhalle Zeche Scherlebeck – Schacht V/VI (winding-engine house of Scherlebeck Colliery – shaft V/VI)(TR 16)*

The Zeche Scherlebeck at Scherlebeck in Herten began production in 1898. Because it was using the latest technology, in 1899 number V and VI shafts became an independent mine, the Zeche Schlägel und Eisen (**206**). This was the third colliery to bear the name of the mining symbol of the crossed hammer and pick. The surface remains of Schacht V/VI are all built of dark-red brick. The pithead baths date to 1899 and the winding-engine house, with large arched windows and round windows above those, dates to 1901. The engine house contains the last, and perhaps the oldest, tandem steam-winding engine in Westphalia, operated by *Koepe* winding (**127**). This was made in 1900 at the nearby Prince Rudolph ironworks (**113**) in Dülmen. The mine closed in 1929 and in 1933 the stores building was converted to a gymnasium for apprentices being trained on the site. In 1987 the colliery buildings obtained national status and funds were granted for restoration of the site. The Friends of Schacht V have restored the winding engine to full working order and converted the winding-engine house into a museum and community meeting room. The pithead baths now house a kindergarten and the other buildings form a socio-cultural centre for the community. The nearby estate, built to house the workforce, still retains much of its original character.

### 154. *Maschinenhalle Zeche Zweckel (winding-engine house Zweckel Colliery)(TR 16)*

The Zeche Zweckel was sunk as a double pit in 1908 in Zweckel, a district in the north of the city of Gladbeck in Recklinghausen (see also **199, 200**), and production began in 1912. The two winding engines were constructed by the Thyssen Company. These were advanced technology for their time, being driven by electricity (designed by AEG), and are now amongst the oldest in the Ruhr. The 1909 winding-engine house was designed to appear as if it were an historical mansion, because construction was by the imperial state of Prussia, which wanted the majestic architecture to be a visible sign of its claim to power. The three-storey building is 126 metres long, with two recessed wings to the main building and the façade divided by pilaster strips. The main storey has a row of semi-circular arched windows above a raised gallery platform.

A saltwater spring was discovered at a depth of 430 metres during sinking of the shafts. So saltwater-hydrotherapy baths, the Royal Solbad Zweckel, were built to utilise this resource. In 1929 the mine merged with the Zeche Scholven and its production was important in the 1930s for the quality of its coal as fuel. The saltwater spring failed in 1950 and the hydrotherapy baths had to close. The mine was modernised after WWII and reopened in 1958. It became unprofitable, however, as the geological conditions were found to need a higher level of mechanisation, so the mine was decommissioned in 1963. Ventilation and drainage were retained to safeguard surrounding mines (**143**).

The site is now in the ownership of the SIG (**228**). The two steel headframes survive, although one was reduced in height in 1984. In 1988 they were given statutory protection, along with the 1909 winding-engine house. The winding-engine house is used as a venue for art and culture in the city of Gladbeck, including the Ruhr Triennale.

**155.** ***Pförtnerhaus Zeche Königsborn 2/5 (gatehouse Königsborn Colliery 2/5)(TR 10)***

Number 2 shaft of the Zeche Königsborn was sunk in 1887 and the number 5 ventilation shaft in 1942. Both shafts continued in use until 1966, after which Zeche Königsborn 3/4 (**186**) took over. The surface structures were all demolished, apart from the former gatehouse, which was built as part of a redesign of the entrance area of the mine in 1910. The gatehouse served as a control point for all employees, visitors, coal merchants and suppliers of the mine. It also housed the telephone exchange, the devices installed in front of the weighbridge for land sales and a common room for the security guards of the mine.

In 1992 the gatehouse was added to the list of recognised monuments of the city of Kamen. It was restored and rebuilt three years later. The significance of the industrial history of the building is shown by the carving on the eastern gable of the crossed hammer and pick (**206**). The western gable carries the date 1910. The building is now used by an organisation dedicated to workers' welfare.

**156.** ***Rheinhauser Bergbausammlung (Rheinhauser mining collection)(TR 3)***

In 1897 the Krupp steelworks opened in Rheinhausen on the left bank of the river Rhine and the Zeche Wilhelmine-Mevissen was sunk in 1914. Workers from many European countries and the eastern part of Germany came to work in the mines and in what was once the largest smelter in the world. The mines closed in 1967 and the not-for-profit Verein Rheinhauser Bergbausammlung (Rheinhauser Mining Collection Association) was founded to preserve mining traditions. The exhibition now has some 850 exhibits, such as documents, minerals, photographs and tools, and an operational model, at a scale of one to 30, of the headgear of the Zeche Diergardt 1/2.

**157.** ***Schacht Gerdt (Gerdt shaft)(TR 3)***

The Zeche Rheinpreussen (**201**)was sunk in 1857. The Schacht Gerdt was completed in 1945 and a new headframe, designed by Fritz Schupp (**108**), was erected in 1955/9. From 1967 the shaft was used only for ventilation and to supply materials underground. The headframe structure is an unusual enclosed-square brick building characterised by narrow horizontal louvres and balconies. The headframe is a striking landmark, located only a few hundred metres from the river Rhine.

**158.** ***Schacht Lerche (Lark shaft)(TR 7)***

The Schacht Lerche was originally a ventilation shaft for return air. The shaft was remodelled and deepened by about 370 metres to a depth of 1,400 metres, enabling the Heinrich Robert and Haus Aden/Monopol mines to be amalgamated as the Bergwerk Ost (**141**), at a total cost of €280 million. The headgear above the shaft, weighing 134 tonnes, was transported at night more than 35 kilometres from the Schacht Romberg in 2001. The last shift came up the shaft in 2011, which was capped shortly afterwards. The headgear has statutory protection. An investor wants to set up a business park, leisure facilities and a small mining museum here, but there is local opposition to this.

**159.** ***Schacht Oberschuir (Oberschuir shaft)(TR 16)***

Schacht Oberschuir was sunk in 1908 as the eighth shaft of the former Zeche Consolidation (**171**) and was named after the operations director of the mine, Johann Oberschuir (1836-1912). It was intended as a manriding and ventilation shaft and continued in use solely as a ventilation shaft until it was filled in 1984. Electricity was used in the mine from a relatively early date.

The buildings around the shaft were all constructed from bright-red brick, interspersed with white plaster and cornices. The gatehouse, pay office, pithead baths and winding-engine house have baroque-style towers, curved gables and windows in the palladian style, together with art-nouveau elements. The headgear with its lattice-girder work (**204**) is a landmark of the area. The buildings have all been restored and in 1996 a modern building of concrete, glass and steel was added, called the Green Tower. The winding-engine house is used for exhibitions and events, while the pithead baths have been converted into residential units.

**160.** ***Schacht Rossenray (Rossenray shaft)(TR 17)***

The sinking of the Schachtanlage Rossenray (Rossenray shaft installation) as a double pit began in the early-twentieth century, but stopped as a result of WWI and did not start again until 1937. The freezing process to enable sinking the two shafts began in 1943. This was interrupted again by the economic crisis at the end of WWII and the company developing the mine went into liquidation. Shaft sinking recommenced in 1957. The mine had three shafts and power and coking plants when it began production in 1962. In 1966 16 miners were killed in a methane-gas explosion. Annual production had reached 935,000 tonnes by 1969, when the mine was amalgamated with the Zeche Pattberg (**196**). It later became part of the Verbundbergwerk West (Combined West Mine), which closed in 2012. Number 2 shaft remained for pumping (**143**).

In 1970 a new cross-shaped 75-metre-high headframe began to be constructed for shaft number 1, but was left unfinished. A similar headgear was planned for shaft number 2, but was never built. The number 1 headgear had lateral wings intended as the entry points for a circulation of mine cars, also never built. The headframe was a prominent landmark in the flat Lower Rhine area where it was located and was nicknamed the 'Cross of the Lower Rhine' (referring to the Christian symbol of the cross on which Jesus Christ was crucified). The number 2 headgear was demolished in 2018 and number 1 in 2019.

In 1999 the conceptual artist Mischa Kuball (born 1959) converted the tower, now called the Westpol (West Pole), into an art installation, known as the Yellow Marker, with the headframe of Zeche Konigsborn 3/4 (**186**), 80 kilometres away, as the Ostpol (East Pole). Opposite corners of the two towers were lit up at night with yellow fluorescent tubes. They were meant to represent the east and west poles or the end points of the industrial area of the Ruhr. The conspicuous spoil heaps of the mine remain and the area is also used for the sorting of gravel and sand from a nearby quarry.

### 161. *Trainingsbergwerk Recklinghausen (Recklinghausen Training Mine)(TR 8)*

The former spoil heap of the Zeche Recklinghausen II (**200**) now accommodates a training mine. The mound was created from the spoil produced during the sinking of numbers 1 and 2 shafts. Over time the material compacted and the heap was colonised by vegetation and trees. Tunnels were dug into the spoil heap during WWII to provide air-raid shelters. In 1975 these tunnels were extended to create a route about 1,400 metres long, with 44 training courses available. All geological, safety and technical conditions can be simulated in a safe environment. The replica underground is equipped with large mining machines and is also used for testing machines and systems before they are approved for use.

### 162. *Zeche Adolf von Hansemann (Adolf von Hansemann Colliery)(TR 6)*

Sinking of the Zeche Adolf von Hansemann began in 1873, but difficult conditions meant that the first coal was not brought to the surface until 1896. The mine was named after Adolf von Hansemann (1826-1903), the chair of the Gelsenkirchen Bergwerks-AG. In 1899 the former boiler house was converted by Dietrich Schulze (**136**) into administrative and changing rooms in the style of gothic brick. This was modelled on the fifteenth-century Uenglinger Gate in Stendal and was one of the few such buildings based on an historical model. On the ground floor there were offices and a laboratory, and baths and changing rooms for the management. The lamproom and pithead baths for the miners were upstairs. The winding-engine houses of numbers 1 and 2 shafts and the 1920s gatehouse and stores building were also constructed in an historical style. The mine closed in 1967 and the buildings, which have statutory protection, were converted in 2000 by the Dortmund Chamber of Commerce into a scaffolding-training centre. Several rooms in the gatehouse contain a small museum for miners, miners' everyday objects and a model of a colliery heading.

### 163. *Zeche Alte Haase (Old Hare Colliery)(TR 16)*

Coal was mined in the area around the city of Sprockhövel in the south of the Ruhr for several centuries. The last mine in the region, the Zeche Alte Haase, closed in 1969. *Haase* is the German for 'rabbit' or 'hare'; it used to be the custom to name a mine from the wildlife of the region. Sinking in the late 1880s of the first shaft of the mine took 10 years. Whereas steel headframes were already in use elsewhere in the Ruhr coalfield, the Schacht Julie (Julie shaft) was provided with a Malakoff tower (**127**), which is now the youngest, smallest and southernmost of the preserved Malakoff towers. In the early-twentieth century the mine had a licensed concession of 47 square kilometres, making it one of the largest mines in Germany. The mine had the advantage of being free from methane gas. The need for heavy capital investment to modernise the colliery meant that ownership passed from local families to foreign investors. When they decided to close the mine in the 1920s, the 1,300 miners were strongly supported by the public. The result was that in 1926 the colliery was taken over by the Vereinigt Elektrizitätswerke Westfalen (United Westphalia Electricity) to supply coal to the power station at Hattingen. The Malakoff tower was restored in the early 1990s to form a symbol for the city of Sprockhövel.

### 164. *Zeche Arenberg Fortsetzung (Successor to the Arenberg Mine)(TR 16)*

The Arenbergsche Bergbau- und Hüttengesellschaft (Arenberg Mining and Metallurgical Company) was set up in 1856, named from the duchy of Arenberg. Various shafts were sunk during the second half of the nineteenth century. The Zeche Arenberg Fortsetzung was sunk as a double mine and began producing coal in 1912, the same year as a coking plant was built adjacent to it. The mine achieved peak production in 1927 of 77,000 tonnes with 2,769 miners. In 1930, however, the coking plant and part of the underground operation were closed down. In 1970 the Ruhrkohle AG (Ruhr Coal Company), now the RAG AG, reactivated the mine and constructed a new steel box-strut headframe above number 1 shaft. The headgear was demolished in 1981 and both shafts were filled and capped.

The main colliery buildings of the blacksmiths' shop, gatehouse, pithead baths and wages office were built in the art-nouveau style and have been preserved, while the remainder of the site has been used for a technology centre and a start-up. The wages building in particular is regarded as a fine example of the industrial architecture of the reign of Kaiser Wilhelm II (1859-1941). The winding-engine house was restored in 2001 as a community-event centre for the city of Bottrop.

### 165. *Zeche Auguste Victoria Schacht 4/5 (Augusta Victoria Colliery shaft 4/5)(TR 7)*

The Zeche Auguste Victoria Schacht 4/5 had a special significance within the Bergwerk Auguste Victoria (**138-40**), because ore was mined in this section as well as coal. Sinking of the shaft began in 1928, with the final depth of 830 metres being reached in 1952. The headgear was built in 1931. A vein of lead, silver and zinc ore was discovered in 1930. In 1937 this section of the mine ceased to produce coal, which was instead brought to the surface via numbers 1 and 2 shafts. The ore was worked from 1938 until 1962, when it became unprofitable, by which time 7 million tonnes of lead, silver and zinc had been produced.

In 1995 the 1930/1 headframe of number 4 shaft was registered in the list of protected monuments of the town of Marl, because of the historical significance of the *Koepe* technology used (**127**). The threatened demolition of the winding-engine house led to the formation of the Erzschachtfreunde (Friends of the Ore Shaft) as a subgroup within the Heimatvereins Marl (Association of the Marl Homeland).

In 2005 it was agreed that the lower level of the winding-engine house should be used as a mining museum, sponsored by the local-history society. The winding engine has been preserved *in situ* on the upper floor, which is used for exhibitions, lectures and various cultural purposes.

### 166. *Zeche Bergmannsglück (Miner's Luck Colliery) (TR 16)*

Sinking of the double pit Zeche Bergmannsglück 1/2 north of Buer began in 1903. The name Bergmannsglück referred to the miners' greeting (**210**). In 1909 a coking plant was constructed, which was decommissioned in 1929 in favour of the plant at the Zeche Westerholt (**217**). The colliery produced about a million tonnes of bituminous coking coal each year, making it one of the most productive in the Ruhr. As a result of the mid-1950s coal crisis, the mine was amalgamated with the Zeche Westerholt, where a new central shaft was sunk. The conveyor belts were removed after the Bergmannsglück 1/2 shafts were closed in the 1970s and 1980s. Some buildings on the site were used as a central warehouse by the Ruhrkohle AG (**164**), but all the buildings were left unoccupied in 2008. Some surviving colliery buildings, including the gatehouse, are preserved as good examples of industrial architecture of the early-twentieth century.

### 167. *Zeche Bonifacius (Boniface Colliery)(TR 17)*

When the Zeche Bonifacius was sunk in 1857, the village of Kray where it was located had only 300 inhabitants, living on 18 farms. The mine was named after Saint Boniface (CE 675-754), an English missionary known as the Apostle of the Germans, who was martyred in Frisia (**198**). By 1870 the colliery was producing 185,000 tonnes of coal a year. The Gelsenkirchen Bergwerks-AG took over the colliery in 1899 and rebuilt it completely. The new surface buildings were constructed around a courtyard between 1902 and 1905. Unlike other major mines of the time in the area, they were designed in the art-nouveau and neobaroque styles. In 1908 a coking plant with 240 furnaces was installed. The mine closed in 1974.

Although now protected as listed buildings, the structures were initially saved by private initiative. They still preserve the original layout. The lattice-girder headframe (**204**) was restored in 1993. In 2004 the former neogothic administrative and wages building was converted into a hotel, the Alte Lohnhalle (old wages building). The winding-engine house still contains a 1913 twin steam-winding engine and a 1914 electric-winding engine made by the Siemens-Schuckert Company. Also preserved are the small pithead baths, with a mortuary for miners killed at the colliery at one end. Guided tours are available.

### 168. *Zeche Brockhauser Tiefbau (Civil Engineering Brockhauser Colliery)(TR 16)*

Coal was already being worked by small adit mines in this area by 1700. The Zeche Brockhauser was formed in 1873 by the merger of the five older drift mines of Diebitsch, Friedrich, Ignatius, Preussischer Zepter and Treue. The name Brockhausen, which originally belonged to a farm in Stiepel, is today a street name in the Bochum suburb of Stiepel. *Brock* used to mean an area of land near a river (in Bochum the river Ruhr) which is flooded regularly. So the name Brockhauser, meaning 'houses next to a river on land that floods regularly', was then transferred to the merged colliery. In 1874 a vertical shaft was sunk to a depth of 150 metres, gaining the colliery name the suffix of the adjective *tiefbau* (civil engineering), and a Malakoff tower (**127**) was erected over the shaft. The colliery was one of the smaller mines in the Ruhr, with a production of 40,000 tonnes a year and 185 miners in 1880. As well as the low levels of production, there were also difficulties in moving the coal out of the narrow valley, so the mine closed in 1912. In 1979 the structures on the site, including the Malakoff tower, were given statutory protection, as the last example remaining in the Ruhr of sandstone buildings. They were, however, left derelict without roofs, so they have suffered from water penetration and vegetation growing on the walls.

### 169. *Zeche Carl Funke (Carl Funke Colliery)(TR 12)*

In 1780 all 16 of the locks on the river Ruhr had been completed, so that coal could be transported along its entire length, and it became economic to mine coal in the area of Baldeneysee, south of Heisingen. There was a rationalisation of seven mines in 1871 and in 1898 a shaft was sunk for the Zeche Carl Funke, named after a prominent mining businessman, Carl Funke (1855-1912), who directed work to increase production. A housing estate for the miners was built on either side of the road to the colliery in 1900. Most of the surface facilities were demolished in 1973 when the colliery closed.

The 1880 winding-engine house, 1920 gatehouse and headframe still survive. The name of the colliery is spelled out in large letters on the existing foundations. In 2012 the headgear was purchased by Mike Schuh, an operator of climbing gyms. Permission was granted by the city of Essen to an application for a change of use for the conversion of the headgear to a climbing frame and to provide a scenic lookout over the area.

### 170. *Zeche Concordia, Schacht 6 (Concordia Colliery, shaft 6)(TR 4)*

Schacht 6 of the Zeche Concordia was sunk in 1912/13 as a ventilation shaft, and survived the 1968 closure of the mine so that it could be used for dewatering the adjacent pits (**143**). The original two-storey 1921 struts-framework headgear was replaced by a small headframe in 1994, demolished in the 1990s. Surviving buildings of historic value were all built between 1911 and 1913 and include the pay office, pithead baths, the building at the pit mouth and the winding-engine house. These were converted in 2006 for use as artists' studios, a musical theatre, offices and a restaurant.

### 171. *Zeche Consolidation 3/4/9 (Consolidation Colliery 3/4/9)(TR 8)*

Development of the Zeche Consolidation began in 1863. It was undertaken by Friedrich Grillo (**147**), together with eight other companies to share the risk and the high investment costs. The name Consolidation refers to the joint nature of the enterprise. Number 3 shaft of the mine was sunk in 1871, followed by number 4 shaft three years later and, after some more shafts, work on number 9 shaft began in 1915. Work was interrupted by WWI, so it was not until 1922 that number 9 shaft acquired its lattice-girder headgear (**204**), which formed a model for those at the Zeche Zollverein XII (**233**) and the Zeche Pluto-Wilhelm.

Zeche Consolidation 3/4/9 escaped damage during WWII. It reached peak production in the late 1950s when the main conveyor for number 4 shaft was extended. Production at the mine ended in 1997 and many of the surface structures were demolished. In 1987, however, the 1874 53-metre-high steel headframe on number 9 shaft had been given statutory protection, so could not be demolished. The associated winding-engine house was given statutory protection in 1992. Together with various other buildings, the 1874 headframe and the 1958 87-metre-high concrete headframe on number 4 shaft have been preserved in the care of the SIG (**228**), which restored them between 2002 and 2005.

The fan winding-engine house of numbers 3 and 4 shafts has been converted into the Consol youth theatre, using the shortened form of the mine's name. A music centre is located in the turbine building of number 4 shaft, with the addition of two extra storeys and a steel staircase. The northern winding-engine house now houses art installations by the artist Werner Thiel (1927-2003), created from his photographs and drawings of demolished mines in the Ruhr coalfield and mining items he had collected. The southern winding-engine house contains an atmospheric-pressure winding engine. This steam-winding engine, built by GHH, was the largest and one of the last to be installed in the Ruhr coalfield. In 1997 an association was founded committed to preservation of the monument. The association offers demonstrations of the operation of the double winding engine.

### 172. *Zeche Erin 7 (Erin Colliery 7)(TR 16)*

The 68 metre-high headframe on number 7 shaft of the Zeche Erin (**147**) was erected in the 1950s. Number 7 shaft was used for raising coal to the surface. The headgear is preserved in the middle of the site as a scenic lookout and a symbol of the industrial park that opened on the site in 1985.

### 173. *Zeche Friedlicher Nachbar (Peaceful Neighbour Colliery)(TR 16)*

The coal seams on the slopes of the Ruhr Valley south of Bochum outcrop at the surface, and then continue underground only a few metres below the surface. So mining here began with adit mines at an early date. One of these drift mines was the Zeche Friedlicher Nachbar, whose first deep shaft was sunk in 1868. From 1850 it had, however, been using a shaft of the Zeche Hasenwinkel-Himmelscroner Erbstolin, 50 metres away. It was from this close relationship that the mine was given the name Friedlicher Nachbar. In 1899 the mine took over the adjacent Zeche Baaker Mulde, which had also developed from a former adit mine. A second shaft was sunk, 500 metres south-east of the first, as well as a ventilation shaft, at the beginning of the twentieth century. In 1950 a new headframe, designed by Fritz Schupp (**108**), was erected to serve number 2 shaft. The Zeche Friedlicher Nachbar closed in 1961, one of the first Bochum mines to cease production. The 1950 headgear was moved to the Zeche Zollverein 1/2 (**221**). The winding-engine house is now used for theatrical performances. Other buildings are part of the Deimketal industrial park.

### 174. *Zeche Friedrich Heinrich (Friedrich Heinrich Colliery)(TR 17)*

Together with the Zeche Niederberg (**192**), the Zeche Friedrich Heinrich formed the western boundary of the Ruhr coal industry. The mine began production in 1861, named after Baron Friedrich Heinrich von Diergardt (1795-1869), a leading Rhenish industrialist. In 1907 the Friedrich Heinrich Company used French finance to start sinking two parallel shafts and constructing the brick baroque-style surface structures and a coking plant. Homes for the workforce were built nearby (**118**). Production began before WWI, after which the mine was subject to continuing litigation about its ownership, until in 1926 it came back into the ownership of the Rheinisch-Westgälischen Kohlen Syndikat. The coking plant was destroyed during WWII and the rest of the site was badly damaged.

In 1958 the colliery became the first in the Ruhr to have full mechanisation underground. Peak production was achieved in 1993 with 4.17 million tonnes of coal. A longwall conveyor system, 430 metres long, was installed in early 1998 to work the Girondelle seam, at a depth of 1,000 metres. In March 1998 this produced 19,938 tonnes daily, believed to be a world record at the time. When the mine closed in 2012 its annual production was 2.5 million tonnes, with 3,366 employees.

The surface structures, and especially the headframe, are considered worthy of preservation, but have not so far been given statutory protection. The town of Kamp-Linfort renovated and converted the mine site for the 2020 Landesgartenschau (garden festival). The Kamp-Linfort campus of the Rhein-Waal University of Applied Sciences is located on part of the site.

### 175. *Zeche Friedrich Thyssen (Friedrich Thyssen Colliery)(TR 3)*

The first shaft of the Zeche Friedrich Thyssen was sunk in 1872, with a rail connection to the station at Neumühl. It was close to the river Rhine, on which the coal was transported to the port of Alsum from 1882. The mine was acquired in the 1880s by the Thyssen brothers, August (1842-1926) and Joseph (1844-1915). They named it after their father, Johann Friedrich Thyssen (1804-77), the founder of the family fortune. Under their leadership the mine became one of the most important collieries of the Ruhr. In 1907 a new headframe was built above number 6 shaft with a steel lattice-girder construction (**204**).

By 1925 the mine had eight shafts in total. With the Great Depression (1929-39) looming, most of the mine was decommissioned in 1928 and number 6 shaft was used for other mines. In 1932 a new solid-wall headgear was constructed above number 2 shaft. During WWII there was damage to the mine, and especially to numbers 4/8 shafts. A new headgear was erected subsequently above number 4 shaft, identical to that over number 8 shaft. After the war the colliery properties of the Gelsenkirchen Mining Company were disaggregated and the Zeche Friedrich Thyssen began to operate as a stand-alone company. The company decided on early rationalisation during the late-1950s coal crisis. Number 1 shaft was filled in 1956. In 1959 the 4/8 system was abandoned, and the shafts and the surface structures were demolished. In 1976 numbers 2/5 shafts were closed and the following year the coking plant shut.

Virtually nothing remains of the structures on the site, which have been replaced by offices and shops. All that survives is the headgear of number 6 shaft, which is protected as an industrial monument representing the first major construction phase of the mine in the early-twentieth century. It was one of the first in the Rhineland to be given conservation treatment and is believed to be the last such headframe in Germany still on its original site. It was also the first industrial plant in Duisburg to be declared a national monument. From 2019 it has been illuminated at night.

### 176. *Zeche Fürst Leopold (Prince Leopold Colliery)(TR 7)*

The Zeche Fürst Leopold was worked between 1913 and 2001. It was named after Nikolaus Leopold (1906-88), prince of Salm-Salm, who held the mineral rights in the area. Peak production was reached in 1997, with 2.4 million tonnes. In 1998 the mine was amalgamated with the Zeche Westerholt (**217**) to form the Verbundbergwerk Lippe (Lippe Combined Mine).

New buildings were erected as required over the time that the colliery operated. The modern headframe was demolished in 2008. The historic headgear of number 2 shaft has been preserved, along with the 1912 and 1915 winding engines, the latter converted in 1972 to high-pressure operation of 4,420 horsepower. In 2015 the SIG (**228**) took ownership of the winding-engine house. Both the modern and the historical buildings have been converted into a cultural quarter for the arts and culture. An information and meeting centre opened in 2015.

### 177. *Zeche Gneisenau (Gneisenau Colliery)(TR 6)*

The Zeche Gneisenau in Derne was linked with the power plant and the Gneisenau coking plant. It was named after August von Gneisenau (1760-1831), a Prussian field-marshal during the Napoleonic wars (1803-15). The headframes were built with steel latticework girders of the *Bock* form. This was a modified version of the English latticework headgear, in which the support for the cage guide ropes (or rails) forms part of the headgear, so cover the shaft opening. The *Tomsom* headgear design allows for a more accessible shaft top, as it has only two vertical legs set away from the shaft, not the more common four surrounding the shaft. It was designed by the Belgian Eugene Tomsom (1842-1905), the plant director at Gneisenau from 1882 to 1900. Gneisenau was the only example in the Ruhr and is believed to be the oldest headgear in this coalfield. The colliery was the largest operated by the Harpen Mining Company, which owned many collieries, the first of which was the Zeche Robert Müser (**204**).

The second headgear was built in 1934 with a height of 58 metres. Because there was only limited space on the colliery site, the two winding-engine houses had to be moved very close to the headframes. This meant that the headgear had to be extremely steep, indeed almost vertical. Coal from Scharnhorst Colliery also came up the same shaft, so the two collieries merged in 1930.

The mine closed gradually between 1985 and 1990. It had been hoped to convert the site into a visitor museum, but this was unsuccessful. The two headframes were, however, listed to protect them, together with the two winding-engine houses and the 4,200-horsepower steam engines that each contain. They are now in the care of the SIG (**228**).

### 178. *Zeche Graf Bismarck 1/4 (Count Bismarck Colliery 1/4)(TR 8)*

The Zeche Graf Bismarck was named after the Prussian statesman, Otto von Bismarck (1815-98), who became chancellor of Germany after unification in 1871. The mine was sunk in 1868, eventually having nine shafts by just after WWI. A coking plant was built in 1913 and a power plant in 1938. The colliery reached peak production during WWII, when 3 million tonnes of coal were mined. By the mid-1960s the colliery was one of the largest employers in Gelsenkirchen with almost 7,000 staff. In 1966 the owner, the German Petroleum Company, closed the mine, filled the shafts and demolished almost all the surface structures. Only the 1905 administration building and pithead baths survived, constructed with red-brick and white-plaster walls and arched windows, similar to those of the Zeche Rechlinghausen I (**199**). The building is now used for concerts, exhibitions and theatrical performances, and houses an organisation offering services to people with disabilities.

### 179. *Zeche Hansa (Hansa Colliery)(TR 6)*

In 1856 the Dortmunder Mining Company acquired several coal deposits in the Huckarder area and combined them under the name Zeche Hansa. The colliery was probably called after the Hanseatic League, which was a confederation of merchants on the Baltic Sea between 1400 and 1800. The company immediately began to sink a shaft and construct a steel mill. Inflows of water into the shaft slowed progress and led to the bankruptcy of the company. In 1866 it was taken over by William Thomas Mulvany (**147**). He used cast-iron segments to seal the shaft, resulting in the first coal being produced in 1869. Five shafts were drilled in total and the first coking plant came into operation in 1895. It was replaced in 1928 by the coke plant (**125**).

The machinery underground at the mine was frequently updated. German coal became too expensive in the latter part of the twentieth century. From 1977 the Ruhrkohle AG (**164**) experimented with removing the coal using hydro-mechanical extraction (high-pressure water jets), but this was not successful. In 1979 a firedamp and coal-dust explosion resulted in seven deaths, and the mine closed in 1980. The original 1905 brick tripartite workshop building designed in the Low German gothic style is now used by Huckarder clubs for events and conferences. The 1930 headframe from number 3 shaft and the winding-engine house, which had been rebuilt after WWII, are now in the care of the SIG (**228**). New fans in number 2 shaft continued to function to protect the remaining mines to the north against water until 2014 (**143**).

### 180. *Zeche Helene Bertha (Helene Bertha Colliery)(TR 17)*

The Zeche Helene Bertha was sunk in 1870 in the eastern part of the Zeche Vereinigte Helene-Amalie (**212**). A coking plant was added next to the Bertha ventilation shaft in 1900. It was renamed the Zeche Salzer-Amalie/Neuack after mergers with other mines. The shafts were named after women from the Krupp family (**127**): Amalie, Barbara and Marie in Altendorf and Bertha and Helene in Altenessen. The two shafts were filled and the surface structures demolished after the final decommissioning in 1965.

The brick administrative and pithead-baths building was designed in 1927 by the Rhenish architect Edmund Körner (1874-1940) in the classical modernist style. It had two wings for pithead baths and for offices and sales, giving it an asymmetrical layout. The most impressive part of the structure is the wages hall, with the inscription 'Helene 1927' outlined in yellow brick. In 1983 the building was given statutory protection and in 1989 it became part of the IBA Emscher Park (**123**). The Helene Park is open to the public .

### 181. *Zeche Holland 1/2 (Holland Colliery 1/2)(TR 16)*

As suggested by the name, the Zeche Holland was developed with Dutch finance. As the form of consortium generally used in German mining was not known outside Germany, the Dutch founded their consortium as a public limited company (a corporation limited by share ownership, whose shares may be traded on the stock market). This was a rarity for a mining company in Germany at this early period, although it subsequently became the norm, described as an *Aktiengesellschaft*, abbreviated to AG. Sinking began in 1856 and the first coal was produced in 1860. From 1926 numbers 1 and 2 shafts were used only for manriding, not for raising the coal, which was done at number 6 shaft (**182**). The Zeche Holland developed into a large colliery, with 15 shafts in operation between the two World Wars (1918-39). Peak production was achieved in 1969, with 1.7 million tonnes and 3,000 employees, even though a number of its shafts had already gone out of operation.

The double Malakoff towers (**127**) of numbers 1 and 2 shafts are believed to be the only ones of their kind to be preserved in Europe. They have now been converted into residential apartments. The 1885 winding-engine house, workshop and 1925 fan building have been restored and converted into office and residential use.

### 182. *Zeche Holland 3/4/6 (Holland Colliery 3/4/6)(TR 16)*

The second major period of investment in the Zeche Holland dated to the period immediately after the Franco-Prussian War of 1870 (see **181** for the first phase). Number 3 shaft was sunk in Wattenscheid in 1872. In 1882 a coking plant was built, the first in Germany with the technology for separating ammonia and tar. Number 4 shaft was sunk in 1900, followed by a ventilation shaft, number 5, and finally number 6 shaft in the 1920s as the main production shaft. In 1921 the leading industrial architects, Fritz Schupp (**108**) and Martin Kremmer (1894-1945), designed a neoclassical administrative and pithead-baths building, with three wings grouped around a courtyard. In 1962 the 1927 headframe from the Zeche Zollverein (**233**) was moved to the Zeche Holland. From the 1950s until its closure in 1983, the colliery was the central production site for a network of mines in the Essen and Gelsenkirchen area, known as the Zeche Holland/Rheinelbe-Alma. The headframe of number 4 shaft and the winding-engine house of number 6 shaft survive on the site today. The 1921 administration building is now a technology centre.

### 183. *Zeche Hugo (Hugo Colliery)(TR 16)*

The Zeche Hugo was developed in 1873 in the period of economic recovery after the Franco-Prussian War (1870-1). The mine was named after its founder, Hugo Honigmann (1833-79). Nothing now remains from that period. By the 1950s Gelsenkirchen was the largest mining city of the Ruhr after Bochum (**108**). In the late 1950s a central extraction shaft was sunk and production was concentrated on Zeche Hugo 2/5/8. In 1952/5 new pithead baths, wages building and winding tower above the new number 8 shaft were constructed, designed by Fritz Schupp (**108**). The mine was amalgamated subsequently with the Zeche Ewald (**231**) and eventually closed in 2000.

The 1950s administrative building, pithead baths and headframe of number 2 shaft and the associated 1974 winding-engine house still survive as a group on the site. The rest of the colliery site is now planted as a biomass park to produce fast-growing timber for use as biomass in power stations.

### 184. *Zeche Julius Philipp (Julius Philipp Colliery)(TR 16)*

The Zeche Julius Philipp was founded in 1863 by the amalgamation of two adit mines, the Glücksburger Erbstollen and the Julius Philipp Erbstollen (see **298** for the *Erbstollen* element of the names). The mine was called after Julius Philipp Heintzmann (1745-94), the *Bergrat* (mine adviser), who owned shares in 65 mines in the Brandenberg coalfield. The mine began production in 1878 after three years of development, and a second shaft and a coking plant were added subsequently. Water problems underground made profitable mining difficult. In 1904 the mine was purchased by a new owner, who really only wanted to acquire its mining licence. This was so the licence could be transferred to other mines owned by the company and so the mine closed the following year. In the 1920s the surface buildings were all demolished, apart from the Malakoff tower (**127**), whose shaft continued to be used until 1962 to provide ventilation for the neighbouring Zeche Prinz-Regent. Since 1990 the Malakoff tower has housed a collection of 10,000 objects belonging to the Department of Medical Ethics and History of Medicine of the Ruhr University. There is a permanent exhibition open to the public on the history and ethics of medicine.

### 185. *Zeche Königin Elisabeth, Schacht Emil (Queen Elisabeth Colliery, Emil shaft)(TR 17)*

The Zeche Königin Elisabeth was called after Queen Elisabeth Ludovika of Bavaria (1801-73), wife of the Prussian king, Friedrich Wilhelm IV (1795-1861). The first shaft was named after her husband, Gründungsschacht Wilhelm (Wilhelm foundation shaft). Starting operation in 1850, this was one of the oldest pits in the area mining fireclay. No buildings survive from this period.

Schacht Emil was sunk in 1910, separated from the Gründungsschacht Wilhelm by a railway embankment. To bring all the functions of the mine under one roof, a new administrative building, with a wages office and a space for the electric-winding engine, was designed by Alfred Fischer (**144**). The single-storey building, with high hipped roofs, had the lamproom, offices and wages office in a slightly recessed central section, with the winding-engine house and the pithead baths occupying the wings. From the turn of the century it had been usual to provide a covered walkway between the shaft and the pithead baths to protect the miners coming up from underground, often in wet clothes. Fischer designed such a walkway, but had to abandon this provision for the comfort of the miners due to the layout of the surface of the mine. Fischer's design is considered to be an excellent example of the move from historicism to classical modernism.

The building above Schacht Emil now houses apartments. An art school and art gallery, called the Zeche Elizabeth, have been established in the winding-engine house, which was given statutory protection in 1988.

### 186. *Zeche Königsborn 3/4 (Königsborn Colliery 3/4) (TR 16)*

Sinking of the Zeche Königsborn began in 1874 and the mine started production in 1880. It was named from the town where it was located. Number 3 shaft was sunk in 1901 and number 4 shaft in 1902 (**144**). Many of the surface structures were badly damaged by bombing during WWII, but the winding tower survived. Zeche Königsborn 3/4 ceased operation in 1981. The tower formed the Ostpol (East Pole) of the Yellow Marker art installation (**160**).

### 187. *Zeche Lothringen (Lorraine Colliery)(TR 16)*

The name of the Zeche Lothringen is a reference to the annexation by Germany in 1871 of the eastern-French/western-German province of Alsace-Lorraine following the defeat of France in the Franco-Prussian War (1870-1). The name of the province in German was Eisass-Lothringen, called after Lothar II (835-69), who was king of the area of *Lotharingia* (Heidecker 2010, p. 1). Sinking of the first shaft of the colliery began in 1872. This was followed by four more before 1914 and a number of estates to house the workforce. Also constructed were coking plants and buildings for the utilisation of by-products of the coking process, such as ammonia and tar. The 1900 winding-engine house for the first shaft has ornate arches, yellow and red tiles and art-nouveau ornamentation. On 8 August 1912 144 miners were killed underground in a major methane explosion, attributed to poor ventilation and lapses in safety standards. Kaiser Wilhelm II (1859-1941) was in the Ruhr at the time, attending the celebration of the one-hundredth anniversary of the Krupp Company

**Figure 187.1. Oil painting of the visit of Kaiser Wilhelm II to the Zeche Lothringen following the methane explosion, by Theodor Heinrich Rocholl, 1914 (© Deutsches Bergbau Museum).**

(**127**). He went to the site to express his condolences and the scene was depicted in a 1914 oil painting by Theodor Heinrich Ludwig Wilhelm Rocholl (1854-1933).

A forced-labour camp was established under the Nazi regime at Zeche Lothringen 3, with several barracks built to house the slave labourers who worked in the mine between 1933 and 1945. After WWII the barracks were used to accommodate properly paid miners. Following the closure of the mine in 1967, the buildings were used to house *Gastarbeiter* (guest workers), the term applied in the 1950s and 1960s to workers from Greece, Italy, Morocco, Portugal, Spain, Tunisia, Turkey and Yugoslavia. From 1961 it referred mainly to workers from Turkey, who did not acquire citizenship rights but were expected to return home at the end of their contracts. In 1983 a student-housing project rebuilt the barracks, but maintained their basic structure. As a result they were registered in 2005 in the list of monuments as one of the very few examples remaining of this type of architecture used in forced-labour camps. Also preserved at Zeche Lothringen number 3 shaft are the administration building, control room, pithead baths and the building where the tallies were kept while the miners were underground. The capping on the filled shaft can still be seen. The administration building, ammonia-factory building and winding-engine houses at Zeche Lothringen numbers 1/2/6 shafts are all preserved.

### 188. *Zeche Minister Achenbach, Schacht 4 - LÜNTEC-Tower (Minister Achenbach Colliery, shaft 4 – LÜNTEC-Tower)(TR 7)*

The first shaft was sunk in 1897 of the Zeche Minister Achenbach, named after the Prussian trade minister, Heinrich von Achenbach (1829-99), and began production in 1900. Number 4 shaft was sunk after WWI and the first coal was raised in 1924. Production ceased in the 1960s, after which it was used as a training centre for apprentices until 1990. The old gatehouse is now a café, while the former administration building and pithead baths have housed the Lüntec technology centre since 1995. The symbol of the technology centre, a large white plastic 'egg' designed by Luigi Colani (1928-2019), has been placed on top of the headgear. Two apartments of the Alten Kolonie are displayed as the Bergarbeiter-Wohnmuseum (**136**).

### 189. *Zeche Minister Stein (Minister Stein Colliery) (TR 6)*

The first three shafts for the Zeche Minister Stein were sunk in 1871, 1891 and 1901. The mine was named after Baron Heinrich Friedrich Karl vom und zum Stein (1757-1831), a Prussian officer and statesman, who was in charge of the areas in France and Germany reconquered from France after the Napoleonic wars (1803-15). A *Hammerkopfturm* (**147**) was erected over number 4 shaft in 1926 and the modern steel headgear became the emblem of the district. The colliery shut in 1987.

A planned conversion of the site into an underground landfill was prevented by extensive community protests. The administrative offices, *Hammerkopfturm* headgear and pithead baths have all been preserved as offices. A modern spandrel office building, in the shape of the former pit spoil heap, has been constructed beneath the headgear, and the former operational areas are now a service-centre commercial area.

### 190. *Zeche Monopol, Schacht Grimberg 1/2 (Monopole Colliery, Grimberg shaft 1/2)(TR 7)*

In 1873 the mining entrepreneurs, Heinrich Grimberg (1833-1907) and Friedrich Grillo (**147**), registered to purchase coal reserves in the Kamen and Bergkamen coalfields. It is said that the Zeche Monopol in Bergkamen was named from Heidsieck Monopole, the German supplier of the sparkling wine imported from Champagne in France, with which the owners celebrated the evening before registering for the licence. The mine was especially profitable because the seams were almost horizontal and so easily worked. Shafts were first sunk in 1890/4, but it was only in the late 1970s and early 1980s that the underground arrangements and machinery were thoroughly modernised. A new headgear was erected above the 73-metre-high Schacht Grimberg 2, which is one of the most prominent headframes in the Ruhr. A new coal-preparation plant was designed with environmental considerations to the fore. The architecture and the tawny-colour scheme were specifically matched to the adjacent centre of Bergkamen. The colliery closed in 1995, with its modern surface installations barely completed and has now been declared a national monument. It is occupied by recycling companies. It is understood that the headframe, shaft building and winding-engine house are to be made publicly available.

### 191. *Zeche Monopol, Schacht Grillo (Monopoly Colliery, Grillo shaft)(TR 10)*

The connection of Kamen to the Cologne-Minden railway (**149**) in 1846 resulted in an economic boom. Work began in 1873 on the first shaft at the Zeche Grillo, named after the industrialist Friedrich Grillo (**147**), although the first coal was not produced until 1879. The mine was subsequently renamed as the Zeche Monopol (**190**), but the Grillo name was retained for the shaft. A second shaft was begun in 1887. It reached a depth of 775 metres in 1892, making it the deepest shaft at that time in the Ruhr coalfield. The colliery was modernised and expanded repeatedly over time. A modern coal-preparation plant was constructed in the 1950s. The mine closed in 1981.

Two administrative buildings have been preserved, one from 1910, the other from the 1950s, along with the headframe of Schacht Grillo 1, as an important feature of the city skyline. A concept for the reuse of the 34-hectare former colliery site and its buildings was developed as part of the IBA Emscher Park (**123**). A business and technology park has been developed on the site. Work by the not-for-profit Verein zur Förderung der Beergbau- und Industriekultur in der Stadt Kamen eV (Association for the Promotion of the Mining and Industrial Culture in the City of Kamen) on the preservation and development of new uses for the remaining colliery buildings is continuing, under the name Monopol 2000.

### 192. *Zeche Niederberg (Lower Mountain Colliery) (TR 17)*

Just before WWI mining began to take over from agriculture and textiles as the main industrial activity in the region of Neukirchen. Together with the Zeche Friedrich Heinrich (**174**), the Zeche Niederberg formed the western boundary of the Ruhr coal industry. Plans for a mine here date back to the nineteenth century, when the mining licence was granted, but the local partners lacked the financial resources to develop the coal reserves. The Krupp Company took over the licence in 1907 and began construction in 1911, with funds from the Berlin bank, Bleichröder. In 1908 the railway crossed the area of the concession, enabling the coal to be moved by rail. The mine was named from the first company to develop it, the Nieder Rheinische Berg.

In 1923 a lattice-girder headgear (**204**), 36 metres high, was erected over number 2 shaft. The mine was mechanised in the 1940s. The mine expanded in the 1950s, including construction of a coal-preparation plant. In the period 1964/9 a new central shaft, number 5, was sunk, eventually reaching a depth of 1,162 metres. The pit started mining anthracite in 1973, the only mine in the area producing coal for domestic use. Most of the coal was processed on-site into low-smoke briquettes. By 1988 the mine had a total area underground of 120 square kilometres.

In 1917 the mine produced 7,443 tonnes of coal with 329 employees. By 1930 the figures were 582,000 tonnes and 1,795 employees, which fell to 307,000 tonnes with 1,515 employees in 1945 at the end of WWII. Production increased to more than a million tonnes for the first time in

1995, when 4,323 employees produced 1.4 million tonnes. Peak production was achieved in 1980 with 2.04 million tonnes of coal and 4,940 employees. With the decrease in demand for coal, production eventually fell to 2 million tonnes and 1,964 employees in 2000.

The mine closed in 2002. Most of the above-ground structures were demolished in 2004, but the buildings from the foundation period of the mine were given statutory status and renovated. These can be viewed from the road alongside the mine curtilage. Still surviving are the gate houses, central winding-engine house, which still contains two winding engines, former fan house, which provided ventilation to the mine, and the two winding towers, one of which is a listed structure.

### 193. *Zeche Oberhausen (Oberhausen Colliery)(TR 4)*

As well as the Zeche Rheinpreussen (**201**) and the Zeche Zollverein (**233**), the Zeche Oberhausen was sunk by Franz Haniel (1779-1868) in the mid-nineteenth century to provide coal for his ironworks, the first such in the Ruhr. No buildings of that date survive. It was first known as Zeche Königsberg (King's Mountain Mine), but was renamed from Oberhausen, the city where it was located. The Great Depression (1929-39) led to the closure of the colliery in 1931. A hostel now occupies the 1911/12 gate house, and the 1922 workshop is used as a meeting hall and recording studio.

### 194. Zeche Osterfeld (Osterfeld Colliery)(TR 4)

The Zeche Osterfeld was sunk in 1853 and operated until 1992, after which the shafts were filled. It was named from Osterfeld, the town where it was located. The gate house and administration building of 1911/12, the 1950 headframe building and the coal-preparation plant are all listed buildings. In 1999 large parts of the colliery and coke works were used for a garden festival, which created a new public garden.

### 195. *Zeche Osterfeld, Schacht IV (Osterfeld Colliery, shaft IV)(TR 4)*

The 1924 headframe on number 4 shaft of the Zeche Osterfeld in Oberhausen-Osterfeld is an example of modern industrial architecture in the Bauhaus style. The headgear stands 43 metres high. It has been preserved, together with the brick administration building. pithead baths and transformer house, also from 1924. Before WWI the miners' trade union had reached agreement with the management that the miners would be paid from the time that they reached the entrance and not, as usual then, from when they arrived at the coalface underground, which could be between one and three hours (Orwell 1937, pp. 25-6); this was not negotiated in the UK until the early 1970s as a settlement of strikes. It was, therefore, in the interests of the management that the distance to reach the coalface should be as short as possible. So a new shaft, number 4, was sunk to reduce travelling times. The spoil from the shaft was used to provide a flat mound on which the headgear was then constructed.

### 196. *Zeche Pattberg (Pattberg Colliery)(TR 17)*

The Zeche Pattberg was named after Heinrich Pattberg (1862-1934), a mine director who pioneered new methods of working and who served on the executive board of the Rheinpreussen Company until 1932. The mine was sunk in 1927 and the above-ground buildings were constructed in 1927/34. In 1969 the mine was amalgamated with the Schacht Rossenray (**160**) and in 1971 with the Zeche Rheinpreussen (**201-3**) to form the Verbundbergwerk Rheinland (Combined Rhineland Mine). The colliery was decommissioned in 1993 and now the only surviving structures are a water tower, with legs supporting a circular container, and the winding-engine house dated to 1912 from the Zeche Rheinpreussen. The winding-engine house is now home to the SIG (**228**).

### 197. *Zeche Prosper II (Prosper Colliery II)(TR 16)*

The Zeche Prosper was sunk in 1856, named after Prosper Ludwig (1785-1861), seventh duke of Arenberg, which in 1815 was divided between Hanover and Prussia. A second shaft was sunk in 1871, 2.5 kilometres north-east of the first shaft. To serve this shaft an especially large and impressive Malakoff tower (**127**) was erected in 1872, with polygonal corner towers and battlements. Zeche Prosper I closed in 1959 and its headgear was demolished in 1976, although the fire station for this part of the colliery still survives in commercial use. In 1982 almost all the old surface buildings of Zeche Prosper II were replaced with new structures. The Malakoff tower and headgear are protected as an industrial monument in the care of the SIG (**228**).

### 198. *Zeche Radbod (Radbod Colliery)(TR 7)*

The Zeche Radbod was named after Radbod (CE 679-719), duke of the Frisians. Frisia is a region on the North Sea coast, most of which is now part of the Netherlands, but a small part belongs to Germany. The mine opened in 1906, but was stopped working by a methane explosion on 12 November 1908 in which 349 miners died (**146**). The accident led to mass demonstrations in many towns in the Ruhr, with 7,000 people alone on the streets in Bochum. As a result of this tragedy the German state enacted that gasoline safety lamps had to be replaced by electrical safety lamps, and the Zeche Radbod was the first mine in which the new lamps were employed. Production was permitted to resume in 1909 with the new lamps, together with restrictions on the use of explosives. Despite these measures, there continued to be fires and explosions underground. In the following years additional facilities were added on the surface. Peak production was reached in 1989 with 1.3 million tonnes of coal and 2,000 miners, but the colliery closed the following year. The surface facilities were taken over by the SIG (**228**) with guided tours available. The headframes and winding-engine houses are now protected as industrial monuments. The main entrance building is now the Radbod cultural centre.

**199.** ***Zeche Rechlinghausen I (Rechlinghausen Colliery I)(TR 8)***

Sinking of the Zeche Rechlinghausen I began in 1869, but was interrupted by the Franco-Prussian War (1870-1), so production only began in 1875. The mine is named from its location, the northernmost city in the Ruhr. In 1900 the Dortmund mining authority forbade the use of the same shaft for turning the ventilation both on and off, which meant that many mines were forced to sink a second ventilation shaft. A new lamproom, magazine and pithead baths were completed in 1901, but the new shaft did not begin production until 1905. As a result of the Great Depression (1929-39), production was transferred in 1931 to Zeche Rechlinghausen II and 826 miners lost their jobs. At the Zeche Rechlinghausen I site only the administration, pithead baths and surface buildings have been preserved, with red-brick and white-plaster facades similar to those at the Zeche Graf Bismarck 3/4 (**178**).

**200.** ***Zeche Rechlinghausen II (Rechlinghausen Colliery II)(TR 8)***

The Zeche Recklinghausen II produced coal between 1901 and 1974. Then until 1988 the shaft was used only for manriding and moving materials. After the demolition of many of the surface structures, an association was formed to campaign to save the remaining buildings. In 1999 the site became part of the IBA Emscher Landscape Park (**111**), with a generous space around the surviving headframe and 1964 winding-engine house. In the eastern part of the winding-engine house is a district leisure centre, including a dance/sports hall and a shooting gallery. One of the winding engines is preserved in the western part of the winding-engine house, along with a small museum run by the association.

The surviving structures, and the now-lost heapstead building, are testament to the architectural trends of the 1960s. The shaft is equipped with two sets of cages worked independently by a pair of 3,800-horsepower *Koepe* winding engines (**127**). These were manufactured in the 1960s by the Prinz Rudolph foundry (**113**), and were some of the last steam-powered winding engines manufactured in Germany, if not in western Europe as a whole. The sheave wheels of the headframe are not parallel to each other, but in two pairs angled such that they open out towards the winding houses in the form of a V. This was done to take account of the distance between both winding engines, which precluded the normal arrangement.

**201.** ***Zeche Rheinpreussen Schacht 1/2 (Prussian Rhine Colliery shaft 1/2)(TR 3)***

The Zeche Rheinpreussen in Homberg was sunk in 1857 by Franz Haniel (**193**) as the first mine on the left (western) bank of the river Rhine, to which the first part of the name refers. The other reference is to the kingdom of Prussia, which subsequently took the lead in the unification of Germany in 1871. It was one of the largest mines in the Ruhr and also the first in the Ruhr designed solely to provide coal, rather than to supply the iron and steel industry with coke. The Zeche Rheinpreussen was expanded in the early-twentieth century. New drilling methods and techniques were developed at Rheinpreussen, which made mining history. The Schacht Gerdt (**157**) was sunk in 1945. In 2006 there was extensive demolition work on the site, but the administration building, workshop and five-storey square winding tower were preserved.

**202.** ***Zeche Rheinpreussen Schacht 4 (Prussian Rhine Colliery shaft 4)(TR 17)***

The sinking of Schacht 4 of the Zeche Rheinpreussen began in 1900, using a new technique developed by Heinrich Pattberg (**196**), and production began in 1904. Because of the size of the Rheinpreussen concession, this was on a site independent from Zeche Rheinpreussen Schacht 1/2 (**201**). Schacht 4 had its own coal-preparation and coking plants and its own style of architecture. The surface buildings, in brick neogothic style, were finished in 1907. The headgear was built to a height of 48 metres and the electric-winding engine was installed in 1906. Schacht 4 closed as an operational shaft in 1964.

The buildings on the site have been given statutory protection as an especially fine example of the layout of a colliery at the beginning of the twentieth century. Beyond the gatehouse at the entrance are the administration building and pithead baths. Behind the workshop and fan building are the headgear and the eastern winding-engine house. The western winding-engine house has not survived. The pithead baths were close to the shaft, so that miners coming up from underground in wet clothes did not have far to go to reach shelter.

The listed buildings have been restored and are used commercially. The eastern winding-engine house still contains the 1906 winding engine, one of the oldest surviving electric-winding engines in the Ruhr. The headgear forms a prominent landmark for the area.

**203.** ***Zeche Rheinpreussen Schacht 5/9 (Prussian Rhine Colliery shaft 5/9)(TR 17)***

Schacht 5/9 of the Zeche Rheinpreussen was developed at the same time as Zeche Rheinpreussen Schacht 4 (**202**). Because of the size of the Rheinpreussen concession, this was on a site independent from Zeche Rheinpreussen Schacht 1/2 (**201**). It had its own coal-preparation and coking plants and its own style of architecture. The headgear, which no longer exists, was similar to that at Zeche Rheinpreussen Schacht 4. Later a large concrete headframe, which has also been demolished, was erected above Schacht 9, which was the main production shaft.

The brick building housing the administration and pithead baths has a central hall laterally flanked by two towers crowned with domes. It has large segmented arched windows set in stone. This structure is now the central building of the Technology Park Eurotec. The two winding-engine houses of the early-twentieth century and a workshop building have also been preserved.

**204.** ***Zeche Robert Müser (Robert Müser Colliery) (TR 16)***

The Zeche Heinrich Gustav was founded in 1856, with a depth of 760 metres. It was acquired by the Harpen Mining Company (**177**) and in 1927 the main shaft was deepened and widened to a diameter of 7.6 metres. In 1928 a 57.2-metre-high lattice-girder headframe was erected, which served as a model for headframes at other collieries, and the colliery acquired its own 20-megawatt power plant. In 1929 the Zeche Heinrich Gustav was amalgamated with the Zeche Amalia and the Zeche Vollmond. The enlarged colliery was named after the former director-general of the company, Robert Müser (1849-1927). The company had many collieries, of which the Zeche Robert Müser was the first and the Zeche Gneisenau (**177**) the largest.

The colliery produced 1.5 million tonnes of coking coal in 1938 and 1.37 million tonnes in 1966 with 4,000 employees. The mine closed in 1968. The headgear and shaft were retained to pump minewater from adjacent collieries to the surface here (**143**) and in 1990 the headgear was made a listed structure.

**205.** ***Zeche Sachsen (Saxony Colliery)(TR 7)***

The Zeche Sachsen, sunk in 1912 with two shafts, was one of the most eastern coal mines of the Ruhr. It is named from the state of Saxony in eastern Germany. The Essen architect Alfred Fischer (**144**) was responsible for the surface buildings. Unlike so many of the collieries of the time, which had magnificent neogothic surface buildings, Fischer initiated a trend towards industrial architecture that was strictly functional. The best example of form following function is the winding-engine house, which was considered the central point of the entire system. The building is a simple, almost cubic, structure with tall narrow rectangular windows (*cf.* **144**).

The mine ceased production in 1976, the shafts were filled and a business park was created on the site with a large proportion of green spaces. The winding-engine house was given statutory protection in 1989 as the only surviving historical structure on the site. As part of the IBA Emscher Park (**123**), the building, from which all machinery had been removed, was converted to an events and exhibition hall with 3,000 square metres of meeting space. The Alfred-Fischer Hall has also been renovated as a venue for concerts and musical and theatrical events.

**206.** ***Zeche Schlägel und Eisen 3/4/7 (Hammer and Pick Colliery 3/4/7)(TR 16)***

The Zeche Schlägel und Eisen was founded in 1873. The name of the colliery refers to the European mining symbol of the crossed pick and hammer (**33.3, 60.2, 206.1**). Number 3 shaft became operational in 1897. The headgear was of the *Promnitz* design, with three legs, rarely found in the Ruhr. This is the oldest surviving example of this type of headframe in North Rhine-Westphalia. The winding-engine house still contains the 1928 double steam-winding engine, built at the Prince Rudolph ironworks (**113**) in Dülmen. Today both headframe and winding-engine house are in the care of the SIG (**228**).

**207.** ***Zeche Sterkrade (Sterkrade Colliery)(TR 4)***

The Zeche Sterkrade was first planned as a ventilation shaft of the Zeche Osterfeld (**194**) and additional capacity for the output from the Zeche Hugo (**183**). The mine is named from Sterkrade, a suburb of the city of Oberhausen. The first shaft of the Zeche Hugo collapsed in 1898 and had to be abandoned. The work was completed in 1903 with construction of the boiler house, office building, pithead baths and shaft house. Sterkrade ceased to be used once the combined mine of Lohberg-Osterfeld had been established in 1991. The 1903 headframe has been preserved with two of its four wheels. The headframe, along with that of the Zeche Vereinigte Carolinenglück 2/3 (**210**), are the only two headframes surviving in Germany of the *Zschetzsche* design. This means that the headframe effectively has three raker legs, with the extra one in the centre, all three being connected by horizontal and diagonal bracing.

**Figure 206.1. Ceramic plaque with crossed hammer and pick and inscription saying 'to commemorate the first trip underground in 1955 Dortmund Mining AG' (© Deutsches Bergbau Museum).**

The *Zschetzsche* design has the traditional diagonal and vertical elements; diagonal raker legs to resist horizontal winding forces and a vertical main structure that deals with the vertical forces. It is the use of fairly lightweight horizontal, diagonal and vertical steel sections to produce either braced lattice or truss structures, which makes for a light yet strong headframe, which is the defining feature of the design. The three-storey brick shaft building is one of the last solid-construction shaft buildings in the Rhineland. The site is now in the care of the SIG (**228**).

**208. *Zeche Trappe (Bustard Colliery)(TR 9)***

The first mention of coal mining in this area occurs in 1662 and there were already three shafts operational at the Zeche Trappe in 1754, when 17 miners produced 3,840 tonnes of coal. The mine is named after the bustard (**163**), the world's heaviest bird that can fly. In 1755 the first horse winch in the Ruhr was installed and by 1799 there were three winch shafts, reaching a depth of 95 metres. In the eighteenth century it was one of the largest coal mines in the Brandenberg Mountains. In 1822 a steam-winding engine was installed at the mine, so that by 1826 the Zeche Trappe was the largest colliery in the Ruhr. The last year for which figures are available is 1847, when there were up to 82 employees and 16,016 tonnes of coal were produced. In 1848 a new winding-engine house was constructed and in 1853 the Zeche Trappe was amalgamated with the Zeche Vereinigte Wülflingsburg to form the Zeche Vereinigte Trappe. The winding-engine house of the Friederika shaft has been restored and is today a private residence.

**209. *Zeche Unser Fritz 1/4 (Our Fritz Colliery)(TR 8)***

The shaft building of the Zeche Unser Fritz 1/4 (**151**) is one of the 14 Malakoff towers (**127**) now considered important enough to be preserved for the history of mining in the Ruhr. It was erected in 1871 above number 1 shaft. Its buttresses and polygonal stair towers, which served as escape routes, resemble the Malakoff tower of the Zeche Ewald (**231**). In 1890 an iron headframe was erected above the tower at the same time as a powerful winding engine was installed. The number 4 double shaft was sunk in 1908/10, after which number 1 shaft was used for ventilation. The Rhine-Herne Canal was finished in 1914, with a loading wharf for Zeche Unser Fritz 1/4. The iron headframe was demolished in 1970 and a new roof constructed. The shaft was last used in 1993 and the shafts were filled in 1995. The Malakoff tower is the only surviving structure on the site, but is now privately owned. Artists' studios, commercial buildings and a secure facility for 90 mentally ill offenders have been constructed on the rest of the site.

**210. *Zeche Vereinigte Carolinenglück 2/3 (United Good Luck Caroline Colliery 2/3)(TR 8)***

The Zeche Glückauf was sunk in 1850. *Glückauf* (good luck) is the traditional greeting between German-speaking miners, originally wishing each other good luck in finding a profitable lode of ore. The technique of piercing the hard water-impermeable marl layer of rock had to be mastered to allow the development of what became the first deep mine in the Ruhr. It was renamed in 1870 as the Zeche Vereinigte Carolinenglück and a new main production shaft came into operation in 1891. In 1898 115 miners were killed in a firedamp and coal-dust explosion, at the time the most serious mining accident that had occurred in the Ruhr. Two years later the Bochum Association took over the colliery and modernised the coke works to supply coal for their cast-steel factory. The following year a cable car was built to take coke to the blast furnaces. The colliery suffered extensive damage during WWII and reconstruction took until 1952. The colliery closed in 1964. Since 1970 this has been the focus of the central water-pumping station of the RAG AG (**143, 164**). The surviving 1912 headframe is of the *Zschetzsche* design (**207**).

**211. *Zeche Vereinigte Gibraltar (United Gibraltar Colliery)(TR 29)***

The Zeche Vereinigte Gibraltar began operating in the 1780s, and the roadways were extended to 2,000 metres in length in 1830. The mine operated only intermittently in the nineteenth century, due to the low levels of coal found, and was decommissioned in 1883. It came back into production after WWI, with new buildings erected in 1922. It again proved unprofitable, however, and closed in 1925.

The Nazis came to power in nearby Bochum in the early 1930s. The colliery buildings were used in 1933 for a training school for leaders of the Sturmabteilung, usually abbreviated to SA and commonly known as storm troopers or brown shirts. This was a violent para-military organisation, which played a major role in the rise of Adolf Hitler (1889-1945) to power in Germany. After that the colliery complex was used as a concentration camp for communists, social democrats and trade unionists until early 1934, when the first purpose-built concentration camps became available in northern Germany. A plaque on the site commemorates the victims of torture here. One person who suffered at the Zeche Vereinigte Gibraltar here was a miner, August Bahrenberg (1878-1933), who had also been chair of the rural Bochum district council of Querenburg. His social-democratic principles led the Nazis to imprison him here, where he died shortly afterwards of heart disease at the age of 53.

After the Kemnader Stausee dam had been constructed on the river Ruhr in 1979, the 1922 buildings were renovated. They are now used as a leisure centre and one serves as a boathouse for Bochum University.

**212. *Zeche Vereinigte Helene-Amalie (Combined Helene-Amalie Colliery)(TR 17)***

The Zeche Amalie was already mining coal in 1843, although no structures survive from that period (**180**). From 1853 the Zeche Vereinigte Helene-Amalie was one of the first two mines in the Ruhr to use pit ponies for transport underground. It was one of the most important mines in the Dusseldorf region in the second half of the nineteenth century. The administration building was designed by the architect Edmund Körner (1874-1940) in 1915. In 1927 the mine was taken over by the Krupp Company, which began rationalisation and modernisation. The main building was designed in the mid-1930s by the Bonn architect, Christian Bauer, with a steel structure clad in brick. The windows were grouped into contrasting horizontal and vertical bands, which aroused the displeasure of the Nazis, who disapproved of this style of architecture. The entrance has four reliefs representing morning, daytime, evening and night-time produced by E. Meisen in 1935. During WWII the mine suffered considerable damage, which was repaired in 1947. The 1915 building was completely rebuilt in brick, but incorporating the entrance and the four reliefs from the original.

The coal-preparation plant was demolished after the mine closed in 1968. The headframe of the Schacht Marie was replaced with a small auxiliary winder in 1982 and the shaft building was destroyed in 1984. In 2008 most of the remaining plant, including the chimney standing 12 metres high, was demolished. The rebuilt 1915 building is now used for offices, and the 1927 administration block was restored and is part of the IBA Emscher Park (**123**). The site is open to the public as the Helenenpark.

### 213. *Zeche Walfisch (Walfisch Colliery)(TR 11)*

The Zeche Walfisch was formed in 1832 by the merger of three smaller mines, which had been operating since the eighteenth century. It was named from a small uninhabited island in the Baltic Sea. Peak production was reached in 1896 with 75,000 tonnes and 388 men. The mine closed in 1925. The winding-engine house was constructed in about 1850 in the late-classical style typical of this region and is now used as a private residence. It is a three-storey building with a gabled roof, quoins and arched windows, which were subsequently bricked up. The slots for the guide for the shaft rope can still be seen on the edges of the building.

### 214. *Zeche Walsum (Walsum Colliery)(TR 3)*

Between 1904 and 1927/30 numbers 1 and 2 shafts of the Zeche Walsum were sunk by the Thyssen Company. The mine was named from the former city of Walsum, now a municipality of the city of Duisburg. The headframe on number 1 shaft was built in 1938/9. The mine was expanded after WWII and by 1949 coal was being mined mechanically with an armoured face conveyor and a compressed-air shearer. The adjacent power plant came into operation in 1960. By 1998 the daily production was 12,500 tonnes, or more than 3 million tonnes per year, with 4,200 employees.

Coal production ended in Duisburg in 2008 and from 2014 the water from all the mines in the area was pumped into the Zeche Walsum. From there pumps installed in one of the two winding towers pumped the water into the river Rhine (*cf.* **143, 221**). The other mining facilities on the 25-hectare site were demolished between 2009 and 2012.

### 215. *Zeche Waltrop (Waltrop Colliery)(TR 7)*

To provide a reliable source of fuel for its railways and navy, the Prussian state bought coalfields in the Ruhr in 1902 and established its own coal industry. Construction by the Royal Prussian mining company of the Zeche Waltrop, named from the town of Waltrop, began in 1903, with brick buildings of a uniform appearance. Production started in 1905 and a coke plant was added to the site in 1909. The peak of production was reached in 1974, when the 2,021 employees extracted 1,112,000 tonnes of coal. In the late 1970s the mine merged with the Zeche Minister Achenbach (**188**) and closed soon afterwards in 1979.

The winding-engine house, which was the largest in the Ruhr after that of the Zeche Zollverein XII (**233**), was listed in 1988, along with eight of the other 10 surviving buildings. The surface structures have all been restored and are now in commercial use. There is a four-cylinder tandem-winding engine preserved in number 3/4 winding-engine house. The adjacent spoil heap covers 6.5 hectares and stands 15 metres high. In 2000 a tower by the artist Jan Bormann (born 1939) was erected on the spoil heap. The tower stands 20 metres high and consists of a truncated pyramid composed of lattice struts, mirroring the struts used in mine headframes (**204**). This has a platform, 3.7 metres above the ground, with views over the eastern Ruhr region.

### 216. *Zeche Werne (Werne Colliery)(TR 7)*

The Zeche Werne is the oldest mine north of the river Lippe. In 1898 flooding led to the closure of the Zeche Piesberg in Osnabrück (**132**). The owners decided to sink a new mine in Werne to supply coal for the new steel and rolling mills they were planning on the edge of the Ruhr coalfield. There were a number of difficulties in the early stages, with shafts having to be abandoned at the 750-metre depth in 1899 and 1903. This was followed by a methane explosion in 1905, which destroyed some of the surface installations. The mine only started production in 1907, a third shaft was sunk in 1914 and a fourth in 1954. The colliery closed in 1975. The headframes above numbers 1 and 2 shafts were demolished, while the remaining buildings were converted to commercial use (see also **150**).

### 217. *Zeche Westerholt (Westerholt Colliery)(TR 16)*

The German state needed large amounts of coal for the nationalised railways and for the imperial navy, which was being expanded considerably in the early-twentieth century in competition with that of the UK. It was, therefore, important to safeguard a secure coal supply. From 1903 the state constructed new collieries, known as fiscal mines, in the northern Ruhr and southern Munster.

The Zeche Westerholt was created as a double pit in 1907 and came into operation in 1910, with the addition of a coking plant in 1912. It was named from a district of the city of Herten. The buildings were of red brick with white recessed cornices and soffits and curved gables with white stepped cornices, arranged around a central courtyard. Westerholt was badly damaged during WWII with the headframe of number 3 shaft being the only headgear in the Ruhr to be destroyed by bombing. In the mid-1950s the mine was amalgamated with the Zeche Bergmannsglück (**166**), with the underground conveyor systems linked together. A concrete closed winding tower was erected above a new central shaft sunk in 1960. Number 1 shaft was deepened in 1987/91 and a new headgear was built in 1989. In 1998 the mine was amalgamated with the Zeche Fürst Leopold (**176**) to form the Verbundbergwerk Lippe (Lippe Combined Mine). Number 2 shaft was filled in 1999 and its headgear was demolished. The Zeche Westerholt closed in 2008.

A number of surface buildings still survive from the early-twentieth century, despite the wartime bombing and subsequent modernisations. In 2010 the Regional Association of Westphalia-Lippe declared that the surviving surface structures had to be preserved as a statutory monument. In 2015 an initiative was launched by the cities of Gelsenkirchen and Herten to renovate the colliery site. By June 2019 renovation was completed of the gatehouse and in December 2020 a development company, Neue Zeche Westerholt (New Westerholt Colliery) was established to implement the 2015 feasibility study in three construction phases to be completed in 2030/2. In 2021 the former training mine of the colliery was transferred to the Zeche Zollern (**129**) to form a simulated underground.

### 218. *Zeche Westfalen (Westphalian Colliery)(TR 16)*

Sinking of the Zeche Westfalen, the easternmost mine of the Ruhr, began in 1909. It was named after the region of north-western Germany now known as North Rhine-Westphalia, centred on Dortmund. Construction began of a coking plant at the same time. Numbers 1 and 2 shafts reached their final depth of 1,050 metres in 1911 and production started in 1912 and 1913 respectively, with a workforce of 1,360 men. Both shafts had headframes of the *Zschetsche* type (**207**).

Peak production was achieved in 1927 with 826,000 tonnes of coal a year and 3,660 miners, just before the Great Depression (1929-39) caused a major drop in production. In 1935 the colliery was linked by rail to a quay on the Datteln-Hamm Canal. The mine was damaged by bombing during WWII, after which number 5 shaft was sunk for ventilation and in 1966 number 6 shaft came into operation. The last shaft, number 7, reached a depth of 1,330 metres with a peak production in 1982 of 2.67 million tonnes. The mine finally closed in 2000.

Many of the buildings around numbers 1 and 2 shafts have been preserved. In 2006 the pithead baths and the wages building were converted to commercial use and for conferences and exhibitions. A path connecting the site with the river Lippe has been constructed along the course of the former railway.

### 219. *Zeche Westhausen (Westhausen Colliery)(TR 6)*

The buildings of the Zeche Westhausen are considered to be amongst the architecturally most significant in the Ruhr. The mine is named after the town of Westhausen in Baden-Württemberg. The Malakoff tower (**127**) over number 1 shaft dates to 1873, the early days of the coal mine. It was fitted with escape stairs, because there was always a high fire risk in the wooden headgear with coal dust, lubricants and open fires in use. Opposite the headgear are the 1906 administration and wages building in the late-art-nouveau style, designed by the architect, Paul Knobbe (1867-1956). The winding-engine house of the 1883/1906 number 1 shaft, the 1905 workshops and the 1908 fan house survive, together with the 1922/7 winding-engine house of number 3 shaft, in which there is a small mining museum. The museum was established by the Westerfilde-Bodelschwingh group of historians and displays the 1926 twin steam-winding engine and other items of mining equipment.

### 220. *Zeche Wohlverwahrt (Well-Guarded/Well-Preserved Colliery)(TR 12)*

The origins of the Zeche Wohlverwahrt go back to the eighteenth century, although the small mine was not very successful. It was one of the smallest private mines surviving in the Ruhr after WWII. By 1956 there were 256 miners, producing 59,000 tonnes of coal annually. The colliery had to be closed in 1962, because of a threat of flooding underground. The 1910 building, which housed the blacksmiths' workshop, coal-preparation plant, cokery, pithead baths and shaft house, has been a listed building since 1989.

### 221. *Zeche Zollverein 1/2/8 (Zollverein Colliery 1/2/8) (TR 2)*

The first coal from the number 1 shaft of the Zeche Zollverein (**233**) in Essen was shipped in 1851. The surviving buildings of Zeche Zollverein 1/2/8 were built either at the beginning of the twentieth century or in the last major-reconstruction phase between 1957 (headframe of number 1 shaft) and 1964 (headframe of number 2 shaft). The headgear over number 2 shaft was designed by the Fritz Schupp (**108**) firm of architects, although for the Zeche Friedlicher Nachbar (**173**) in Bochum in 1950. When this mine closed, the headframe was brought in pieces to Zollverein and rebuilt in 1964. Number 2 shaft and that of Zeche Zollverein XII (**233**), were used for dewatering the mine with pumps set at the 1,000-metre level to pump the water to the surface and then through a pipe to the river Emscher (*cf.* **143, 214**). PACT Zollverein (Performing Arts Choreographic Centre NRW Tanzlandschaft Ruhr), founded in 2002, is housed in the pithead baths of numbers 1/2/8 shafts. This combines an international theatre, a residency for artists and an institute for research and development.

### 222. *Zeche Zollverein 3/7/10 (Zollverein Colliery 3/7/10)(TR 2)*

Working of the Zeche Zollverein (**233**) meant that eventually numbers 1 and 2 shafts placed the miners underground on roadways too far from the working coalface and there were also difficulties in ensuring the circulation of fresh air. So further shafts were sunk: number 3 in 1880, number 7 in 1899 and number 10 in 1913. These were all placed in relation to the Cologne-Minden railway (**149**). When number 12 shaft was sunk in 1932, these three shafts continued to be used as well. The 1913 single-storey headframe above number 10 shaft, which has a lattice-girder framework (**204**), has been restored. The shaft building has not survived, but the 1913/20 former machine and transformer building, with three naves of brick with arched windows, cornices and decorative artificial stone pilasters, has been preserved and is used as a science centre and a venue for events.

### 223. *Zeche Zollverein 4/5/11 –Triple Z (Zollverein Colliery 4/5/11– Triple Z)(TR 2)*

Number 4 shaft of the Zeche Zollverein was completed in 1894, number 5 in 1891 and number 11 in the late 1920s, so the site has adjacent buildings from different construction phases, the last before the major rationalisation phase of number XII shaft (**233**). The modernisation doubled the amount of coal produced. These shafts accessed seams that were very steep. In the 1960s the introduction of compressed-air picks and new shearers failed to increase production and the shafts were decommissioned in 1967. The buildings have been restored as a centre for entrepreneurs and young companies, and artists are offered six-month scholarships to work here.

### 224. *Zollverein (Zollverein)(TR 2)*

The Zollverein coking plant was built in 1958/61 to produce coke for use in the steelworks. It was designed to complement the existing building of Zeche Zollverein XII (**233**), both designed by Fritz Schupp (**108**). The coke ovens are 6 metres high, aligned on the tracks of the Cologne-Minden railway (**149**). The ovens reached their maximum capacity in 1964, with 4,920 tonnes produced per day. An extension was built in 1971 to give a total of 304 ovens over a length of 650 metres, making it one of the largest coking plants in the world. Over 10,500 tonnes of coal were converted into 8,000 tons of coke daily, along with 3 million cubic metres of gas. The plant employed 1,000 workers. Demand for coke declined in the 1980s and 1990s with the crisis in the steel industry. The plant closed in 1994 and in 2010 was taken into the care of the SIG (**228**) as one of the largest industrial monuments in

**Figure 224.1. Coking plant, opened in 1961 and closed in 1994, with modern ferris wheel and photovoltaic installation (© Stefan Bruggerhoff).**

the world. It is used for exhibitions of contemporary art. At night the monument is illuminated in red and blue (**225.2**). The RAG Montan Immobilien opened its corporate headquarters here in 2012.

### 225. Ruhr Museum (Ruhr Museum; formerly Ruhrlandmuseum: Museum of the Ruhrland)

The Ruhr Museum is now located in the former coal-preparation plant of the Zeche Zollverein XII (**233**). The Ruhrlandmuseum was founded in 1904, concentrating on art, ethnology, history and natural history, later adding botany, geology, mineralogy, prehistory and zoology. From the beginning samples were collected from the coal seams of the Ruhr, including typical fossils, minerals and rocks. There was extensive damage during WWII, including the loss of the African collection, and the museum did not reopen until 1954. In 1984 the museum moved to a new building, with exhibitions linking the geology of the Ruhr with the social history of industrialisation. In 2006 the museum was transferred from the ownership of the city council of Essen to an independent foundation, funded by sponsorship from the city of Essen, the Rhineland Regional Association and the state of North Rhine-Westphalia, and was renamed the Ruhr Museum. Four million objects were moved to the coal-preparation plant of the Zeche Zollverein XII. The museum opened to the public in its new premises in 2010 with 7,500 square metres of exhibition space. The museum has 35 members of staff, two volunteers and 10 interns. The museum averages 250,000 visitors a year.

Three floors of the main building were converted for the Ruhr Museum, which is not an industrial museum, but forms the regional museum for the cultural and natural history of the Ruhr. These floors, as well as the Besucherzentrum Ruhr (Ruhr visitor centre), are reached by a 58-metre-long escalator. The museum describes the pre-industrial history of the region, followed by the story of its rapid and extensive industrialisation, which made the Ruhr one of the largest industrial areas in the world, and its post-industrial current situation. The displays enable visitors to follow the path of coal in the Ruhr coalfield, covering the themes of history, memory and the present. The museum interweaves the structure and mining contents of the building itself into its story.

The museum also has extensive archaeology, geology, history and photography collections. The largest part of the collections is written material, such as mine books, work and pay slips, promotional materials ranging from children's books that address mining and graphics to artistic drawings and prints, documents, historic illustrative charts, mine plans and postcards. There are also decorations, medals and plaques. The fabrics collection includes banners of protests against pit closures, miners' clothing and personal clothing of working men and women. There are larger workshop machines and tools from between the eighteenth and twentieth centuries, such as bay lights, hatchets and hammers, pit telephones, signage and wedges. The collections range from furniture from both colliery buildings and miners' apartments to

**Figure 225.1. Exterior of the Ruhr Museum in the former coal-preparation plant of the Zeche Zollverein XII (© Ruhr Museum; photographer Brigida González).**

**Figure 225.2. External access to the Ruhr Museum illuminated at night (© Ruhr Museum; photographer Rainer Rothenberg).**

objects of everyday life and the working world, including mining devotional art.

The photographic archive of the museum holds some 3 million photographs of the history of the Ruhr coalfield as well as of the present Ruhr, including works by eminent artists such as Ruth Hallensleben (1898-1977), Willy van Heekern (1898-1989), Bernd Becher (1931-2007) and Hilla Becher (1934-2015). The nearby mineral museum in Essen is a branch of the museum displaying items from the geological collection, including an exhibition on coal. The mineral museum is the starting point for the Baldeneysee geological trail, on which surface outcrops of coal can be seen.

*References*

ERIH, *s.n.* Essen

https://de.wikipedia.org/wiki/Ruhr_Museum

https://www.ruhrmuseum.de

personal communications, Ulrike Stottrop, head of collections, 2016; Philipp Bänfer, head of communications, 2021

### 226. Saarländisches Bergbaumuseum Bexbach (Bexbach Saarland Mining Museum)

The Saarländisches Bergbaumuseum Bexbach is located in Bexbach in the Saarland. Coal mining began here as early as the Roman period, when the Saarland was part of the Roman province of *Belgica*, but was never very profitable because the seams were only between 1 and 2 metres thick. Mining only intensified when the demand for coal increased with the Industrial Revolution **(379)**, making the deeper seams worth exploiting. In 1879 the mine in the Bexbach suburb of Frankenholz (close to the Saarländisches Bergbaumuseum Bexbach) was expanded into the largest colliery in south-western Germany, with over 3,000 miners at its height. The longest cable railway in the world at the time carried the Frankenholz coal to the railway at Bexbach and then onto the first railway line in the Saar region to Ludwigshafen, which opened in 1849. Mining in this area ceased in the 1970s, although there are still reserves sufficient for several hundred years.

The museum is located in a former water tower and opened to the public in 1934. It is situated in parkland, where large mining machines are on display. Adjacent to the flower garden is a grassed spoil heap, Monte Barbara (Mount Barbara), with a statue of Saint Barbara (**96**). The museum deals with the history, safety and technology of lignite mining from its beginnings in the fifteenth century up to the present day. The museum is run by the Verein Saarländisches Bergbaumuseum Bexbach e.V (Association of the Bexbach Saarland Mining Museum). It has eight floors, with about 700 square metres of exhibition space in the museum building. The museum

**Figure 226.1. Underground roadway with haulage engine, pump range and compressed-air venturi (© Saarländisches Bergbaumuseum Bexbach e.V).**

**Figure 226.2. 1873 banner of the Royal Merchweiler Reden Coal Mine (© Saarländisches Bergbaumuseum Bexbach e.V).**

currently has two part-time employees and around 10 volunteers.

The visit starts at a height of 40 metres on the viewing platform on the seventh floor of the Hindenburg tower. This has a panoramic view over the Saar landscape, which has been shaped by over 250 years of coal mining. Each floor deals with a specific topic on Saarland coal mining, with aerial photographs of former Saarland mining facilities on two of the stairwells. The displays include information about the administration of the Saar mines; the conversion of coal to coke; mine safety with personal protective equipment; technology used in the mines, such as measuring devices; and transport systems for people and coal. Mining society includes festivals and miners' associations and the social life of the miners and their families. A miner's house has been recreated with kitchen, upstairs bedroom and a stable with a goat, known as 'the miner's cow'. As well as photographs and documents, there are face masks, coal plant fossils, mining lamps and surveying equipment. A replica Carboniferous forest has been created with plant fossils to explain the formation of coal (*cf.* **99**, **241**). A lamproom and pithead baths are displayed, together with the original desk where miners were assigned their tasks. A film explains how a mine works, how coal is processed and how it is transported to the end user.

An underground facility was created underneath the museum tower, and then the parkland was formed by dumping stone debris from mining activities over the artificial roadway. The roadway was extended in 1957 and again in 1986 to a final length of 190 metres, dealing with mining over the last 100 years. In front of the entrance is a simulation of a rope haulage, devised to give the acoustic and optical illusion of descending in a cage 1,276 metres into the pit. There a stairway descends to a reconstructed 1.2-metre thin seam, and then down another stair to a tunnel with a higher roof showing heavy twentieth-century mining machinery. Some of the machinery can be operated by visitors under the supervision of a museum guide.

The collections include the 1873 banner of the Royal Merchweiler Reden coal mine. Merchweiler is a municipality in the district of Neunkirchen in Saarland and Friedrich Wilhelm von Reden (1752-1815) was a famous mining expert. The museum also displays the 1928 banner of the Catholic Association of the Bexbach Apprentices. This shows the crossed hammer-and-pick mining symbol (**206**) and the miners' greeting *glück auf* (**210**) on the front. The image of Saint Barbara and the Latin exhortation *ora pro nobis* (pray for us) appear on the reverse.

*References*

ERIH, *s.n.* Bexbach

https://www.bbm-bexbach.de

personal communications, Rolf Bongard, director, 2012, 2021

showcaves

**227. Städtische Sammlungen Freital (Freital City Collections)**

The Städtische Sammlungen Freital is located in Freital in the district of Sächsische Schweiz-Osterzgebirge in the Free State of Saxony, south-eastern Germany, 8 kilometres south-west of Dresden. In the nineteenth century the administrative centre of the coal mines of the Freiherrlich Burgk (Burgk Baronial) Company was Burgk Castle. This is a manor house rebuilt in 1846, although its cellar, foundations and part of the ground floor date back to 1580. In 1819 Burgk Castle and the title of Baron Dathe of Burgk were inherited from his aunt by Karl Friedrich August Krebs (1791-1872). The baron was a mining entrepreneur who founded the Freiherrlich of Burgker Coal and Ironworks the same year, which became one of the leading companies in Saxony. In 1821 he introduced the first steam engine and by 1830 he employed between 700 and 800 miners producing 800,000 bushels (25,276 tonnes) of coal a year. He was active on social issues and introduced health insurance for his employees.

In 1828/32 a 1,300-metre adit was driven from the grounds of the Schloss Burgk to provide ventilation to the mine and a route underground for the miners to reach the coalface. This took advantage of a steep section of the herb garden, which still exists today. By 1835 the drift had reached the bottom of the Fortuna shaft and afterwards the bottom of the Neuhoffnungsschacht (New Hope shaft), which had been sunk in 1837. The Segen Gottes Schacht (God's Blessing shaft) was sunk in 1856.

On 2 August 1869 a methane-gas explosion occurred in the roadways of the Neuhoffnungsschacht and the Segen Gottes Schacht, which were interconnected. The toll of 276 miners killed was the highest in mining disasters in the Döhlen coalfield. As the explosion happened at 5 a.m., not all the workforce on the morning shift had yet gone underground and three miners were able to escape up the drift unharmed after the explosion.

On 6 August Crown Princess Carola of Saxony (1833-1907) visited to express her condolences. An appeal launched to provide pensions for the families and the few survivors raised donations from throughout the whole of Europe. Black-and-white and colour prints were sold to help raise funding. One print, entitled 'Der Gang zur letzten Schicht' (the walk to the last shift) shows the miners setting out for the colliery on the fatal morning. They had first gathered for prayers in the prayer room on the ground floor of the Huth Haus, which is shown in the background. Six vignettes around the central image show Burgk Castle, the Huth Haus and the buildings around three of the shafts, including the Segen Gottes Schacht and the Hoffnungs-Schacht (Hope shaft).

A memorial was inaugurated on the first anniversary of the disaster in the Kleinnaundorf district in a park in the Am Segen settlement. It consists of a stele with semi-circular

**Figure 227.1. First electric underground locomotive, exhibited in 1879 (© National Coal Mining Museum for England).**

panels bearing the names and ages of all the victims and the inscription:

> Dedicated to the memory of the officials and workers who, loyal to their profession, were crushed in the depths of the earth on August 2 1869.

The company based in the Schloss Burgk and responsible for the coal mine went bankrupt in 1930. In 1945, during the land reforms of 1945/6, the manor house was expropriated by the communist state from the last owner, Elisabeth von Boxberg, Baroness Dathe of Burgk (1889-1971).

In 1924 the council of Freital, which had gained city status in 1921, established a municipal museum. Since 1946 this museum has been based in Burgk Castle and is now amongst the largest non-state museums in Saxony. The museum employs five staff, attracts 20,000 visitors each year, and is funded as the municipal museum of the city of Freital. There are 4,000 square metres of exhibitions about art history, regional history and mining history.

There are a number of statues in the park of Burgk Castle. These include one dating to 1861 of Rotkopf Görg who, legend says, was given a hat full of glowing coals by a mountain spirit in return for playing music to which the spirit could dance. Rotkopf threw away the coals, which in the morning had turned to ash, but one small piece which remained in his hat had become gold.

The adit in the park is the only visitor mine to document the history of the coal industry of Saxony, concentrating on the period from 1945 to 1989. It is also the only place in Saxony where visitors can see a coal seam underground. The entrance to the underground display is by a steep flight of stairs. The displays trace the formation of the Döhlener coalfield back to their origins, 296 million years ago. They also demonstrate the hard working conditions of the miners in the early collieries of this region. On display are wall and roof timbering and ventilation piping. There is also a four-wheel electric mine locomotive, displayed in context with the pantograph engaged with the overhead line, dating to 1945. On the surface a technology garden gives an open-air demonstration of mining in the period immediately after WWII.

In 1879 the Siemens and Halske Company produced the first electric underground locomotive in the world for public demonstration. This is now preserved at the Verkehrszentrum (Transport Museum) of the Deutsches Museum (**228**). The exhibition at the Städtische Sammlungen Freital includes Dorothea, the first electric locomotive ever actually to work underground. This was built in 1882 and worked underground here between 1882 and 1927. In 2000 it was loaned back to its original location by the Siemens Forum in Munich.

*References*

https://de.wikipedia.org/wiki/Schlagwetterexplosion_im_Segen-Gottes-_und_Neuhoffnungsschacht

https://de.wikipedia.org/wiki/Schloss_Burgk_(Freital)

http://www.museum.de/museen/städtische-sammlunge-schloss-burgk-freital

personal communication, Juliane Puls, museum administrator, 2016

### 228. Stiftung Industriedenkmalpflege und Geschichtskultur (Foundation for Industrial Monuments and Historical Culture)

The SIG was set up in North Rhine-Westphalia, with membership from the state government of North Rhine-Westphalia, RAG AG (**164**) and representatives of various state and local organisations. Its aims are to provide a secure future for the most important listed industrial monuments of North Rhine-Westphalia by preserving them, identifying new uses, exploring their scientific importance and making them publicly available. Since 1997 its office has been in the former administrative building of the Kokerei Hansa (**125**) in Dortmund.

The profession of engineer was highly regarded in Germany and its members were interested in preserving the records of their work. In 1903 the Deutsches Museum (German Museum) was founded in Munich, which in 1915 established the definition for 'technical cultural monuments' as dealing with installations such as bridges, cranes, forges, mine facilities, waterwheels and windmills. The first monument to be preserved was the seventeenth-century Frohnauer forge, which was restored in 1907/25. From 1910 conservationists and engineers especially devoted resources to saving contemporary industrial buildings, the first being the 1818/30 foundry of Sayner Hütte. All efforts were devoted to rebuilding the economy in the period immediately after WWII and in fact more attention was paid to industrial monuments in communist East Germany than in democratic West Germany. By 1952 an institution for technical monuments had been established in Dresden and industrial monuments were recorded from 1961 throughout the whole of East Germany.

Impetus for dealing with the industrial heritage of West Germany came from TICCIH, which had been founded in 1973 at the Ironbridge Gorge Museum Trust (**379**). The second TICCIH conference took place in 1975 in Germany, as a result of which the buildings of the Zeche Zollern II/IV (**129**) were saved from demolition.

In 1970 the state government of North Rhine-Westphalia adopted a five-year plan to preserve industrial monuments in the state, at a cost of 2 million Deutschmarks (€1,022,580). These were now seen as integral and important historic elements of cities in the regeneration projects of the late 1970s. Due to their location and size and the often very-high quality of their architecture, former industrial buildings were particularly suitable for cultural uses. First workers' settlements and later industrial plants, such as the surface buildings of collieries and in particular their prominent headframes, came to be symbols for the identity of the areas where they were located. The results of this change in sentiment can be seen especially in the very large number of colliery and other industrial sites preserved on the ENERGIE-Route Lausitz Industriekultur (**111**) and the Route der Industriekultur (**135**), together with the museums of the LWL-Industriemuseum (**126-9**).

A monument topography of the Federal Republic of Germany has been published ever since 1981, providing an overview of the local and regional monument inventory and its historical and geographical context. North Rhine-Westphalia has an extensive network of industrial-history monuments, including the Zeche Zollverein (**233**). By 2003 there were about 76,400 registered monuments in the state, corresponding to around 2.55 *per cent* of the entire building stock. About 3,500 monuments of these belong to the fields of technology and industry. Some of the 25 anchor points of the Route der Industriekultur are in the care of the SIG, which is responsible for the management of 12 historic coal-mining sites that are described in more detail elsewhere in this publication (**125, 135, 154, 171, 176-7, 196, 197-8, 206-7, 232**).

*Reference*

www.industriedenkmal-stiftung.de

### 229. Strasse der Macht (Road of Power)

The Strasse der Macht is located in North Rhine-Westphalia, close to the border with Belgium and the Netherlands. Some 100 million tonnes of lignite are mined each year by RWE Power from three opencast mines. The Inden opencast mine, named from the village of Inden, itself named from the river Inden, extends across the administrative districts of Düren and Aachen. It produces 21 million tonnes every year, supplied to the Weisweiler power plant, which produces 18 billion kilowatt hours of electricity every year. The Inden seams are 230 metres below the surface and up to 45 millimetres thick. Some 1.6 billion cubic metres of overburden will need to be removed to allow the remaining 502 million tonnes of coal to be worked until the planned closure in 2030. From 1996 the course of the river Inden was moved to allow all the coal to be extracted. The resultant floodplain was planted with 400,000 trees and shrubs, and between 2030 and 2060 a 1,100-hectare lake will be created with water from the river Ruhr. To the west a new lake, the 100-hectare Blausteinsee, already fills the void left by the Zukunft-West opencast mine.

The brick renaissance-style Paffendoft Castle in Bergheim was built in 1531/46 by William of the Bongard. In 1958 the last owner, Marietta Baroness of the Bongard, sold the castle to a predecessor company of RWE Power, because the Fortuna Garsdorf opencast mine had reached the edge of the castle grounds. There is an RWE information centre in the castle itself and a permanent exhibition on the first floor about Rhineland lignite and its exploitation.

The castle is the starting point for the Strasse der Macht, which runs for 30 kilometres across the flat plain east of Aachen, through an area from which the villages were removed to permit opencasting. The route crosses a number of points of industrial interest, including a glass-and-steel platform near Inden-Lucherberf that gives a view over the whole area. Features along the route include the restored opencast landscapes of Fortuna-Garsdorf and Peringsmaar; the Fortuna-Nord factory, which makes lignite products; the Hambach opencast mine, which is one of the largest in Europe; the Niederaussem power station, fuelled by Rhineland lignite; and the Wiedenfelder Heights opencast-mining landscape.

*References*

ERIH, *s.n.* Inden

https://de.wikipedia.org/wiki/Schloss_Paffendorf

### 230. Tagebau Welzow-Süd (Welzow-South Opencast Mining)

The Tagebau Welzow-Süd is an enormous lignite mine in the south-east of the state of Brandenburg in north-eastern Germany. Underground mining in the Grube Clara (Clara Pit) began here in 1866. Opencast mining started with drainage works in 1959 in the most easily accessed area of coal. As the thickness of the overburden increased, an F60 (**106**) overburden conveyor bridge was installed in 1969 and became fully operational in 1973. To enable the mine to expand, 17 villages were relocated between 1968 and 2011. The mine was operated by Vattenfall Europe Mining AG, owned by the Swedish state, and produced 130 *per cent* of Sweden's carbon-dioxide emissions. When the Social Democrat-Green coalition won power in Sweden with a promise to leave coal in the ground, instead of closing the mine, as required by the legally binding commitment they made six months earlier at the 2015 Paris climate conference, in 2016 they sold it to a consortium of Czech businessmen (Butler 2021, p. 9).

The mine is working the second-thickest lignite seam in the region at a depth of up to 130 metres and produces up to 90,000 tonnes each day. Welzow-Süd is the main supplier for the Schwarze Pumpe power station and the Schwarze Pumpe refining plant. About 8,000 people are employed in mining and converting the lignite; for example, into briquettes.

The Welzow-Süd former railway station now forms the visitor centre for excursions to the mine. In preparation for visits, information about the changing landscape is provided in the station, including films such as 'Eighty years of environmental change in eight minutes'. To see the technology at first hand, guided tours are offered to the opencast mine on foot, by land rover or transported in a manrider. The F60 overburden conveyor bridge is some 600 metres long and has three chain-and-bucket excavators. It removes excavated material from one side of the mine and then carries it across the void to dump it on the other side. There are altogether seven further excavators of varying sizes working on the site. Visitors can see all the machines in operation.

Vattenfall uses the removed overburden to create new landscapes. For every tonne of coal obtained, 7 cubic metres of overburden and 7 cubic metres of water have to be removed. The Tagebau Welzow-Süd is part of the ENERGIE-Route der Lausitzer Industriekultur (**111**).

*References*

ERIH, *s.n.* Welzow

https://de.wikipedia.org/wiki/Tagebau_Welzow-Süd

museum leaflet

www.bergbautourismus.de

### 231. Zeche Ewald (Ewald Colliery)

The Zeche Ewald is located in the south of Herten in North Rhine-Westphalia. The mine is named after Ewald Hilger (1859-1934), who was one of three mining contractors who sank the first shaft in 1871. This was at some distance from any other settlements, as in the UK (**382**), presumably to ensure that the miners could not move easily to other jobs if they were discontented with their wages or working conditions. Production ceased in 1931 during the Great Depression (1929-39), resulting in considerable unemployment and hardship in the settlement. Production resumed in 1938 and it became the most productive pit in the Ruhr after WWII, employing over 4,000 miners.

The mine closed in 2000. Many of the buildings were demolished, but the most imposing were incorporated into the 750-hectare Hoheward landscape park. These included the 1888 Malakoff tower (**127**), similar to that of the Zeche Unser Fritz (**209**), and the double-strut headgear built in 1955 above the central, number 7, shaft. The heapstead building over number 2 shaft was constructed in 1926, with winding done from the south side. A new headgear had to be built in 1986 because of subsidence, with winding done after a rotation of 90 degrees (*cf.* **277, 376, 386**) and all the remnants of winding removed from the south side. The new headgear was a refinement of the German A-frame style of headgear (*cf.* **108**, **233**). The duplex-cylinder steam-winding engine serving number 3 shaft was built in 1919 by the Prinz Rudolph foundry (**113**). It was rebuilt in 1940 and was converted to run on compressed air in 1985.

**Figure 231.1. General view of the Zeche Ewald (© RVR-Besucherzentrum Hoheward).**

**Figure 231.2. Former heating plant, now the Revue Palast Ruhr, with 1926 heapstead building and 1986 headgear to the left (© RVR-Revue Palast Ruhr).**

**Figure 231.3. Top of Hoheward spoil heap and the horizon observatory (© RVR-Adler).**

The 1986 scaffolding stands 40 metres high above number 2 shaft. In 2009 the heating plant of the colliery was developed as the Revue Palast Ruhr (the Ruhr Palace revue theatre) by the theatrical entrepreneur Christian Stratmann (born 1951). In 2010 the lamproom and wages office were converted into the Hoheward tourist information centre for Herten, which is now the RVR visitor centre Hoheward.

The site also accommodates more than 20 international companies and the Hydrogen Competence Centre, h2 Herten, with a total of some 1,000 employees. During 2019 the visitor centre employed five staff and the park had about 143,600 visitors. At the centre of the park are the Hoheward and Hoppenbruch spoil heaps, which at 220 hectares together form the largest spoil-heap landscape in Europe. The top plateau of the Hoheward spoil heap supports two curved steel arches that form the 'horizon observatory'.

*References*

https://de.wikipedia.org/wiki/Zeche_Ewald

https://www.hoheward.rvr.ruhr/

http://www.landschaftspark-hoheward.de/hoheward/

https://www.ruhr-tourismus.de/de/das-ruhrgebiet/staedte/kreis-recklinghausen/zeche-ewald.html

personal communications, Linda Feige, deputy manager, Besucherzentrum Hoheward, 2016; Eberhard Lantz, photographer, 2016, 2017; Alina Guss, tourism assistant, 2021

### 232. Zeche Sophia-Jacoba (Sophia-Jacoba Colliery)

The Zeche Sophia-Jacoba is located in Hücklehoven in the very west of North Rhine-Westphalia in the Aachen anthracite coalfield. Exploratory drilling for coal began in this region in 1885. Mining only became economically viable after the opening in 1908 of a railway line between Baal and Roermond, with a stop at Hücklehoven. Sinking of the Zeche Sophia-Jacoba began in 1911 and it opened in 1914, although with only eight miners. Sophia and Jacoba were the names of the wives respectively of the owner of the Dutch company, Frits Fentener van Vlissingen (1882-1962), and the mining director, Isaac Pieter de Vlooys (1875-1955). The workforce had grown to 2,161 employees by 1926 and 4,500 in 1936. The Schaufenberg settlement was built in 1921/6 for the miners and their families and a mining-vocational school opened in 1929. Number 3 shaft was deepened in 1927 and its two-storey headframe was erected in 1934, 42 metres high. The first diesel locomotives were introduced in 1937, although the last pit pony did not finish work until 1951.

Much of the above-ground infrastructure was destroyed by Allied artillery fire in 1944. The result was that 1 million cubic metres of water had to be pumped out of the mine after WWII. In the mid-1950s it was decided to make the Zeche Sophia-Jacoba the focal point of a large central mine, as it was at the centre of a circle, with a radius of 7.5 kilometres, of the main coal seams. The workforce peaked at 5,660 employees in 1959 and production at 420,000 tonnes in 1984. From 1960 it was one of the most modern coal mines in Europe. With the decline throughout Europe of the use of coal, however, the mine was decommissioned in 1997, even though there were still coal reserves for a further 100 years.

The former administrative building was designated as a listed building, but had to be demolished in 2007 because of mining subsidence. The briquetting plant was demolished in 2009, although several conspicuous spoil heaps have been preserved in the area, grassed over and used by hikers. Former miners formed a not-for-profit association, the Förderverein Schact 3 e.V. (Association of Friends of Shaft 3) to preserve the memories of the mining history of the region. In 2007 they opened a mining museum in the mining-vocational school on the site of the former colliery, set up with contributions from the members. The headgear, cage and winding machinery, and two other buildings have been preserved. Since 2010 the original headgear has been in the care of the SIG (**228**). The associated shaft house, winding-engine house and 1966 winder are not listed in the inventory of monuments, and remain in the care of the Förderverein. As the shafts had been filled with concrete, a replica roadway of about 60 metres has been constructed next to the pithead, with all the appropriate machinery and equipment in full working order. Special events are held, otherwise access is by application to the Förderverein.

*References*

https://de.wikipedia.org/wiki/Zeche_Sophia-Jacoba

https://www.industriedenkmal-stiftung.de/en/monuments/zeche-sophia-jacoba/

museum guidebook

### 233. Zeche Zollverein XII (Zollverein Colliery XII)

The first Zollverein shaft was sunk in 1847 by Franz Haniel (**193**) and three more before the end of the century (**221**). The name Zollverein refers to the customs union agreed in 1833 between the independent German states to coordinate tariffs and economic policies across the whole area. Officially it continued in force until 1919, but was actually subsumed into the German empire with the unification of Germany in 1871. Number XII shaft of the Zeche Zollverein was sunk in 1932 by the company Vereinigte Stahlwerke AG because the other four areas of the mine had not been modernised since the end of the nineteenth century. At the time the mine was producing 12,000 tonnes of coal a day and employing 5,000 men, making it the largest coal mine in Europe. Shaft XII was designed by the Schupp (**108**) and Kremmer firm of architects in the simple Bauhaus style. The buildings were the first structures to use a steel-skeleton construction, with red-brick walls and

**Figure 233.1. Exterior of the Zeche Zollverein XII, showing red-brick walls and steel trelliswork (© Stefan Bruggerhoff).**

**Figure 233.2. Pressing machine at the coking plant, illustrating the extensive area of derelict structures behind the main buildings of the Zeche Zollverein XII (© Stefan Bruggerhoff).**

steel trelliswork. This was extremely practical and became a model for buildings at mines throughout the Ruhr. Coal was moved by conveyor belt from all underground areas of the mine to number XII shaft, so that when the modernised colliery began operation it was able to produce 11,000 tonnes of coal a day, believed to be the largest output of any contemporary European coal mine and to be the most advanced in the world. Peak production was reached in 1937, with an annual output of 3.226 million tonnes and 6,900 employees.

The colliery suffered only minor damage during WWII and by 1953 was producing 2.4 million tonnes a year. The mine stopped production in 1986. Its outstanding architecture, described as the 'most beautiful coal mine in the world' (*cf.* **129**) and dominated externally by a massive German A-frame style of headgear (*cf.* **108**, **231**), meant that the whole site was immediately given protected-monument status. Two years after it closed former miners started organising informal tours of the site. In 1993 the industrial monument was opened officially to the public, with weekend tours attracting more than 5,000 visitors. In 2001 the 35 hectares of the Zeche Zollverein XII, the Zollverein (**224**) and Zeche Zollverein Schacht 1/2/8 (**221**) were together designated as a World Heritage Site. From 1998 the Entwicklungsgesellschaft Zollverein mbH (Zollverein Development Association), the Stiftung Zollverein (Zollverein Foundation) and the SIG (**228**) were tasked jointly with developing the full potential of the site. In 2008 the roles of the three organisations were combined under the Stiftung Zollverein.

IBA Emscher Park (**123**) planned the restoration and conversion of the site by the company, Bauhütte Zeche Zollverein Schacht XII GmbH. This was completed in 1999, with some of the machinery – the conveyor belts, screens, tipplers and tub-run – preserved. The premises of the Ruhr Museum (**225**), the treasures of Essen cathedral and the headquarters of the Stiftung Zollverein and the Route der Industriekultur (**135**) are located in the coal-preparation plant. A spectacular escalator, adjacent to the existing conveyor belts and 68 metres long, takes visitors 55 metres up to the 24-metre level to the Zeche Zollverein visitor centre, with a ride lasting 90 seconds. There are also conference, event and exhibition spaces and a restaurant. The largest presentation in the world of contemporary design, the Design Zentrum NRW (North Rhine-Westphalia), is housed in the Red Dot Design Museum in the redesigned boiler house of the mine, converted by the British architect Norman Foster (born 1935). The landscaping of the park around the coking plant and the Zeche Zollverein XII was completed in 2012. The site is now a status symbol of the Ruhr and a model throughout the world for regeneration using industrial heritage. Behind the main buildings, however, there are many hectares of derelict structures, for which no use has yet been identified.

*References*

ERIH, *s.n.* Essen

https://de.wikipedia.org/wiki/Zeche_Zollverein

http://www.zollverein.de

museum guidebooks

# Hungary

Hungary has four major lignite coalfields in the east and four in the north-west, including the Ajka and Oroszlány coalfields (Jaskó 1989, p. 37). The Mecsek anthracite coalfield lies in the south-west, north-west of Pécs (euracoal). Hungary is especially famous for its wealth in gold and silver. Agricola (**108**) often refers to Hungary and analysis of the clothes and weapons of gold washers and soldiers in woodcut 173 shows that they are Hungarian (Szemán 1994). Ore-mining techniques informed those used in coal mining in the country. The first coal mine was opened in Brennbergbánya near Sopron in 1759 and in 1789 2,000 Viennese *glazes* (1,016 tonnes) of coal were produced there. Priority was given to finding coal near the western border with Austria to serve the needs of Austrian industry, so in 1766 the offer of 50 gold sovereigns by the queen of Hungary, Maria-Theresa (1717-80), resulted in a large number of coal deposits being discovered. Because of a lack of demand, however, major mining activity only began with the industrialisation of Hungary in the nineteenth century, mainly by Austrian companies. Between 1846 and 1896 Hungarian railways grew from 35 kilometres to 14,878 kilometres, offering transport for the coal while increasing the demand.

Hungary was on the losing side in WWI. Under the 1920 treaty of Trianon, Hungary lost two-thirds of its territory to Austria and Czechoslovakia, including much of its mineral deposits. This included the anthracite of Banat and the lignite Zsil coalfield, both now in Romania, although the miners continued to be Hungarian. Coal imports were considerably restricted by inflation and the worsening economic situation, leading to increased interest in exploration and in exploiting local reserves. The leading geologist, István Vitális (1871-1947), realised that drilling in 1900/2 in the Nagyegyháza coalfield, north-west of Budapest, had not reached the main coal seam. The 64 million tonnes of reserves there were reached in 1923. Between 1918 and 1938 prospecting identified a definite total of up to 690 million tonnes of additional reserves (summarised in Vitális 1939), and the likelihood of some 1,000 million tonnes.

The communist Hungarian People's Republic was established in 1949 and survived until 1989. The 1956 popular uprising against the regime was supported by the Miners' Brigade, four of whose leaders were executed in 1957. Large-scale exploration from 1959 identified further considerable lignite seams, making up two-thirds of Hungarian total coal reserves. These lay close to the surface, so they could be exploited by fully mechanised large-scale opencast working (see Fejér 1989 and Jaskó 1989 for the history of coal mining in Hungary). Since 2014 all production has been from lignite opencast mines and Hungary still has 10.5 billion tonnes of coal reserves (euracoal). The 2018 National Climate Change Strategy aimed to reduce greenhouse-gas emissions by 40 *per cent* by 2030, and by 2020 only two opencast mines were left in operation (*ibid.*).

### 234. Ajka Bányászati Múzeum/Alsó-Csingervölgy Bányászati Múzeum (Ajka Coal Mine Museum/Lower Csinger Valley Coal Museum)

Coal seams in the Lower Csinger Valley, near Ajka, slope downwards from the north-east to the south-west. The railway from Graz to Budapest was constructed in the 1870s and coal mining started in 1872. Basalt, bentonite, clays, gravel, manganese and especially bauxite were mined in the area as well as coal. The economy here was a mixture of mining and farming. The shaft of the Armin Akna Colliery was sunk in 1903 to a depth of 128 metres. A fire underground in 1909 killed 55 miners. The colliery shut in 1959 and the last coal mine in the area closed in 2004, so the region is now completely rural. The Ajka Bányászati Múzeum, also known as the Alsó-Csingervölgy Bányászati Múzeum, opened in 1965.

In 1977 an adit, 54 metres long, opened to illustrate working conditions underground. The tunnel is supported by arched girders, with timbering behind. On display are mining tools ranging from medieval axes and picks to modern pneumatic drills and pumps. When the mine closed, the surface machinery was all left *in situ*, including coal cutters, conveyors, mine cars, shaft-sinking equipment and underground locomotives. The buildings were also all retained, including the blacksmiths' shop, transformer building and winding-engine house. There are displays of archival material in the main building, such as cross-sections of the coalfield and geological maps of the area.

*References*

https://www.ajka.hu/?page_id=1886

https://www.termeszetjaro.hu/en/poi/museum/ajkai-banyaszati-muzeum/51047203/

showcaves

### 235. *Soproni Múzeum Központi Bányászati Múzeum (Sopron Museum Central Mining Museum): Központi Bányászati Múzeum (Central Mining Museum)*

The Központi Bányászati Múzeum opened in 1957 in the baroque Esterhazy Palace in the centre of the city of Sopron in the north-west of Hungary near the border with Austria. The museum occupies an area of more than 1,000 square metres, and attracts 10,000 visitors a year. It has outposts at the Oroszlány Bányász Múzeum (**236**), which presents local coal mining, and the Balás Jenő Bauxitbányászati Kiállítás in Gánt, which deals with local bauxite mining. In 2020 the Központi Bányászati Múzeum was modified and became part of the Soproni Múzeum, so is now called the Soproni Múzeum Központi Bányászati Múzeum. It has one curator, one cultural organiser and five other staff. This is the only mining museum in Hungary designated to hold a national collection. The museum deals with all Hungarian mineral mining, including in the territories lost in 1920 as they are considered to be an integral part of the history of Hungarian mining.

The permanent exhibition, of about 600 square metres, includes metal mining such as of gold and silver, as well as coal. The first section deals with mining art, including that of Sándor Gaspár (1916-2011), who had worked at the Dorogi Colliery in Komárom-Esztergom County. Of particular note is a table decoration in the form of a model of the surface of a mine, made of silver, minerals and ores such as amethyst, copper, marcasite and pyrite. It was made in *Selmecbánya* (now Banská Štiavnica in Slovakia) for presentation to the queen, Maria Theresa (**316**), when she visited the town in 1751. It shows the horse-drawn car, pump, vertical shafts, *etc.*, to demonstrate mining techniques of the time.

The museum displays a number of simulated undergrounds. One shows a typical underground setting of a 1930s/1940s roadway supported by timbers with a trapezoid section. Fixed turbo-electric lamps provide lighting. The simulation also has a canvas ventilation tube with wire reinforcement, compressed-air hand-boring machine tools, an iron tub on rails, and a safety lamp hanging from the cross-bar to alert the miners to any gas present: changes in the colour of the flame of the lamp reveal the presence and composition of any dangerous gases. Transport facilities occupied most of the space in narrow galleries so as much as possible was hung from the timbers out of the way. Acetylene lamps were used in flame-safety pits, such as the Brennberg Mine in the Sopron Mountains. The pit props carry electric cables. A compressed-air tube with branches and rubber hoses attached enables operation of machinery such as Flottmann cutters and drilling equipment, without generating sparks that could trigger an explosion. A separate section shows a freshly fired roadway in which coal has been brought down, with displays of hand tools, such as a crowbar, hatchet, leaf-shaped shovel, miner's rake and pickaxe.

Longwall working began to dominate in the second half of the twentieth century, using the Várpalota-type VOB-HP-108 overhead Hungarian shield chock and the Soviet KS-IK combine with a double-cutter head.

**Figure 235.1. Entrance to the Központi Bányászati Múzeum (© Soproni Múzeum Központi Bányászati Múzeum).**

Another simulation shows an ESA-60-L-type German milling cutter-shearer equipped with a VOB-HP-108-type Hungarian hydraulic shield chock and cable carrier. It was expected that working a coalface of 100 to 200 metres in length should make 3 to 5 metres progress each day. It was not, however, possible to use longwall working in medium-depth deposits of shallow slopes, such as found at Oroszlány and Várpalota. Manual labour continued to be essential, as shown by the display of hand-getting tools, although to an ever-decreasing degree.

There are a number of models, such as a 1628 pump and horse-driven whim gin (**372**) to dewater the *Selmecbánya* Mine. There is also a model of the Newcomen atmospheric steam engine (**371**), installed in 1722 by Isaac Potter at the *Újbánya* gold mine (**21**). In Hungary it was, therefore, known as the Potterian fire engine. There is also a model of the first Hungarian skip-transportation shaft at Várpalota Colliery, which operated between 1938 and 1984. This was important for longwall coal production, because the Hungarian hydraulic shield chock was the first of its type.

**Figure 235.2. Table decoration in the form of a model of a mine made in *Selmecbánya* for presentation to Maria Theresa in 1751 (© Soproni Múzeum Központi Bányászati Múzeum).**

*References*

https://sopronimuzeum.hu/kozponti-banyaszati-muzeum/

https://www.google.hu/search?q=k%C3%B6zponti+b%C3%A1ny%C3%A1szati+m%C3%BAzeum&client=firefoxa&hs=Wjh&rls=org.mozilla:hu:official&source=lnms&tbm=isch&sa=X&ei=krs-UufqFcestAbO5YDgCg&ved=0CAkQ_AUoAQ&biw=1280&bih=683&dpr=1

http://www.kbm.hu/

personal communications, Erzsebet Bircher, director, 2013; Attila Szemán, curator, 2021

## 236. *Soproni Múzeum Központi Bányászati Múzeum (Sopron Museum Central Mining Museum): Oroszlány Bányász Múzeum (Oroszlány Mining Museum)*

The Oroszlány Bányász Múzeum was established in 1979 in Majkpuszta, Oroszlány. It was set up to house the Oroszlány mining collection and concentrates on the history of mechanised Hungarian coal mining in the second half of the twentieth century. With the closure of the coal mines, the Vértesi Thermal Power Corporation (which was working in cooperation with the Oroszlányi Bányász Múzeum Foundation) was keen to preserve the characteristics of a group of buildings and equipment. The result was the opening of the museum on the outskirts of the Majk Carmelite hermitage. The museum was run by

**Figure 235.3. Simulated underground exhibition showing shield chock with a shearer and cable carrier (© Soproni Múzeum Központi Bányászati Múzeum).**

Figure 236.1. Steel headgear at number 20 shaft at the Oroszlány Bányász Múzeum (© Soproni Múzeum Központi Bányászati Múzeum).

the corporation until 1994 and the foundation merged with the Központi Bányászati Múzeum (**235**) in 2004. In 2017 the Központi Bányászati Múzeum handed over operation of the museum to the municipality of Oroszlány. Care of the collections and professional supervision remain the responsibility of the Központi Bányászati Múzeum. The Oroszlány Bányász Múzeum occupies 1,000 square metres. It has one curator and four other staff and attracts 12,000 visitors a year.

The museum's buildings and equipment are presented in their original state as a modern mining concern of surface facilities and operations. Items were collected from various mines as they closed, including banners, geological specimens, machinery, miners' clothing and tools. The museum is located in the former number 20 mining plant and preserves the compressor house, headframe and winding engine, switchgear and ventilation building. The first exhibition provides an overview of Oroszlány coal mining starting in 1937, with instruments used by geologists and mine surveyors, miners' tools and work clothes, as well as the brass bands that are typical of mining communities throughout Europe. The operation of the control room and monitoring equipment are also dealt with here. Outside the building are tracked heading machines from the late 1950s, known as F machines, which were developed by Hungarian engineers. There is

Figure 236.2. F4 small boom-type heading, cutting and filling machine with caterpillar track (© Soproni Múzeum Központi Bányászati Múzeum).

also underground transport with tubs to convey coal to the surface.

For safety reasons it was not possible to preserve the original underground levels and the shafts were capped. A simulated underground exhibition space, based on number 20 shaft, opened to the public in 2001. This shows the coalface, the headgear and the system for raising and lowering full and empty tubs of coal. This section also deals with problems such as the dangers of explosion, mine lighting, minewater drainage, roof collapses, roof supports and ventilation. The pit headgear is used as a lookout tower and a circular route can be undertaken in mine cars pulled by a mine locomotive.

*References*

http://www.banyaszatimuzeum.hu/

http://www.museum.hu/m/Oroszlanyi_Banyaszati_Muzeum/aktualis?f

http://www.1hungary.com/oroszlanyi_banyaszati_kiallitas/1309321

personal communications, Erzsebet Bircher, director, 2013; Attila Szemán, curator, 2021

showcaves

### 237. Mecsek Bányászati Múzeum (Mecsek Mining Museum)

The Mecsek Bányászati Múzeum is located at Pécs near the border with Croatia. The first coal mine in the area opened in the eighteenth century. The museum housing the Pécs Mining Collection opened in 1977 in the original headquarters of the Dunagőzhajózási Company (DGT), which began as a steam-boat company. The DGT had its own exhibition of mining equipment, but went into liquidation and the collection was sold in 1993. The Mecsek Bányászati Múzeum took over the building in 2004 and it was part of the Központi Bányászati Múzeum (**235**). In 2014 the coal-mining collection was taken over by the Janus Pannonius Museum, which is the regional museum for the county of Baranya, funded by the Pécs local authority. The museum occupies 1,100 square metres, and has one member of staff responsible for the coal-mining collection of some 600 items, all of which are on display. It attracts about 13,000 visitors a year and three volunteers assist with the special 'museums at night' events.

An exhibition on the history of Mecsek coal mining was set up in 2004 and covers geology, mine lighting, monitoring equipment, movement of coal underground, pneumatic drills, shaft sinking, underground locomotives and mine cars, and ventilation. The main feature used to be a 440-metre-long underground tour, which opened in 1981 in a former wine cellar used to create an artificial tunnel. This displayed mining machinery, as well as a mineral collection, including fossilised dinosaur prints. Archival photographs, models and reconstructed scenes illustrated working conditions underground, including hydraulic supports, movement of machinery and coal, protective clothing, safety equipment, shaft sinking, transport for the miners, tunnelling machines and ventilation. Unfortunately the underground exhibition had to close in 2015 due to a severe fungal infection of *Serpula lacrymans* (dry rot). All the underground exhibits had to be removed and the museum has so far been unsuccessful in obtaining funding to treat the fungus.

*References*

jpm.hu

personal communications, Andrea Pásztor, curator, 2015, 2021

showcaves

Small amounts of anthracite are found in Jammu and Kashmir, the northernmost part of India. Bituminous coal reserves are located in Jharkhand in the north-east, Odisha and West Bengal on the Bay of Bengal in eastern India, and Chhattisgarh and Madhya Pradesh in central India. Lignite is found in Rajasthan in the north-west, Lakhimpur in Assam in the north-east and Tamil Nadu on the southernmost point of India (CMMCP 2020, fig. 16.2).

Coal was first mined in 1774 in West Bengal, but was always hindered by having to be moved by bullock cart to its markets, often taking many months (Islam 2013, p. 194). So it was easier to import coal; for example, 150 tonnes from Australia in 1801. The Geological Survey of India was established in 1851. The major coalfields had all been identified by about the middle of the century (*ibid.*, p. 195; Ranestad and Sharp 2020, p. 5), and the transport difficulties were resolved with the introduction of railways in 1855. The British monopoly was broken with Indian involvement in coal mining from 1894. By 1900 India was producing 6.12 million tonnes annually, so that coal imports declined (Islam 2013, pp. 194-5), and local production reached a peak in 1920 of 18 million tonnes. WWI increased demand for coal, leading to mechanisation and the establishment of coking plants, coupled with price controls, but demand plummeted from 1920. In the early-twentieth century some 60,000 women were working underground, until banned in 1929, although women had to be employed underground again during WWII (Khaitan 2020). Demand increased in and after WWII, resulting in the development of opencast mines and further mechanisation, as underground mines went deeper to reach better-quality coal (Islam 2013, pp. 195-6). The mines were nationalised in 1971/3, but denationalised in 2015/18.

The colonial authorities' main emphasis was on eradicating tropical diseases, such as cholera, malaria and smallpox, because epidemics led to labour shortages (Chatterjee 2020, p. 190). In 1901 it was falsely claimed that Indian mines were safer and healthier than agricultural work (because underground they were working protected from the fierce Indian summer heat) and also than their European counterparts (*ibid.*, pp. 190-1). Yet a report in 1950 identified still-rampant work-related illnesses, including bronchitis, hookworm, nystagmus, phthisis and pneumoconiosis, resulting from insufficient lighting, ventilation and modern equipment in comparison with British mines (*ibid.*, pp 191-3).

There are still considerable difficulties in India. A coal mafia has developed running illegal mining (*ibid.*, p. 197), in which it is alleged that the senior trade-union leadership is involved. Most of the coalfields are in tribal areas, leading to the displacement of tribal populations, and there is exploitation of tribal women and children in coal-mining activities (*ibid.*, pp. 196-7). Equally serious are the appalling health problems, both in the workforce and in people living close to the mines, as well as in areas where lignite is being burned.

In 2017 coal generated over 70 *per cent* of electricity. India is the second largest producer of coal in the world, with 739 million tonnes in 2018. It does, however, have to import high-quality coking and thermal coal from Colombia and Australia, making it the second largest importer of coal in the world (Hui 2021) and one of the three largest polluters, along with China and Australia.

### 238. Bidala Audyogik Tatha Praudyogikee Sangrahaalay (Birla Industrial and Technological Museum)

The Bidala Audyogik Tatha Praudyogikee Sangrahaalay is located in the south of Kolkata, the capital of West Bengal, on a site that had housed the ancestral bungalow of the Tagore family. In 1919 the land was sold by Surendra Nath Tagore (1872-1940) to the industrialist, Ghanshyam Das Birla (1894-1983), who was usually known as G.D. Birla. It formed Birla Park, the home of the Birla family from 1919 to 1956, when the park was converted into a museum, after the bungalow used by the Tagore family had been demolished. The present museum building was constructed by the architects N. Guin and Company and opened to the public in 1959.

The initiative to start the museum project was taken by G.D. Birla; Bidhan Chandra Roy (1882-1962), Chief Minister of West Bengal; and Jawaharlal Nehru (1889-1964), Prime Minister of India (1950-64). The main objective of the museum is to enhance public understanding of science and technology through educational programmes, outreach activities and permanent exhibitions in the museum. When set up by the federal government in 1959, it was the first science museum opened in India and its popularity spurred the science-museum movement in India. Today there are 60 science museums and centres in various parts of India and their numbers continue to increase. Since 1978 the museum has been funded by grants from the government of India and internal revenues and has been run by the National Council of Science Museums, the parent body of science centres and museums, which is part of the Ministry of Culture of the government of India.

The museum occupies an area of 10,352 square metres and attracts about 230,000 visitors a year, with a further 5

Figure 238.1. Exterior of museum building (© Bidala Audyogik Tatha Praudyogikee Sangrahaalay ).

Figure 238.2. Single-ended ranging-drum shearer (© Bidala Audyogik Tatha Praudyogikee Sangrahaalay).

**Figure 238.3. Single-strand armoured flexible conveyor on left and steel roof supports on right (© Bidala Audyogik Tatha Praudyogikee Sangrahaalay).**

million people involved in the outreach programme. The museum employs 103 members of staff and has about 20 volunteers. The exhibitions deal with various branches of the physical sciences, industry and technology, many of the displays and exhibits having been designed and fabricated in its own workshop. There are several galleries with models demonstrating biotechnology, electricity, mathematics, mining, motive power, physics and transport. There are an interactive science gallery and a gallery specifically for children.

The museum opened its simulated underground in 1983, which is the only one of its kind in the country. Because 80 *per cent* of India's coal production came from the nearby Bengal-Bihar coalfield, the museum aimed to give visitors an idea of what going underground in a coal mine is like and demonstrating how miners work. It was also intended to serve as a practical training facility for mining-engineering students, who had hardly any access to a real coal mine.

Visitors are taken into the mine at fixed times in a cage, which descends a few metres, but gives the impression of rapidly going down hundreds of metres. They pass through a double-door airlock system into the mine-safety demonstration and an orientation room, then into the pit bottom in groups. The simulated coal mine depicts pillar-and-stall mining (**47**), as well as longwall systems and drilling for shortwall drivage. Equipment was collected from various coal companies in eastern India, in particular the Bengal Coal Company. On display are the machinery and tools used in the different systems, including single-strand armoured flexible conveyors, caving systems, a continuous miner, disc shearer, friction props, power loader, rotary drill, shortwall coal cutter and single-ended ranging-drum shearer. Mine-safety devices and practices are also highlighted.

*References*

personal communications, E. Islam, director, 2012, 2016, 2021

www.bitm.gov.in

# Indonesia

Indonesia is estimated to have 18.8 billion tonnes of mineable reserves and 95 billion tonnes of potential reserves: 58.6 *per cent* lignite, 26.6 *per cent* sub-bituminous, 14.4 *per cent* bituminous and only 0.4 *per cent* anthracite (https://globalenergycertification.org/indonesian-coal-quality-coal-reserves/). The main coalfields are in Kalimantan, which occupies the southern three-quarters of the island of Borneo, and in Sumatra, in the centre of the main island of Indonesia (CMMCP 2016, fig. 17.1).

Small quantities of coal were produced from just a few mines in Kalimantan and Sumatra during the Dutch colonial period (1849-1945), mainly to serve the shipping industry. Production peaked in 1941 at 2 million tonnes, after which it declined steeply (Lucarelli 2010, p. 14). Indonesia (then the Dutch West Indies) was occupied by Japan from March 1941 until September 1945, initially to obtain strategic raw materials, such as coal, oil, petroleum, rubber and tin. To ship out the coal, the Japanese constructed a port at Bayah on the south coast using 15,000 forced labourers, known as *romusha*, 400 to 500 of whom died each day. They also connected Bayah to the existing coal railway using 120,000 *romusha* with a similar deathrate (Thee 2013). From mid-1943, however, attacks by Allied submarines prevented the export of these raw materials.

The Japanese occupation caused enormous destruction of the physical infrastructure, especially bridges, ports and railways. This exacerbated the problems caused by the struggle for independence, the anti-western policies of Sukarno (1901-70), president from 1945, and the use of oil for shipping and rail transport, so that by 1972 Indonesia was producing only 200,000 tonnes of coal. In 1967 Soeharto (1921-2008) replaced Sukarno and aimed to stabilise the economy to make the country attractive to multinational corporations. In 1967 the Mining Law allowed foreign companies to export coal that they had mined and the Foreign Investment Act allowed foreign investors to repatriate their earnings. From 1989 to 1999 production increased from 4.53 to 80.89 million tonnes. Between 2000 and 2009 majority ownership of the largest coal companies transferred to local investors, and provincial governments took over regulatory control of the mining sector, including issuing new mining licences (Lucarelli 2010, pp. 11-15). In 2020 550 million tonnes were produced, 25 *per cent* of which by law had to be sold domestically, mainly for electricity generation. The remainder is exported, to India, Japan, South Korea, Taiwan and especially to China, boosted by China's ban on imports from Australia (https://www.argusmedia.com/en/news/2183405-indonesia-coal-exports-rise-in-december). In the first seven months of 2021 exports to China increased by 45.2 *per cent*, from 76.4 million tonnes to 110.9 million tonnes. Indonesia is the world's largest exporter of thermal coal (Hui 2021).

**239. Lubang Mbah Suro (Mbah Suro Tunnel)**

Sawahlunto is located in the area of Bukit Barisan in the province of West Sumatra. Coal was discovered here in 1867 by a Dutch engineer, Willem Hendrik de Greve (1840-72), leading to the settlement being known as the 'town of the black pearl'. In 1882 the Dutch established the settlement of Sawahlunto. The Dutch Government invested 5.5 million *gulden* (€163 million) in developing coal mining here and the first mine opened in 1891. In 1894 the government invested a further 17 million *gulden* in developing a rail service from Sawahlunto to Padang, the capital of West Sumatra and the largest city on the west coast of Sumatra, in order to move the coal to *Emmahaven* (now the Teluk Bayur seaport). The railway used rack rail on the steepest mountainous stretches. This enabled production to increase from tens of thousands to hundreds of thousands of tonnes of coal each year. Production peaked in 1930 at 610,000 tonnes a year. The original locomotives were Swiss, but they did not survive WWII and were replaced by Japanese rack-rail locomotives of similar appearance. Further mining was developed following the independence of Indonesia, and production peaked again in 1976 at 1.183 million tonnes a year. An opencast mine operated until 2002, after which only underground mining continued.

Coal mining ended in Sawahlunto in 2004, leaving the settlement as a ghost town. Since 1990, however, the area had been developed as a tourist destination with an emphasis upon its landscape and handicrafts, including artefacts made by local craftsmen from local coal, and from 2004 coal mining was an added attraction. The old town has been preserved as a living example of a colonial-era mining settlement. The area covers some 250 hectares and preserves more than 100 early buildings. The hotel built for Dutch scientists and geologists is still available for accommodation for visitors. The Societeit was built in 1910 for Dutch workers to play pool and bowls. After use as a bank, it has been restored and is now the town's cultural centre.

The Padang Panjang to Sawahlunto Colliery railway closed in 2000 because of the decline of the mining industry. It now, however, operates as a tourist line related to the railway museum, which was established in 2005 at the railway station jointly by the local authority and the railway company to record the history of the railways in

West Sumatra. It displays carriages, historic photographs, locomotives and signals, as well as rides across the city on a train pulled by an historic steam locomotive built in Germany and known as the Mak Itam (black lady). The 1918 Gudang Ransum Museum (Rations Warehouse Museum) is located in the former mess hall where some 2,000 meals a day were provided for the indentured coal miners. The museum preserves the fireplace, 4 metres high, and cooking pots with a diameter of 132 centimetres and a depth of 62 centimetres.

In 2019 Sawahlunto was given World Heritage Site status by UNESCO on the basis that it is the oldest coal-mining town in south-east Asia with a surviving large colliery complex. Tour operators in Padang now offer specialised mining tours, visiting the Gudang Ransum Museum and the Lubang Mbah Suro, named after a colliery foreman who died about 1930.

The mining museum is located in the centre of Sawahlunto and includes 186 metres of underground roadway, mined between 1898 and 1932. A number of seams, one above the other, were worked by the mine and only the uppermost roadway has been used for the tour. The tunnel was renovated by the local authority to enable safe visits, while maintaining the authenticity of the roofs and the sides of the roadways. Work was carried out between June and December 2007 and the mine opened to visitors in April 2008. Lighting has been installed, as well as a fan to provide ventilation. Visitors are equipped with boots and a helmet and descend a stairway to the roadway, 15 metres below ground. Contraband items like matches or cigarette lighters that could cause gases to ignite, must not be taken underground, as in a working mine.

The museum displays and interprets mining technology and working conditions and is currently planning a new coal-mining gallery. This will deal with the coal-mining history of Sawahlunto, together with the economic, infrastructure, political and socio-cultural developments now underway in the region. Outside the museum there is a life-size statue of two colliers pushing a tub full of coal, while being observed by a European supervisor. The museum explains the history of the use of forced labour, known as *narapaidana*, by the *orang rantai* (chained people), who were prisoners who came from all over the Dutch colonies, including China, Java, Madura and Makassar. They were overseen by armed guards wielding whips, their hands and feet were chained while they worked, hence the name, and they had to drag iron balls weighing between 5 and 10 kilograms to stop them trying to escape. In 1908 the *orang rantai* were paid 18 cents a day, while free contract miners received 32 cents a day, with meals provided in the *gudang ransum* plus housing. At the time the usual wage for a labourer was 62 cents a day, although without housing and meals.

*References*

https://en.wikipedia.org/wiki/Sawahlunto

http://whc.unesco.org/en/list/1610

http://www.thejakartapost.com/news/2011/03/23/from-coal-mine-tourist-spot.html

showcaves

# Ireland

The anthracite Kanturk coalfield in the south-west of the Republic of Ireland ceased production by the 1870s. The Leinster (**241**) and Slieveardagh (**242**) coalfields in the south-east have high-quality anthracite coal with a low-sulphur content, but in three very narrow seams. The Connaught coalfield (**240**) in the north-west has seams of semi-bituminous coal, only 45 centimetres thick, which had to be worked by miners lying on their backs and then turning onto their sides to shovel it out.

The first recorded coal mining began in the Leinster coalfield in 1638, with the first steam engine to pump water in a mine in 1740. In the Slieveardagh coalfield it was recorded in 1654 that coal being mined was 'suitable for blacksmiths', with the mines being unlicensed private enterprises, and 35 mines in production employing about 1,000 men through the mid-1800s. Mining is recorded as starting in the Connaught coalfield in 1765; the coal was used for the smelting of local iron ore from 1795 until the ironworks closed in 1838.

In the 1860s there were 73 collieries in Ireland, each year producing about 136,000 tonnes of anthracite, mainly for brickmaking, domestic fuel, lime-burning and malting. In the nineteenth century coal was used to manufacture town gas, as well as for domestic cooking and heating. As Irish seams were very thin compared to British mines, from the mid-nineteenth century coal was imported, which continues to the present day. Between 1904 and 1924 the output of coal gas increased and railways with coal-fired steam engines peaked at about 3,500 kilometres of track. Extra pressure was placed on coal in the 1930s and competition with coal from Britain and oil made mining of indigenous coal unprofitable. A state mining company, Mianrai Teoranta (Minerals Limited), established in 1941, took over many of the Irish mines, but by 1950 the Slieveardagh mines were operating at a loss and closed in 1970 (https://www.nmrs.org.uk/mines-map/coal-mining-in-the-british-isles/ireland/slieve/). In 1990 the Air Pollution Act banned the use of bituminous coal in the Dublin area and beyond. All coal mines had ceased production by 1990 (http://ireland2050.ie/past/coal/ for the history of coal mining in Ireland). Moneypoint, built in 1987 and the only power station fuelled with coal (mainly imported from Colombia, totalling 1.3 million tonnes in 2018), is scheduled to close in 2025 (euracoal).

### 240. Arigna Mining Experience

The Arigna Mining Experience is located in County Roscommon, close to Carrick-on-Shannon. The Arigna area was associated with coal mining for some 400 years and is located in a scenic area of north Roscommon overlooking Lough Allen. Much of the agricultural land in this area is of relatively poor quality and there was considerable poverty and much emigration to America and Australia in the nineteenth and twentieth centuries, particularly during the years of the Irish Potato Famine (1845-52). Only in Arigna was there stable, reasonably paid employment in the coal mines. The coal was used in the nineteenth century for ironworking and subsequently as fuel at the Arigna power station, which opened in 1958 and burned 55,000 tonnes a year. The power station closed in the late 1980s and, without their main market, the last mines closed in 1990. There is still a coke works in Arigna, which employs former miners and is dependent on coal imported from Northern Ireland; some coal dust is also imported from Poland to make eco-briquettes for stoves.

The initiative to open a museum came from the local community, which raised more than €125,000 itself, with funding also coming from the Arigna Enterprise Fund and the Department of Communications, Marine and Natural Resources. Altogether €1.5 million were invested. The museum opened in 2003 in a striking building designed by Wejchert & Partners Architects of Dublin. It was the first of its kind in the Republic. The museum's running costs are partly self-financed and part-funded by the state. Together the museum and visitor centre occupy some 280 square metres and employ 14 staff with support from a couple of volunteers. The museum is open seven days a week, 360 days a year. By early 2020 there had been over 450,000 visitors.

The centre deals with the history of coal mining in the Arigna Valley from the 1700s until 1990, as well as the geology and formation of coal deposits in the area. The displays in the museum include explanatory panels, mining artefacts, photographs and wage slips. A digital-video-disc presentation of the development of the mine into a visitor attraction includes footage of the mining methods used in Arigna, recorded weeks before the closure of the mines. The exhibition demonstrates the impact the industry had on the local communities over the years. On the site is an original miner's cottage, which has been restored and is available for letting as self-catering accommodation. The museum also explains the concepts and operation of different energy systems, past, present and future. Some of the new renewable-energy systems can be seen in operation in the Energy Centre and in the nearby windfarms of the Kilronan Mountains. The museum is situated on the route of the Arigna Miner's Way and Historical Trail.

To illustrate the methods used in extracting coal from the Arigna mines, the centre has a simulation of an underground explosion. This shows how a main roadway

Figure 240.1. Building housing the Arigna Mining Experience (© Arigna Mining Experience).

Figure 240.2. Entrance to simulated underground with guide, former miner, Seamus Lehany, at its mouth and a statue of the Virgin Mary with the Christ child above (© Arigna Mining Experience).

**Figure 240.3. Tunnel in the simulated mine with stone walls and arched girders supporting the roof (© Arigna Mining Experience).**

would be driven, with low secondary roadways branching off every 5 to 6 metres to follow the thin seams. There is a working internal haulage system on a looped wire rope which the visitor can see in operation. Above the entrance to the simulated mine is a statue of the Virgin Mary with the Christ child. The tours last 45 minutes and are guided by a former miner. The tunnel is level-entry and fully accessible, with authentic lighting and sound effects. The visitor can see a coalface and the very low roadways, stone walls and arched girders supporting the roof.

*References*

http://en.wikipedia.org/wiki/Arigna

http://www.arignaminingexperience.ie/

museum leaflet

personal communications, Valerie Stenson, manager, 2014, 2021

## 241. Castlecomer Discovery Park

The 32-hectare Castlecomer Discovery Park is located in the grounds of the former Wandesforde estate in County Kilkenny. Coal appears to have been worked here from about 1600 where it outcropped at the surface. Mining of some 10 million tonnes from the Three Foot coal seam began about 1640 using bell pits (**395**), some of which were later exposed during opencast mining. From the beginning in 1637 the Wandesforde family owned and, except for brief periods of leasing, ran the mines.

About 1740 the very profitable Jarrow seam was identified and worked by a succession of shafts. One advantage was that the coal here produced very little methane, so could be worked with a naked flame for lighting. By the early 1930s there were five major mines in the Castlecomer coalfield: the Deerpark, Jarrow, Monteen, Rock Bog and Vera Pits. About 1,000 men were employed in mining-related activities between WWI and WWII (1918-39). In 1924 the Deerpark drift mine was sunk to the high-quality Skehana seam, where eventually some 18 kilometres of roadway were worked at a depth of 213 metres. The water make, the frequent cases of pneumoconiosis and the thinness of the seams made working conditions difficult and resulted in frequent union disputes and strikes. It was not until 1939 that pithead baths were provided for the miners, the first in Ireland.

From about 1945 all mining in the Deerpark Pit was mechanised with coal cutters, electric haulages,

**Figure 241.1. Exterior of the Castlecomer Discovery Park (© Castlecomer Discovery Park).**

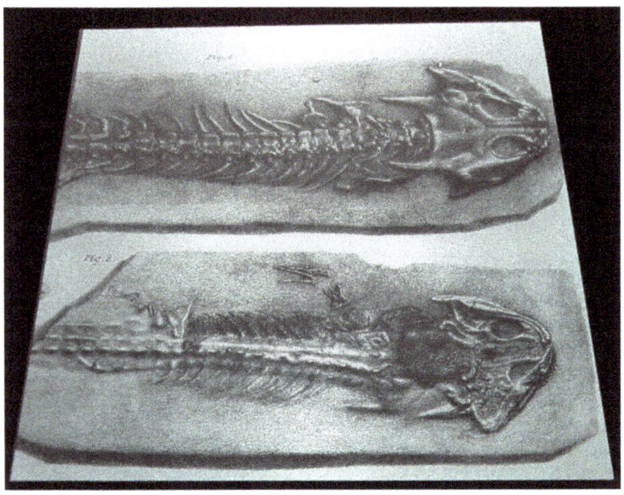

**Figure 241.2. Replica of fossilised remains of *Keraterpeton galvani* (© Castlecomer Discovery Park).**

underground conveyor and pumps, required as some 272,765 litres of water had to be pumped every hour to keep the pit dry. In about 1952 water problems underground increased and the seam became more difficult to mine, so that by 1969 the mine was making a loss of about £2,000 a week. The government had been subsidising the business while seeking advice on how to improve the situation, but it was finally accepted that the pit was uneconomic and it closed in 1969.

Castlecomer Discovery Park began as a community project to rejuvenate the town of Castlecomer following the closure of the mines, as the local community wanted a museum as a permanent commemoration of the coal-mining heritage of the area. After many years of planning, research and fundraising, for example from Kilkenny County Council, the museum opened in 2007. It was launched with a coal-mining exhibition, a visitor centre in the former farmyard and kitchen gardens, and design-craft studios in the former stable yard. It is run as a social-community enterprise on a not-for-profit basis. Funding is raised partly from the operation itself and partly from the Pobal Community Employment Scheme. The museum shares a receptionist with the discovery park and has a number of part-time education tutors, who do guided tours for pre-booked parties. The park attracts approximately 140,000 visitors a year, about 3,000 of whom visit the museum.

The coal-mining exhibition is a 45-minute tour through the mining past of Castlecomer using an interactive multimedia display, moving from the formation of coal 300 million years ago, through the 300 years of the history and heritage of coal mining on the Castlecomer plateau up to the closure of the mines. There are five rooms of displays. Life-sized reconstructions of prehistoric animals and plants illustrate life in the swamps of the Carboniferous period (*cf.* **99, 226**), when most of the coal was formed. One room displays fossilised plant remains from the shale, whose seams are found above the coal seams.in the Leinster coalfield. There are scale models of the local mines, together with mining equipment, such as a kibble (**355**), information panels and oral-history video interviews with former miners and their families.

In 1865 miners at the Jarrow Pit started to find strange animal fossils in the coal seam. Various eminent academics and scientists, including Edward Percival

**Figure 241.3. Image from the collections of the Deerpark Pit when operating (© Castlecomer Discovery Park).**

Wright (1834-1910) and Thomas Henry Huxley (1825-95), came to the area and recognised them as the remains of ancient amphibians. Altogether 10 genera of amphibian fossils were identified in Castlecomer, five of which were previously unknown: for example, *Keraterpeton galvani*, an extinct, relatively small, vertebrate amphibian from the Carboniferous era, similar to a modern newt, was first identified in Castlecomer. The Jarrow amphibian fossils are now held at the National Museum of Ireland in Dublin, but replicas are on display at the Castlecomer Discovery Park, along with other key specimens from the time.

*References*

http://www.askaboutireland.ie/reading-room/environment-geography/physical-landscape/castlecomer-plateau/coalmining-in-castlecomer/

https://www.discoverypark.ie/visitor-info/our-history/

http://www.irishtimes.com/news/science/intrigue-and-egos-in-a-tussle-over-irish-amphibian-fossils-in-1866-1.2044173

personal communications, Nichola Salmon, education officer, 2016, 2021

### 242. Clonmel Museum

The Slieveardagh coal-mining district is located in the Slieveardagh Hills on the border between County Tipperary and Kilkenny in the south-east of Ireland. The Mining Company of Ireland was set up in 1824 and sunk a mine at Mardyke. The company built 25 terraced houses for their workers and a Cornish engine, whose engine house remains, together with four ventilation-furnace brick chimneys, 28 metres high. By 1841 the company was operating three mines employing 400 people. The company mined at a loss during the Irish Potato Famine (1845-52), but kept its workers in employment.

The Slieveardagh Heritage Centre, located in a nineteenth-century former church in Killenaule, County Tipperary, opened in 1995. The centre told the story of the area, including coal mining. The centre has recently closed down its displays about coal mining, and transferred the historical items to the Clonmel Museum. The centre does still list historical items on its website, but the actual building now houses a library and a printing and training facility.

*References*

ERIH, p. 6, *s.n.* Glengoole

http://www.slieveardagh.com/coalmining/ballingarry-coal-mines/

http://www.slieveardagh.com/coalmining/coalmining-in-slieveardagh/

http://www.slieveardagh.com/places/mining-remembered

personal communication, Corina Bolger, office manager, 2012

# Italy

Coal is found in only two regions of Italy. There are small amounts of lignite in Tuscany in central north-western Italy. The two regions of Sulcis and Iglesiente in the island of south-western Sardinia, off the western coast of southern Italy, are rich in sub-bituminous reserves.

Theophrastus (*c.* 370-287 BCE) records the use of lignite by smiths in Liguria, in north-western Italy (Travis 2008, pp. 5-6). Pliny the Elder (CE 23-79) states (Rackham 1952, XXXIV, 96) that copper *in multis partibus Italiae*, 'in many parts of Italy', was smelted *carbone*, 'with coal', *propter inopiam ligni,* 'because of the scarcity of wood', *i.e.* wood from which to make charcoal, the usual fuel for smelting metals.

Coal mining certainly occured in Sardinia in the Roman period, but modern mining did not start in Sardinia until the 1850s, with the national Geological Committee being established in 1867 (Ranestad and Sharp 2020, p. 5). Coal mining only really developed as an industry when it became possible to move coal by rail and the mines were mechanised. Initially coal was very successful in supplying the electricity industry, especially during WWI (Musa 2016), but in the 1920s the electricity companies replaced coal for generation with water power from large artificial lakes. In 1920 seven miners were shot dead during a labour dispute. The fascist regime of the dictator, Benito Mussolini (1883-1945), needed indigenous energy and to increase supplies of coal leading into WWII, so a state company, the Società Mineraria Carbonifera Sarda (SMCS), was formed. In 1937 Mussolini ordered the construction of new mines and of the regional town of Carbonia, in south-western Sardinia, which was inaugurated in his presence in 1938. The town was named from *carbone* (coal). This was to form the centre for a revived coal industry, together with a port at Sant'Antioco from which minerals could be shipped. Miners came from all the regions of Italy, so Italian rather than Sardo was used as a common language amongst the many different dialects spoken. The coal industry continued in the period after 1945, when the economy of Italy was being revived, but became uneconomic with the development of the European Coal and Steel Community in 1952.

The SMCS was always in state ownership. With the nationalisation of electricity in 1964, ownership of the coal mines and power stations was transferred from SMCS to Ente Nazionale per l'Energia Elettrica (ENEL), which was also state-owned. In 1972 ENEL ceased all mining activities. In 1976 the Carbosulcis SpA (Sulcis Coal) was established with the aim of reopening the coal mines. The company is owned by the Sardinian Government and is the only coal-mining company left in Italy. The new project was focused on the Nuraxi Figus (**246**) and Seruci mines in the Monte Sinni coalfield, 13 kilometres north-west of Carbonia. The aim was to produce 1,722,000 tonnes a year, although this was never achieved.

There were frequent demonstrations, marches and strikes by the miners of the Iglesiente and Sulcis coalfields during the period from the 1970s to the 1990s. There were sit-ins underground at the mines of Nuraxi Figus and Seruci in 1984 (**246**). This resulted in the Mining Law of 1985, which allocated 505 billion Italian lire (€259.375 million) to restart coal production in 1988 at Seruci, but nevertheless it closed in 1991. Domestic coal production finished in Italy in 2001, except for the Nuraxi Figus mine. This closed in 2018, when saleable production was an estimated 243 tonnes, although for economic reasons this was left underground. Under the 2013 national energy plan, repeated in 2019, it is intended that all Italian coal-fired power plants should close by 2025 (euracoal).

### 243. Archivio Storico Minerario (Mining Historical Archives)

The Archivio Storico Minerario is located in the municipality of Iglesias in the province of South Sardinia. The archives are housed in the former warehouse of the now-closed Monteponi lead mine. The archives are managed by the Interventi Geo Ambientali (IGEA SpA), The Geoenvironmental Operations Company is a public company charged with the preservation and restoration of old mining plans and records. The archives constitute the Sardinian industrial record, regarded as of national and international importance. Records on mining and enrichment of minerals in Sardinia and in several Italian regions (Friuli Venezia Giulia, Lombardy and Trentino Alto Adige) have been collected from companies that have gone into liquidation or have merged with other companies, with corporate archives ranging from as far back as 1850 up to the recent period of closures (*cf.* the more narrow focus in the UK: **384**).

Qualified archives personnel of the IGEA SpA are working on the identifying, sorting and reorganisation of the entire documentary record in preparation for producing inventories. The aim is to make the records widely accessible through on-line platforms such as the National Archival System and the Unified Information System for Archive Superintendents. The documentation consists of thousands of mining and technical plans that show the development of the underground workings, and plans and sections of the development of extraction shafts and underground roadways, from different time

**Figure 243.1 Exterior of the Archivio Storico Geominerare (© IGEA SpA).**

phases. Equally important and extensive are the technical cartographic records, covering the machinery, plants, and industrial and civil buildings, and the glaciation records, often accompanied by explanatory reports and detailed studies and technical reports.

There is a considerable amount of administrative and technical-administrative documentation, relating to budgets, internal-management reports, personnel, production, purchasing and sales, and correspondence between the mining companies and other institutions, such as consultants, lawyers, notaries, partners, personalities, schools, suppliers, technical visits, trade unions and various experts. The library is comprised of magazines, newspapers, photographs, printed matter, publications, technical books and university theses showing the technological adaptations and cultural aspects of mining.

The archives are fitted out with the most modern storage facilities and ancillary equipment. The documents are housed in a compact filing system with roller-racking, while books are housed on antique bookshelves. On the walls there are paintings, photographic reproductions and some nineteenth-century photographs of groups of very proud-looking men in traditional dress and armed with rifles. These formed the bodyguard who accompanied the mine owners as they travelled from the city with money to pay the workforce. Sardinia is notorious for its bandits, who date back to as early as the Roman period, and it was essential for mine owners transporting money or valuable goods to have armed protection.

The centre has a thermostatic bath for conservation of paper and vellum documents, with a low-pressure vacuum table to wet the paper with water and/or chemicals and to soak the fluids out of the paper after a short contact time. The centre also possesses a state-of-the-art very large flatbed scanner for digitising the enormous documents produced to record underground levels and roadways, especially maps and plans.

In 1994 the Archival Superintendency of Sardinia designated the archives as being of high historical importance. This statement was updated in 2006, recognising the documentary heritage and historical mining as having an important role in the conservation of a large section of the cultural, industrial and social history of Sardinia.

*References*


http://san.beniculturali.it/web/san/dettaglio-soggetto-produttore?id=5714

https://sa-sardegna.beniculturali.it/index.php?id=488

https://siusa.archivi.beniculturali.it/

personal communication, Michele Caria, president, IGEA SpA, 2021


### 244. Centro Italiano della Cultura del Carbone (Italian Centre of the Culture of Coal)

The Centro Italiano della Cultura del Carbone is located on the site of the Grande Miniera di Serbariu (Great Coal Mine of Serbariu) in Carbonia, in the province of South Sardinia. The mine operated from 1937 to 1964 and was the most important energy resource for Italy and the largest colliery in the country throughout that period. It employed 18,000 workers, 12,000 of whom were miners. The surface site encompassed more than 33 hectares, with an additional 32 hectares of spoil heaps, and the mine had about 100 kilometres of roadways at depths of up to 317 metres. In 1991 the town council of Carbonia purchased 25 hectares of the derelict site and began restoration work in 2002, after a period of detailed planning. The museum occupies about 15 hectares of the site and opened in 2006 as a centre for education and tourism and part of the Parco Geominerario (**247**). Restoration of the site included alternative uses for some of the colliery buildings; for example, accommodation; catering facilities; the Centre for Research on Clean-Energy Technologies; a convention

**Figure 244.1. Simulated underground gallery with arched girders and machinery (© Centro Italiano della Cultura del Carbone).**

centre; handicrafts workshops; historical, demographic and anthropological archives; the Edouard Alfred Martel Civic Museum of Palaeontology and Speleology; and university training. The colliery's two remaining headframes are important landmarks for the town.

The museum operates as a not-for-profit association between the municipality of Carbonia, the Parco Geominerario and the provincial administration of Carbonia-Iglesias. It has a director and four employees, together with three seasonal guides. Visitors to the museum progress from the lamproom to the winding-engine house, then the simulated underground gallery. The lamproom not only housed the miners' lamps, but also the lockers where they left their clothes. The lamproom is now a large open space, of about 2,400 square metres, with displays about the everyday life and work of the miners, the history of coal mining in Sardinia and the story of the colliery itself. On display are documents, mining lamps, mining tools, photographs, segments of period films and video-interviews with individual miners. The displays also deal with the establishment of the industrial town of Carbonia.

The winding-engine house still contains the original 1937 winding engine of 1,600 horsepower made by Demag-Brown Boveri of Switzerland. Guides escort groups of up to 25 visitors on a tour lasting about 45 minutes of the simulated underground gallery. This shows how mining methods changed in the period from 1937 until the closure of the mine.

The museum also has an archive of historic photographs. These show how coal mining in Sardinia either differed from or resembled that in areas elsewhere in Europe. For example, whereas women worked in screens buildings in the UK until 1972 (**264**), in Sardinia women sorted coal from dirt only until the second half of the 1950s, when machinery was introduced in the screen buildings. In 1950 the Italian photographer, Federico Patellani (1911-77), produced four illustrated articles on Sardinia for the magazine *Tempo,* especially about Barbagia, Cagliari (the capital) and Carbonia. As part of the series entitled 'L'inchiesta in Sardegna' (investigation in Sardinia), Patellani took striking images of women working in the screens building of the Grande Miniera di Serbariu.

*References*

ERIH, *s.n.* Carbonia

personal communications, Mauro Villani, director, 2012, 2017, 2021; Patrizia Medas (**247**), 2012

www.museodelcarbone.it/

### 245. MINE Museo delle Miniere e del Territorio di Cavriglia (MINE Mining Museum and the Territory of Cavrigli)

The MINE Museo delle Miniere e del Territorio di Cavriglia is situated in Castelnuovo dei Sabbioni, a hamlet of the municipality of Cavriglia in the province of Arezzo

**Figure 244.2. Close-up of women sorting coal from dirt at the Grande Miniera di Serbariu in 1950 (© estate of Federico Patellani; image courtesy of Centro Italiano della Cultura del Carbone).**

**Figure 245.1. Opencast lignite mine of Castelnuovo dei Sabbioni in the 1970s, Santa Barbara power plant in the background (© Mine Museum Documentation Centre, Municipality of Cavriglia).**

**Figure 245.2. Deserted village of Castelnuovo dei Sabbioni (© Mine Museum Documentation Centre, Municipality of Cavriglia).**

in Tuscany. The museum documents and illustrates the history of the area of Cavriglia, including the opencast lignite mine of Castelnuovo dei Sabbioni, which closed in 1994, and how it affected the local economy and culture. The old village of Castelnuovo dei Sabbioni was abandoned to allow unimpeded mining. There are still some deserted houses in the lower part of the village, together with a monument to the 74 civilians shot on 4 July 1944 by Nazi soldiers in reprisal for partisan activities in the area. The opencast void is now a lake, although the Santa Barbara (**96**) thermoelectric power plant is still operating. These mines occupied a third of the municipal territory of Cavriglia, which covers an area of about 600 hectares. The low-quality lignite was used mainly for powering blast furnaces and producing electricity, unlike the high-quality coal produced in Sardinia (**247**).

In 2006 a survey of scientific places in Tuscany, called 'Itinerari Scientifici in Toscana' (scientific itineraries in Tuscany), was carried out by the Museo Galileo. Part of the project was the creation of a documentation centre here with an archive, books and photographic library, which includes photographs of both underground and opencast mining. This occupies one floor of a house dating to the early-twentieth century, whose other floor is used for educational activities and a laboratory. The Tuscan region was characterised by sharecropping, from 1200 to the end of the nineteenth century. The photographic archive with more than 11,000 digitised and catalogued images illustrates the transformation of the area over 150 years. This runs from sharecropping through underground mining to opencast working up to the closures, the end of mining in 1994 and subsequent reclamation works. Key features of the archives are photographic images of the transformation of settlements and the destruction of the surrounding landscape following the opencast excavations.

The documentation centre developed into the mining museum, which opened in 2012 as a facility of the city of Cavriglia and attracts between 3,000 and 4,000 visitors a year. The municipality funds activities and the salary of the two museum professionals. The museum also has some private sponsorship. The museum director and management are responsible to a cooperative selected by tender. The museum is recognised by the region of Tuscany as a museum of regional significance and is, therefore, eligible to obtain financing from specific regional programmes.

The museum occupies about 600 square metres with three separate buildings in the upper part of Castelnuovo dei Sabbioni. The main museum building has two storeys of exhibits, while the former Castelnuovo village church of San Donato (Saint Donatus of Arezzo) has been restored and transformed into a multifunctional space. The museum also has an external space for outdoor events in summer.

The museum has seven rooms dedicated to the history of local mining, beginning with the first evidence for the working of lignite, then the development of mining technology and the social life of the miners, including early labour struggles. The next section deals with the 1944 massacre, the long post-war trade-union struggles and changes in mining techniques, moving from underground roadways to opencast mining. The displays end with an installation called 'Linfa', a tree made of corten steel that represents the life cycle of life, death and regeneration, which permeates every part of the mine territory.

The galleries use tactile and olfactory experiences, sounds and smells to bring the history alive. The exhibits include minerals, and equipment and tools needed for mining lignite. The most impressive exhibit is a huge block of brown coal, the last evidence of the exhausted lignite seams.

*References*

personal communications, Gianfranco Molteni, director, 2016; Paola Bertoncini, director, 2021

showcaves

www.minecavriglia.it

### 246. Nuraxi Figus

In 2006 Carbosulcis SpA agreed to supply 1.1 million tonnes of coal to ENEL over a three-year period. At that time it was estimated that Sardinia had coal reserves of 600 million tonnes. The Nuraxi Figus mine was employing 500 people, had an annual production capacity of 1.5 million tonnes and one of the largest reserves in Europe of sub-bituminous coal. It had been open for some 40 years, but was costing €40 million a year to maintain, because the poor-quality coal had high sulphur levels. The reaction to the announcement of the closure of the Nuraxi Figus mine was that in 2012 120 miners occupied the mine, 396 metres underground, along with 350 kilograms of dynamite, for a week.

The government had provided funding to convert the mine to a carbon-capture and storage (CCS) facility, which would have cost €200 million. As a result of the financial crisis of 2008, this money was cut off under Mario Monti (born 1943), prime minister from 2011 to 2013. Under directives of the EU, coal production had gradually to be reduced over a four-year period, with final closure in 2018. The programme for developing CCS is, however, still alive. Carbosulcis and Sotacarbo SpA, both publicly owned and focused on clean technologies, are carrying out major research activities, despite the scheduled cessation of coal mining.

As the Nuraxi Figus is located in the Parco Geominerario (**247**), national rules for the park oblige the structures of the mine to be preserved. There are two headgears at the colliery. The winding engines for both shafts are combined in a single winding-engine house, a relatively common arrangement in a number of European countries. On the surface there are skip conveyors and disposal points

**Figure 246.1. Headgear of the Nuraxi Figus (© Parco Geominerario Storico e Ambientale della Sardegna; photographer A. Monteverde).**

discharging contents onto a 180-degrees dispersal belt, from where it was put through the coal-preparation plant. There are photographs of roadways underground with steel arching and timber infill at a junction, with the arches diverging from the same point; an electrically driven endless-haulage engine underground; and a return wheel of a pulley for directing the cable of the endless-haulage cable, without tensioning gear because it is on the engine itself. It is likely that some at least of the underground infrastructure will also survive.

*References*

https://transterramedia.com/collections/2079

http://www.ibtimes.com/sardinian-workers-occupy-italys-last-coal-mine-demand-government-funds-759161

personal communications, Patrizia Medas (**247**), 2016; Salvatore Cherchi, mining engineer/former president, provincia di Carbonia (province of Carbonia), 2016

### 247. Parco Geominerario Storico e Ambientale della Sardegna (Sardinian Historical and Environmental Geological Mining Park)

The island of Sardinia has extraordinary mineral wealth and a wide variety of natural landscapes. It first became important in prehistory when there was trade throughout

**Figure 247.1. Administrative centre of the Parco Geominerario in Iglesias (© Parco Geominerario Storico e Ambientale della Sardegna; photographer A. Monteverde).**

the Mediterranean of its obsidian (solidified volcanic glass with sharp edges used before the introduction of metals for cutting implements. including weapons). The island is rich in antimony, aragonite, azurite, barite, bauxite, cadmium, calcite, cerussite, chalcopyrite, cinnabar, coal, cobalt, copper, corundum, dolomite, fluorite crystals, germanium, gold, hornblende, iron, lead, magnetic hematite, magnetite, malachite, manganese, marcasite, molybdenite, nickel, pyrite, pyrolusite, rock salt, siderite, silver, tungsten and zinc. This was a mixed blessing, as the island was fought over and passed from the control of one powerful state to another, starting as early as 1,000 BCE with the Phoenicians, followed by the Carthaginians, Romans, kings of Spain, dukes of Savoy and finally the kingdom of Italy, formed by the union of the various Italian states in 1861.

Mining of the various minerals was a major source of income for the people of Sardinia during the nineteenth and first half of the twentieth century, with continuous technological innovations. It was, though still cheaper to import than to mine coal: Sardinian coal in the second half of the nineteenth century was used only in processing lead and zinc locally, before those minerals were shipped to Belgium, France, the German Rhineland and the UK (Musa 2016, p. 212). The major boost came in 1938 with the construction of the town of Carbonia. The exhaustion of some seams, decline in the quality of extracted material and rising production costs led, however, to the closure of almost all of the mines.

The Parco Geominerario Storico e Ambientale della Sardegna was established in 2001 to preserve the most important sites and to promote the significance of the environmental, geological, historical and mining sites of south-western Sardinia. The UNESCO definition of a geopark is that it is a territory that has a geological heritage of particular importance and a development strategy supported by a suitable programme. A geopark must also include a number of geological sites of special scientific quality, aesthetic appeal, educational value and rarity. Already in 1997 the UNESCO General Conference recognised the Sardinian geopark as the first symbolic example of international importance to be included in both the global and European geoparks networks.

The geopark was established in 2001, with a total area of 380,000 hectares in which the mining heritage is protected. The geopark is organised into eight different core areas, each distinguished by unique geological deposits, history of mining activity, mineralogical characteristics and the various different mining techniques used. Particular attention is also devoted to the archaeological and historic-cultural aspects relating to mining, such as documentary archives, extraction shafts, historic railways, industrial plants, thousands of kilometres of underground roadways and workers' villages. The geopark contains a number of important coal-mining sites (**244**, **246**).

The geopark has been charged with the development, dissemination of information, protection, research and restoration in these fields. In 2014 the Autonomous Region of Sardinia additionally charged the geopark with the promotion, protection and recovery of all Sardinian geosites, even outside the park boundaries. The Sardinian geopark now also works outside Italy through a network of international collaborations.

The activities of the geopark are overseen by a consortium made up of all interested parties, such as Italian national departments, regional institutions, universities and 81 municipalities. It is funded mainly by the Italian Ministry for the Environment, Land and Sea, supplemented by several EU activities and projects. The Sardinian Regional Government finances some activities in the territory of the geopark. The administrative centre is in Iglesias, a major town in the province of South Sardinia. There are 12 staff managing several multisectoral teams in multiple locations, municipalities and sites. As the geopark is so large and completely open, it is not possible to state the total number of visitors each year, but in 2019 the three main sites attracted over 82,000 visitors.

*References*

ERIH, *s.n.* Iglesias

https://m.youtube.com/watch?feature=youtu.be&v=XNidT-aqlp8

http://www.parcogeominerario.eu/

http://www.parcogeominerario.eu/images/archivio_documenti/documenti_consorzio/VERTENZA%20PARCO%20GEOMINERARIO%20vs%20GEOPARKS/006%20RIVALIDAZIONE%202011%20Parcogeominerario%20report%202011%20rivalidatori%20EGN.pd

personal communications, Francesco Muntoni, manager of environmental technology, 2012; Patrizia Medas, head of culture, events and international relations, 2012; Roberto Rizzo, geologist, environmental and technical department, 2021

# Japan

Bituminous and sub-bituminous coal, but not anthracite, is found in all three islands of Japan. There are seven coalfields in Hokkaido, the northernmost island of Japan; three in Honshu, the central and largest island; and four in Kyushu, the south-western island. In 1854, after 250 years of isolation, the American Commodore Perry (1794-1858) forced the opening up of Japan to American commerce, partly to ensure a supply of coal for foreign vessels. Inflation in the nineteenth century forced the government to begin industrialisation. The 1872 Mining Law prohibited non-Japanese exploitation of underground resources. Railways requiring coal for steam were developed by the government and annual production increased from 210,000 tonnes in 1874 to over 10 million tonnes in 1903 (Xue 2015, p. 218). The coal mines were developed mainly by *zaibatsu* (financial corporations), in particular Mitsubishi, Mitsui and Sumitomo. The workforce included convict labour (kept separate from 'respectable' miners), who endured appalling working conditions until Japan became a signatory in 1932 to the International Labor Organization's 1930 Forced Labor Convention (Takashi 2017). The use of convict labour had already been restricted by the 1899 Prison Law, which had led to a rise in husband-and-wife teams. Women who worked in the mines previously had often been the widows of miners killed underground. Legislation up to 1933 gradually prohibited women working underground, although women continued underground illegally in the small mines of the Chikuhō region of Kyushu (Burton 2014; Palmer 2021, pp. 4-5, fig. 2). Japan's mines mechanised much more slowly than those in America and Europe, only moving from pillar-and-stall (**47**) to longwall working in the late 1920s (Palmer 2021, p. 7).

Following victory in the Russo-Japanese war (1904-5), Japan annexed Korea in 1910, giving it access to large supplies of anthracite. Japan then developed its armed forces in preparation for war against America, beginning with the attack on Pearl Harbour on 7 December 1941. Large quantities of steel were required for gun batteries, planes, rifles and ships. After WWII coal was an essential source of energy until it began to be replaced by oil and cheaper coal from Australia and China in the 1960s, when most of the mines closed (Lim 2017). The last deep shaft in Japan closed in 2001 (**260**). The 1970s oil crisis forced Japan to look to cheaper fuels, such as nuclear energy (*ibid.*), but the public reaction to the 2011 disaster at the Fukushima nuclear station, means that Japan continues to be heavily reliant on imported coal (Thompson 2021; see https://archive.unu.edu/unupress/unupbooks/uu36je/uu36je0b.htm and Palmer 2021, pp. 7-10 for the history of coal mining in Japan).

In 2015 the coal-mining, iron, steel and ship-building heartland of the 23 Sites of Japan's Meiji Industrial Revolution on Kyushu were inscribed on the UNESCO World Heritage Site list, including Hashima Island (**248**), the Miike Colliery complex (**255, 265**) and Miike port (**265**).The designation was, however, dependent on a promise to include a 'full history' of the sites, in particular dealing with the extensive use of prisoners-of-war, especially Americans, Australians, Britons, Dutch and New Zealanders; for example, Miike Mine used almost 2,000 Allied prisoners-of-war, hundreds of whom died (Kotler 2021). There was also forced labour, especially by Chinese and Koreans, but including also men from Arabia, Argentina, Egypt, India, Ireland, Jamaica, Norway and Portugal, a total of more than 13,000 prisoners from 16 different countries (Kotler 2021; Palmer 2021). This is also the case with individual sites; for example, many of the workers on Hashima Island (**248**; see also **259**) were conscripted Chinese and Korean labourers. Despite objections by the Korean Government, they are not mentioned on trips to the island or in the brochures describing the history and lifestyle of those who lived on the island (Thompson 2021). In July 2021 UNESCO issued a reprimand stating that Japan had still not complied with the commitment to a full history, which must be implemented by 1 December 2022 if the sites are to retain their World Heritage Site status (*ibid.*). This is unlike museums and publications in Germany, which fully acknowledge the use of forced labour and prisoners-of-war in WWI and WWII (**26, 69, 97, 115, 127, 129, 187**).

### 248. Gunkanjima Dejitaru Myūjiamu (Gunkanjima Digital Museum; formerly Gunkanjima Hakubutsukan: Gunkanjima Museum)

Hashima is a small island, 6 hectares in size, 15 kilometres south-west of Nagasaki in the East China Sea (see also **259, 260, 261**). It lies in Fukuoka Prefecture in the northern part of Kyushu. Viewed from a distance, Hashima Island resembles a battleship, so the island came to be called Gunkanjima (Battleship Island) colloquially. Originally uninhabited, coal was officially discovered on Hashima about 1810. Coal exploration by the Fukahori family (**261**) began in 1870 and the first shaft was sunk to the underground coal seams in 1887, when the first permanent inhabitants came to live on the island. The Mitsubishi Company purchased the island in 1890. Their engineers extended the island with land reclamation using spoil from the underground mine. They added high retaining walls for protection from the sea and the works were completed in 1907. There was no permanent quay on Hashima because the waves are too strong for a ship to dock. So coal was loaded onto ships using a dolphin dock, formed by piles

**Figure 248.1. View of island of Hashima, with the coal-mine site in the centre (© Nagasaki-Shi).**

driven into the seabed and joined to the land by a short bridge.

From 1916 onwards accommodation for the workforce was provided in steel-framed high-rise cement blocks (the first of their kind in Japan), able to withstand typhoons and linked by covered walkways. At its greatest extent, in 1959, there were more than 30 accommodation blocks and 5,259 people living on the island. A passenger ship, the Yugao-maru, connected the islands of Hashima and Takashima to Nagasaki on the mainland. The residents had access to facilities such as bars, a Buddhist temple, cinema, doctor's surgery, gymnasium (building 71), restaurants, schools, swimming pools, theatre and 25 shops. Working conditions were extremely dangerous, and *go anzen ni* (be safe) was the miners' common greeting. Miners prayed at the Shinto Hashima shrine to the island's guardian deity for protection while working underground. The shrine, whose remains still survive, was built on a small hill and was dedicated to four deities, including Ōyamazumi no Kami (**272**).

The high-quality coal from Hashima was extracted from seams up to 1,000 metres deep and extending some distance out under the sea, reaching a peak of production of 371,946 tonnes in 1941. The coal was used especially at the Yawata steelworks in Kitakyūshū. Coal usage declined from the 1960s with the increase in the use of oil and competition from cheaper coal from Australia and China. The mine closed in 1974, leaving the island once more uninhabited. In 2002 Mitsubishi transferred ownership of the island to the municipality of Takashima, now part of the city of Nagasaki.

The concrete buildings still survive, but in a ruinous state. Some of the buildings have collapsed, although the exterior walls of some have been reinforced. None of the interiors are accessible, but Google Earth recorded the whole of the island in 2013, including inside the derelict buildings. In 2002 the Swedish film maker Thomas Nordanstad (born 1965) produced a short documentary on Hashima Island.

A 220-metre-long walkway has been installed, with three viewing points. This enabled a small part of the island to be reopened in 2009 for people to learn about the coal-mining industry. Tourists are taken by ferry from Nagasaki (**260, 261**)and landed using the dolphin dock. The tours take about 45 minutes and run when the weather is favourable, on about 160 days a year. Tourism now brings in about 600 million yen each year.

When the museum opened in 2015 it was originally known as Gunkanjima Hakubutsukan, but the name has now been changed to Gunkanjima Dejitaru Myūjiamu, with a

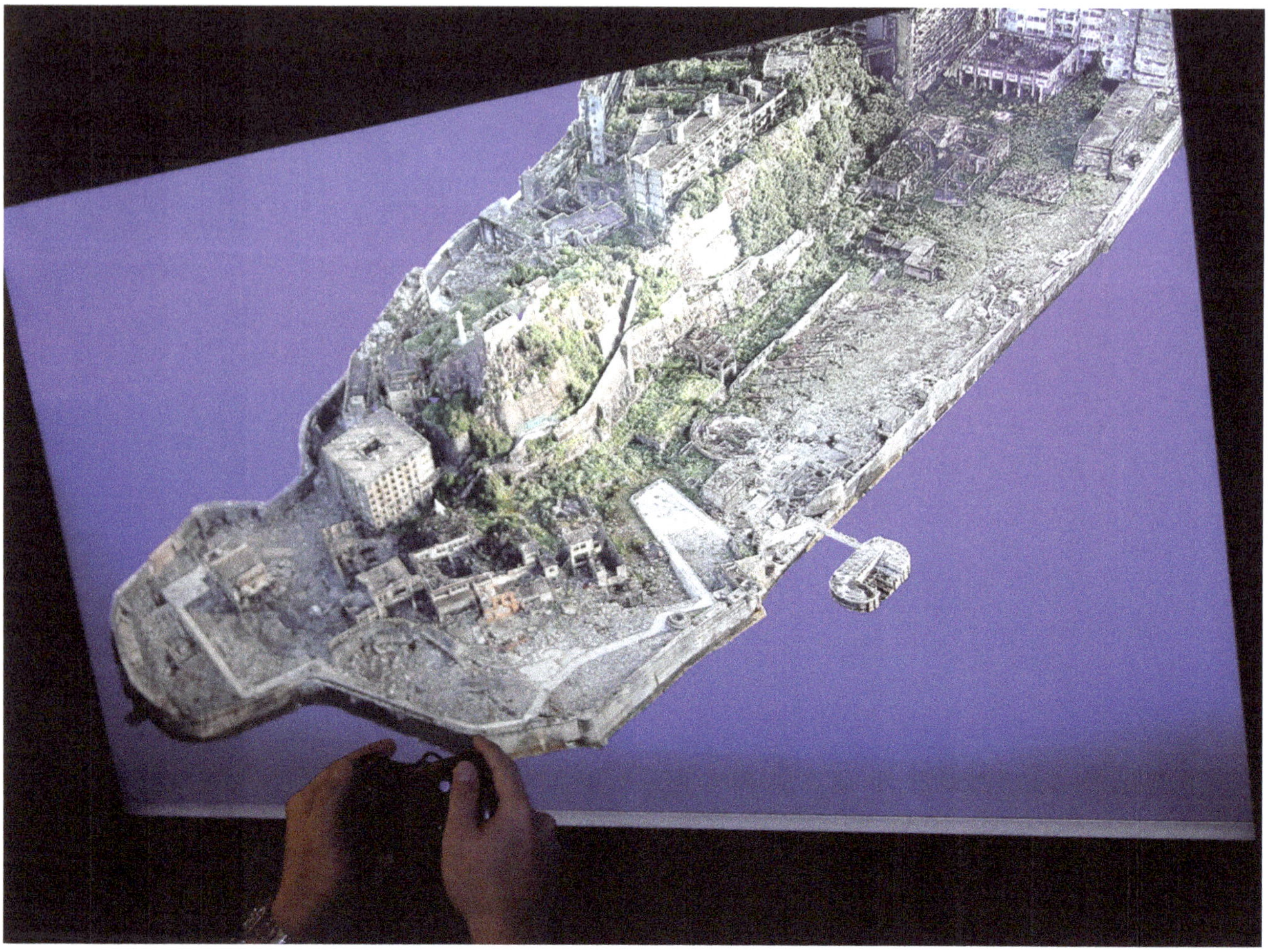

**Figure 248.2. Three-dimensional model of Hashima Island, showing the dolphin dock on the right-hand side (© Kabushikigaisha Yunibāsaruwākāzu).**

**Figure 248.3. Replica of the Hashima shrine (© Kabushikigaisha Yunibāsaruwākāzu).**

new website homepage in English. The museum has 10 members of staff and is run by a private limited company, the Kabushikigaisha Yunibāsaruwākāzu (Universal Workers Company Limited).

Artefacts on display include implements used in the mine, such as a crowbar and a pickaxe, together with interviews with the miners who worked with them. Virtual-reality technology allows visitors to explore the roadways of a coal mine by riding in a mine car in a colliery roadway. The life of the families is shown with the reconstruction of a room in one of the apartments in the late 1950s. Another replica is of the Hashima shrine.

There are three-dimensional models of the island, building 30, the gymnasium and the Yugaomi Maru passenger ship. In a video, complemented by interviews with five former residents, Yoichi Nakamura, an architect who grew up on the island where his father was the head of mining, considers a possible future for building 30. This was the first apartment block to be built in 1916 and so is of architectural significance. Finally there are videos of interviews with researchers into the preservation plan for the buildings, especially those in danger of collapse.

*References*

https://en.wikipedia.org/wiki/Hashima_Island

https://www.greenshinto.com/wp/2018/04/17/world-heritage-shrine-gunkanjima/

https://www.gunkanjima-museum.jp/data/en/

http://www.uwosh.edu/home_pages/faculty_staff/earns/hashima.html

http://www.ikjeld.com/japannews/00000010.php

http://www.japan-guide.com/e/e4414.html

personal communication, Kuwaoka Fumi, director, 2021

showcaves, *s.n.* Hashima Coal Mine

### 249. Hoshi no Furusato Hyakunen Kinenkan (Hoshi no Furusato Centennial Museum)

The Hoshi no Furusato Hyakunen Kinenkan is located in the city of Ashibetsu in Sorachi, a western subprefecture of Hokkaido. The name means literally 'the Centennial Museum of the Hometown of the Stars', but the museum prefers the English translation to be 'Hoshi no Furusato Centennial Museum'. The museum, which opened in 1993 and is funded by the municipal authority, attracts about 8,300 visitors every year. There are four members of staff and the museum occupies an area of 1,500 square metres.

The museum's permanent displays deal with the fauna, flora, history and literature of Ashibetsu, including the coming of Japanese settlers into the territory of the Ainu, the original indigenous people of northern Japan. The development of agriculture, forestry and mining is illustrated by the specialised tools used in each area of activity. Particular attention is given to the rise and then decline of the Mitsui Ashibetsu coal mine, together with the associated Mitsui Ashibetsu railway. An eight-minute film is shown in a mini-theatre about a day in the life of a community of coal miners living in a housing terrace set in 1953, complemented by the diorama model of the 1953 row of coal-mine houses made for the film.

The museum's stores have a wide variety of items, including miners' clothing and mining equipment, such as breathing apparatus. The most important element is a document and photographic archive, with more than 3,000 items related to coal mining, including a miner's certificate issued by the Tomoko organisation (**276**). A model shows the surface facilities and underground roadways used at a mine.

The museum also offers a large hall for events for the community, a special-exhibitions hall and a training or workshop room. There is a library open to the public, with books relating to the displays. The museum has published three volumes on the history of Ashibetsu.

*Reference*

personal communications, Haseyama Takahiro, director, 2016; Daisuke Yamaguchi, curator, 2021

### 250. Iwaki-shi Sekitan Kasekikan 'Hororu' ('Hororu' Iwaki City Coal and Fossil Museum)

The Iwaki-shi Sekitan Kasekikan is located in the city of Iwaki in Fukushima Prefecture on the central-eastern side of Honshu. Large deposits of coal were discovered during the second half of the nineteenth century in what became known as the Jōban coalfield (see also **256**). Coal mining was the main industry in Iwaki until the 1960s. A committee was set up in 1982 to develop a museum devoted to the history of coal mining in Iwaki, with the intention of encouraging both tourism and academic research. The museum first opened in 1983 and reopened in 2010 after renovation. The museum has an extensive display of fossils from Iwaki, especially of ammonites and *futabasaurus*, a type of plesiosaur from the Late Cretaceous of Fukushima, first described and named in 2006. The fossils are mainly from the Early Permian, Jurassic, Cretaceous and Quaternary Cenozoic periods.

The museum building occupies 4,499 square metres and the museum has in total 4,900 square metres when the gardens are included. Next to the entrance is the headgear from the East Pit of Jōban Colliery. The museum employs 11 staff, has 12 volunteers and attracts about 75,000 visitors a year. It is run by a not-for-profit foundation, the Iwaki-shi Sangyō Shinkō Kōsha. The term Hororu in the name is a nickname for the museum, derived from the Japanese verb *horu* (to dig).

The museum has two floors of exhibits, with displays of large coal-mining machinery in the grounds outside.

**Figure 249.1. Exterior of Hoshi no Furusato Hyakunen Kinenkan (© Ashibetsu-shi).**

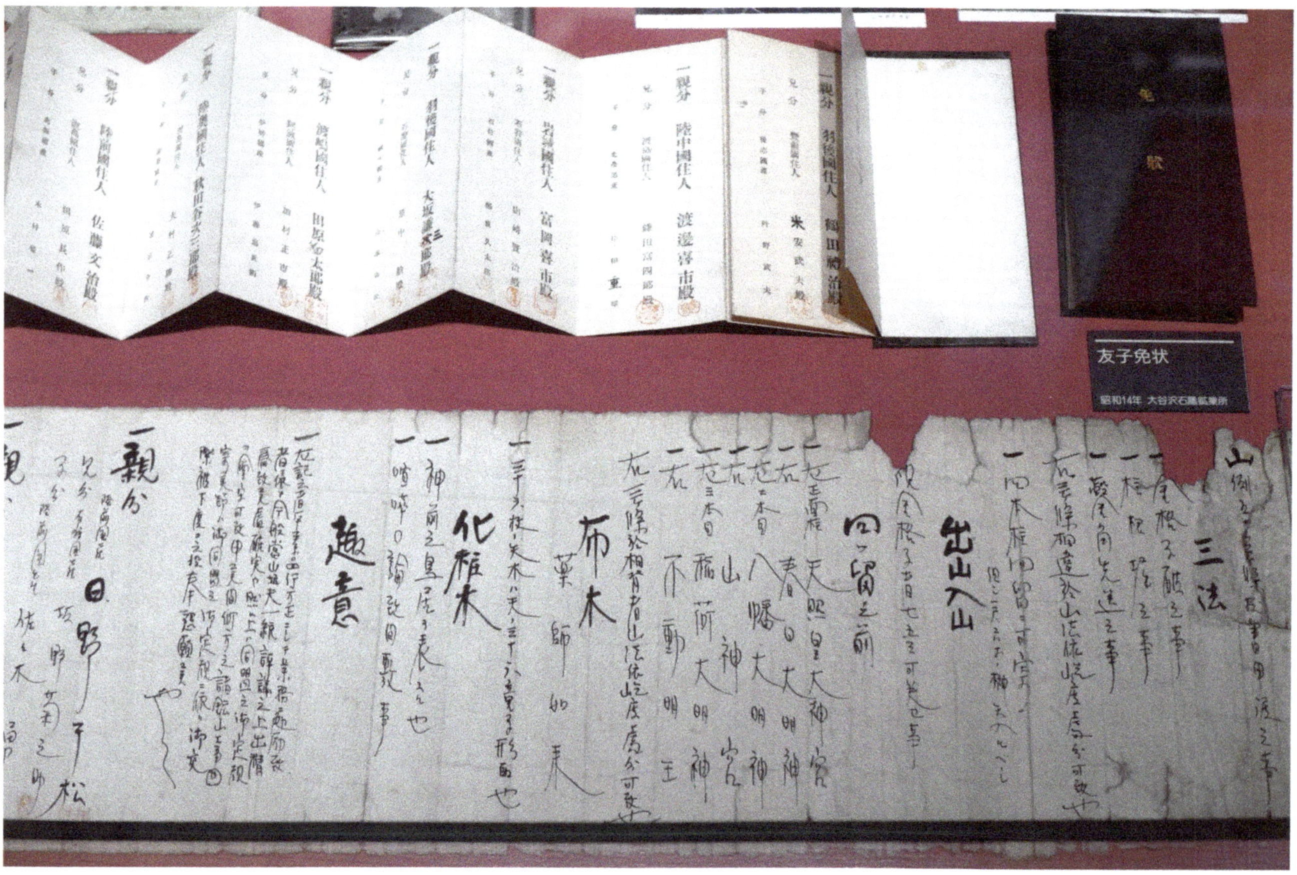


**Figure 249.2. Miner's certificate (© Ashibetsu-shi).**

**Figure 250.1. Entrance to the Iwaki-shi Sekitan Kasekikan 'Hororu' with re-erected headgear from the East Pit of the Jōban Colliery in the background (© Iwaki-shi Sekitan Kasekikan).**

**Figure 250.2. Pumphouse with miner taking a cold bath underground, with box on right containing salt (© Iwaki-shi Sekitan Kasekikan).**

In the late-nineteenth century unmarried miners lived in extremely cramped communal dormitories, while married miners lived with their families in tiny separate rooms. A number of these accommodation scenarios are shown in the museum. Cooking was done by the wives in a communal kitchen, also illustrated by a recreated scene. The mines employed forced labour and severe punishments (Palmer 2021, p. 7), including execution, were inflicted on those who tried to escape. As the museum points out, this did not encourage worker loyalty or efficiency, so was gradually replaced from 1914 by a direct-control system. This meant that the miners no longer had to live in dormitories. Miners were paid monthly and worked either day shifts or night shifts. As well as formal displays of mining equipment and machinery, there are also videos of underground coal mining in 1955.

A simulated colliery underground uses life-size models of miners and replicas of tools to show how coal was worked and the way in which new technology was gradually introduced. The simulation is located in the basement, 600 metres deep, reached by a replica cage in what would have been the intake shaft. The earliest scene is of a family unit in 1856 – brothers, parents and children – using pickaxes to work the coal, which was then moved to the surface in baskets carried on the backs of workers, who were usually women (*cf.* **273**). The method of mining was by pillar-and-stall (**47**), known locally as 'the digging of the raccoon dog' (**253**), which it resembled.

The next scene is of mining in 1944 with miners bending rails to fit the curve of the roadway. From the early 1960s miners are shown setting an extension bar to existing 1950s-style prop-and-bar metal roof supports on a longwall face equipped with a double-drum shearer, and inserting packing with wedges. This illustrates the problem of flowblow, in which the roadways rise up towards the roof of the tunnel. Also shown is the site office where safety and production rates were monitored, wages for each miner were calculated and the surveyors had their base.

The Jōban coal mine went down to a depth of 600 metres, where the temperature reached 42 degrees Celsius and the humidity was 90 *per cent* or more. To counteract the danger of dehydration and heat stroke, the miners had to consume salt regularly and take a cold bath every 30 minutes to lower their body temperature. Pumps were installed in the 1960s to reduce the danger of inflows of hot water from the walls of the mine by lowering the water level of the hot spring. The water was pumped to a traditional Japanese *onsen*, or large communal bath filled with hot spring water rich in minerals. The underground section has a display of the borers used to install the pumps.

From the late 1960s there are coal-cutting machines shown in place, including a single-boom roadheader displayed as though on development work driving an arch-shaped roadway. There is also a side-dump rotor loading into a mine car in a development heading being cut to access the seam using an Eimco side bucket, with conveyor belts to move the coal and powered roof supports to keep the roof in place above the coalface. The machinery on display came from the Jōban Coal Mining Corporation. Safety was ensured by the use of staff representatives and security staff, each responsible for a specific area of the mine. One scene shows a rescue after an explosion by a fully equipped rescue team. The museum also possesses a life-size statue of a miner (**251**).

*References*

http://www.japan-guide.com/e/e7779.html

http://www.sekitankasekikan.or.jp/

personal communications, Watanabe Fumihisa, curator, 2016; Takehiko Owada, curator, 2021

### 251. Kamisunagawa Tankōkan (Kamisunagawa Coal Mining Museum)

Kamisunagawa is located in Sorachi, a central subprefecture of Hokkaido. Mitsui Sunagawa Mine was sunk in 1914 to a depth of 660 metres. It was the first mine sunk by Mitsui on concessions that had been granted to the company in 1898 in various parts of Hokkaido. This concession lay in the northern part of the Ishikari coalfield.

After the mine closed in 1987, successful efforts were made to develop construction, industrial, manufacturing and service industries and other new enterprises in the town, including the Kamisunagawa Tankōkan, which opened in 1994. The museum was closed between 2005 and 2013 and, on its reopening in 2014, it attracted 1,071 visitors and then 639 in 2015. The museum is managed and operated by the municipality of Kamisunagawa, under the control of the regional director of the Kamisunagawa Municipal Planning Department. There are no full-time members of staff, but four volunteers provide support. The museum occupies a reinforced-concrete one-storey building with a floor area of 739 square metres. The entire site is 790 square metres. The museum does not have its own website, but features on the website of the municipality.

There is a life-sized statue of a miner with a pneumatic drill outside the museum. This is one of 11 statues made in the last two years of WWII to encourage miners to increase production to support the wartime economy, with two further examples preserved at other museums (**250**, **277**). The museum is entered by a tiled ramp under a ceiling designed to imitate a coal-mine roadway with arched girders and suspended industrial-style lamps. There are display cases containing mining equipment and tools, as well as the memorial inscription of the second shaft of the Mitsui Sunagawa Mine. A display shows the method of extracting coal using a high-pressure water jet. This blasts the coal out of the seam; it is then washed into the gully and pumped out by a gravity pump. This is not a method used in Europe for extracting coal, although the china-clay industry in Cornwall in south-western England uses an identical extraction method. The technique was developed in the Union of Soviet Socialist Republics. It was introduced by the mine manager, Fujii Nobuya, in 1964 in an attempt to reduce coal dust and so prevent explosions underground and to mine steeply sloping coal seams more easily. Mitsui Sunagawa Mine was the first in Japan to implement this method of mining coal.

*References*

http://aribaka.blog112.fc2.com/blog-entry-356.html?sp

personal communication, Aoyama Masahiro, assistant to the regional promotion director, Kamisunagawa Municipal Planning Department, 2016

### 252. Kikuchi Hiromi Kinenkan Takahagi Tankō Shiryōkan (Kikuchi Hiromi Memorial Hall Takahagi Coal Mining Museum)

The Kikuchi Hiromi Kinenkan Takahagi Tankō Shiryōkan is located in Takahagi, a city in Ibaraki Prefecture, in the Kantō region in the centre of the eastern coast of Honshu. This coal-resource centre is located on the site of the former Takahagi Colliery and is made up of both a museum and an art museum. The museum is surrounded by an open green area in which various coal-mining machines and a return wheel are displayed, as well as stumps of fossilised wood from the Carboniferous period,

**Figure 252.2. Daruma stove fired by coal (© Kikuchi Hiromi Kinenkan Takahagi Tankō Shiryōkan).**

**Figure 252.1. Fossilised Carboniferous tree stumps with small return wheel in background (© Kikuchi Hiromi Kinenkan Takahagi Tankō Shiryōkan).**

including an intact tree trunk. There is also a large stone plaque inscribed with the words *Takahagi kou Tsurekou*. This used to stand at the entrance to the second coal-mining shaft (*tsurekō*) of Takahagi Colliery, which was used for ventilation and as an emergency exit if there were problems with the main shaft. The stone was moved to the museum after the mine closed.

The museum concentrates on the history of Takahagi Colliery, which played an important role in the development of key industries in Takahagi. Objects and photographs of the mine are on display, together with plant fossils from the Carboniferous era. There is a diorama model, at the scale of 1 to 600, of the surface of Takahagi Colliery. This shows the communal pithead baths and the entrance to the underground, together with the adjacent spoil heap and housing for the workforce and their families. One domestic item on display is a daruma stove, which came into use in the mid-1920s. The daruma was initially a firewood stove, but came to be fired by coal. The term *daruma* refers to a typical Japanese puppet with a round belly, referring to the shape of the stove.

*References*

http://www.ibaraki.biz/takahagitankou.html

personal communication, Tsugito Shikada, management committee member, 2016

www.takahagitanko.org

### 253. Kurate-machi Rekishi Minzoku Hakubutsukan (Kurate Town History and Folklore Museum)

The Kurate-machi Rekishi Minzoku Hakubutsukan is located in the town of Kurate, in the Kurate district of the prefecture of Fukuoka in the northern part of Kyushu. The Chikuhō coalfield in the north-western part of Fukuoka Prefecture was the largest coalfield in Japan, although being inland was hampered by having to send coal a considerable distance by rail. The development of the coal industry goes back to the Chikuho Mine, intended to develop the salt marshes of the Seto Inland Sea (Xue 216, p. 218). Its mines were usually small, privately run and less mechanised than those in the south of Kyushu, such as Miike Colliery (**255, 265**). At its height in 1940 the Chikuhō coalfield was producing 18.6 million tonnes a year. Over the 100 years that it was operating altogether about 726 million tonnes were mined. The coal industry flourished in Kurate from the Meiji era (1868 -1912) to the Shōwa era (1926-89), supporting the growth of modern industries in Japan. The development and subsequent decline of the Chikuhō coalfield greatly influenced the history and culture of the town and the lives of the local people.

From 1955 information had been collected from all areas of the town and temporarily stored in the local school building. Construction work in the second half of the 1960s on the Sanyō Shinkansen (high-speed) railway uncovered material

**Figure 253.1. Entrance at the Kurate-machi Rekishi Minzoku Hakubutsukan to the Kurate-machi Exhibition Hall dealing with coal mining in the Chikuhō coalfield (© Kurate-machi Kyōiku Iinkai).**

from the culture of earlier periods. This was taken to the historical museum of Kyushu. Local volunteers established an association to campaign for the return of the material and for a museum in which to house it. Construction began in 1984 with a subsidy from the national treasury and the museum opened in 1985. Takashima Colliery (**261**) in the Nagasaki coalfield closed in 1986 and donations of material were received from the mine.

The Kurate-machi Rekishi Minzoku Hakubutsukan was established as a museum of the history and folklore of the town of Kurate and the lives of its inhabitants. The exhibitions have three sections: a folklore area, dealing with festivals and agricultural and everyday tools; the history area, dealing with material from the Jōmon era (14,000-300 BCE) to the Edo era (CE 1603-1868) excavated from the town; and an area dealing with the achievements of Itō Tsunetari (1774-1858), a national academician and educator who wrote 82 volumes of the topographical book *Dazaikannaishi* over 37 years. As well as exhibiting material about the local culture, the museum also aims to be a centre of research and to preserve and develop traditional arts. The museum has a library with collections about the life of the local people. It serves as an educational and information centre for regional history. There are two members of staff and five volunteers. The museum has about 3,000 visitors each year.

**Figure 253.2. Male and female miners working underground in the Sensui Mine in the town of Kurate sometime between 1912 and 1930 (© Kurate-machi Kyōiku Iinkai).**

**Figure 253.3. Replica underground with a battery-powered locomotive and exhaust ventilation bagging above (© Kurate-machi Kyōiku Iinkai).**

One aim of the museum was for people in the region to understand the coal culture that has greatly influenced the town's modern history. The Kurate-machi Exhibition Hall for coal-mining material was opened in 1991. This section of the museum occupies an area of 445 square metres and the reinforced-concrete building is 240 square metres. The exhibition hall is closed at present and is scheduled to reopen in 2023. The contents of the exhibition are currently under consideration, but it is expected that the displays will not be changed significantly.

The displays aim to enable visitors to understand what life working underground was like by showing the development of coal mining from the hand-getting of coal in the Edo era up to the mechanisation of the late 1960s and early 1970s. The exhibition is divided into two sections: the first covers coal mining in the period of small and middle-sized mines in the Chikuhō coalfield and the second deals with coal mining in the modern period. The method of early mining was by pillar-and-stall (**47**), known as 'the digging of the raccoon dog' (**250**).

The exhibition uses commentaries, drawings and photographs of both male and female miners working underground. There are models of women working naked except for a loincloth and images of women attached to tubs of coal by a rope running over their shoulders to enable them to pull the tubs on their hands and knees (*cf.* **378**). For the modern period there are full-size exhibits, mining machines and a replica underground roadway, with a battery-powered locomotive and exhaust-ventilation bagging above.

*References*

https://www.youtube.com/watch?v=LaYdYOVYMDo

http://www.town.kurate.lg.jp/sisetu/map_sekitan.html

personal communications, Hasegawa Tomie, curator, 2017, 2021

### 254. Kyū Taiheiyō Tankō Tankō Tenjikan (Taiheiyō Colliery Exhibition)

The Kyū Taiheiyō Tankō Tankō Tenjikan is located in Kushiro in eastern Hokkaido. The Taiheiyō Coal Mining Corporation was formed in 1920 with the merger of the Kimuragumi Kushiro Coal Mining Company and Mitsui Mining Company Limited. In 1946 drilling started under the sea at Harutori Pit. In 1949 the company was listed for the first time on the Tokyo stock exchange. With the decline in coal mining in Japan, the coal-mining division was spun off from the main company in 1970.

Kushiro was an important port for Japan because it was ice-free throughout the winter, becoming especially important during the 1920s with the growth of commercial fishing. Taiheiyō Colliery was linked to the harbour at Kushiro by the Rinkō-Sen (portside line) railway. A 6-kilometre line was opened from Harutori Pit to Irifune wharf in 1925. A 3-kilometre line connecting the mine to Higashi-Kushiro opened in 1928 and was reduced by 2 kilometres in 1966 when a new coal-loading wharf opened. This line closed in 1986. The mineral railway has not carried passengers since 1966 and was the only railway operated by the Taiheiyō Coal Services and Transportation. The local community decided to keep the mine alive by establishing the Kushiro Coal Mine Company Limited, which began small-scale mining in 2002, with Taiheiyō Colliery renamed as Kushiro Colliery. This is the last deep coal mine operating in Japan, and it continues to use the railway, the only private railway on the island of Hokkaido and the only surviving mineral line anywhere in Japan.

In 1980 the Taiheiyō Coal Mining Corporation established the Kyū Taiheiyō Tankō Tankōtenjikan to preserve and display the history of Taiheiyō Colliery and of the company itself. The company owns and funds the project. The museum occupies a total area of about 590 square metres. It employs one member of staff and attracts 1,500 visitors each year.

The first part of the museum is 180 square metres of exhibition dealing with various aspects of the history of coal and coal mining. At the entrance is a 6-tonne lump of coal, believed to be the largest in Japan, and a continuous miner, or heliminer. This was used on development work, especially for driving roadways. The cutting head can be raised and lowered to cut the roadway. The skirt collects the cut material with its gathering arms and pushes it for disposal onto an armoured conveyor that passes through the centre of the machine.

The creation of coal is explained, including displays of fossils, as well as the uses of coal in trade products and the future forecast for coal. There are examples of coal liquefaction and information about the size of coal reserves around the world, with displays of bituminous and lignite coal in America, Australia, China and Russia. The history of the mechanisation of the Taiheiyō Colliery and of the relationship of the mine to the city of Kushiro uses information panels, maquettes of underground machinery and photographs. A sectioned model of a relatively modern drift mine working mechanised longwall faces has a console with descriptions and buttons for indicator lights on the model.

The museum has a simulated adit of a coal mine of 410 square metres. This is displayed with machinery that came entirely from Taiheiyō Colliery. There is a modern fully automated longwall coalface equipped with an armoured face conveyor and computer-sequenced four-leg roof supports, a double-ended ranging-drum shearer using rack-and-pinion haulage, and the motor for the hydraulic powerpack. There is also a continuous miner with a shuttle car behind to take the material off the back of the heliminer. Also on display is a bo-bo electric flameproof mine locomotive, equipped with a pantograph (*cf.* **344**) for electrical supply from an overhead line. The locomotive is displayed with two mine cars of around 5-tonne capacity. The museum also holds an archive of documents from Taiheiyō Colliery.

*References*

https://en.wikipedia.org/wiki/Taiheiy%C5%8D_Coal_Services_and_Transportation_Rink%C5%8D_Line

personal communications, Takaori Ishikawa, vice-director, Kushiro City Museum, 2016, 2021

### 255. Mandakō Sutēshon (Manda Pit Station)

The Mandakō Sutēshon is located in the city of Arao in the north-west of Kumamoto Prefecture, which lies in the centre of Kyushu. During the Meiji era (1868-1912) the Manda Pit was the centrepiece of the Miike Colliery complex on the border between the cities of Arao and Ōmuta (see also **265**, **276**), although a larger pit was developed during the Taisho era (1912-26). The first two shafts were sunk here in 1897 and 1898, to a depth of 274 metres, making this amongst the deepest and most modern mines in Japan at the time. A 49-horsepower double-acting twin-cylinder air compressor, made in America, was installed in 1899. The first coal was produced in 1902 from the first shaft. The headgear of the second shaft was installed in 1908 at the same time as the introduction of a railway underground, with manriding cars from 1928. Maximum coal output was reached in 1944 with 10.092 million tonnes. Production ceased in 1951 and the facilities of the first shaft were dismantled, although the shaft is still unfilled, while those of the second shaft were maintained for pumping until 1997.

The headgear, pumping building, warehouse, winding-engine house and the pithead baths with their wooden lockers were saved. Miners prayed for their safety each day to the god of the mine, Ōyamazumi no Kami (**272**), at his shrine, which has also been preserved. This is the largest coal mine to be preserved that played a major role in the modernisation of Japan, from the Meiji and Taishō periods of the Miike Mining Company Limited.

Figure 254.1. Exterior of the Kyū Taiheiyō Tankō Tankō Tenjikan with continuous miner to the left (© Takaori Ishikawa).

Figure 254.2. Sectioned model of relatively modern drift mine working mechanised longwall faces, with console for information on right (© Takaori Ishikawa).

**Figure 255.1. Latticework headgear, with two-storey winding-engine house in background (© Arao-shi Sangyōshinkōka Kankōsuishinshitsu).**

**Figure 255.2. Barnsley Main Colliery pithead today (© National Coal Mining Museum for England).**

**Figure 255.3. Colliery workshops with centre lathe in foreground and line shafting in background (© Arao-shi Sangyōshinkōka Kankōsuishinshitsu).**

The site was designated as a national historical site in 2000, because it shows the technology of a major coal mine of its period. Conservation of the headgear of the second shaft and the winding-engine house were completed by 2009 and the site opened to the public in 2010. The technology used at the colliery was fairly standard in Japan from 1910 onwards. The most interesting story with this site is the adoption of western architecture and technology, rather than the technology itself. In the 40 years following the end in the late-nineteenth century of Japan's self-imposed isolation, much use was made of western expertise in engineering, industry and transport. In fact the layout of the heapstead, headgear and two-storey winding house of the Manda Pit is almost directly equivalent to the surviving installation at Barnsley Main Colliery (**355**), even down to the electric-winding engine. The latticework headgear is directly paralleled by those used on the English and Welsh coalfields. The headframe has a topmost platform (above the winding wheels), which features two sheave wheels used for the raising and lowering of pitwork in the shaft; the capstan for this survives in a building adjacent to the shaft. The colliery workshops still retain the original line shafting and belt drives.

The site occupies about 261 square metres. There are six members of staff and eight volunteers. A manager is assigned to the museum by the Arao City Tourism Office. In 2019 the museum attracted just under 40,000 visitors. Visitors to the Manda Pit begin their tour at the nearby road station, where a model is displayed of the colliery as it was in 1939. There are also old pictures, panels and sculptures relating to the mine, as well as items such as helmets and parts of the second headgear of the pit. The mechanism of the coal mine is illustrated in a diorama.

*References*

http://blog.goo.ne.jp/musshu-yuu/e/9f5564725a370a764daa5bb0f902f214

http://www.city.arao.lg.jp/q/list/385.html

http://www.sekitan-omuta.jp/legacy/12.html

personal communications, Ōshima Masatomo, assistant manager, Industrial Promotion Division, 2016; Hirose Tatsuya, manager, 2021; Nakano Hiroshi, planning staff, 2021; Sakuma Atsushi (**276**), 2021

### 256. Miroku-zawa Tankō Shiryōkan (Miroku-zawa Coal Mining Museum)

The Miroku-zawa Tankō Shiryōkan is located in the city of Iwaki in Fukushima Prefecture on the central-eastern side of Honshu. Coal mining in the Jobān coalfield (**250**) was the main industry in Iwaki until the 1960s, when it was displaced by oil and competition from cheaper coal from Australia and China.

The Miroku-zawa Tankō Shiryōkan is a privately owned museum consisting of archives, books, objects and photographs collected by Tameo Watanabe over a number of years. A document entitled 'Cultural heritage register' sets out three pages of detailed description of the nature of the heritage designated by the city of Iwaki and preserved in the museum, followed by two pages listing everything in the collection of about 90 items.

The museum occupies an area of about 50 square metres and attracts around 3,000 visitors a year. It opened in 1989 and is open seven days a week. A sound guide is also available from the Iwaki city hall. Items on display include a map of an underground and pictures and paintings of mineworkers and about coal mining. The museum was run by Tameo Watanabe, with the assistance of eight volunteers and, with prior reservation, Tameo Watanabe would give a guided tour of the items on display. Tameo Watanabe died in 2020 and his family has yet to decide what to do with the museum.

*References*

http://4travel.jp/domestic/area/tohoku/fukushima/souma/iwaki/museum/11364356/

personal communications, Tameo Watanabe, curator, 2016; member of staff, Iwaki City Cultural Promotion Division, 2021

### 257. Mitsubishi Bibai Kinenkan (Mitsubishi Bibai Memorial Museum)

Bibai is located in approximately the centre of the Ishikari Plain and of the Ishikari coalfield in Hokkaido. There are a number of remains of the coal mining. First is the former Mitsui Bibai Mine, which was sunk in 1915 as the Numaki Colliery by the Tanaka Steamship Coal Mine Company. It was bought by Mitsui in 1928 and closed in 1963. Elements that survive, such as the bunkers in which the coal was stockpiled and the colliery office, have been preserved as part of the industrial heritage of the area. Second is the Mitsubishi Bibai Colliery, which was sunk in 1913, to a depth of 170 metres. It was bought by Mitsubishi in 1915, eventually reaching annual production of a million tonnes and peak production of 1.86 million tonnes in 1944. The mine closed in 1972, but the two headframes have been preserved in the Coal Mine Memorial Forest Park. In 2007 the Ministry of Economy, Trade and Industry selected these headgears to be preserved as symbols of the modernisation of the industrial heritage. Also preserved are 103 examples of the housing provided at Minami-Bibai for the mining staff of Mitsui Bibai. There were four classes of housing, dependent on the status of the employee.

The Mitsubishi Bibai Kinenkan was set up to preserve the history of coal mining and the coal-mining heritage of the region. It was opened and donated to the city council in the grounds of what is now the Mitsubishi Heavy Materials Company. The museum is housed in a purpose-built, single-storey building of steel construction with an area of 162 square metres. There is one member of staff and in 2014 the museum attracted 612 visitors. The public parks division of the local authority manages the museum and it is funded completely by the city.

**Figure 256.1. Exhibition with equipment used in the underground (© Miroku-zawa Tankō Shiryōkan).**

**Figure 257.1. Housing at Minam-Bibai for employees of the Mitsui Bibai Colliery (© NPO Hōjin Yama No Kioku Suishin Jjigyōdan).**

**Figure 257.2. Headframes of the Mitsubishi Bibai Colliery (© NPO Hōjin Yama No Kioku Suishin Jjigyōdan).**

Conservation is carried out in the museum of coal-mining objects before they are displayed. The exhibitions include the history of the Mitsubishi Bibai Colliery, the mining utilities, and pictures and books depicting mining work, as well as local events and utilities used by miners in and out of the pit. A model of a miner shows how rescue equipment was worn. The museum has an account book from the housing company responsible for the provision of housing for the workforce.

*References*


https://images.search.yahoo.com/search/images;_ylt=AwrE1xlwr65gAcIAfp5XNyoA;_ylu=Y29sbwNiZjEEcG9zAzEEdnRpZAMEc2VjA3Nj?p=Mitsubishi+Bibai+Kinenkan&fr=mcafee

personal communication, Hasegawa Hirokazu, officer responsible for museums, Mitsubishi City Council, 2016


## 258. Miyawaka-shi Sekitan Kinenkan (Miyawaka City Coal Mining Memorial Museum)

The Miyawaka-shi Sekitan Kinenkan is located in the city of Miyawaka in the prefecture of Fukuoka in the northern part of Kyushu in the Chikuhō coalfield (**253, 274**). The museum is organised and funded by the municipality of Miyawaka. It occupies a two-storey building of 830 square metres dating to 1888 and was restored by the local authority in 1964. The mine closed in 1976 and the museum opened the following year. There are two members of staff and the museum attracts 1,800 visitors a year.

By the time the Kaijima Coal Mining Company opened the Ōnoura Mine in 1885, it had become the leading producer of coal among the local mining entrepreneurs. In 1903 Kaijima's net production ranked fourth nationally. The Kaijima Company has been both praised and criticised for its paternalistic management policies. Forced labourers were few, unlike many contemporary mines. Kaijima strove to encourage docile well-behaved labourers; for example, it was the first company to provide the children of its workers with a private elementary school, in which the museum is now located, and to offer scholarships for higher education. Another project provided vocational training for unskilled workers. Critics claim such programmes merely camouflaged the company's exploitation of workers, appeasing them with benefits, but not addressing their legitimate concerns.

When it was sunk in the 1950s, the total cost of the creation of the Shinsugamuta Mine, belonging to the Kaijima Coal Mining Company, was about 3 billion yen. Around 1960 the mine shaft was 530 metres deep, and the mine had the highest spoil heap in Japan. Shinsugamuta Mine had two cages going down to a depth of 500 metres. The total area belonging to the Kaijima Company was

1,708 hectares, stretching for 8 kilometres on an east-west axis and 4 kilometres north-south. During its peak production year in 1961, the mine employed 12,000 men and women. When it closed in 1976, the last mine of the Chikuhō coalfield to do so, the company had been operating for 91 years.

The collections include historic photographs of the mine and a wide variety of mining tools and equipment, such as pneumatic percussion coal picks, which are on display in the museum. The most important item in the museum's collection is an American Locomotive Company (ALCO) 22 locomotive, with a 0-6-0 steam tank engine and American-style bar frames. This was manufactured in 1919, probably in Japan under licence from ALCO using some parts supplied by ALCO. The locomotive operated between 1919 and 1976 in the coal-mining area, hauling trains of waste rock for backfilling areas of subsidence caused by underground coal workings. This locomotive has been designated as a tangible cultural property by the city of Miyawaka.

*References*

https://www.pref.fukuoka.lg.jp/uploaded/attachment/118177.pdf (no. 25, p. 26)

personal communications, Tatsuhiko Funakoshi, head of General Affairs Division, Miyawaka City, 2012, 2016

### 259. Nagasaki Gunkanjima Shiryokan (Nagasaki City Warship Island Museum)

The Nagasaki Gunkanjima Shiryokan is located in Nomozaki, which in 2005 merged with the expanded city of Nagasaki in Fukuoka Prefecture in the northern part of Kyushu. The museum was opened in 2003 by the Nagasaki Minami Shōkōkai Seinenbu (South Nagasaki Youth Group of Commerce and Industry) and in 2005 the museum merged its organisation with that of the Nagasaki city museum. When Nagasaki City Council started taking tourists on guided tours of Hashima Island (**248**), it used the museum for those visitors who could not get to the island; for example, because of bad weather or infirmity.

The Section of Culture and Tourism of the Nagasaki City Department of Economy is responsible for the running and funding of the museum. There are no members of staff, just one volunteer. The museum occupies about 90 square metres. The visitor figures increased from 16,580 in 2006 to 27,580 in 2014.

The exhibition was renewed in 2008 and again in 2011. The Japanese Meiji Industrial Revolution in the city of Nagasaki is examined, with the development from iron, steel and ship building to coal mining. The display does not, however, deal with the 26 camps that provided more than 13,000 prisoners from 16 different countries to five of these heritage sites (Kotler 2021; Palmer 2021). The museum has 220 panels of information, 30 photographs, a film about Hashima Island and a scale model of the island. A telescope is available for visitors to look at Hashima Island in the distance.

*References*

https://www.nagasaki-tabinet.com/guide/61067

http://www.at-nagasaki.jp/spot/61067/

personal communication, Naijima Daisuke, general office clerk, 2016

### 260. Nagasaki-shi Ikeshima Tankō Taiken Shisetsu (Nagasaki City Ikeshima Coal Mine Experience)

Ikeshima Island is located 30 kilometres to the north of Nagasaki. Like Hashima Island (**248**) it is situated in Nagasaki Prefecture of Kyushu. During WWII Ikeshima was identified as having large reserves of coal, extending out under the East China Sea. The Mitsui Matsushima Company purchased more than half of the island, which had a radius of only 600 metres and a circumference of only 4 kilometres. The name Ikeshima means pond island, but the pond from which it took its name was dredged to form the harbour, so a desalination plant had to be built in 1966 to provide fresh water. Originally there were several small islands, but the spoil from the mine was dumped in the water between them, so that they were eventually joined up to form one large island. Development work began in 1952, but coaling did not start until 1959, just as oil was beginning to displace coal. Ikeshima Kōdō (Ikeshima Mine) was the last deep mine to close in Japan, in 2001, caused by rising labour costs. At its peak of production, in 1970, there were 8,000 people living in large concrete apartment blocks, as on Hashima Island, but by 1971 there were only 720 left. Between 2002 and 2007 mining trainees were brought in from Indonesia, so notices are in both Japanese and Bahasa Indonesian.

Apart from a few apartment blocks by the shore, all the buildings associated with the mine are still intact. Because of the eerie atmosphere, the site is often used for cinema and television films. Guided tours are now offered by the residents of the island, which is reached by ferry (**248**). Visitors are shown the apartments by a local resident, including climbing to the top of one for a panoramic view of the island, entering a typical coal miner's house of the 1960s and visiting the second shaft for a demonstration of the machinery of that time.

Transport is then provided on an authentic mine car into the mine workings under the sea. The Mitsui Matsushima Risōshisu Kabushiki Gaisha (Mitsui Matsushima Resources Corporation) opened the underground to visitors in 2003. Since 2016 the underground experience has been the responsibility of the city of Nagasaki. There are six members of staff, including one guide for the above-ground tour and three underground guides, who are all former coal miners. The visitor numbers increased from 993 in 2011 to 5,057 in 2014, and are expected to increase to 8,600.

The underground experience on Ikeshima is the only place in Japan with a genuine original coal-mining underground. The mine is reached by an adit and the guides take visitors on foot through the 350 metres of roadways. Visitors are able to see underground machinery operating, such as a single-ended ranging-drum shearer and a drilling machine, as well as a safe haven (a fresh-air base for use in case of an explosion underground) and a static exhibition of a roadheader.

*References*


http://abandonedkansai.com/2011/07/13/ikeshima-in-the-morning/

http://archive.unu.edu/unupress/unupbooks/uu36je/uu36je0b.htm

http://en.japantravel.com/nagasaki/ruins-of-ikeshima-coal-mine-island/3965

http://ikeshima.info/

**Figure 260.1. Abandoned apartment blocks on Ikeshima Island (© Nagasaki-shi).**

**Figure 260.2. Coal-mining train hauled by a battery electric locomotive for transporting visitors into the mine (© Nagasaki-shi).**

**Figure 260.3. Interior room of a miner's dwelling (© Nagasaki-shi).**

https://en.wikipedia.org/wiki/List_of_coal_mines_in_Japan

https://spikejapan.wordpress.com/2010/08/31/ikeshima-goodbye-to-old-king-coal/

http://traverseworld.com/Nagasaki/Ikeshima.html

http://www.kanko-sotome.com/

personal communications, Yuta Fuji, industrial curator, 2012; Hirayama Tomohisa, chief investigator of the Planning and Finance Department Regional Development Section, Nagasaki City, 2016

### 261. Nagasaki-shi Takashima Sekitan Shiryōkan (Nagasaki City Takashima Coal Mining Museum)

The Nagasaki-shi Takashima Sekitan Shiryōkan is located on the eastern side of Takashima Island, which is 1.34 hectares in size, 15 kilometres south-west of Nagasaki in the East China Sea and 3 kilometres north-north-east of Hashima Island (**248**). It is situated in the Nagasaki Prefecture of Kyūshū. The museum opened in 1988 and renewed its displays in 2004. The museum is only a short walk from the depot for the ferry between Takashima and Nagasaki (**248**). It attracts about 70,000 visitors a year and there are two members of office staff. The museum occupies an area of 511 square metres and is managed and funded by the city of Nagasaki.

For hundreds of years people on Takashima are supposed to have collected coal from where it outcropped on the surface for use as a household fuel. A man called Goheita was said to have lit a bonfire on an exposed outcrop and so discovered its use as fuel, so coal was known as *goheita*. When transport links improved in the eighteenth and nineteenth centuries, Takashima coal began to be sold further afield, mainly to salt makers on the coast of the Seto Inland Sea. At the time Takashima Island belonged to the area administered by the Fukahori family (**248**), who took over the management rights and made the islanders become subcontractors and provide the labour force. Thus coal profits became a staple of the local economy, as was still the case when Japan opened up to the world from the late 1850s.

Nagasaki gained new importance as the closest port to China and a stopover for foreign commercial ships and naval vessels, at a time when sail-rigged ships were being replaced by steam-driven vessels, increasing the demand for coal. Nabeshima Naomasa (1815-71), lord of the Nabeshima clan, decided to increase production at Takashima Mine with assistance from a Scottish merchant, Thomas B. Glover (1838-1911). At this time coal was still simply gathered from where it outcropped. Glover imported modern mining equipment from the UK and hired British mining engineers to drill a vertical shaft on the island. Takashima Mine was the first in Japan to have a steam engine. In 1869 a coal seam was reached some 45 metres below the surface, thus creating the first modern coal mine in Japan and the first mine worked by roadways extending out from a deep shaft. This was one of the islands from which the present single island was created (**260**). The enormous success of Takashima Mine enriched Nagasaki with foreign currency and led to the development of mines on nearby islands, such as Hashima. The Mitsubishi Company was established as a shipping firm by Yatarō Iwasaki (1835-85). During the Pacific War (1937-45) the Nagasaki shipyard built 72 warships and 15,000 torpedoes (Palmer 2021, p. 7). In 1881 the company acquired Takashima Mine, using the coal to power their fleet of modern steamships, and working the pit until its closure in 1986.

The museum collected machinery from the mines. On the ground floor there are photographs of Takashima Mine in the past, models of the mine and the underground adit network, films and bricks of konjac, or glucomannan, a starchy food used in local cooking. The archives and library are also situated on the ground floor. On the first floor there are fossils, items from the coal-mining trade union and about the life of people of Takashima and a section on Hashima Colliery, including the shrine on Hashima Island (**248.3**). In the outer courtyard of the museum there is a large-scale model of the whole of Hashima Island, including Hashima Colliery, the accommodation for the workforce and the dolphin dock.

*References*

http://archive.unu.edu/unupress/unupbooks/uu36je/uu36je0b.htm

http://media.virbcdn.com/files/a7/FileItem-11144-HashimaTheGhostIsland.pdf

https://en.wikipedia.org/wiki/List_of_coal_mines_in_Japan

https://www.pref.fukuoka.lg.jp/uploaded/attachment/118177.pdf (p. 4)

http://www.aya.or.jp/~marukimsn/kikaku/2013/2013sakube.html

http://www.city.nagasaki.lg.jp/kanko/820000/828000/p000836.html

personal communications, Kuriwaki Yoshiro, director, Nagasaki City World Heritage Promotion Office, 2012; Kurata Noriko, Nagasaki City Office, 2016

### 262. Nishishoro Komyuniti Sentā-nai Sekitan Shiryōshitsu (Coal Museum in the Nishishoro Community Centre)

The Nishishoro Komyuniti Sentā-nai Sekitan Shiryōshitsu is located in Kushiro in eastern Hokkaido. The museum opened in 1986. It occupies one room, 66 square metres in size, in the Nishishoro Komyunitisentā (Nishishoro Community Centre). The museum had 41 visitors in 2015.

A large piece of coal is displayed beside the entrance to the museum, which is ringed with a blue imitation of an arched girder. There are displays of magneto-powered

**Figure 261.1. Large-scale model of the whole of Hashima Island in outer courtyard of the museum, dolphin dock to the left (© Nagasaki-Shi).**

telephones, gas-warning indicator, workwear and miners' personal equipment, including cap lamps. There are also hand tools used underground, ranging from a bamboo coal-sorting riddle, carrying baskets, early picks and rakes, through to modern compressed-air borers and picks, and heavy-duty drills for boring shotfiring holes. Testing for gas underground is dealt with, for example, with hand-held methanometers. The major exhibit is a reconstruction of an advancing roadway with a life-size model of a coal miner using a compressed-air pick beside panels giving instructions for the layout of the working areas and support rules for the roof of the roadways.

*References*


http://blog-imgs-33-origin.fc2.com/g/a/b/gabinshiranuka/DSCF0228.jpg

http://blog-imgs-33-origin.fc2.com/g/a/b/gabinshiranuka/DSCF0230.jpg

personal communication, Shōji Yukinori, General Planning Department, Shiranuka Town Hall, 2016


### 263. Nōgata-shi Tankō Kinenkan (Nōgata City Coal Mining Memorial Museum)

The city of Nōgata is located in the north of Fukuoka Prefecture, which is situated on the northern shore of Kyushu. The Nōgata-shi Tankō Kinenkan opened in 1971 to preserve the history of coal mining in the region, as the number of mines in the Chikuhō coalfield (**253, 273**) decreased. It is operated by the Kōeki Zaidanhōjin Nōgata Bunka Seishōnen Kyōkai (Nōgata Youth Cultural Association), a not-for-profit foundation established by the Board of Education of the city of Nōgata.

The museum occupies 4,119 square metres of land, with 665 square metres of buildings. The main building, which is built in the traditional style in wood, measures 246 square metres, the steel-framed annexe is 369 square metres and the brick coal-chemical house is 50 square metres. At the back of the museum building is a surface gallery, with a re-creation of a typical underground roadway, used for training mines rescuers. The museum attracts about 10,000 visitors a year. Free open days are held on topics related to coal, such as combustion experiments. It employs five staff, although there are a further 150 volunteers, who work mainly on maintenance of the steam locomotives.

Coal-mining equipment and details of the history of the Chikuhō coalfield are displayed in the three buildings, including the largest piece of coal in Kyushu, weighing 1.8 tonnes (*cf.* **254**). Large-scale mining machines are displayed at the front of the museum, together with two steam locomotives formerly used by Japan National Railways. Behind the museum are also several *metasequoia glyptroboides* trees (dawn redwoods), which are native to Lichuān in the Chinese province of Húběi. Until their rediscovery, these were known only from fossils and were believed to have become extinct in the Miocene period. The fossils were first described in 1941 by Shigeru Miki (1901-74), a Japanese paleobotanist at Kyoto University. A living specimen was found in 1943, forming part of a local shrine in Lichuān. In 1946 the Chinese professors Zhèng Wànjūn (1908–87) and Hú Xiānsù (1894-1968) connected this tree to the fossil remains. In view of the threat of deforestation to the existence of the small number of trees identified, Harvard University collected seeds in 1948 and distributed these to arboreta around the world. Local people at the shrine in Lichuān called it *shuǐshān* (water fir), but in Japan they are known as coal trees, because coal seams often derive from compacted metasequoia trees.

*References*

http://en.wikipedia.org/wiki/N%C5%8Dgata,_Fukuoka

https://en.wikipedia.org/wiki/Metasequoia_glyptostroboides

personal communication, Yoshiyuki Yamamoto, member of staff, 2015

### 264. Numata-chō Tankō Shiryōkan (Numata Town Coal Mining Museum; formerly Numata-chō Tankō Shiryōkan ((Numata-chō Furusato Shiryōbunkan)): Numata Town Coal Mining Museum ((Numata Hometown Annexe Museum))

The Numata-chō Tankō Shiryōkan is located in Numata-machi in the Rumoi coalfield of the district of Uryū on Hokkaido. The Numata-chō Nōgyō Shiryōkan (Numata-chō Agricultural Museum) opened in 1984. In 1999 the name was changed to Numata-chō Tankō Shiryōkan (Numata-chō Furusato Shiryōbunkan). In 2010 part of the displays were transferred to the Hotaru Gakushūkan (Firefly Museum), which deals with the history of the town of Numata. The building of the old Numata-chō Furusato Shiryōkan was closed and the Numata-chō Tankō Shiryōkan opened.

The museum occupies an area of 553 square metres and in 2016 had 1,511 visitors. A detailed exterior view of the museum can be seen on Google Maps. There are no staff allocated specifically to the coal-mining annexe of the museum, but there are four permanent staff whose duties include responsibility for this annexe. The museum is run by the Numata Kyōiniinkai (Numata Board of Education). It does not have a website.

The museum deals with coal mining in the region. A diorama shows the surface of Shōwa Colliery in the mid-twentieth century, based on the landscape around Numata about 50 years ago. The model shows a coal-preparation plant linked to the rail system and a large supply of roof-support timber props. Another diorama depicts the conditions underground. The site of Shōwa Colliery survives as a derelict ruin in the mountains, shown on the livedoor blog.

Tools used underground in coal mining are exhibited, including three self-rescuers, showing the developments in design during the 1960s and 1970s, and a larger Kawasaki breathing apparatus that would have been used by mines-rescue teams in emergency situations underground. A single-shot exploder using a shotfiring key to charge it and the button to fire the shot would have been used for typical driving of roadways underground. Safety equipment used by miners underground include a yellow Japanese safety helmet of a smooth rounded shape with a small brim. The electric cap lamp that would be used with the helmet has an alkaline battery made by the Japan Storage Battery Company. This company began production in 1895 as the first lead-acid battery manufacturer in Japan and produced their alkaline range from 1938.

Photographs on display show a range of mining images, including mines-rescue men in overalls, an underground locomotive dating to around the 1950s and surface locomotives, including passenger trains, presumably for the workers. The locomotives, one of which is displayed outside the museum, were vital for transporting the coal from the relatively isolated mine in the mountainous area around Numata. The display has an image of women sorting the coal from the dirt in the screens building (*cf.* **244.2, 378**). Photographs cover the history of three coal mines, including the Shōwa Mine, and a case contains the banner of the Tomoko Seidō (see **275** for the Tomoko organisation) of Shōwa Colliery. This has crossed picks underneath the name of the mine. To the right is written *Shōwa Tankō Ryō Tomoko Ichidō*, meaning something like 'All of us, both Tomoko [organisations] of Shōwa Colliery'.

*References*

http://blog.livedoor.jp/darker916/archives/12123870.html

http://www.town.numata.hokkaido.jp/section/kyouiku/ujj7s30000001o7e.html

personal communication, Shinohara Satoshi, curator, 2017

### 265. Ōmuta-shi Sekitan Sangyō Kagakukan (Ōmuta Coal Industry and Science Museum)

The Ōmuta-shi Sekitan Sangyō Kagakukan is located close to the centre of Ōmuta in Fukuoka Prefecture, in Kyushu on the shore of the Ariake Sea. The coastal location made Ōmuta ideal for shipping coal out from the local mines, together with the large-scale mining machines produced in factories in the city. Coal was so important to Ōmuta that a large piece is on prominent display at the entrance to the railway station. The museum occupies a modern purpose-built structure, with powered roof supports on display outside. The ceiling of the entrance has replica coal and stone seams above the reception desk.

The first section of the displays deals with how coal was formed and then the methods of winning the coal. These range from early hand-getting, after coal was first found here in 1469, through to modern machinery. Cutaway models, with explanatory panels above, show various aspects of typical twentieth-century Japanese collieries, with winding systems, conveyors for moving the coal and pumping and ventilation arrangements. One, for example, shows the surface of a typical coal mine of the mid-twentieth century with vertical-winding and drift-conveyor methods of mineral transport and pumping and water arrangements. Panels outline the development of Ōmuta as it grew to be Japan's largest coal-mining city. In 1873 all the collieries in the Miike coalfield were nationalised in order to introduce modern technology and the largest colliery, Miike Mine (**255, 265**), sunk in the 1860s, was named at that time. Convicts were used as labour in the mine from 1873 until 1931 (Takashi 2017). Ōmuta reached its peak in 1959 and then declined as industry switched to oil, although Miike Mine survived until 1997. The many roles played by coal are explored, from simple lamps to fuel for generating electricity.

The second section focuses on understanding modern technology. Full-scale dioramas with large modern coal-mining machines are displayed; for example, an underground locomotive and the driving of an underground roadway using a continuous miner, with models of coal miners alongside. Smaller displays deal with lighting, including flameproof underground items such as a klaxon, an electric signal bell, faceline lighting and surface lighting. Complementary to these displays is a video of the headframe of Miike Mine (**255**), and the railway from the colliery to Miike port, from which the coal was shipped. The third section of the museum deals with general science, mainly with interactive experiments aimed especially at schools. Eight of these cover the question of energy, including the energy supplied by coal and other aspects, such as electromagnetic waves.

The museum has an archive of videos of the testimonies of people involved in the Miike Mine. A panel introducing the historically important sites is on display, alongside a model of a local colliery's winding installation with headgear and winding-engine house, made by local people so that children could understand the mechanism of the headgear and the winding of the cage in the shaft.

*References*

http://www.sekitan-omuta.jp/

personal communications, Emiko Gohonmatsu, director, 2012; Nakano Hiroshi, planning staff, 2021

### 266. Sasebo-shi Sechibaru Tankō Shiryōkan (Sasebo City Sechibaru Coal Mining Museum)

The Sasebo-shi Sechibaru Tankō Shiryōkan is located in the city of Sasebo in the prefecture of Nagasaki in the northern-western part of Kyushu. The museum opened to the public in 1978. It does not have any members of staff and is run by two volunteers. It occupies 221 square metres and attracted 627 visitors in 2019. The costs of running the museum are borne by the Sasebo-shi Kyōiku Iinkai (Sasebo City Board of Education).

The museum occupies the former office building of the limited-liability partnership of the Matsuura Colliery, constructed in 1912 of sandstone excavated from the nearby river Saza and in use until the closure of the mine in 1970. This is the only stone building constructed in the western-European style in the north of Nagasaki Prefecture. As such, it has been designated as part of the prefectural cultural heritage under the classification of important mining heritage.

The exhibition focuses on the history of Matsuura Colliery. The museum displays machinery used in the mine, such as an hydraulic-powered roof support, as well as photographs. The collections include fossils found during operations in the mine and mining tools from the period before mechanisation. Near the museum is a pithead from 1906, with a coalface supported with replica wooden pit props.

*References*

http://www.city.sasebo.lg.jp/benrimap/shisetsu/kyoiku/073.html

personal communications, Taro Nagamoto, superintendent, Sasebo City Board of Education, 2012; Kawachino Atsushi, chief of Cultural Properties Division, Sasebo City Board of Education, 2016, 2021; Shinya Nishimoto, superintendent, Sasebo City Board of Education, 2021

**Figure 265.1. Large piece of coal on display on Ōmuta railway station (© Margaret L. Faull).**

**Figure 265.2. Exterior view of entrance to the museum (© Margaret L. Faull).**

**Figure 265.3. Coarse-scale model of a local colliery's winding installation with headgear and winding-engine house (© Margaret L. Faull).**

Figure 266.1. Exterior of the museum (© Sasebo-shi Kyōiku Iinkai).

Figure 266.2. Example of a coal face with wooden pit props supporting the roof (© Sasebo-shi Kyōiku Iinkai).

### 267. Sorachi Yama no Kioku Manejimento Sentā (Sorachi Coal-Mine Heritage Management Centre)

Iwamizawa is the capital of the subprefecture of Sorachi in Hokkaido. The Sorachi Yama no Kioku Manejimento Sentā opened in 2009 with the purpose of the regeneration and revival of the Sorachi coal-mining region by using its local coal-mining history and providing an information centre for the coal-mining heritage of the region. It functions as a regional think-tank and also organises the planning and maintenance of the Sorachi Coal Mine Heritage Route (**268**).

The centre is located in a 1927 stone building with a 1909 warehouse attached. It opened in 2009. There were 3,670 visitors in 2010, rising to an average now of about 5,000 a year. It is run by a not-for-profit organisation, which raises funding from donations, membership fees and income from various sources, such as projects. There are two members of staff and about 20 volunteers. The centre occupies some 200 square metres, with books, booklets, files and leaflets, as well as some coal-mining artefacts on display. The former warehouse is employed as an area for temporary exhibitions. The warehouse can also be used for conferences, live piano performances and private hire to raise funding. As another way of raising funds, in a separate area visitors can buy drinks, attracted by the free wi-fi, which is still rare in Japan.

*References*

http://www.mc.soratan.com/

http://www.soratan.com/

personal communication, Yoshioka Hirotaka, board chair, 2016

### 268. Sorachi Yama no Kioku no Rūto (Sorachi Coal-Mine Heritage Route)

During his journey in 1857 through Hokkaido, the explorer Matsuura Takeshirō (1818-88) identified coal seams on the banks of the river Sorachi near Akabira. In 1868 carpenter Kimura Kichitarō discovered coal in Horonai, Mikasa. Six years later the mining engineers, Benjamin Smith Lyman (**277**) and Takeaki Enomoto (1836-1908), prepared a report showing that there were coal seams here between 3 and 7 metres thick. As a result the Meiji Government decided to sink the first coal mine of the Ishikari coalfield in Horonai. In 1889 the government sold the mine and its railways to Hori Motoi (1844-1912), who founded the Hokkaidō Tankō Tetsudō Kaisha (Hokkaido Colliery and Railway Company), abbreviated as Hokutan.

At its height, the Sorachi region was one of the largest coal-mining areas in Japan, with some 100 mines. Later cheap imported coal and the shift from coal to oil led to mine closures until all of Sorachi's coal mines had disappeared. Today the region is still characterised by a variety of reminders from the era, ranging from aspects of local lifestyles to production facilities, effectively forming an open-air coal-mining museum. Many aspects of coal-mining culture also remain, such as the Akabira Fire Festival and the Hokkai Bon Dance.

In 1974 the outcropping coal seam at Yūbari (**277**) was designated as a natural monument of Hokkaido. Colliery-related facilities and aspects of the Sorachi culture and lifestyle were given designation in 2001 by the Hokkaido Heritage Concept Promotion Council as part of the heritage legacy of Hokkaido to be preserved for future generations. A total of 52 properties had been designated as of 2012. The management of the Sorachi Yama no Kioku no Rūto is carried out by the Sorachi Yama no Kioku Manejimento Sentā (**267**). The route was inspired by the designated route in the German Ruhr (**135**), but does not have an official name. The entries below deal in alphabetical order with some major points on the tour that do not have separate entries in this publication.

*References*

http://blog.goo.ne.jp/sorachitabinikki

https://ja-jp.facebook.com/zuriyama

personal communication, Yoshioka Hirotaka, board chair, 2016

### 269. *Akabira*

The explorer Matsuura Takeshiro (**268**) recorded the presence of coal seams in Akabira in 1857. There are still many colliery-related structures in and around the former coal-mining cities of Akabira and Ashibetsu. These include the remains of the Hokutan Kitasumi Akama coal-preparation plant and of the headgear and winding-engine house of the Sumitomo Akabira Colliery. The mine, 600 metres deep and constructed in 1963 at a cost of 2 billion yen, closed in 1994. The headgear is 43.8 metres high. The Tantan Society operates tours in and around the mine several times a year, including the interior of the winding-engine house. The remains of the pithead baths, which were separated from the headgear by being located on the opposite side of the road, were originally reached by an underground passage. There was a dressing room and five tubs using the nacelle system, two large bathtubs, 20 metres in length, and three small ones. The spoil heap of the mine, with an average slope of 18 degrees, survives as a 130-metre-high outlook point above Akabira, reached by the Zuriyama stairs with 777 steps. The development, operation and closure of the mine are dealt with at the Akabira City Historical Museum.

*Reference*

https://ekinavi-net.jp/destinations/akauta/

### 270. *Kamisunagawa*

The first coal mining in Hokkaido developed along the river Penkeutashinai, which runs through the town of Kamisunagawa (see **251** for the museum in Kamisunagawa). After the closure of the Mitsui Sunagawa

**Figure 267.1. Exterior of the Sorachi Yama no Kioku Manejimento Sentā (© NPO Hōjin Yama No Kioku Suishin Jigyōdan).**

**Figure 268.1. Outcropping coal seam, 7 metres thick, at Yūbari, designated as an Hokkaido natural monument in 1974 (© NPO Hōjin Yama No Kioku Suishin Jigyōdan).**

**Figure 269.1. Headgear and winding-engine house of Sumitomo Akabira Colliery (© NPO Hōjin Yama No Kioku Suishin Jigyōdan).**

**Figure 270.1. Headgear and associated building of Mitsui Sunagawa Colliery in Kamisunagawa (© NPO Hōjin Yama No Kioku Suishin Jigyōdan).**

Colliery in 1987, the headgear and shaft were used for zero-gravity experiments. After these ended in 2003, the headgear and the associated building were preserved as industrial monuments.

*Reference*

https://images.search.yahoo.com/search/images;_ylt=A2KLfSwed69gF5IAQktXNyoA;_ylu=Y29sbwNiZjEEcG9zAzEEdnRpZAMEc2VjA3Nj?p=Mitsui+Sunagawa+Colliery&fr=mcafee

### 271. *Mikasa*

Mikasa was the location of the first railway in Hokkaido (and only the third in Japan) constructed specifically for moving coal from the mines. The first colliery in Hokkaido, the Honorai Mine, was sunk by the state company in 1879. In 1966 a two-level cage was installed that could transport 50 miners each time, and a new shaft, 915 metres deep, was sunk as an air intake. The mine closed in 1989. The winding-engine house and the intake shaft were dismantled in 2004 and all the winding machinery removed. Also in 2004 the Hokutan Horonai Coal Mine Landscape Park was established, containing a spoil heap, with another spoil heap and the mine headgear on the other flank of the mountain about 1 kilometre away.

The Sumitomo Ponbetsu Mine was sunk in Mikasa in 1880. In 1960 what was claimed to be the largest shaft in Asia was constructed, with a headgear. This was 51 metres high, above a shaft 750 metres deep with a diameter of 6.4 metres. The mine operated a system of skip winding. It closed in 1971. The derelict headgear, fan building, screens building, winding-engine house and other colliery buildings, all constructed of concrete, still survive.

The Mikasa Shiritsu Hakubutsukan (Mikasa City Museum), which is 1,295 square metres and opened in 1979, has one gallery devoted to the history of the coal mines of the area. The museum has six staff, supported by 16 volunteers, and attracts 32,000 visitors a year. The museum displays miners' tools and equipment used in the Hokutan Honorai and the Hokutan Ikushunbetsu Mines.

*References*

personal communication, Manabu Kano, curator, Mikasa City Museum, 2016

www.city.mikasa.hokkaido.jp/museum

### 272. *Utashinai*

The Hokutan Sorachi Colliery (**276**) at Utashinai, which had a depth of 248 metres, was founded in 1890 and the railway to Iwamizawa opened the following year. The mine company had not, however, yet prepared a shrine to protect the miners. After constant accidents in the pit, in 1894 the miners set up stones that they found in the mine, in which the goddess could live and before which they could pray for safety.

In Japan's oldest history and sacred book *Kojiki* (records of ancient matters), completed in 712, there is a myth that two gods, Izanagi and Izanami, cooperated to give birth to the land of Japan and then to the gods related to nature. One of these was Ōyamazumi no Kami, the god of mountains, sea and war. There is, however, no myth in which Ōyamazumi no Kami plays the leading role, and his daughter, Konohanasakuya-hime, is more famous as the goddess of Mount Fuji. Ōyamazumi no Kami has been worshiped as a guardian deity of mining since the time of the metal mines of the Edo period (1600-1868). In 1899 the Hokutan company (**268**) built a shrine dedicated to Ōyamazumi no Kami, his wife Kaya-no-hime (the goddess of vegetation and protector of fields) and Ōkuninushi, his descendant. A *kami-dana* (a miniature altar to worship a god at home or at work) was also set up on the surface next to the shaft at the deployment station, where the miners of each of the three shifts were assigned their tasks for their shift. The mineworkers prayed there every day for safety before entering the mine. When the new shrine was made, the stone where the goddess dwelt broke into two, perhaps because she thought that her role had finished.

The colliery festival of the Utashinai shrine, known as the Tanzan festival, was held every year on 11 and 12 May, and an *Omikoshi* (a portable Shinto shrine) was paraded around the residential area of the colliery. Employees of the coal mine were given a holiday on 12 May. Many stalls were lined up near the shrine and crowds of people from the neighbourhood gathered at the festival. Now that there are no more colliers, there is nobody to carry the *Omikoshi* and there are no more stalls; the festival survives only as a ceremony at the shrine, which is now run by a religious corporation.

There are still stones in the precincts of the Utashinai shrine that are said to contain the name Sanjin (Mountain Goddess), although the inscription has weathered away. The heapstead building also survives.

*References*

https://power-spot.me/oyamazuminokami/

personal communication, Sakuma Atsushi (**276**), 2021

### 273. Tagawa-shi Sekitan Rekishi Hakubutsukan (Tagawa City Coal Mining Historical Museum)

The city of Tagawa lies close to the centre of Fukuoka Prefecture, which is situated on the northern shore of Kyushu. The Tagawa Coal Commemorative Park, with a total area of 4.8 hectares, was created to stress the important role that coal played in the development of the region, highlighted by a statue of a coal miner. In the park and behind Tagawaita station is the site of the former Mitsui Tagawa Colliery, on which the Tagawa-shi Sekitan Rekishi Hakubutsukan is located.

**Figure 271.1. Headgear of the Hokutan Honorai Mine in Mikasa (© NPO Hōjin Yama No Kioku Suishin Jigyōdan).**

**Figure 271.2. Derelict headgear and winding engine house of the Sumitomo Ponbetsu Mine in Mikasa (© NPO Hōjin Yama No Kioku Suishin Jigyōdan).**

**Figure 272.1. Surviving buildings of Hokutan Sorachi Colliery at Utashinai (© NPO Hōjin Yama No Kioku Suishin Jigyōdan).**

**Figure 272.2. Shrine from the Hokutan Sorachi Colliery at Utashinai (© Sakuma Atsushi).**

The Tagawa Coal Reference Library, specialising in the history and business of coal, was established in a modern building, of 3,400 square metres, in 1983. In 2005 this was transformed into the Tagawa-shi Sekitan Rekishi Hakubutsukan. A retired secondary-school history teacher, Tatsuo Asō, who had written a history of coal mining in Kawara, Kawasaki, Soeda and Tagawa, was appointed as curator in 2006. His first main task was ensuring the preservation of the physical remains of the coal-mining industry, including the large machinery used in the mines.

The museum now employs 13 staff. One display room deals with the local history and folklore of Tagawa district, while the other two rooms concentrate on coal mining. The first explains the origins of coal and how it has been mined. This is illustrated by digging tools, mining machinery, a model of the Mitsui Tagawa Colliery and a diorama showing underground workings with wooden roof supports and a woman carrying coal in a basket on her back (*cf.* **250**). There is also information about the living conditions of the coal miners and videos of the coal miners who worked at Mitsui Tagawa Colliery and about the history of the Chikuhō coalfield (**253**). The other room deals with paintings and literary works inspired by the lives of coal miners, including the arduous labour in the mines and the accidents suffered by the workers. There is a reproduction of a coal miner's house during the early-

Showa period (from 1926), as well as a steam train that used to pull waggons of coal.

Behind the museum building are the two remaining chimneys from the Mitsui Tagawa Colliery, standing 45.45 metres high. The steel headgear of the Ita shaft of the mine has been re-erected nearby. With a height of 28.4 metres, completed in 1909 in a form used in the British coalfields, it is the only surviving headgear in the Chikuhō coalfield. The chimneys and the headgear are often described as the symbols of the city of Tagawa and of the Chikuhō coalfield. All three were registered as national tangible cultural properties in 2007. Also registered earlier, in 1996, were the 584 historic documents and illustrations of the Yamamo to Sakubei Collection showing the life of the mine and of the mineworkers during the Meiji period (1868-1912). These are owned by Fukuoka Prefectural University and held by the museum. In 2011 this collection was registered by UNESCO as the first example in Japan of a collection of World Memory Heritage status on the basis that the collection describes Japan at a time when most surviving records are official documents. So a private record created by one of the workers is very rare.

An attempt was also made to have the Chikuhō coalfield, which accounted for half of the coal production in Japan and so supported the modernisation of Japan, registered on the UNESCO World Heritage List. But so little now survives, and the Ita shaft headgear is not in its original position, that the bid was unsuccessful.

Tagawa is proud of its coal-mining heritage. Nakamura Museum, with a collection of some 1,300 pieces of ceramics, glasswork, inkstone, sculpture and traditional and modern paintings, is presented as 'the museum of artwork from the period of coal mining'. Since 2006 a Tagawa coal-mine festival has been held every year outside the museum. Some 2,000 people participate in the Tankōbushi Sōodori, which is described as a traditional local coal dance. It is followed by the Tankōbushi parade and celebratory fireworks.

*References*

https://blog.goo.ne.jp/tchm-sekitan/c/df2e409d4d8d0686e0a29ae135aaf09b

personal communication, Kazumi Asagarasu, curator, 2015

### 274. Tankō to Tetsudōkan, Yūdzuru-eki (Coal-Mine and Railway Hall, Yūdzuru Station)

The Tankō to Tetsudōkan,Yūdzuru-eki is located in Kushiro, the capital of the sub-prefecture of Kushiro, on the south-eastern coast of Hokkaido. The Yūbetsu railway line was opened in 1921, branching off the Sekihoku main line to run south-east from Nayoro to Kitami. This branch line served Yūbetsu Colliery.

After the line closed, the Akan Town Folk Museum opened in 1972 on the site of the former station of Akan. In 1987 it moved to Tanchō no Sato, also in the Akan region, with the revised name of Tankō to Tetsudōkan, Yūdzuru-eki, and reopened in 1988. The museum occupies about 170 square metres, including the offices for the manager who runs the museum. It is open annually from late April until the end of October, and attracts about 1,500 visitors annually. The museum is owned by the city of Kushiro and is funded and managed by a public-private partnership.

The intention of the museum is to present the history of Yūbetsu Colliery, as coal mining was the core industry of the town of Akan between 1919 and 1970. Akan village was granted the status of a town in 1957, but was merged into the expanded city of Kushiro in 2005. As well as mining tools, the museum displays material relating to memories of the town itself and of the Yūbetsu railway line.

The geology and types of coal in the region are explained, with coal and coke test samples alongside a collection of core sections in individual cases. A strata plan above these shows the results from an analysed core section. Two models show the full gear worn by miners of boots, cap lamp, helmet and self-rescuer, and by mines-rescue staff, including breathing apparatus with full mask. Material for breaking ground is illustrated by compressed-air tools, such as long drill rods with a compressed-air drill and air leg to show how drilling and firing was used to drive roadways underground. This is complemented by a photograph of a miner driving a trepanner coal-cutting machine. In the centre of this display is a white banner with the red logo of the company, the Yūbetsu Tankō Tetsudō; this uses the letters YC from the English translation of the name, Yūbetsu Coal Railway.

Other photographs show a miner with a bandaged hand eating rice during a break underground and scenes of Yūbetsu Colliery, built around the base of a hill, with the railway leading to the large coal-preparation plant. A boiler house with a very tall chimney stood in the centre of the colliery, surrounded by mine buildings such as workshops. A walkway from the mine led over the river to the adjacent hill.

Displayed under open cover outside is locomotive 65, a class C11 2-6-4 steam locomotive. The Japanese Government and Japanese National Railways built 381 examples of this type of locomotive between 1932 and 1947, based on the C10 2-6-4-type locomotive manufactured in 1930. The code C11 65 refers to the numbering system introduced when Japan's railway system was nationalised in 1907 and updated in 1928. The letter refers to the number of axles, so C means three axles. The first number is the class of locomotive, with 10-49 designating tank locomotives and 50-99 tender locomotives. Therefore class C11 engines were a specific class of three-axle tank locomotives. Locomotive 65 was produced by Kawasaki Heavy Industries at their Hyogo works in 1935 (manufacturer's number 1533). This locomotive appears to have operated first on mainline work at Sakura near Tokyo, then at

**Figure 274.1. Exterior of the Tankō to Tetsudōkan, Yūdzuru-eki (© Kushiro-shi).**

**Figure 274.2. Locomotive 65 displayed under cover outside (© Kushiro-shi).**

Katsuura, Shizuoka and Takaoka, before being transferred to the colliery railway in 1961.

The importance of locomotives to collieries is emphasised in the displays. An image shows locomotive 65 being lifted by crane, possibly when it was being delivered to the colliery. The YC company logo is displayed again, both on a banner and as a cast-metal plaque. The exhibition continues with a series of shunters' lamps, including an acetylene/carbide lamp between three electric lamps and two railway lanterns. Above them are nameplates from locomotives, including two from America made by the Baldwin Locomotive Works, Philadelphia, in 1898 and 1918. There is also a plaque of a C56 1001 locomotive. The number of C56s (steam-tender locomotives) produced only went up to 164, so the number 1001 comes from the numbering system used by the Japanese Government following the nationalisation of the railways in 1909.

*References*

http://c5557.photoland-aris.com/C1165.htm

http://hokkaidobicycle.web.fc2.com/kankou/29_19.html

https://en.wikipedia.org/wiki/Japan_Railways_locomotive_numbering_and_classification

personal communications, Komura Hiromi, chief planning clerk, Tourism Promotion of Akan Town, Akan-chō Kankōshinkōkōsha Public Corporation, 2017; Mark Carlyle, curator of industry, NCMME, 2017

### 275. Ube-shi Sekitan Kinenkan (Ube City Coal Mining Memorial Museum)

Ube is located in Yamaguchi Prefecture, which is on the south-west side of Honshu. Ube lies on the Seto Inland Sea, which meant that coal could be shipped out easily from its port. All mines of the Ube coalfield had closed by 1967. In 1969 the first Ube-shi Sekitan Kinenkan was created with donations from citizens, the city, mineworkers and the prefecture. The headgear on the site originally served the Higashi Mizome Colliery. It has been painted red and white and has an observation deck at the 37-metre level. The headgear now functions as the symbol of the museum.

The museum occupies a two-storey building of reinforced concrete with a total floor area of 942 square metres. It is under the direct management of the municipality of Ube. Approximately 30,000 people visit each year. The museum employs one curator and has support from about 50 volunteers.

The displays include equipment, maps, photographs, workers' clothing and a large piece of coal. The technology of coal mining is explained using a reconstruction of a colliery underground in the basement. This demonstrates how the roof has been supported at different eras, from wooden pit props through to steel powered supports. The living conditions of the workforce are illustrated with a reconstruction of the interior of a typical coal miner's house. The museum has a collection of about 3,000 items, ranging from large machinery to fossils and folklore elements. Large machines and a steam locomotive are displayed outside the building.

*References*

https://commons.wikimedia.org/wiki/Category:Ube_Coal_Museum

personal communication, Nakamura Tomoyo, curator, 2012

### 276. Utashinai-shi Kyōdokan Yumetsumugi (Utashinai City Local Municipal Museum for Spinning of Dreams)

The Utashinai-shi Kyōdokan Yumetsumugi is located in the city of Utashinai in Sorachi district, at the north-eastern end of the Ishkkari Plain in approximately the centre of Hokkaido. The Hokutan Sorachi coal mine (**272**) closed in 1995, the fourth-last underground coal mine in Japan to do so, and the museum opened in 1997. The building has three floors with a total area of 1,600 square metres. There is one member of staff, supported by 70 volunteers. The museum is managed and funded by the city and in 2020 attracted 1,599 visitors.

The economy and history of Utashinai were based on the coal-mining industry. The Utashinai-shi Kyōdokan Yumetsumugi museum was established by the city to commemorate this heritage and to celebrate the centennial anniversary of Utashinai achieving city status, despite having the smallest population of any city in Japan. The museum displays methods of coal production, the lives and jobs of coal miners and the tools used in the many coal mines of Utashinai. There is a small simulation made by an experienced coal miner where visitors can experience the darkness of a mine. There are also bird-cage lockers, which were open baskets in which the miners' clothes were stored. The cages were suspended from the ceiling by ropes. The remainder of the exhibitions deal with utensils from the miners' homes and a model of a miner's tenement home of the Taishō era (1912-26), built by the Hokutan Sorachi Colliery.

Of special importance are the original manuscript and illustrations of the novelist Takahashi Kiichirō (1928-2007) for *Nobuyo* (1978), for which he won the Akutagawa Prize, which is one of the most prestigious literary prizes in Japan. The previous year he had been a candidate with *Kannon Riki Shissō*, a literary work describing the life of a coal miner. On display are illustrations of scenes that Takahashi Kiichirō saw as a child, of a coal miner wearing a cap with a cap lamp, a coal-mine tenement house and a line of mine tubs filled with coal under covered accommodation.

The museum also possesses original documents and the banner of the Tomoko organisation of the Hokutan

**Figure 276.1. Exterior of the Utashinai-shi Kyōdokan Yumetsumugi (© Utashinai-shi Kyōin Iinkai).**

Sorachi Colliery. The Tomoko organisation was a nationwide mutual society of coal miners, similar to the German Knappschaftskassen (**297**). This developed from nineteenth-century recruitment practices, whereby mines used to hire a *hambakashira* (foreman), who then hired Tomoko workers in their turn. Most coal miners during the Edo period (1603-1868) died before they were 30 years of age of the occupational disease known as *yoroke*, literally 'staggering', but actually pneumoconiosis (**92**). The basic principle of Tomoko was to establish a parent-child relationship of a young miner with an older collier in order firstly to pass on mining techniques and secondly to provide support for each other. This was known as the Tomoko alliance. The younger miner lived with his 'parent' and obeyed his orders, rather than those of his natural parents and the younger miners supported the older miners in their old age as one might a parent. The 'child' constructed the tomb for the 'parent', whose name was engraved on the front of the tomb and the name of the 'child', on the back.

The Tomoko system effectively licensed miners as being competent in their skills (**249.2**). There were two kinds of Tomoko members: *Watari Tomoko*, a group of miners who went to various coal mines and were highly skilled, and *Jikofu* or *Ji-Tomoko*, miners who worked in just one area.

Mechanisation of the mines was introduced to deal with the effects of the recession following WWI. This eliminated the need for such skilled workers and so the practice of licensing miners. The first labour trade union, Yuaikai, founded in 1912, had a coal-mining section. The miners at the Mitsui Miike coal mine (**255, 265**) were defeated by the company after a five-month strike in 1912 and again after a four-month strike in 1960. With the development of the trade unions and of the social-security system in the 1920s, Tomoko became a shell of its former self, with a focus only on family-style relationships.

In Utashinai a local dish called *nanko* was introduced by miners of the Akita Prefecture during the Meiji era (1868-1912). It was not customary to eat meat in Japan. But in the Akita Prefecture, silver and copper mining were very important to support the finances of the lord. So he recommended eating nutritious horse meat to build up the strength of the miners and because it was believed to be a remedy against *yoroke*. Work for pit ponies underground was very heavy and their life expectancy was short, after which they were eaten. At some point the custom changed to eating the large intestine of a horse, cut into bite-sized pieces, simmered for a long time and then seasoned with miso (fermented soybeans). *Nanko* is still eaten on special

**Figure 276.2. Example of bird-cage lockers that would have been suspended from the ceiling to store the miners' clothes and possessions while underground (© Utashinai-shi Kyōin Iinkai).**

occasions by some households, such as a gathering of relatives.

*References*

https://en.wikipedia.org/wiki/Akutagawa_Prize

https://www.tripadvisor.com/Attraction_Review-g1122378-d10448024-Reviews-Local_Museum_Yumetsumugi-Utashinai_Hokkaido.html

personal communications, Sakuma Atsushi, chief examiner of the Social Education Group, Utashinai City Board of Education, 2016, 2021

### 277. Yūbari Sekitan Hakubutsukan (Yūbari Coal Mining Museum)

Coal was discovered in 1888 in the Yūbari region, located in Sorachi, a central subprefecture of Hokkaido. The coal seams, which are up to 7 metres thick, come to the surface in this mountainous region (**268.1**), so could be worked easily and economically by adit mines following the coal into the hillside. The first coal mine was opened at Yūbari in 1890. Peak production of coal from more than 20 mines was reached in the early 1960s, at 3.6 million tonnes. As a result of competition from oil, coal became uneconomical in the 1980s, when most of the mines shut. The last coal mine in the Yūbari region closed in 1990. The population dropped to its present level of about 9,000, from a peak of 108,000 when the mines were operating.

The Yūbari Sekitan Hakubutsukan was opened in 1980 by the municipality as a major enterprise, with considerable investment in the 3,526 square metres of displays. This was in an attempt to develop tourism as an alternative source of employment in a region that had been very dependent on the coal-mining industry. The museum was located in the Coal Mining History Park, which hosted various other activities that had no real relationship with coal mining, such as a leisure park. Central government contributed subsidies to this attempt to rebalance the economy, but the municipality incurred considerable debts as well.

Unfortunately Yūbari is not a recognised tourist destination in a country that has many alternative tourist attractions and where rail fares are high. Also Yūbari is served only by a single-track very-slow railway, so visitor numbers were low. The museum attracted only about 25,000 visitors a year, which was insufficient to support such a major enterprise. The result was that in 2007 Yūbari became the only municipality in Japan to go into liquidation, as a result of the running costs of the many attractions that had been created (*cf.* **363**). Civic services were cut drastically, the amusement park was demolished and the museum and some 20 other tourist attractions were put up for sale in 2008. All the attractions were mothballed, apart from the museum, which remained open, although its website was taken over by a pornographic website.

At present the museum belongs to a private company and has five members of staff. Now active steps are being taken to reinvigorate the museum, whose orange headgear stands out 15 metres high from the landscape. The headgear has its winding wheels one above the other, rather than the more usual side by side (*cf.* **231, 376**, **386**). The museum deals first with the exploration of the area by Benjamin Smith Lyman (1835-1920), an American geologist commissioned in 1872 by the Japanese Government to explore for coal; then, after his return to America, the identification in 1888 by his assistant Ban Ichitarō (1854-1920) of the thick coal seams of Yūbari. The displays continue with the history of the development of mining here and the technology used, including rescue apparatus. The ground-floor displays deal with metasequoia trees (**263**), as the thick Yūbari coal seams were formed from Hokkaido dawn redwood trees buried by mudslides or volcanic eruptions.

On the second floor there is a drawing of a dawn redwood tree with examples of how coal can be used. shown in pictures in its branches. This illustration used to be included in Japanese school textbooks. The second-floor displays deal mainly with the history of coal mining in Hokkaido. A lift from the second floor travels very slowly down about 12 metres, giving the impression of travelling down hundreds of metres. This leads to a simulated underground roadway with reduced levels of lighting. Single lights are switched on and off by motion sensors as visitors move through the exhibition. Headings off the roadway cover mining at various periods using dioramas, starting in the 1890s when both male and female workers wore only light clothing and scanty footwear (*cf.* **250.2**, **253.2**) and the roof was supported by wooden pit props. Models of pit ponies show how they were used for haulage. The later dioramas have modern machinery, such as an Eimco tracked rocker-shovel, roadheader machine, ventilation systems and longwall faceline equipment, including an armoured face conveyor, powered roof supports and a single-ended shearer.

The adjacent buildings display about 250 fossils found during mining and replica dioramas about the mining community in the 1960s. These include a scale model of Yūbari, showing the houses of the coal miners, with overhead lighting simulating the course of the day, including a red sunset and, at night, the windows glowing with light. The houses for the management staff each had a bathroom, whereas the miners lived in buildings for 10 or 20 families, with a shared bathroom. Each family had just one room, measuring 3.7 by 4.6 metres. The museum also possesses a life-size statue of a miner (**251**). A section through an original 7-metre seam is on display in the grounds and the museum also possesses a piece of coal weighing 3 tonnes.

*References*

http://archive.unu.edu/unupress/unupbooks/uu36je/uu36je0b.htm

http://lothes.blogspot.co.uk/2006/04/yubari-coal.html

https://en.wikipedia.org/wiki/Ishikari_coalfield

https://en.wikipedia.org/wiki/List_of_coal_mines_in_Japan

http://talk-hokkaido.blogspot.jp/2012/05/coalmine-museum-in-yubari-city1.html

http://talk-hokkaido.blogspot.jp/2012_05_01_archive.html

http://www.amateras.com/trip/japan/hokkaido/yubari-e.htm

http://www.aya.or.jp/~marukimsn/kikaku/2013/2013sakube.html

http://www.yubari-resort.com/en/attrac.html#coalmine

personal communications, Kazuaki Ikeda, director, Regional Policy Division, Hokkaido Government, 2012; Kenichi Takahashi, curator, 2015

showcaves

# Malaysia

The largest deposits of coal, comprising 98 *per cent* of the coal in Malaysia, are in northern Borneo: in central Sarawak, especially in Merit Pil; and in south-central Sabah, particularly the Meliau coalfield (Chen 1993). The total reserves of the country are 1.7 billion tonnes, small amounts of which are anthracite and lignite, but mostly bituminous coal. In Sarawak the Sadong Colliery was active between 1874 and 1931, producing coal for both export and local consumption (*ibid.*). Although only 2 *per cent* of Malaysian coal is located in Peninsular Malaysia, the only coal ever mined economically was by the Malayan Collieries in Batu Arang (charcoal) in Hulu Selangor.

Coal was first identified here in 1908 by Haji Abdul Hadi while he was prospecting for tin. The Geological Survey Department confirmed the presence of 1,295 hectares of two seams (0.64 and 12.19 metres thick) of sub-bituminous coal, estimated at 9 to 18 million tonnes. This was at a time when the railways were having to rely on imported coal and timber. Between 1915 and 1960 the area produced 13 million tonnes of coal, mainly for the power stations, railways and tin mines. All mining in the country began while Malaysia was under British control.

Batu Arang was a company town, with cinemas, hospital, markets, schools, shops and other facilities supplied by the company. There was segregation between the housing and the recreational areas of the Chinese, the Europeans and the Indians. There were four categories of staff: European technical and administrative staff; Asian technical and clerical staff; the skilled artisan class; and the labourers, who were divided into overseers, coal hewers, haulers, shotfirers and timberers. Tin mining was preferred by the miners because there they were relatively free of close supervision and there was profit-sharing, so they earned more. The company, therefore, purchased coal-cutting machinery to overcome labour shortages. Batu Arang was well known for industrial activism. The first major strike occurred in 1936, quickly resulting in a 10 *per cent* wage rise, and the next in 1937; after police action in which one person was killed, the miners achieved a further 10 *per cent* rise, while the 1939 strike ended linkage of sales tonnage to wages.

The Batu Arang mines were evacuated at the start of the Pacific War (1937-45), but the ship taking the records to Australia was sunk. The company had always kept a balance between underground and surface mining. As soon as the Japanese landed they resumed production, taking 1.240 million tonnes between 1942 and 1945. They concentrated on the easily worked surface deposits, resulting in destruction, fires and flooding underground, leading to heavy operational costs when the company resumed operations in June 1946. The destructive Japanese mining methods meant that restoration was slow.

In November 1946 and January 1947 there was considerable labour unrest, so the government took over the running of the mine under emergency legislation. The miners achieved rations of rice, seven days' pay for six days' work and 12 paid public holidays. In 1954 consultants reported that production costs could never become low enough to compete with oil or imported coal, which was also of superior quality. It was not until 1955 that women were prohibited from working underground. Production ceased in 1960 when the company's lease expired. In its heyday the population stood at 18,000, but by 1960 it was only 4,300 (see Kaur 1990 for the history of coal mining in Batu Arang). In 2011, the centenary of its foundation, Batu Arang was granted heritage-town status and is now an important tourist attraction in Selangor.

### 278. Pulau Labuan Muzium Chimney (Labuan Island Chimney Museum) and Muzium Labuan (Labuan Museum)

Pulau Labuan is a small island in the China Sea just off the coast of Sabah in Borneo in eastern Malaysia. Coal on Labuan was first identified in 1844 by Admiral Sir Edward Belcher (1799-1877), at that time a captain in the British navy and an important explorer. The island was attractive to the British because of its plentiful coal supplies and became one of the Royal Navy's network of international coaling stations established to maintain the fleet at sea.

The Chimney is a local landmark, standing 32 metres high and constructed of 1,000 red bricks, imported from England in the 1880s and strengthened with iron tie bars. It is located in the very north of the island, about 13 kilometres from the town. It stands in the Tanjung Kubong coalfield, which occupies the northern tip of the island and where the first shaft was sunk in 1847. The mine was operated by various British companies for 64 years. A railway was installed in 1893 by the New Central Borneo, the last company to operate the mine. It took the coal 13 kilometres to Victoria port to facilitate its export. Most of the mineworkers were from Borneo, Brunei and Sarawak or else Chinese immigrants, brought from Hong Kong and Singapore, as well as prisoners from Hong Kong and India. This use of unskilled labour led to reduced productivity and a series of accidents which, together with disease and water ingress into the mine, resulted in its closure in 1912. Next to the Chimney is a shaft, known as Lubang Gantang (first tunnel) and there are seven tunnels, which have been explored by staff of the Department of Museums Malaysia.

Figure 278.1. Entrance to the Muzium Chimney on the left with the Chimney on the right (© Department of Museums, Malaysia, Ministry of Tourism and Culture, Malaysia).

Figure 278.2. The Chimney (© Department of Museums, Malaysia, Ministry of Tourism and Culture, Malaysia).

Figure 278.3. Museum diorama of miner shovelling coal into a mine tub (© Department of Museums, Malaysia, Ministry of Tourism and Culture, Malaysia).

It used to be thought that the Chimney was a ventilation shaft for the adjacent coal mines, but excavations by the Department of Museums and Antiquities between 1997 and 2000 showed no tunnel under the Chimney that could have linked it into colliery roadways to ventilate them. There was also no evidence of smoke or burning on the inner walls. Because of its location near the sea and the existence of a port called Raffles Anchorage in this area, it had been suggested that it was a lighthouse with a beacon for passing ships, but the design makes this extremely unlikely. Mining engineers in the UK are agreed that it is unlikely to be a ventilation shaft, as it is much taller than would normally be required, and that the lack of signs of burning suggest that only hot air was released from the Chimney. Their view is that the design at the bottom looks as if it is to accommodate a steam boiler or possibly a very large oven. This is supported by the fact the top looks to be some sort of diffuser for hot air. Moreover the Chimney is almost identical to chimneys of the same date in the UK serving colliery boilers powering winding engines. It is quite likely that the Chimney was built to accommodate a winding-engine boiler but that for some reason, possibly finance, the rest of the work did not proceed, leaving the Chimney standing without a purpose.

Conservation work was carried out on the Chimney in 1997 by the Department of Museums and Antiquities to stabilise the structure. The Muzium Chimney, located in a colonial-style building, opened in 2002 and attracts 50,000 visitors each year. It has five members of staff. The museum occupies an area of 325 square metres, while the area of the Chimney itself is 2,292 square metres. The site is managed and funded by the Department of Museums Malaysia, part of the Malaysian Ministry of Tourism and Culture.

The museum has displays showing the history of coal mining on Labuan Island on the first floor and of the Chimney on the floor above. Also displayed are coal-mining dioramas, mining equipment, the history of the Labuan railway and specimens of coal.

The Muzium Labuan, the main museum in Labuan itself, is opposite Labuan Square. The museum has two floors of displays dealing with the history and culture of the island, especially the role of the island in WWII and the importance of coal mining.

*References*

http://goseasia.about.com/od/malaysiastopattractions/tp/labuan_what_to_do.htm

http://labuanisland.net/historical-places/labuan-chimney/

https://en.wikipedia.org/wiki/Edward_Belcher

https://en.wikipedia.org/wiki/Fuelling_station

http://www.jmm.gov.my/en/museum-directory

personal communication, Salmiahwati Djafar, curator, Muzium Labuan, 2016

# Mongolia

Mongolia has extensive coalfields, with bituminous coal in the north-west, west and south and sub-bituminous coal and lignite in the north, north-east and east. Only the Saikhan-Ovoo coalfield in Central Mongolia has anthracite in addition to bituminous coal. Mongolia's estimated reserves stand at 173.3 billion tonnes (CMMCP 2016, 22).

Coal mining in Mongolia only began in the early 1900s. In 1990 Mongolia exported 0.5 million tonnes of coal and used 6.6 million tonnes internally. There was a major slump when the previous financial and technical support from Russia was withdrawn. The move from a communist planned economy, based on rural agriculture, to one based on mining exports led to foreign investment, in particular from Australia and Hong Kong (*ibid.*). Many Mongolians have left rural areas to move, especially to the capital, Ulaanbaatar (often anglicised as Ulan Bator), or to work at the Tavan Tolgoi bituminous mine in the South Gobi desert of southern Mongolia. Tavan Tolgoi has the largest untapped coking-coal and thermal-coal deposits in the world, with an estimated reserve of 6.4 billion tonnes, a quarter of which is high-quality coking coal. They were first identified in 1945 by Soviet geologists working with the first Mongolian geologist, Jalte Sevyn Dugersuren. There has been a six-fold increase in coal production in Mongolia since 2000. Mongolia is now a major exporter of coal, replacing Australia as a leading supplier of coal to China. In the three-month period January to April 2021, 5.8 *per cent* of exports went to Singapore and 93.5 *per cent* went to China, giving an annual amount of 26.8 million tonnes (https://www.mysteel.net/article/5022611-0504/Mongolian-Q1-coal-exports-soar-114-on-year.html).

A railway from Tavan Togoi, planned for 2021, has failed to materialise, so coal exports to China have to go by truck, leading to massive road gridlocks. There has also been insufficient investment in the infrastructure for settlements around the mines. Three-quarters of Mongolia's energy comes from burning bituminous coal, which has led to considerable pollution in the capital; in 2019 there were an estimated 2,260 deaths from air pollution (Carrington 2021).

### 279. Mongolyn Gyeologi Erdes Bayalgiin Muzyei (Museum of Geology and Mineral Resources of Mongolia)

The Mongolyn Gyeologi Erdes Bayalgiin Muzyei is located in the Sukhebaatar district of Ulaanbaatar, in the valley of the river Tuul in north-central Mongolia. This geology museum is part of the Mongolian University of Science and Technology and deals with the geological history and mineral resources of Mongolia, including coal and petroleum, which was first identified in the country in the 1950s. The museum was established in 1967 when the geological school was set up at the university. It occupies an area of 400 square metres and attracts about 10,000 to 15,000 visitors each year. There are two members of staff and six volunteers.

The nucleus of the museum was formed from donations of samples collected by Mongolian geologists doing exploration work. More than 7,000 deposits of over 400 different minerals have been identified in Mongolia, of which 3,000 are on display in the museum. Exhibits are divided into geological types in relation to the provinces from which they come. A large map of Mongolia shows the location of the mineral deposits of the country, with especial attention given to fluorspar, of which Mongolia has about 600 deposits, forming the fourth largest reserves in the world. As well as practical minerals, such as tin, Mongolia is also rich in minerals used in jewellery, such as chalcedony, crystal, garnet, green jade, sapphire, sardonyx and topaz. Items made from these are on display, along with newly discovered minerals found only in Mongolia, such as armstrongite, mongolite and ongonite.

Images and graphics are used to explain how coal was formed from compressed vegetation in the Carboniferous period and how the seams are found in layers. There are explanations of the differences between coking, rock and thermal coals, graphically illustrated by a very large oval lump of coal. A model demonstrates the Tavan Tolgoi coking-coal deposits.

*References*

https://en.wikipedia.org/wiki/Tavan_Tolgoi

personal communication, Uondon Majigsuren, curator, 2015

# Netherlands

The main coalfield of the Netherlands, in Limburg in the south close to the border with Belgium and Germany, has both anthracite and lignite. The Augustinian abbey of Rolduc employed local miners to extract coal for sale as fuel from the beginning of the sixteenth century. Extensive mining only began in the south-eastern corner of Limburg when boreholes near Eygelshoven revealed a major coal seam, at a depth of 154 metres. This led eventually to 12 anthracite deep mines here, which extracted a total of 600 million tonnes. This changed south Limburg from an agricultural regime of small villages to an industrial landscape with miners from many different countries (https://en.wikipedia.org/wiki/South_Limburg_coal_mining_basin).

Initially exploitation of the coal was by foreign companies, but from 1903 the state financed the development of a number of mines. The first was the Dominiale Mine (1815-1969) in Kerkrade (**97**). Of the six shafts, only the Nullandschacht survives. In 2013 the Koempels van de Domaniale started providing tours for pre-booked groups of the renovated shaft building (https://vakantieinlimburg-nl.translate.goog/schacht-nulland-kerkrade/). The Maurits State Mine was the largest mine, both in Limburg and Europe. Sunk in 1926 it was the last to come into operation. Meanwhile from 1915 lignite began to be extracted by opencasting around Eygelshoven and Hoensbroek (euracoal). Before WWII coal provided 85 *per cent* of energy (Gales and Hölsgens 2017). During WWII the Netherlands were occupied by the German army and Dutch miners were forced to continue to produce coal, which contributed to the German war effort. On one occasion the miners threatened to go on strike for better conditions and were informed that the Germans would simply shoot any men refusing to go underground, which was the end of that strike (personal communication, Joe Dezelak, former Dutch miner and later mine manager, Caphouse Colliery, West Yorkshire; died 2009: **386**). A strike did actually take place between 28 April and 6/7 May 1943, which was ended when seven civilians, including three miners, were killed by the Germans (Luyten 2018, pp. 118-22).

In 1965 the minister for economic affairs announced the closure of all Dutch mines, as their coal was too expensive in comparison with the energy from natural gas found in the northern province of Groningen, so that by 1975 coal provided only 4.3 *per cent* of energy (Gales and Hölsgens 2017). The first deep mine to be closed was the Maurits State Mine in 1967 and the last was the Oranje Nassau Number 1 Colliery in Heerlen in 1974. Lignite mining ended in 1968 with the closure of the Carisborg site. The result was widespread unemployment of former miners, and economic depression and closure of related industries, apart from the chemical industry (*ibid.*). There have been attempts to develop tourism in the area, as well as an innovative project to use hot underground minewater to heat buildings in Heerlen (Roijen 2011; *cf.* **113, 286**). Although there is no longer any coal mining in the Netherlands, the port of Rotterdam has become the main point of entry for coal imports into Europe. In 2015 30.7 million tonnes passed through Rotterdam, decreasing to 17.3 million tonnes as a response to the 2015 Paris climate agreement. Of these imports, mainly from America, Australia and Colombia, 10 *per cent* is used in power stations in the Netherlands and the remainder goes to Germany, where 60 *per cent* is used in the steel industry and 40 *per cent* in power stations. Both of these countries have pledged to end the use of coal in their power plants by 2030, although not for steel making, which requires high temperatures (stationshttps://www.portofrotterdam.com/en/logistics/cargo/dry-bulk/coal).

## 280. Nederlands Mijnmuseum (Dutch Mining Museum)

Heerlen in Limburg was once the centre of extensive coal mining and, as a result, became one of the richest cities in the Netherlands. Since the mines closed, almost all traces have now disappeared, including what, at 155 metres high, was once the tallest mine chimney in Europe: the 1953 Lange Lies (Tall Lies, short for Tall Elizabeth) chimney of the boiler of the Oranje Nassau Number 1 Colliery. Lange Lies and the nearby 1938 Lange Jan (Tall John), 135 metres high, were once major landmarks of the area. In 2003 there were plans to build skyscrapers in Heerlen called Lange Lies and Lange Jan, but these schemes have been abandoned. To absorb the economic impact of the mine closures on Limburg, €120 million were invested in the region each year. Most of the spoil heaps have been removed or else converted into residential or recreational areas as part of an operation called *van zwart naar groen* (from black to green). The Eastern Mining Area is now known as Parkstad Limburg (Limburg Park City), with many parks and ample space for new activities. The new offices for the Central Bureau of Statistics were built on the site of the Oranje Nassau Number 1 Colliery. Rising up through the centre of the building is a glass column on which are marked the various local coal seams and the depths at which they occur.

Amongst the few remaining mine buildings are those of the number 2 shaft of the Oranje Nassau Number 1 Colliery. These now house the Nederlands Mijnmuseum, run by the Stichting Carboon (Carboniferous Foundation). The workforce consists partly of professional staff and partly of volunteers, including some retired miners. The

**Figure 280.1. Exterior of the Nederlands Mijnmuseum at night (© Stichting Carboon).**

**Figure 280.2. The shaft top of Orange Nassau 1 Colliery, now conserved, within the original colliery heapstead (© Stichting Carboon).**

**Figure 280.3. Steam-winding engine built in 1897 by the G. Mahler Company of Aachen, Germany (© Stichting Carboon).**

museum receives an operating grant from the municipality of Heerlen, members of the Friends of the Museum pay an annual contribution and other income is raised from donations and entrance fees. The museum opened in 2005 and attracts 12,000 visitors a year.

The museum is housed in an elegant square red-brick building, standing 14 metres high, with rectangular blind windows on the ground floor, three sets of double-arched windows on the first floor and three large arched windows on the second floor. The building is 225 square metres in size and stands adjacent to the steel headframe, 30 metres high. The winding-engine house occupies 100 square metres and now houses the museum's collections.

Each floor of the museum has photographs and displays of mining machines and equipment, including a large collection of fossils and mine lamps. The building houses the top of the shaft and the cage for going underground (*cf.* **87.1**). This is the only building in the Netherlands where the cage is still intact. The winding engine dates to 1897 and is now in operation, following repairs. In the grounds of the museum, there are two four-wheel diesel mechanical flameproof mine locomotives with mine cars for transporting miners underground. There is also an original 0-6-0 tank locomotive, Oranje Nassau 14, with two coal waggons, for moving coal on the surface. This was constructed in 1924 by S.A. Ateliers de Construction de La Meuse (often abbreviated as La Meuse) at their Liège, Belgium, works in 1924. It worked until 1969 shunting coal waggons. A year or so later it went into conservation by the Stichting tot behoud van Mijnlocomotieven (Foundation for the Preservation of Mine Locomotives), a Dutch railway-preservation group specialising in restoring mining locomotives. The locomotive was returned to Heerlen for static display in April 2015, in the ownership of the Nederlands Mijnmuseum since December 2015.

The Pand Kneepkens (Kneepkens building) was designed in 1939 by the architect Frits Peutz (1896-1974) for the Kneepkens family as a department store at a time when Heerlen was flourishing because of the mining industry. It was known as the 'small glass palace' and was designated as a national monument in 1999, but has been vacant since the 1990s. It is now being transformed into a new museum by the Stichting Carboon, due to open in April 2022. It is being financed by the municipality of Heerlen, the province of Limburg and donations from companies, including major Dutch cultural funds and the Nederlandse Staatsmijnen (Dutch State Mines) Company, which is now a producer of food supplements. Renovation of the building will cost €2.3 million and fitting out a further €1.5 million. The museum will tell the story of the mines, including the history of the development and decline of the mining industry, as well as the impact the coal mines have had on the population of Limburg. There will be an emphasis on the life of the miner, above and below ground. This will complement the displays in the Oranje Nassau Number 1 Colliery.

*References*

http://en.wikipedia.org/wiki/Heerlen

http://nederlandsmijnmuseum.com/

personal communications, Erwin Roijen, consultant, Remining-Iowex/Cauberg-Huygen, 2011; Jan de Wit, secretary to the Board, Stichting Carboon, 2016, 2021

### 281. Steenkolenmijn (Coal Mine)

Valkenburg, 14 kilometres west of the Nederlands Mijnmuseum (**280**) and in the same coalfield in south-eastern Netherlands, is the site of the Steenkolenmijn. This was founded in 1917 by Emile Caselli (1873-1961) to demonstrate the nature of work underground. A drift was driven into the soft marl clay of the region to create a simulated underground fitted out with mining machinery. Marl is soft underground, but when exposed to the air becomes hard, so was a popular material for buildings and was often mined in the same vicinity as the coal mines. It is estimated that there are more than 300 kilometres of underground tunnels in the Valkenburg/Maastricht region, totalling some 20,000 individual roadways.

Visits begin in the underground cinema with a film from the mid-1960s of the state mines in Limburg. Tours of the mine, lasting about 75 minutes and available in either Dutch or English, are guided by one of 14 former miners, who demonstrate the running of the machinery and recount their own experiences of working below ground. Underground there are also a heated café and a fossil museum, with exhibits from the Cretaceous and Tertiary periods, including the large lower jaw of a mosasaur, a giant marine lizard. Also available locally are trips through the South Limburg Hills with carriages hauled by a coal-fired steam train.

*References*

ERIH, *s.n.* Valkenburg

http://www.steenkolenmijn.nl

# New Zealand (known by Māori as *Aotearoa*)

Coalfields with 16 billion tonnes of reserves are found in New Zealand: 80 *per cent* of the reserves are of lignite in the South Island. The remainder is sub-bituminous in both islands and a lesser amount of high-quality bituminous coal, mainly on the West Coast of the South Island (https://www.nzpam.govt.nz/nz-industry/nz-minerals/minerals-statistics/coal/).

The indigenous Tainui Māori (who had settled New Zealand in 1320/50) of the central North Island used coal for cooking. British mining began in the South Island in 1849 in Dunedin on the south-east coast and in 1864 at Brunner Mine in the western part (**285**). This was the most productive coal mine in the nineteenth century in New Zealand, supplying most of the country's coal requirements. Bonds were issued in the 1870s to fund railway construction, whose trains needed coal. Many early miners brought a tradition of trade-union activism from Britain, with some union leaders going on to become members of parliament (**287**).

Mining was deemed an essential industry from 1942 onwards during WWII, when Australasia was cut off from the rest of the world, so that miners could not change jobs and large-scale opencast mines were developed. Coal use declined in the 1960s and 1970s with more oil and hydroelectric power available, but revived with the 1980s increase in oil prices (Sherwood and Phillips 2021). In some deep mines high-pressure water jets were used to cut the coal (*cf.* **179, 251**), but by the 2000s most mining was opencast, with 18 in production in 2017, producing 2.9 million tonnes. Coal is used mainly for electricity generation and industrial processes, including export for steel making to Brazil, China, India, Japan and South Africa. The 2010 Pike River Mine disaster, in which 29 men died, was one of nine major underground methane explosions, in which 211 miners have been killed, leading to the setting up of a national independent workplace-safety organisation.

The Department of Conservation is a government department responsible for looking after the cultural and natural heritage of public conservation lands within New Zealand. The department now administers and maintains the surface sites of many of the abandoned underground mines (see http://www.doc.govt.nz/our-work/heritage/heritage-topics/mining-coal/ for all sites in the care of the Department of Conservation).

### 282. Alborns Coal Mine

The site of Alborns Coal Mine in the West Coast region attracts about 1,000 visitors a year. The site is located on the Alborns Coal Mine Track, developed for walkers near Reefton (**289**) in the Victoria Forest Park beech woodland. The mine dates back to 1880 and was family-run from

**Figure 282.1. 1915 Leyland lorry, used as the last winch on the site, under lean-to roof covering, and coal skip (© NZ Department of Conservation, photographer Jim Staton).**

**Figure 282.2. Incline up to site of 1915 truck and part of the walkway around the site (© NZ Department of Conservation; photographer Les Wright).**

1923. The old surface workings still survive, including the remains of makeshift winch equipment and a 1915 Leyland lorry, which was used as the last winch on the site and now has a lean-to roof covering. Alongside the lorry is a rebuilt 3-ton (2.7-tonne) coal skip.

*References*

https://coastingnz.wordpress.com/2013/10/09/reefton-alborns-mine-walk/

personal communication, Jim Staton, ranger, New Zealand Department of Conservation, 2021

### 283. Benneydale Coal Mine

The site of Benneydale Coal Mine, in the north of the King Country region of the North Island, attracts about 5,000 visitors a year. Coal was not discovered here until 1931. In 1940 the mine was purchased by the government, which built the township of Benneydale to house the workers at Mangapehi State Coal Mine. This is one of the few towns that did not have a Māori name, the name being formed by joining together the surname of Charlie Benney, the under-secretary of state for mines, and Tom Dale, the superintendent of the mine. In 2019 the New Zealand Geographic Board approved a change to a dual name of Maniaiti/Benneydale, the Māori element meaning a small slide or landslip.

**Figure 283.1. Rail-mounted Samson coal cutter, fitted for overcutting, made by Mavor and Coulson, Glasgow, Scotland (© NZ Department of Conservation; photographer Neville Ritchie).**

The mine closed after a major fire in 1962, although another part of the coal seam nearby was worked between 1978 and 1998, but on a much smaller scale. The 1962 fire had a major impact on the town, which only existed to serve the mine. While the mine was working the population was some 2,000, but is now only about 200. In the early 2000s Benneydale still had a dairy, petrol station, police station, primary school and public house, but empty houses and shops were conspicuous.

The site still has a large coal hopper and a rusty rail-mounted Samson coal cutter, fitted for overcutting. This was made by Mavor and Coulson in Glasgow, Scotland, and shipped out to New Zealand.

*References*

http://www.teara.govt.nz/en/king-country-places/page-4

http://www.teara.govt.nz/en/speech/34684/benneydale-township

### 284. Blackball Coal Mine

The site of Blackball Coal Mine, in the West Coast region of the South Island, attracts about 5,000 visitors a year. The Blackball site is one of several heritage places that collectively comprise the coal-mining heritage of the Blackball area. Blackball town itself is the only original British-funded mining town in New Zealand still to be occupied.

The two seams at Blackball were opened in 1890, at depths of 1.2 metres and 3.7 metres. Initially from 1893 coal was moved by an aerial ropeway, 5.5 kilometres long, running through the town and across the Grey River to Ngahere. The ropeway moved at 6.4 kilometres an hour and could carry 363 tonnes. It continued in use until the railway, which had reached Ngahere in 1889, was extended to Blackball in 1909. The tubs of coal were transported without their chassis directly from the mine entrance.

At first the mine was ventilated by furnace (*cf.* **386**), with three chimneys built between 1891 and 1897. One, 10.7 metres high, was originally in wood, but the timber was replaced by brick in 1898. All three were made redundant in 1904 by the installation of a double-inlet capell fan. This was in turn superseded by a Sirocco fan in 1923, installed in a fan house constructed in 1921. Two of the chimneys are still in good condition, but are not within the area maintained by the Department of Conservation.

This section of the mine closed in 1937. A new southern section of the mine was opened up and worked until the mine closed in 1964 due to major water issues, with up to 4.546 million litres a day having to be pumped (*cf.* **386**). The Mines Department took over ownership of the mine in 1941.

The foundations for the conveyor, the main chimney and the drainage tunnels are still in good condition, unlike most of the other historic features at the site. The early return-airway roof and parts of the fan house are on the verge of collapse, 60 *per cent* of the Sirocco fan has corroded away and the dam has collapsed. There is, however, enough to show how the mine surface operated. The tunnel and

**Figure 284.1. Blackball Coal Mine chimney prior to repairs in 2020 (© NZ Department of Conservation; photographer Jim Staton).**

return-airway entrances, concrete foundations of the fan engine and water-diversion tunnels remain intact, in an area of about 8,000 square metres. The main drainage tunnel continues underground for about 250 metres. The main physical features are the remains of the Sirocco fan and the 1909 conveyor foundation, of which only the front face is totally visible, and the chimney, 19 metres high. Scaffolding was erected around the chimney to enable it to be repointed, which was completed in 2020. Volunteers have cleared part of the site and the local community has erected interpretation panels.

*References*

https://www.doc.govt.nz/search-results/?query=Blackball%20 Mine

personal communication, Jim Staton (**282**), 2013

### 285. Brunner Mine

Brunner Mine, named after the explorer, Thomas Brunner (1821-74), was one of the earliest large-scale industrial sites in New Zealand, where coal was mined and coke, firebricks and other products were manufactured from the 1860s until the 1940s. Coal was initially taken by barge 11 kilometres to the port at Greymouth, the largest town in the West Coast region, until a railway was built to the site by 1877. An endless ropeway hauled filled coal tubs out of the mine, with the empties returning in the opposite direction. Once the mine workings extended beyond the Brunner fault into the dip workings in 1884, a 50-horsepower horizontal-winding engine was installed in 1886.

At first the mine relied on natural ventilation, but as it was extended underground, a Schiele fan, with a diameter of 2.9 metres, was installed in 1888. This was driven by a steam engine able to displace between 453 and 510 cubic metres of air per minute. The fresh-air intake was the main mine-entrance tunnel. An air-compressor plant was introduced in 1888, but was not used for long, as electricity was introduced at Brunner in 1899.

The intake and return airways ran parallel underground for about 1 kilometre, then extended down into the dip workings beyond the fault line. The worst industrial disaster in New Zealand was the explosion in these dip workings in 1896, in which 65 men and boys were killed. A memorial was erected here, 100 years later, to commemorate those who died at Brunner, as well as all victims of other mines in the district of Grey.

The Department of Conservation has provided detailed interpretation on site and on their website of the surface remains of the colliery. It is estimated that the site attracts about 40,000 visitors a year. There are brick remnants of the boilers dating to before 1883 and the iron base survives of the tall chimney of the boilers which provided steam power from the 1880s until the 1920s to power the mine machinery. The chimney was bricked around in 1900. Two kilns down by the river were built in the 1860s. The Saint Kilda fireclay mine worked until 1921, starting in 1907, after the main Brunner coal mine closed in 1906. The clay was transported by tramway to the brickworks.

*Reference*

https://en.wikipedia.org/wiki/Brunner_Mine

### 286. Coaltown Museum

The Coaltown Museum, opened in 1977, is located in Westport in the northern part of the West Coast region of the South Island, at the mouth of the Buller River. The Buller region had an important role in coal mining in New Zealand. Westport's Coaltown Trust was established in the early 1970s as a voluntary body dedicated to the local coal-mining heritage. In 2013 the trust moved the museum from its original home in Westport's former Morley and Company Brewery building into a new centrally located facility, owned by the Buller District Council and open seven days a week.

The trust received operational funding from the district council, but in 2021 a new company, Te Ara Pounamu, took over management of the museum. While there will be no further funding from the council, the Coaltown Trust has continuing responsibility for the care and maintenance of its specimen collection, which comprises more than 6,000 objects, photographic specimens and primary documents donated since 1977.

The story in the museum currently begins with the Gold Rush of the mid-nineteenth century when the isolated area began to be opened up. Westport was the main town of the area, because transport of the coal mined at all the local mines was possible only through the harbour there. The museum shows the work necessary to turn the river into a usable port where it reached the sea, including a triple-expansion steam engine, two storeys high, from the bucket-dredger steam ship, Mawhera.

The main displays deal with the formation and geology of coal, the development of the communities, transport of the coal, the working life of the coal miners and the miners' trade unions. There is a simulated coal mine fitted out with machinery and tools, with the technology explained in a video. The display of the working conditions of the miners is augmented by photographs showing the hard life involved of the miners and their wives and families in a remote undeveloped area. The main exhibit is the machinery used on the Denniston Incline (**291**). The 8-tonne Q Waggon is displayed at an angle of 45 degrees, as when it was working on the steepest part of the Incline, together with the 20-tonne brake drum used to slow down and hold the waggon when it was full of coal. Te Ara Pounamu is to undertake a programme of major gallery renewal and expansion, plans for which should be revealed in 2022.

**Figure 285.1. Site of Brunner Mine (© NZ Department of Conservation).**

**Figure 285.2. 1864 coke kilns (© NZ Department of Conservation; photographer John Edwards).**

**Figure 286.1. Bottom-discharge skip waggon used on the Denniston Incline (© Coaltown Museum).**

*References*

http://www.coaltown.co.nz

personal communication, Ralph C. Rataul, collections cataloguer, 2021

showcaves

### 287. Granity Incline Track

Millerton is located in the north-west of the South Island in the Papahaua Ranges of the West Coast region, about 25 kilometres north-east of Westport. The Millerton bathhouse is one of the few traces remaining of the abandoned colliery site. Millerton Coal Mine began production in 1896, after the Westport Coal Company built what is known as either the Granity Incline Track or the Granity Dam Track in 1891. This was to enable the coal to be moved down the steep slope from the mine. The incline was a narrow-gauge incline (self-acting, but controlled by a brake at Millerton), 61 centimetres wide, with two parallel tracks. The coal tubs travelled along the tracks to the bins at Granity, which at the time were the largest coal-loading wooden bins in New Zealand. In 1904 William McIntyre (1881-1949), came to New Zealand from Scotland. He and his brother worked at Millerton Mine, where together they set a New Zealand record for the amount of coal mined in one shift. He became actively involved in the coal-mining union, leading to his political career as a member of the Legislative Council. Mining ceased at Millerton in the late 1960s and today the settlement is effectively a ghost town.

Solid Energy Limited put in a new road to their Stockton Coal Mine, bypassing the road where the site is situated. To get to the incline site it is necessary to backtrack and then go down to it via Millerton township, down a road that is hardly ever used, so the site is now no longer maintained by the Department of Conservation. The site previously attracted about 8,000 visitors a year.

The boilers, stone walls and concrete shell of the pithead baths still survive on the site of the mine, together with tracks, trucks and other installations. The remains of the Incline might be compared to those at Denniston (**291**), although those at Denniston are much more impressive.

*References*

https://westcoast.co.nz/news/granity/

https://en.wikipedia.org/wiki/Millerton,_New_Zealand

personal communication, Jim Staton (**282**), 2016

## 288. Lankey Creek Coal Mine

The site of Lankey Creek Coal Mine, near Reefton (**289**) in the West Coast region, attracts about 2,000 visitors a year. The remains on the site are winches, some of the 1877 tramway and an 1883 battery for extracting gold from cement conglomerate.

*References*

https://images.search.yahoo.com/search/images;_ylt=A2KLfSYZv69gVb8AzS5XNyoA;_ylu=Y29sbwNiZjEEcG9zAzIEdnRpZAMEc2VjA3Nj?p=Lankey+Creek+Coal+Mine&fr=mcafee

personal communication, Jim Staton (**282**), 2021

## 289. Reefton

Reefton is located in the West Coast region, in the western part of the South Island, 80 kilometres north-east of Greymouth (**285**). The area is rich in both coal and gold, which was first discovered here in 1866, although it was not until 1892 that Reefton was connected to the rail network.

The Reefton School of Mines opened in 1886 to teach skills applicable to both coal and gold mining (see also **290**). Local children also came to the school to learn chemistry. A 1930 poster advertised the skills and occupations then being taught to certificate level: assaying, battery managers, electricity, enginemen, firemen deputies, gas testing, prospecting for gold, mathematics, mechanics, metallurgy of gold and silver, mine managers, mining (coal), mining (gold), surveying, theoretical chemistry, underviewers and ventilation, with notes that special subjects could also be arranged and that assaying and analytical work could be undertaken, presumably by the trainees. This being in the middle of the Great Depression (1929-39), fees were low, with special consideration given to unemployed and slack-time miners. All teaching was done by the director alone.

The school was housed in a simple gabled building, dating to 1880. This is now on a New Zealand Department of Conservation historic reserve, but managed by an elected board. The extensive mineral collection of items from around the world and the large technical library have been preserved, along with the crucibles used in teaching gold assaying. Also in Reefton is a replica miner's hut, constructed by Reefton Rotary Club in the 1990s from slabs of local cedar wood. This is furnished as it would have been in the 1860s and there is also a working blacksmith's forge. The hut is staffed by two bearded re-enactors wearing 1860s clothing. The hut is meant to be that of a gold miner, but a coal miner's hut would have been identical and many gold mines also had their own coal mines to provide fuel for their boilers.

*References*

http://www.doc.govt.nz/documents/parks-and-recreation/tracks-and-walks/west-coast/reefton-walks-brochure.pdf

http://www.reefton.co.nz/reefton/reefton-s-history

http://www.teara.govt.nz/en/ephemera/9251/poster-for-reefton-school-of-mines

personal communication, Jim Staton (**282**), 2016

**Figure 289.1. Exterior of Reefton School of Mines (© Jim Staton).**

## 290. Thames School of Mines Mineralogical Museum

The Thames School of Mines Mineralogical Museum is not a specialised coal-mining museum, but interprets the school, which provided training and other support for coal mining in New Zealand. It is located in Thames at the south-western end of the Coromandel Peninsula in the north of the North Island. The museum has three to four part-time employees and between two and four volunteers, and is visited by 2,500 visitors each year.

The mineralogical collection of the Thames School can be traced back to 1870 and the establishment of a mechanics' institute at Grahamstown, close to the present school site. Established in part to meet the need for adult education on the Thames goldfield, the mechanics' institute had a wider scope than just gold mining, as is illustrated in the fossil collection from the institute preserved in the mineralogical museum.

Set up in 1885 as one of about 30 such schools of mines in New Zealand (**289**), the Thames School was at one time the largest in the country. As well as training, the school was involved in practical matters, such as producing activated soda lime for the breathing apparatus used by the rescue teams at the disaster at the Huntly Coal Mine in 1914, in which 43 men were killed.

The collection from the old mechanics' institute was moved into the premises of an old Wesleyan Sunday school around 1886. This site is unique in being a formally recognised *Waihi Tapu* (a place sacred to the Māori people in the mythological, religious, ritual, spiritual or traditional sense), being the location of an *urupa* (burial ground). So it is of special significance to the local Māori people and has a history dating back around 400 years.

The Thames School of Mines complex, including the Mineralogical Museum, is owned and funded by Heritage New Zealand Pouhere Taonga, which is largely financed by the government, but also has a fee-paying membership in the tens of thousands. The School of Mines complex was bought from the local council in 1979 by Heritage New Zealand to preserve this unique collection of buildings. Although the collections have grown considerably, the museum itself remains much as it was when it opened in 1901, with classic nineteenth-century wall cabinets and upright cases. From the beginning it was open to the public as well as to the students.

The museum has one of the best mineral collections in Australasia, including a number of coal and related mineral specimens, such as amberite in coal. The inclusion of coal in the school and museum in an overwhelmingly gold-dominated local mining industry reflects the philosophy

**Figure 290.1. Thames School of Mines Mineralogical Museum building (© David Wilton).**

of the school in looking wider and government policy in the late-nineteenth century of sponsoring training and qualifications that were applicable to both hard-rock and coal mining. While local coal in the Coromandel Peninsula/Hauraki goldfields area was generally of extremely poor quality, as is shown by the specimens, there was good coal quite close by in the Waikato coalfield. The collection also contains examples of other hydrocarbon minerals but coal dominates the hydrocarbon collection, in keeping with the importance of coal, particularly during the second half of twentieth century when gold mining in the North Island just about died out. Items on display in the school buildings also include domestic-fuel briquettes made from coal dust and early coal-mining helmets and lights.

*References*

https://nzmuseums.co.nz/collections/3309/thames-school-of-mines-mineralogical-museum

https://www.heritage.org.nz/places/places-to-visit/coromandel/thames-school-of-mines

https://www.heritage.org.nz/the-list/details/132

personal communication, John Isdale, curator, 2013

### 291. The Denniston Experience (formerly the Banbury Tourist Mine Railway)

The Denniston Experience is located approximately 25 kilometres north of Coaltown, Westport, in the north of the South Island. Denniston was the main township of a collection of mining settlements servicing the nearby coal mines. For many years Denniston was the largest underground coal mine in New Zealand, producing an estimated 12 million tonnes of premium-quality coal between 1879 and 1967. The coal was loaded into railway waggons, weighing in total about 12 tonnes. The terrain was such that the conveyor technology of the period was unable to deal with the height and the gradient involved. So the waggons were lowered by cable, at speeds of up to 80 kilometres per hour, down the extremely steep Denniston Incline. This was regarded as a remarkable feat of engineering, sometimes referred to as the eighth wonder of the world. The incline fell 510 metres over 1.7 kilometres in two sections, taking about 14 waggons full of coal each hour down from the plateau and returning empty waggons back to the top. The 8-tonne Q Waggon and the 20-tonne brake drum are now preserved and displayed at the Coaltown Museum (**286.1**). The incline ceased operating in 1967 and an earthquake the following year caused considerable damage.

The Denniston Heritage Charitable Trust was formed in 2007 and the Department of Conservation restored the rail at the brim of the incline in 2009. The Denniston Experience opened in 2012 and operated on a purely commercial basis with trips into Banbury Mine. The experience was forced to close to the public in 2015 with the introduction of more stringent mining regulations following the 2010 Pike River Mine disaster. Plans are, however, underway to make the experience compliant with the legislation, so the experience has been included here.

The hands-on Denniston Experience was the only underground coal-mine experience in New Zealand. Visitors had to get dirty and participated in the coal-mining techniques of the 1880s. The mine has natural ventilation, so a fan was not necessary. Visitors reported to the mine office, to receive a miner's helmet, cap lamp, raincoat and union card, then walked 500 metres to the train boarding point. There they were given a safety talk and received a miner's tally, before joining the narrow-gauge train in groups of a maximum of 17 with two mining guides. In the coal roadways holograms and audio-visuals provided the experience of an 1880s mine in a temperature of about 10 to 12 degrees Celsius.

While underground visitors could take on a number of different roles: clipper attaching tubs onto the haulage; hewer, or coalface miner; shiftman installing pit props and laying rail tracks; shotfirer setting mock explosives and firing the shot; surveyor responsible for mine design and recording the roadway layout; trucker moving full and empty mine tubs; underviewer in charge of the underground management; or union representative. Back on the surface, visitors reported back to the mine office to return their tally and protective equipment. The entire experience took 150 minutes.

When operating, there were five members of staff and it attracted about 2,500 visitors a year. On the surface there are 360 metres of the 2-foot (0.61-metre) mineral train line running along the Waimangaroa Gorge (operating under the Banbury Tourist Mine Railway licence) and a further 240 metres of track underground. Remnants still survive of the mine and town, which is listed as a category 1 historic place on the Heritage List for its outstanding historical significance. Most of the industrial and domestic buildings have disappeared, with only two houses still occupied.

*References*

personal communications, John Gurney, general manager, the Denniston Experience, 2015, 2021

showcaves, *s.n.* Banbury Tourist Mine Railway

www.denniston.co.nz

### 292. Waikato Coalfields Museum

The Waikato Coalfields Museum Society is based in the middle of residential Huntly, in the heart of the North Waikato coalfields region in the Upper North Island, where the coal was of high quality. The society has a collection in storage of about 25,000 to 30,000 items, independently assessed as being of significance nationally, although most of the objects relate to the North Waikato coalfields.

The collection comprises around 4,000 photographs, mining gear, including banjo shovels, canary-resuscitation

**Figure 291.1. Railway and waggons on the surface (© Stelvio Ltd Trading as The Denniston Experience).**

**Figure 291.2. Visitors riding in mine car on the surface (© Stelvio Ltd Trading as The Denniston Experience).**

**Figure 292.1. Old railway station building to be relocated to close to its original site for use as museum (© Waikato Coalfields Museum).**

unit (**327**), detonating gear, an endless rope-haulage winch, coal cutters, coal crusher, coal sizer, helmets, main-and-tail machine, miners' lamps, mines-rescue gear, multiple drills, pickaxes, Pikrose winch, scales, skips and tallies. The museum has the complete library inherited from the old State Coal Department. There is a comprehensive array of archives: extraction records going back to the very early days of mining in Huntly, miners' letters, oral-history recordings, shift reports and activities of the union, which was very strong in the coal-mining communities. The archives relating to coal mining across New Zealand hold artefacts, maps and photographs as well as reports relating to the two major disasters of the North Waikato coalfields, in 1914 and 1939.

The Waikato Coalfields Museum was formerly housed in a two-storey colonial homestead, dating to the 1890s. The building was previously a residence for mine managers and their families and was converted to a museum, which opened in 1980. The museum was governed by an elected committee of community members. The museum building was sold in 2017. The committee is working with Waikato District Council on the relocation of the old railway-station building back to close to its original site near the town centre, with the intention of developing this into a small museum.

*References*

http://www.coal.net.nz/

personal communications, Jane Ferrier, registrar, 2012; Julie Grut, administrator, 2021

**Figure 292.2. Contemporary postcard of the 1914 disaster at Ralph's Mine, Huntley, from the Waikato Coalfields Museum collections (© Waikato Coalfields Museum).**

**Figure 292.3. Funeral cortege following the 1914 disaster at Ralph's Mine, Huntley, from the Waikato Coalfields Museum collections (© Waikato Coalfields Museum).**

# Nigeria

Nigeria has anthracite, lignite and bituminous and sub-bituminous coal in 10 coalfields in the Anambra region in the south of Nigeria and one in eastern Nigeria (CMMCP 2016, 24). While exploring for silver in 1909, the British geologist, Sir Albert Kitson (1868-1937), identified sub-bituminous coal deposits in the Enugu region. The Udi and Nkanu chiefs signed agreements in 1915 and 1917 for mining at Enugu. The first rail line linking the Udi coalfields to the coast at Port Harcourt was completed in 1916, at the same time as the shafts were sunk at Ogbete Mine and Udi Mine, and the Iva Valley Mine opened the following year. Two settlements were constructed for the coal miners. Enugu Colliery reached its peak production during WWII, producing 658,451 tonnes of coal in 1944.

From 1944 the Colliery Workers Union had been demanding recognition of their union and an increase in wages to deal with inflation, so in 1949 the miners started a work-to-rule. The authorities believed that this was a communist-inspired insurrection for national independence and were concerned that the miners could access gelignite underground. So they moved in 900 soldiers, and on 18 November 21 miners and a bystander were killed and 51 seriously wounded. The Ivo Valley massacre led to protests in Aba, Onitsha and Port Harcourt.

Initially the collieries were officially part of the Nigerian Railway Department, from which they were separated in 1950 when the railway department was transferred to the Nigerian Coal Corporation. In the 1950s commercial quantities of oil were discovered in Nigeria and the two main purchasers of coal, the Electricity Corporation of Nigeria and the Nigerian Railway Corporation, changed to diesel-powered machinery. At independence in 1960 coal made a major contribution to revenue to pay for the development of national infrastructure. The collieries were also badly affected by the civil war of 1967, so that production at Enugu Colliery had fallen to 260,030 tonnes by 1972. The Nigerian Coal Corporation lost its monopoly in 1999 and the colliery assets were sold by the Bureau of Public Enterprises (see Baiyewu-Teru 2015 for the history of coal mining in Nigeria). There are still an estimated 2 billion tonnes of reserves, including the largest lignite deposits in Africa. Its low-sulphur sub-bituminous coal is exported to Egypt and Ghana (CMMCP 2016, 24). In November 2021 the president, Muhammadu Buhari (born 1942), signed the Climate Change Act setting a deadline of 2060 to reach net-zero emissions, but there are also plans to revive the coal sector (Maja-Pearce 2022, p. 33).

### 293. National Museum of Unity Enugu

The National Museum of Unity is located in Enugu, the capital of Enugu State in south-eastern Nigeria. This was the heartland of Igbo culture and cooking materials, elaborate woven clothes and iron ore were exported from Enugu to the rest of the Igbo region. The museum opened in 2006 and is managed and funded by the government of Nigeria. It has 126 members of staff, occupies an area of 800 square metres and attracts about 26,000 visitors a year. The museum displays tools and other equipment relating to coal mining and explains the various uses of coal. Of particular importance is the British steel ropeway. This transported buckets containing coal, suspended in midair, from the Enugu coal mines to the Oji River power station, a distance of 45 kilometres.

In 2013 the headquarters of the Coal Corporation and the 1915 coal-preparation plant at Ogbete Mine were declared to be national monuments in celebration of Nigeria's centenary anniversary. The headquarters, comprising six buildings of a colonial-architectural design made of sandcrete blocks and corrugated-iron sheets, each of which housed a different department, is now the archives for coal-mining records for the whole of Nigeria. Plans are underway to develop the Enugu Colliery assets into an eco-museum under the auspices of the National Museum of Unity Enugu.

*References*

https://enugumetro.com/city-directory/national-museum-of-unity-enugu/

personal communication, Onyejekwe Ijeoma, monument, heritage and sites officer, 2015

# Norway

There are no workable coal reserves in mainland Norway. The Norwegian archipelago of Svalbard, in the Arctic Ocean about 950 kilometres west of continental Norway and halfway between mainland Norway and the North Pole, has reserves of 90 million tonnes of good-quality high-calorific anthracite.

When the coal began to be exploited, it formed the basis for the economy of the communities on Svalbard. Coal mining is concentrated on the town of Longyearbyen, where exploitation started in 1903 with the sinking of Trønder Mine. The first year-round mining on Spitsbergen took place through the winter of 1905/1906 at Advent City on Revneset, a small headland in Adventfjorden. Advent City did not last long as coal production was poorer than had been expected. In 1906 an American company, the Arctic Coal Company (ACC), started to mine coal in Longyearbyen, named after the main shareholder in the company, John Munro Longyear (1850-1922). The hope of financial gain from the mineral resources in the archipelago led to a brief Klondike-mood. The archipelago was no man's land, divided up between private partnerships. Everyone who found an area where it looked as if there could be resources like coal would go ashore and put up a sign warning others that the area was taken. More than 100 signs were put up between 1898 and 1920 to mark claims on all the areas annexed. They had different appearances and texts, and had to be renewed every year to remain valid. No regulations existed to control the land division, which often led to arguments and conflicts. In 1916 the state-owned company, Store Norske Spitsbergen Kulkompani AS (SNSK), bought the construction site from ACC and started mining operations. Subsequent mines were not named but numbered.

In 1925 Svalbard became part of Norway under the 1920 Svalbard Treaty. This made Svalbard a visa-free zone, regulated the demilitarisation of the archipelago and gave the right of establishing settlements and mining there to all 39 of the signatories. At present the only countries active on Svalbard are Norway and Russia, which has a mining community at Barentsburg, 40 kilometres south-west of Longyearbyen, although the population has dropped from 2,500 to under 500 people (see euracoal and Avango, Hacquebord, Aalders, *et al* 2011 for the history of coal mining in Spitzbergen).

**294. Gruve 3 (Mine 3)**

The coal in Gruve 3, on the northern slope of Platåberget, was identified in 1913, and mining first began in 1917. The town was destroyed in 1944, but restored after WWII. There was a break in mining between 1949 and 1970 and it ceased completely in 2020. There has been a major reclamation project of the landscape of Svea, in the southern part of Spitzbergen. All buildings and other infrastructure later than WWII are being removed, after Svalbard Museum (**295**) has documented the remains and collected artefacts, history and information.

Gruve 3 closed in 1996. Previously visitors were allowed to go underground in Gruve 3 and Gruve 7, but this was stopped in 2009 for safety reasons. In 2012 SNSK started to develop plans to convert Gruve 3 into an underground mining museum. Meanwhile in April 2015 the government agreed to provide 500 million Norwegian kronor in financial support to the mining company. In September 2015, however, the company announced that coal world prices had made the enterprise uneconomic and that it intended to mothball Lunckefjell Mine, leaving only the much smaller Gruve 7 in Longyearbyen still producing coal for use on the island and for export to the iron and steel industry

Gruve 3 was opened to the public in 2015, owned and mainly funded by SNSK, but with advice from Svalbard Museum, which has also catalogued the remaining objects. Visitors, averaging 15 a day, can crawl along a mining shaft about 700 metres into the mine itself. Innovation Norway and Svalbard Miljøvernfond, a government fund, paid for roof bolting by a couple of coal miners from Gruve 7 of about one-fifth of the main tunnel H1 to preserve Gruve 3 and the mining history of Svalbard. This work was completed in 2019.

Basecamp Explorer, a local tourist company, guided visitors in the mine until Store Norske Gruve 3 AS was established in 2016 as a subsidiary of SNSK. Store Norske Gruve 3 AS is now running all tours of the mine and is responsible for the daily maintenance of the museum and the logistics of guiding visitors. The site is staffed by five people, including a booking manager, a daily manager and guides working both full-time and on a casual basis.

Gruve 3 is effectively a time capsule, as very little has changed since it closed. Visitors can experience part of the mine, the workshops and the section where the coal was brought out of the mine by locomotives to be taken to the harbour for shipping. Underground there are about 700 metres of roadway open, as roof bolting allowed more of the mine to be accessible to visitors, including one mining shaft.

There is no machinery underground. The surface area of about 11,000 square metres displays cutting machinery, drilling equipment, hand-held shot-blasting equipment, personal and protective equipment, pit props and Jeffery locomotives as used in the low-seam mine, with faces only between 40 and 110 centimetres thick. There are also some pieces of old equipment, including colliery tubs, and the skip and A-frame from an aerial ropeway to lower tubs into the mine off the track.

*References*

https://en.visitsvalbard.com/things-to-do/coal-mine-3-visit-gruve-3-as-p2523693

https://www.gruve3.no/

http://www.reuters.com/article/norway-coal-idUSL5N11A12I20150904

personal communications, Terje Carlsen, head of communications, SNSK AS, 2012, 2016; Ragnhild Utne, booking manager 2021; Herdis Lien, curator, Svalbard Museum, 2021

**Figure 294.1. Gruve 3 (© Store Norske Spitsbergen Kulkompani AS).**

**295. Svalbard Museum**

Svalbard Museum, whose name is the same in English as in Norwegian, opened in 1981 in the oldest part of Longyearbyen in a former pig barn. The present museum opened in 2006 in a larger and newer building measuring 750 square metres. It is run by a private foundation with some public support. There are eight members of staff and in 2019 the museum attracted 56,000 visitors.

The exhibition covers the geology of Svalbard, natural history and 400 years of cultural history: from mineral extraction and the first whalers of the seventeenth-century to present-day society, based on modern mining, research and tourism. The themes of the displays range from the interplay between nature and people to the basis of life on Svalbard. The museum has 10 zones, one of which deals with geology and mining.

The geology of Svalbard is especially distinctive and very visible, due to the sparse vegetation. The landscape looks almost as it did at the end of the Ice Age. The land is barren and more than half is covered by glaciers. Millions of years ago the area had lush vegetation and was populated by dinosaurs on land and reptiles in the sea. The landscape was worn away and in some places coal beds, formed in a warm climate, were exposed at the surface. The displays include fossils found in Svalbard, a geological timeline, and minerals and rocks that have been mined or quarried locally.

The method of mining on Svalbard was that a main roadway was driven first; smaller roadways, called cross-cuts, or gains, were then driven branching off the main gallery to reach the actual coal seam. Another roadway, the stope, or working face, was driven between the cross-

**Figure 294.2. View from Gruve 3 of colliery tubs and skip and A-frame from an aerial ropeway to lower tubs into the mine off the track (© Store Norske Spitsbergen Kulkompani AS).**

**Figure 295.1. Exterior of Svalbard Museum (© Stiftelsen Svalbard Museum).**

cuts into the coal seam itself, where the coal was actually mined. The area between two cross-cuts was called the production panel. The cross-cuts were driven just beneath the coal seam so that the entrance to the stope was a short way up the wall, at the actual face. To keep the mine operating continuously, new cross-cuts had to be driven all the time to prepare the way for those who actually mined the coal. Until 1996 the coal was hand-worked by miners lying on their sides in the thin seams (**294**). One of the displays includes a full-scale model showing this, and a stope is simulated for children to crawl through to see what it was like to be a miner before 1996 (*cf.* **385**). Modern highly technological machinery operating in Svea Mine is shown on a film in the displays.

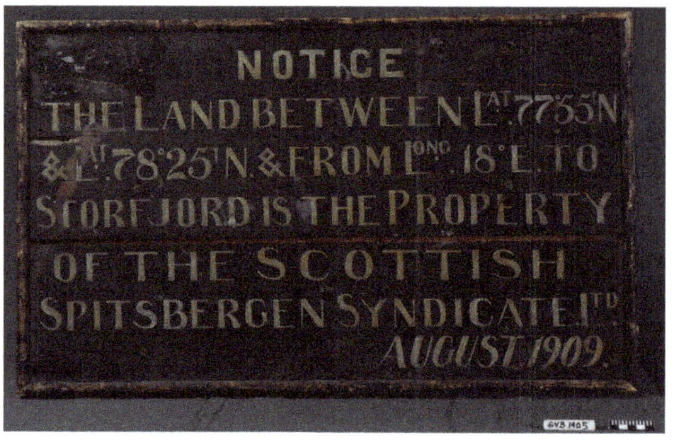


**Figure 295.2. Wooden claim sign dated 1909 (© Stiftelsen Svalbard Museum).**

The museum has about 140 claim signs, of different shapes and materials, such as wood or iron. The museum also has footprints preserved in stone from a suborder of herbivorous mammals. The pantodonta were one of the first groups of large mammals to evolve after the end of the Cretaceous period, when Svalbard was still located near the Equator. In 2008 miners found some stones in the roof of Gruve 7. They were about to kick them down, because such stones in the roof can fall and cause serious injuries (*cf.* protection in Germany from 1976: **105**). Then they realised that these were not ordinary stones, and a palaeontologist identified them as footprints from an extinct animal, which had lived in the valleys of Svalbard about 60 million years ago. Most remains of pantodonts are found in Asia, North America and South America, and these footprints appear to be the only traces found in Europe. Three of the footprints are on display.

**Figure 295.3. Pantodont footprint (© Stiftelsen Svalbard Museum).**

*References*

http://digitaltmuseum.no/info/owners/SVB

http://svalbardmuseum.no

personal communications, Herdis Lien, curator, 2016, 2021

# Poland

Poland has reserves of 22.3 billion tonnes in two large anthracite coalfields, the Lublin in the east and the Upper Silesian in the south. There are five lignite coalfields in the west and central regions, with reserves of 24.3 billion tonnes (euracoal). In 1742 Silesia was annexed by Prussia and the first permanent coal mine was established in Szczakowa in 1767. The Kopalnia Królowa Luiza (**298**) was sunk in 1791 as the first state coal mine in Upper Silesia. Mines continued to be developed throughout the nineteenth century (**299-305**); for example, by 1800 there are already 50 pits in the Wałbrzych region (**301**), employing more than 1,000 miners with an output of approximately 128,000 tonnes per year.

After WWI the Prussian owners of the Upper Silesia coal mines were ousted. Following three Polish uprisings Upper Silesia was divided, with Poland receiving only 28 *per cent* of the territory, but 75 *per cent* of the industrial areas, including 53 out of 67 coal mines. Between WWI and WWII (1918-39) British, French and German capital was invested in these mines until, in 1939, the German army took all of them over to supply the German war industry (Ponty 2016, p. 223). The miners included Russian prisoners-of-war. Until 1945 the miners' greeting was *Glückauf* (**210**), but it then became *Szęść Boże*, 'God bless' (personal communication, Beata Piecha-van Schagen (**297**), 2022). The region returned to Poland in 1945, under communist rule with a state-directed economy. Three-shift working was introduced, but with no real investment in the mines. In 1949 a Miners' Statute attempted to attract more young people into mining by offering special privileges, including a retirement age of 55 for those who had worked underground for 25 years (Szmidtke 2011). Inefficiencies in mining techniques and declining coal prices reduced the number of mines from 70 in 1990 to 25 today and the number of miners from 388,000 to 80,000. Poland is, however, still heavily dependent on coal for generating energy, 74 *per cent* of which came from coal in 2020. In 2019 Poland published an energy policy to 2040, which anticipated anthracite and lignite plants remaining stable until 2030 (euracoal). Coal usage is due to be phased out by 2049.

### 296. Muzeum Gornictwa Podziemnego w Nowej Rudzie (Nowa Ruda Underground Mining Museum)

The Muzeum Gornictwa Podziemnego w Nowej Rudzie is located in Nowa Ruda, in Lower Silesia in south-western Poland near the border with Czechia. The town is situated in the Klodzko Valley of the river Włodzica in the central Sudetan Mountains. Iron ore was mined here in the Middle Ages from about CE 1000 to about 1500. Coal mining came later and continued until 2000. Underground workings from the nineteenth and twentieth centuries and mining machinery, together with training adits of the former school of mining, have been converted into the museum. The underground tour, over roadways extending for 700 metres, takes 40 minutes. On the surface there is a museum with coal-mining exhibits, such as mining implements and plans, as well as a tour of the surface lasting 30 minutes.

*References*


http://kopalnia-muzeum.pl/

showcaves


### 297. Muzeum Górnictwa Węglowego w Zabrzu (Zabrze Coal Mining Museum)

The Muzeum Górnictwa Węglowego w Zabrzu is located in Upper Silesia in Zabrze, the capital of Silesia. The 4,500-square-kilometre Górnośląskie Zagłębie Węglowe coalfield used to be one of the largest producers of anthracite in the world, but is currently being intensively restructured. The city was originally called Zabrze, but in 1915, while Silesia was part of Prussia, Zabrze was renamed as *Hindenburg*, called after Paul von Hindenburg (1847-1934), a German field marshal during WWI. It regained its previous name when the region returned to Poland in 1945. The museum is owned by the city of Zabrze and the province of Silesia, which together fund the museum, along with support from the Polish Government. The museum has over 300 members of staff and attracts more than 200,000 visitors a year.

The museum possesses some 55,000 objects, not all of them on display. These include documents from Tarnowskie Góry attesting to membership of the Upper Silesian Knappschaftskassen. This was a miners' association that represented the employment interests of the members and provided social support, similar to the Japanese Tomoko (**276**). The earliest records date back to the Late Middle Ages (CE 1250-1500), while the organisation developed further in the nineteenth century and laid the basis for benefits for dependents, health insurance and modern pensions. The first Knappschaft miners' hospital in present-day Poland opened in 1832 in *Waldenburg* (**301**) and in the early-twentieth century a number of large and beautiful Knappschaft hospitals were built in Upper Silesia, now operated by official organisations.

The museum has seven different sites in Zabrze, including the main museum building of 932 square metres; the buildings and underground workings of the

**Figure 297.1. Main museum building (© Muzeum Górnictwa Węglowego Zabrzu).**

Sztolnia Królowa Luiza (**298**); the Główna Kluczowa Sztolnia Dziedziczna mouth in the centre of Zabrze; the former water tower; and the Zabytkowa Kopalnia Węgla Kamiennego Guido (**299**), which merged with the Górnictwa Węglowego in 2013. The Guido hostel is a modern self-catering accommodation and conference centre located at the Kopalnia Guido. The Sztolnia Królowa Luiza and the Kopalnia Guido are both on the Industrial Monuments Route of Silesia Province.

### 298. *Muzeum Górnictwa Węglowego w Zabrzu: Sztolnia Królowa Luiza (Zabrze Coal Mining Museum: Queen Louise Adit)*

The Kopalnia Królowa Luiza was so important to the economy of the Prussian state that it was named in honour of one of the most respected women in early-nineteenth-century Europe, Louise of Mecklenburg-Strelitz (1776-1810), queen of Frederic William III of Prussia (1770-1840). By 1816 it was producing 25 *per cent* of the coal of Upper Silesia. In 1846 railways enabled coal to be moved over distances and by 1898 it was producing 3 million tonnes and employing 8,500 miners. Its presence contributed to the evolution of the region from an agricultural to an industrial economy and Zabrze from a small settlement into a large city. Many technological innovations, such as powered roof supports, were introduced first at the Kopalnia Królowa Luiza, which was regarded as one of the most modern and efficient coal mines in Europe and did not stop production until 1998.

The Sztolnia Królowa Luiza complex, which has statutory protection, consists of the surface buildings of the Carnall and Wilhelmina shafts and more than 5 kilometres of underground roadways. The tour is enhanced with the sounds of mining machines and pickaxes and three-dimensional presentations of the shortwall method of mining and displays of equipment, machines and tools used in the nineteenth and twentieth centuries. It includes the wharf where coal was loaded onto boats using a timber crane. The steam-winding engine, which was installed at the Carnall shaft in 1915, is believed to be the oldest steam-winding engine still operational in Europe (see also **314**). The complex was restored between 2009 and 2019. The Sztolnia Królowa Luiza is on the Industrial Monuments Route of Poland.

A drainage channel, 14.25 kilometres long, was constructed underground at the Kopalnia Królowa Luiza between 1799 and 1863 (*cf.* **301**). The channel is known as the Główna Kluczowa Sztolnia Dziedziczna, which is a direct translation of the German *Hauptschlüsselerbstollen*, of which an exact translation is the 'Main Key Hereditary Adit'. The 'main' element refers to its central role in draining all the mines of the area, and the 'key' element to the fact that it was essential, while its 'hereditary' nature consisted in its inheriting of water. In hydrology-technical terms, an 'hereditary' adit/canal/river would be called the 'mainstem' or 'trunk' adit/canal/river. So strictly speaking the *Erbstollen* does not refer to the adit that inherits or takes over water from the mines, but one that takes

**Figure 298.1. Steam-winding engine servicing the Carnall shaft at the Luiza Mine (© Muzeum Górnictwa Węglowego Zabrzu).**

**Figure 298.2. Mine car carrying visitors in the Luiza underground (© Muzeum Górnictwa Węglowego Zabrzu).**

over the water from more than one canal or watercourse underground. The Główna Kluczowa Sztolnia Dziedziczna was the longest hydrotechnical construction in Europe and connected Upper Silesia with the ports of western Europe via its outlet at the Klodnica canal (Jurkiewicz 2013). Powered boats are used on the canal for 1,100 metres of the visit underground. A nineteenth-century wooden boat, 4 metres long, was found while removing mud from the underground corridors of the adit. This will be displayed in the adit after conservation.

### 299. *Muzeum Górnictwa Węglowego w Zabrzu: Zabytkowa Kopalnia Wegla Kamiennego Guido (Zabrze Coal Mining Museum: Historic Guido Coal Mine)*

The Zabytkowa Kopalnia Wegla Kamiennego Guido was named after Prince Guido Henckel von Donnersmarck (1830-1916), who was responsible for establishing the Guido Coal Mine in 1855 to provide the nearby Donnersmarck mill with coal. The nineteenth-century heapstead building below the headgear was renovated in the 1930s and contains a two-drum 560-kilowatt BB-550 electric-winding engine manufactured in 1927 by the Berlin AEG Company. When its coal was exhausted, the colliery was used to dewater the surrounding mines until just after WWII. In 1982 an open-air museum was set up on the surface of the site, which was listed in 1987, but the museum closed in 1996.

The museum reopened to the public in 2007, this time with an underground component, 170 metres below the surface, using the Kolejowy shaft. Level 170 exhibits geological specimens, mining tools and rescue equipment in the original late-nineteenth-century roadways. The pit ponies' stables still exist underground in their original state. Level 179 is also known as the Saint Barbara level because it also exhibits books, sculptures and other objects associated with Saint Barbara (**96**) and legends about her life using multimedia. The geological structure of the mine itself is unusual, with layers of calcite formations, crystals and tectonics.

**Figure 299.1. Guido headgear (© Muzeum Górnictwa Węglowego Zabrzu).**

**Figure 299.2. Level 170 of the Guido underground with arched girders and mine transport (© Muzeum Górnictwa Węglowego Zabrzu).**

**Figure 299.3. Level 320 of the Guido underground with arched girders and mine cars (© Muzeum Górnictwa Węglowego Zabrzu).**

At the end of 2008 a second tour was opened at a depth of 320 metres, at the level that in 1967 was used for testing mining machinery. The roadways here are long and for much of their course the roofs do not require support. Level 320 demonstrates mining methods in the twentieth century, including a 246-tonne coal container, conveyor belts, hydraulic roof supports, an Alpina tunnelling machine in full working order, a seismograph chamber and various transport systems, as well as a simulated mining accident. Four underground conference rooms are available, including an area licensed to sell alcohol.

Level 355 offers tour to what was the last active longwall coalface, 100 metres long and cut in the second half of the twentieth century. This has been preserved in the same condition as when the miners left it.

*References for **297, 298, 299***

ERIH, *s.n.* Zabrze

https://de-m-wikipedia-org.translate.goog/wiki/Knappschaft?_x_tr_sl=de&_x_tr_tl=en&_x_tr_hl=en&_x_tr_pto=sc

https://wydawnictwo.muzeumgornictwa.pl/images/Katalog-Sztolnia.pdf

http://www.muzeumgornictwa.pl/

personal communications, Grzegorz Buchała, Zabrze Department of International Documentation and Research Centre on Industrial Heritage for Tourism, 2015, 2016; Aleksandra Margas, head, Development and Projects Department, 2015, 2021; Magdalena Szczypkowska, deputy head, Development and Projects Department, 2021; Beata Piecha-van Schagen, scientific activity project manager, 2022

www.kopalniaguido.pl

showcaves

### 300. Muzeum Miejskie Sztygarka w Dąbrowie Górniczej (Dąbrowa Górnicza Mining Foremen Municipal Museum)

The Muzeum Miejskie Sztygarka is located in Dąbrowa Górnicza in south-western Poland. It occupies a building constructed in 1842 and designed by the Italian architect, Francesco M. Lanci (1799-1875). There are 16 exhibition rooms, two multimedia rooms and, about 250 metres from the building, an underground facility used for training. The museum has 29 members of staff with 10 volunteers and attracts 10,000 visitors a year. It is funded by the city of Dąbrowa Górnicza.

In 1889 a school of mines was founded in the 1842 building. A chemical laboratory, exhibitions of mineralogical and geological collections, a library and workshops were established, in addition to the departments of mining and metallurgy. After WWI the system for training miners was reorganised, with the classrooms and lecture halls expanded and the mineralogical and geological museum restructured. In 1927 shallow coal deposits were exposed by drilling to produce a training coal mine to give students practical mining experience.

The geological museum resumed activities after WWII and operated until 1968, when the mining school closed together with the museum. In 1985 the museum was reactivated as having been an integral part of the mining school. The Muzeum Miejskie Sztygarka was established in 1997. The mining school was commonly referred to as Sztygarka, because *sztygarów* (mining foremen) were educated there. The term was included in the name of the museum to provide historical continuity with its predecessor. The museum has displays on archaeology, ethnography and history. and organises lectures and

**Figure 300.1. Entrance to training coal mine (© Muzeum Miejskie Sztygarka).**

**Figure 300.2. Longwall shearer installed in the training coal mine (© Muzeum Miejskie Sztygarka).**

field trips, including one specifically to the traces of coal mining in the area of Dąbrowa Górnicza. Every June it also participates in the organisation of Industriada, an annual event promoting the heritage of technology in Silesia.

Coal mining has been practised for more than 220 years in the region of Dąbrowa Górnicza, with almost 120 coal mines having worked at some time during that period. The training coal mine at the museum is now the only one left, having been converted into an underground experience. Visitors descend 25 metres below ground to 800 metres of roadway. As well as mining machinery, such as mine cars, mining transport and shearers, there is a longwall face and natural coal deposits are on display. The route is equipped with authentic coal-mine notices, such as one saying, 'carelessness at work results in death or disability', and with audio-stories and videos that describe the mining world as if seen through the eyes of students of the mining school.

*References*

http://new.muzeum-dabrowa.pl

http://www.zabytkitechniki.pl/Poi/Pokaz/1706

personal communications, Arkadiusz Rybak, director, and Andrzej J. Wójcik, deputy director, 2015; Arkadiusz Rybak, 2021

### 301. Muzeum Przemyslu i Techni w Walbrzychu (Walbrzych Museum of Industry and Technology)

The Muzeum Przemyslu i Techni w Walbrzychu is situated on the site of a former lignite mine in Wałbrzych in Lower Silesia. There has been coal mining since 1536 in the region, at that time known as *Waldenburg*, although exploitation of the coal reserves only really began in the nineteenth century. The first use of boats to move coal began in 1791 (*cf.* **298**) on underground canals, which both drained minewater and gave level transport in a mountainous area (Szewczyk 2009, p. 176). The coal enabled the development of chemical plants and steelworks (*ibid*., pp. 178-9). What is believed to be the first china factory in the world to be coal-fired was built in 1834. Railways reached the area in the second half of the nineteenth century.

At the end of WWII the city became part of Poland, renamed as Walbrzych. The late-twentieth century saw coal decline in importance. Julia Mine, which had operated since the seventeenth century, ceased production in 1996. The site was converted into a museum, which opened in 2014. At the same time, there has been restoration of the polluted industrial landscape (*ibid*., pp. 181-2).

A complex of 11 mine buildings was restored and became a centre for the arts, culture, education and science, with the aim of connecting the past and the present. The museum deals with coal-mining technology, including displays of machinery, geology and the working lives of miners and their living conditions. Some displays consider other aspects of the history of Wałbrzych, such as the chinaworks. The complex also houses an art gallery, ceramic workshops and a concert hall, known collectively as Old Mine. Contemporary Polish art is displayed in the former boiler house and a choral group, cultural centre and dance company are housed in the former pithead baths.

*References*

ERIH, *s.n.* Walbrzych

https://pl.wikipedia.org/wiki/Muzeum_Przemysłu_i_Techniki_w_Wałbrzychu

http://www.starakopalnia.pl/

showcaves

### 302. Muzeum Śląskie w Katowicach (Katowice Silesian Museum)

The Muzeum Śląskie w Katowicach is located on the site of Katowice Colliery, which had opened in 1823 as Ferdinand Colliery, in Katowice, in Upper Silesia. The Towarzystwo Muzeum Ziemii Śląskiej (Society of the Silesian Museum) began in 1924 by collecting folk costumes, handicrafts, paintings and pieces of sacred art. The Muzeum Śląskie (Silesian Museum) opened to the public in 1929. Initially the museum was housed on the fifth floor of the Silesian parliament building. In 1936 work began on an impressive new building for the museum, designed by the architect, Karol Schayer (1900-71) and completed in 1939. When the German forces invaded Poland in 1939, they demolished the building with the aim of destroying this symbol of Polish identity. Much of the collections were also destroyed; those that survived were moved to the Landesmuseum in Bytom, now called the Upper Silesian Museum. In 1984 the museum moved its collections back from Bytom into a four-storey building in Katowice and opened all its displays in 1992.

The colliery closed in 1999, with the buildings at the front of the colliery being demolished to make room for additional museum buildings. The museum acquired ownership of the 7-hectare site in 2004 and construction of a new purpose-designed building for the museum began in 2013, which opened in 2015. It displays a large collection of Polish paintings and archaeological, ethnographic, historical and theatrical exhibitions. The displays about Upper Silesia throughout history include the development of the coal-mining industry and of Katowice Colliery. The Department of Amateur Fine Arts collects work created by former miners, some of which is on display. The museum occupies 2.5 hectares, with an exhibition space of 7,500 square metres. The museum attracted 246,069 visitors in 2019 and has 180 employees and 70 volunteers. It is funded by the Ministry of Culture and National Heritage and the Silesia Regional Government.

Much of the museum has been constructed below ground to provide sufficient space, while avoiding as far as possible encroaching on the historic structures of the colliery. The former clothing storehouse now houses the Centre for Polish

**Figure 302.1. Renovated former winding-engine house of the Warszawa shaft of Katowice Colliery in front of legs of observation tower (© Muzeum Śląskie; photographer Sonia Szeląg).**

**Figure 302.2. Interior of winding-engine house converted to restaurant and bar (© Muzeum Śląskie; photographer Sonia Szeląg).**

Figure 302.3. 1892 steam-winding engine, installed in 1893 (© Muzeum Śląskie; photographer Jacek Mężyk).

Scenography and the offices of the education section. The winding-engine house of the Warszawa shaft is occupied by the museum's restaurant and bar. The structure of the restaurant has been suspended so that none of the new material actually touches the fabric of the historic building and that the original brickwork, large windows and tiled walls can still be seen (*cf.* **304**). Stairs and a lift have been added to the Warszawa headgear, so that visitors can reach a platform at the top, from which there is a panoramic view over the city. Other buildings preserved on site are the electrical workshop, which has been adapted to serve as a technical building for the museum, and the carpenters' workshop, which is used for education, entertainment and exhibitions. The entrance to the museum exhibition, 'The light of history; Upper Silesia over the ages', recreates the entrance to the historical Katowice Colliery.

The complex of buildings relating to the Bartosz shaft – the pithead buildings, winding-engine house and others adjacent – have been entered in the Register of the Monuments of Silesia. The buildings are not open to the public except for special events taking place during the spring and summer. The pithead baths are going to be occupied by museum departments previously housed elsewhere. The winding-engine house still contains the original steam-winding engine, which was manufactured in 1892 in Isselburg in North Rhine-Westphalia in Germany and installed in 1893. The engine worked until 1996, making it the oldest and longest-functioning winding engine in the Polish mining industry. The rope of

Figure 302.4. Former clothing storehouse of Katowice Colliery after renovation, now premises of the Centre for Polish Scenography (© Muzeum Śląskie; photographer Sonia Szeląg).

the horizontal engine has a diameter of 7,000 millimetres. The diameter of the cylinders is 1,100 millimetres, with a 2,000-millimetre stroke.

*References*

http://www.muzeumslaskie.pl

personal communications, Barbara Willak, promotion specialist, 2016; Magdalena Czerny-Kehl, coordinator of artistic projects, 2021

### 303. Szyb Bończyk (Bończyk Shaft)

The Szyb Bończyk is located in Mysłowice in Upper Silesia, close to the boundaries of Katowice, Mysłowice and Sosnowiec. In 1837 the Danzig shaft was sunk at the Mysłowice Kopalnia Wegla (Myslowice Coal Mine). In 1901 the mine was at the forefront of technological development in using hydraulic propellant to backfill the voids left underground by the extraction of the coal. In 1919 the First Silesian Rebellion against German rule was sparked by the deaths of seven people when the Grenzschutz soldiers (border guards) attacked miners here during a labour dispute. A coal-dust explosion in 1987 killed 18 miners. The mine was connected to the KWK Wesoła Mine to form the Mysłowice-Wesoła Mine, which still produces some 3 million tonnes of coal every year and has reserves of 232 million tonnes.

The Bończyk shaft was first a ventilation shaft, originally named the Otto shaft, as this was sunk while Silesia was part of Germany. It had two cages for carrying materials and ventilators, with a *Koepe* winding mechanism (**127**) of 160-kilowatt power. The shaft ceased to be used in 1984. The machinery and the materials of the buildings above and around the shaft started to be salvaged for scrap 10 years later; for example, the winding engine and the shaft building were cut up for sale. The remainder of the machinery was stolen. In 2007, however, work started on converting the former transformer building (constructed in 1911) into a privately owned banqueting, concert, conference and exhibition centre. The transformer building has been transformed into four separate halls, with a total area of 900 square metres and able to accommodate 160 wedding guests. The original large round-headed windows and brick walls have been retained. The building opened in 2012 and highlights local Silesian customs; for example, providing Silesian food and a welcome from children dressed in traditional Silesian costumes.

*References*

http://szybbonczyk.pl/index.php/en/view-szyb-bonczyk

personal communication, Maciej Sikora, director, 2016

### 304. Szyb Maciej (Maciej Shaft)

Szyb Maciej is located on the site of the former Concordia Colliery in Zabrzu-Maciejów, a district in the western part of Zabrze in Upper Silesia. The Maciej shaft was sunk 220 metres deep to work the coal seams in the western part of the underground area of Concordia Mine and subsequently came into the ownership of Pstrowski Mine. A complex of buildings, including the 25-metre-high headgear, was constructed around the shaft. Concordia Mine was a very wet colliery. The difficulties caused by the excessive inflow of water meant that the sinking of the shaft, which had started in 1922, was not completed until 1928, much later than had been expected. The shaft had to be equipped with a very complicated pump system to keep it dry. The plans for this are still kept at the site. Water was drained from the shaft by tubular drains positioned on the vertical walls; it was then pumped into an underground tank and later pumped to the surface. In 1928 the owner of Concordia Mine at that time, the Castellengo-Abwehr Guild, began utilising this abundant water supply and established municipal waterworks. The water was distributed by a system of iron pipelines to supply towns and workplaces along the railway line from Gliwice to Bytom.

Underground mining finished here in 1978 and in 1992 the Pstrowski Company transferred ownership to the present mining company, Demex Limited, to use the shaft to pump water at a temperature of about 8 degrees Celsius. Two ground-source heat pumps, each with a 42-kilowatt heating capacity, provide domestic hot water, and heating and cooling for the buildings on the site, which have been converted to a bar, office and restaurant. The restaurant, with an industrial atmosphere, is located in the former heapstead building and concentrates on providing high-quality international cuisine, alongside traditional Silesian food. The former electrical workshop now houses a bistro. The water pumped to the surface is still used to supply high-quality potable water to the local community. The site opened in 2014 and attracts 16,000 visitors a year. The museum is 1,000 square metres in size and employs three members of staff, as well as tour guides.

The historic buildings and their machinery are presented to visitors in one-hour guided tours from Tuesdays to Sundays throughout the whole year. As Maciej shaft was not a coal-production site, but a ventilation and dewatering shaft, the museum does not display any coal specimens. The first item of especial interest is the right-hand cylinder of a pair of cylinders for a steam-winding engine, together with associated valve gear with drop valves. This was manufactured in 1914 in Breslau by Linke Hofmann, one of the ancestor companies of the present-day French multinational company, Alstom, now mainly producing locomotives and associated railway equipment. This is one of the two engines that used to drive the lifting shaft of Władysław Mine in Jadwiga, then later Pstrowski Mine in Zabrze-Biskupcach.

The second is the medium-duty double-drum electric-winding engine that provided power to raise and lower the two cages in the shaft with miners and materials. This was built in 1925 by Siemens Schuckertwerke, who produced heavy-duty electrical equipment, and became part of the larger Siemens group in the 1960s. The winding engine

**Figure 304.1. Right-hand cylinder and associated valve gear with drop valves, from Linke Hofman steam-winding engine from Wladyslaw Mine beside the Szyb Maciej building with headgear beyond (© Szyb Maciej).**

**Figure 304.2. View of headgear from below, seen through glass insert in the ceiling of the banqueting suite (© Szyb Maciej).**

was restored in 2009, together with the interior of the winding-engine house. It is still in its original position and is believed to be the oldest Siemens-Schuckertwerke electric-winding engine still operating.

The outstanding feature of the site is the attention that has been paid to preserving the historic fabric (*cf.* **302**), while converting it for modern purposes. Immediately below the historic headgear, at the point where coal, men and materials entered and left the shaft, is the banqueting suite. This has a glass ceiling to provide a full view of the headgear above. The walls have been lined to ensure modern levels of insulation, but with glass so that visitors can see the original steel supports. On the floor beneath is a second restaurant and below that the bar area. A viewing platform provides views over the surrounding area of Gliwice and Zabrze.

The restoration methods and the standards of the conservation works were all approved by the General Conservator, a government office based in Katowice. The conservator is responsible for the listing of buildings of special architectural or historical importance for preservation for the nation. In 2011 the complex won an award for the conservation work and conversion of the industrial monument for tourism and in 2015 for the restoration and adaptation specifically of the heapstead building. The result is that this is the only restaurant in Silesia that is on both the Industrial Monuments Route of Poland (**298**) and the Route of Silesian Tastes.

*References*

https://www.szybmaciej.pl/pl/

https://www.youtube.com/watch?v=WteRT9KaVwc

personal communication, Aleksandra Nemś, managing director, 2016

www.slaskiesmaki.pl/en-US

**305. Szyb Wilson (Wilson Shaft)**

The Szyb Wilson is located in Katowice in Upper Silesia. Coal mining began in the late-eighteenth century in Janów, a village now part of Katowice. The Szyb Wilson was sunk in 1864 and the two shafts of the pit were initially called Richtofen and Hulda. In 1935 the pit was renamed after Woodrow Wilson (1856-1924), president of America (1913-1921). He was one of the main influences on the setting up of the League of Nations in 1920, having set out 14 principles for world peace in a speech in 1918 to the United States Congress. The thirteenth point was that there should be an independent Polish state. The renaming of the mine was a compromise between the Polish workforce and the American owners of the mine.

After WWII, when Silesia became part of Poland, the complex became part of Wieczorek Mine, which had been sunk in the early-nineteenth century in Nikiszowiec, a district of Katowice. It had been renamed in honour of a miner, Józefa Wieczorka (1893-1944), who was a labour activist and a member of the Silesian parliament, who had died in the Auschwitz concentration camp. The red-brick pithead baths and shaft building of the Szyb Wilson, which date back to 1918, were designed by the Zillman brothers, Georg (1870-1937) and Emil (1871-1958), who were also responsible for the housing in the Giszowiec and Nikiszowiec districts. The Szyb Wilson closed in 1995 and the last miners came out in 1997.

The site now houses a hotel. In 1998 the Fundacji Eko Art-Silesia (Eko-Art Silesia Foundation) and the Pro Inwest Sp. (Pro Invest Company), together with a group of enthusiasts, started adapting the pithead baths and shaft building into a gallery of modern art within the hotel complex. The renovation and restoration were accompanied by the social revitalisation of the surrounding area and the provision of modern offices in the former boiler house, pithead building and winding-engine house. There is a total area of 2,500 square metres of display space. The small gallery occupies the building where the miners used to collect their tallies. The medium gallery is located in the shaft building. The former pithead baths have been converted to form the large gallery. The main aim of the Szyb Wilson gallery is to promote young and innovative artists, graphic designers, performers and photographers. An association of amateur painters, the Grupy Janowskiej, includes retired miners amongst its membership. Since 2008 the Szyb Wilson has also held a festival of naïf art, exhibiting the ceramics, metalwork, paintings and sculptures of naïf artists from all over the world, which is believed to be the largest event of its kind in Europe. The large gallery is used for balls, concerts, conferences, fashion shows, festivals, lectures, performances and shows and the site has also served as a setting for artistic photographic shoots. films and music videos. The gallery employs 10 staff, with five volunteers, and attracts 20,000 visitors a year. It is funded by grants and its own resources.

The derelict Pulaski mineshaft survives behind the gallery buildings, a *pulaski* being a tool with a head that has an axe blade on one side and an adze on the other. The sympathetic conversion of the industrial buildings has enabled the Szyb Wilson to be included on the Industrial Monuments Route of Poland (**298**). It is also on the Silesian Industrial Monuments Route list.

*References*

http://jura.slaskie.travel/en-US/Poi/Pokaz/15922/1951/kopalnia-wegla-kamiennego-wieczorek

https://en.wikipedia.org/wiki/Fourteen_Points

http://www.inyourpocket.com/katowice/wilson-shaft-gallery_65097v

https://www.poland.travel/en/tourist-route/industrial-monument-route-in-silesia

http://www.szybwilson.org

personal communications, Monika Pac-Bros, president, Fundacji Eko Art-Silesia, 2016; Johann Bros, president, Pro Inwest Sp., 2016

# Romania

Romania has anthracite in the Jiu Valley coalfield of the county of Hunedoara in south-western Transylvania, western Romania, and lignite in the Oltenia field in the south-west and the Ploiesti field in the south-east (euracoal). Coal was first found in the Jiu Valley in 1782 by a mineralogist, Janos Benko, when an outcrop was accidentally ignited. The first mine was sunk in 1790 after coal was identified by a woodcutter in Staierlac in the south-west on the border with Serbia. The first systematic geological prospecting was undertaken in 1835, finding considerable reserves of coal, and large-scale exploitation by Czech, German and Polish companies began in 1848. The largest mine in Romania, Petrila, was sunk in 1859; it reached a depth of 940 metres and production peaked at 1.2 million tonnes in 1983 when 4,500 people were employed.

Once construction of the railway from Petrosani to the important railway junction of Simeria was completed in 1870, the Jiu Valley became a major coal-mining region for the second half of the nineteenth century and during the twentieth century. Transylvanian coal powered railways and ships and the industries supporting the economic unification of Greater Romania, created in 1918 (see Bulearcă, Neagua, Simaa and Mărguş 2014 for the history of coal mining in Romania up to 1948). The high level of industrialisation of the Jiu Valley drew many people from other regions of Romania during the communist period (1946-89). Many mines were closed down in the 1990s in the years following the collapse of communism. In 1997 the government began a restructuring of the coal industry, with a compensation scheme for redundant miners, whose numbers dropped from 55,000 under communism to fewer than 7,000. This has left Hunedoara County with an unemployment rate of 9.6 *per cent*, in comparison with the Romanian national average of 5.5 *per cent*. Possible regeneration schemes were considered, including a feasibility study for an underground tourist mine carried out by the NCMME (**386**), but the promises of regeneration were not kept. Difficult geological conditions and outdated machinery led to many accidents, including the deaths of 13 miners in 2008 in two explosions at Petrila, which closed in 2015 (https://www.systext.org/node/696).

There are reserves of 10 million tonnes of anthracite, still produced by four deep mines, and 281 million tonnes of lignite, 80 *per cent* of which is mined in 12 opencast mines. These are licensed for another 50 years and are adjacent to two power stations. In 2018 the government published an energy strategy to 2030, under which it will modernise its low-emission power plants with investment in gas, lignite and oil (CMMCP 2016, 29; euracoal).

### 306. Muzeul mineritului Petroşani (Petroşani Mining Museum)

The Muzeul mineritului Petroşani is located in the city of Petroşani in the Jiu Valley. The museum opened in 1961 and is the only museum in Romania that focuses on coal-mining technology and history. At first it was located in a house, on a site now occupied by the central park. Its limited size meant that the displays were very small, as most of the building was used for storage of the collections. In 1966 the museum moved into a building that had been erected about 1900 and used for various purposes, including the headquarters of the S.A.R. Petroşani Company and housing for the company's employees, the first tenant having been Chief Inspector Joseph Iancu. The building was appropriate for use as a museum because it had been designated as an historical monument by the Culture Ministry. Extensive renovations were made to the museum between 1996 and 1998.

The museum's collections include some 1,500 items related to mining, including clothing, documents, equipment, fossils, insignia, sketches, tools and objects unique to the region. Also included are axes, breathing apparatus, drill bits, picks, steel flints for firing explosives and shovels. Mining technology and models are displayed by period – 1944-65, 1965-8, 1968-77 and 1977-89 – to illustrate how technology developed through time in areas such as automated cutting machinery, high-performance drilling machines and submersible pumps. For the period 1918 to 1944 there is material on different types of energy to drive machinery, beginning with steam, then compressed air and finally electricity.

The museum also deals with social history. At the beginning of the exhibitions there is a painting of 'Sfânta Mare MucenIţă Varvara, ocrotitoarea minerilor' (the Great Martyr Varvara, protector of miners), known elsewhere in Europe as Saint Barbara (**96**). The displays use documents and photographs to deal with the contribution made by miners to the 1918 creation of Greater Romania and their participation in the workers' strikes and struggles in 1919-20, 1933 and 1941. Special attention is given to the Lupeni miners' strike in 1929, when troops fired on the miners, leading to a number of deaths and many wounded. There is also a collection of old photographs of mining settlements in the valley.

*References*

http://ghidulmuzeelor.cimec.ro/id.asp?k=383

https://en.wikipedia.org/wiki/Lupeni_strike_of_1929

https://en.wikipedia.org/wiki/Petro%C8%99ani_Mining_Museum

http://valeajiului.blogspot.com/2007/06/muzeul-mineritului-petroani.html

museum leaflet

# Russian Federation

Russia has the second largest coal reserves in the world, 157 billion tonnes, behind only America (Aden, Fridley and Zheng 2008, p. 9), 80 *per cent* of which are in Siberia (in the centre of the country) and 10 *per cent* in the Far East region. Approximately one-third is anthracite or bituminous and two-thirds sub-bituminous or lignite (CMMCP 2016, 30). In 1721 Michael Volkov, who discovered the Kuznetsk coalfield of south-western Siberia in the south of Russia, found his first samples of coal in Kemerovo (**309**), east-north-east of Novosibirsk. The Kuznetsk coalfield still produces about 150 billion tonnes of coal annually.

Coal was first found in the Pechora coalfield in 1828, but mining did not begin until 1930 when Georgy Chernov (1906-2009) identified the coalfield's coal resources at Vorkuta, together with the Bolshezemelskaya tundra petroleum deposits. The Vorkuta museum (**313**) displays a photograph taken sometime after 1957 of Georgy Chernov and his father, Alexander Alexandrovich Chernov (1887-1965), both geologists. The Pechora coalfield, situated on the western slopes of the northern Urals, is about 90,000 square kilometres in size. There are nine geological industrial regions in the Pechora coalfield, mainly with thin (up to 1.3 metres) and medium-sized seams (1.3 to 3.5 metres), with estimated reserves of 312 billion tonnes. The first experimental mine was sunk on the right bank of the Vorkuta River in 1932 and the first coal for industrial use was produced in 1934, mainly by prisoners between 1937 and 1942. The Number 1 Kapitalnaya Mine was developed on the left bank of the river and the region developed further with the building of the Pechora railway in 1942. By 1970 there were 13 mines in operation and in 1975 the giant coal mine, Vorgashorskaja, began producing 4.4 million tonnes a year. By the 1980s there were 218,000 inhabitants in the Vorkuta coalfield, including the adjacent mining villages. Pechora is the second largest source of coking coal in the European half of the Russian Federation, now used mainly for coking at the Cherpovets metallurgical plant in Vologda Province, for industry in Saint Petersburg and for powering the railways.

The major coalfield of Neryungri (**312**) in the extreme south of the Sakha (or Yakutia) Republic, one of the 10 autonomous Turkic republics situated in the eastern part of the Russian Federation, was only identified in 1970. The town of Neryungri, on the main Baikal to Amur railway, was founded in 1975. To fund the development of the mines, half of the 850 million tonnes of coking coal were mortgaged to Japan for $450 million, much of which was used to purchase 10 American-designed Marion Super Front excavators built by Sumitomo in Japan. Unfortunately, when they began to arrive in 1979, they could not cope with the extreme cold and frequently broke down, as their hydraulic hoses kept bursting. The mines are now operated by Yakutugol OAO, one of the largest coal companies in Russia. The company owns three large opencast mines and one underground operation, with a total of reserves at the beginning of 2016 of 206.8 million tonnes.

The coal industry became an important part of the Russian economy in the 1930s and by 1950 coal formed 59 *per cent* of the fuel used. In the 1960s discoveries of massive reserves of oil and natural gas, followed by development of nuclear power, began to displace coal, so that by 2009 only 14.7 *per cent* of power generation came from coal. In 2014 Russia exported 136.7 million tonnes of coal, and 41 *per cent* of coal imports into Europe in 2017 came from Russia. Russian coal mines remain dangerous, because 70 *per cent* of miners' income is based on production rather than a wage: in 2007 36 miners were killed in a methane explosion near Vorkuta and at least 108 at the Ulyanovskaya Mine in a similar explosion (Drexler 2011, p. 244).

### 307. Gukovskiy Muzey (Gukovo Museum)

Gukovskiy Muzey is located in the Rostov region of the Russian Southern Federal District in the south-west of the Russian Federation on the border with Ukraine. It was registered as a museum in 1964 and became a museum of miners' labour in 1999. Its full name is GBUK RO Gukovskiy Muzey Shakhterskogo Truda im. L.I. Mikulina (GBUK Gukovsky Mining Museum named after L.I. Mikulin), with GBUK standing for Gosudarstvennoe Byudzhetnoe Uchrezhdenie Kultury (State Institution of Culture of the Rostov Region). It is, however, usually known just as Gukovskiy Muzey. It is managed and funded by the Rostov Region Ministry of Culture.

In 1950 a local teacher of history and geography and a veteran of WWII, Mikulin Leonid Ivanovich, encouraged his students to collect items to form a school museum. In 1964 this achieved the status of an urban folk museum, with Mikulin Leonid Ivanovich as director, and the museum opened in 1965 as a not-for-profit organisation in the palace of culture (where ideological lectures, and concerts and other creative activities were held). In 1992 the museum acquired the GBUK status of a cultural and educational institution and in 1998 its own purpose-built museum with 24 galleries, by which stage it had some 11,000 objects in its collections. The museum's permanent exhibitions occupy 1,600 square metres, with a temporary exhibition area of 150 square metres, situated in a 0.9 hectare park. There are 22 employees, six of whom are

**Figure 307.1. Entrance to the Gukovskiy Muzey Shakhterskogo Truda (© Gukovskiy Muzey Shakhterskogo Truda).**

**Figure 307.2. Bust of Vladimir Ilyich Lenin in coal, on display in the Exhibition of Mine Builders (© Gukovskiy Muzey Shakhterskogo Truda; photographer VN Plekhanov).**

scientific staff, and it attracts about 25,000 visitors each year.

A collection of the national museum on the history of the city and the region, together with the collection gathered by Mikulin Leonid Ivanovich, form the basis of the museum's collections. From 2002 the museum has been actively collecting items of mining equipment relating to the history of the coal industry in the region. The museum deals with the history, life and work of the miners' settlements in the eastern Donbass from the eighteenth century to the present day. It also deals with the campaigns during WWII which liberated the area from the Germans. This is known in Russia as the *Velikaya Otechestvennaya Voyna*, usually translated as the 'Great Patriotic War', but which really refers to a war fought actually on the soil of Russia (the first *Otechestvennaya Voyna* being the 1812 Napoleonic invasion).

The first of the main coal-mining displays is the Exhibition of Mine Builders. This includes a 1.2-metre-high 1-tonne coal bust of the politician and leader of Russia, Vladimir Ilyich Ulyanov, usually known as Lenin (1870-1924). The other display is the Complex of Mining Halls. An early item is a Bakhmutka open-flame wicked oil lamp. This type of lamp was very common in Europe from the Middle Ages (CE 500-1500) onwards and imitated ceramic Greek and Roman designs. It was the only means of lighting underground in Russia at the end of the nineteenth century. This gallery is laid out like a full-scale underground roadway with cage and face machinery, including a gathering-arm loader used for coal and waste clearance in roadway headings, powered roof supports, shearer and tunnelling machine, together with an underground railway, a VL-10 mine car with seats for 12 miners and shaft linings of various kinds.

These displays are complemented by galleries on subjects such as mining engineers, the work of miners during WWII and miners' art, encompassing fiction writing, paintings, photography, poems, sculptures, songs and weaving. There are over 22,600 objects in the collections, 50 *per cent* of which are on display. These encompass archaeology, coins, household clothing, material objects, mineralogy, natural history, pictures, sound recordings, watches and written sources associated with the development of the coal industry of the eastern Donbass (see **348** and **349** for the western Donbass) from the late-nineteenth century up to modern times. They also include material on the work of Donbass miners in the pits and the way of life and leisure activities of their families. The museum has a collection of archival documents and books, which are catalogued as exhibits.

*References*

http://gukovo-museum.ru/.

http://www.russianmuseums.info/M862

http://www.tenderguru.ru/companyzakupkinfo.php?id=1526860

personal communications, Garkushev Yevgeny Nikolayevich, director, 2013, 2016, 2021

### 308. Korkinsk Shakhtnogo (Korkinsk Coal Mine)

Korkinsk Shakhtnogo is located at Korkino in the province of Chelyabinsk, on the eastern slopes of the southern Ural Mountains on the border between Europe and Asia. It is the deepest and largest opencast coal mine in Europe and second in the world in size only to El Cerrejón (**61**). The mine measures 3 kilometres in length, 3 kilometres in width and 600 metres in depth. As with all opencast mines, it is worked by enormous excavators and earth movers (*cf.* **72.2, 106.1, 114.1-3**) with long spiral descents and ascents. For visitors to the site there is a stairway, 1.5 kilometres long, going down into the mine, although it is understood that it is sometimes possible to ride down on one of the earth movers.

*References*

http://englishrussia.com/2006/11/27/biggest-hole-in-the-ground/

showcaves

### 309. Krasnaya Gorka (Red Hill)

The full official name of this museum is Munitsipalnoye Avtonomnoye Uchrezhdeniye Muzey-Zapovednik «Krasnaya gorka» (Municipal Independent Institution, Museum-Reservation, Red Hill), but it is generally known as Krasnaya Gorka. It was founded in 1991 and opened in 2001 in the city of Kemerovo. It is located on the high right bank of the river Tom on the site of a former coal mine, next to the Gorelaya (Burnt) Mountain. Along the right bank of the river Tom stretch structures associated with different stages of the history of Kemerovo Mine, which was sunk in 1907, close to the stone dock on the river. Between 1922 and 1927 the area became the centre of a unique international experiment, the Kuznetsk Autonomous Industrial Colony, in which the working people of 30 nationalities from around the world participated in mining and civil-engineering activities.

The museum is the responsibility of the municipality of Kemerovo, with funding from charitable donations, the city budget, grants, and income-generating activities such as consultancies and organised excursions. There are 33 members of staff, 13 of whom are curators, and up to 150 volunteers, including interns, students and participants in the environmental activities in the reserve collections. The museum attracts about 60,000 visitors each year. There are 400 square metres of exhibition and 92 square metres of stores for the 18,650 items in the collections.

Various monuments have been erected on Red Hill, including one unveiled in 2007 to the memory of the Kuznetsk miners to commemorate the centenary of the sinking of Kemerovo Mine and another to Saint Barbara (**96**). The exhibitions in the museum describe the history of these monuments and of the autonomous colony, the development of Kemerovo Mine and the story of the Kopikuz mining company, twentieth-century underground production and the safety and rescue equipment used by the Kuznetsk miners and mines rescuers.

At first the museum occupied only the house built in 1916 for the Kemerovo mine manager, but Dutch specialists worked with the museum on the preservation of the area's cultural heritage to create a new Kuznetsk coal-mining museum. This presents the main elements of the mine – the Gorelaya Mountain roadway, the quay, the supporting pillars and a larger replica mine with tunnels and shafts, created on the basis of an existing old roadway. There is an exhibition in a specially constructed building and a park displaying machinery used in opencast mining.

The main purpose of the museum is to preserve the historical and cultural heritage of the coalfield, but it has also become an important cultural centre for the city, hosting chess tournaments, conferences, theatrical performances and weddings. There are other mining remains in the vicinity. These include the miners' houses. together with a school and a public sauna, and the pillars supporting the aerial system by which the coal was transported.

*References*

http://www.redhill-kemerovo.ru

http://www.russianmuseums.info/M814

personal communication, Natalia Shelepova, director, 2013

### 310. Muzey Istorii Razvitiya Gornogo Proizvodstva (Museum of the History of Mining Development)

The full name of the museum is MBUK «Muzey Istorii Razvitiya Gornogo Proizvodstva im. A. Demedova» (MBUK Museum of the History of the Development of Mining Production named after A. Demidov), MBUK standing for Munitsipalnoye Budzhetnoye Uchrezhdenie Kultury (Municipal Budgetary Institution of Culture). The institution is generally known as the Muzey Istorii Razvitiya Gornogo Proizvodstva. The museum is located in Zmeinogorsk in the south of the province of Altai Krai in the far south of the Russian Federation.

The museum opened in 1985 and commemoration of the industrialist, Akinfiy Nikitich Demidov (1678-1745), was added to the name in 2010. It is a not-for-profit organisation run by the municipality. There are five employees and the museum attracts about 18,000 visitors a year. The museum has about 17,000 items in its collections. There are 286.5 square metres of permanent galleries, 300 square metres for temporary exhibitions and the museum's store occupies 30 square metres. The museum building stands in 2.3 hectares of grounds and it deals mainly with the history of mining in western Siberia.

*Reference*

http://www.russianmuseums.info/M1360

### 311. Sankt-Peterburgskiy Gornyy Universitet (Saint Petersburg Mining University)

The mining museum of the Sankt-Peterburgskiy Gornyy Universitet is located in Saint Petersburg in the very west of Russia. The mining university was founded in 1773 by decree of Empress Catherine II (Catherine the Great; 1729-96), as a strong engineering profession was considered to be essential for maintaining Russia's status as a great power. It was the first institution of higher-technical education to be established in Russia and the fourth-oldest mining-technology college in Europe (**316**). The mining school was initially housed in two stone buildings on the banks of the river Neva and began with 19 students. Until 1899 the university was the only mining and technical institution of higher learning in Russia.

The museum was established in 1774 at the same time as the school of mining, with an annual allowance of 2,250 roubles for the purchase of metal ores and minerals from around the world. The decree also stated that 'proceeds from ore-mining enterprises and also money remaining from students' allowances' was to be used 'for the establishment of a mining cabinet, a metals cabinet, and a mineralogical cabinet'. The superintendents of all Russia's metallurgical works, mines and pits were instructed to send the museum the most remarkable specimens of factory products, minerals and ores that they encountered.

**Figure 311.1. Exterior of the Sankt-Peterburgskiy Gornyy Universitet (© Sankt-Peterburgskiy Gornyy Universitet).**

**Figure 311.2. Hall with display of individual minerals (© Sankt-Peterburgskiy Gornyy Universitet).**

**Figure 311.3. Model of 1950s tracked roadway-tunnelling machine (© Sankt-Peterburgskiy Gornyy Universitet).**

In 1777 Gustav III of Sweden (1746-92), an advocate of the Enlightenment (1685-1815) and a patron of the arts, visited Catherine the Great, who was his first cousin, at Saint Petersburg. He donated 202 items of Swedish 'ores, salts and stones' to the museum.

The main building of the mining school was designed by Andrej Nikiforovich Voronihin (1759-1814) in the neoclassical style and was constructed between 1806 and 1811. The interior of the museum was laid out between 1821 and 1826 by A.I. Postnikov in classical style and occupies the most impressive ceremonial halls of the school of mining. With the closure of boarding dormitories for students from outside Saint Petersburg in 1830, the cadets' now-redundant sleeping quarters adjacent to the conference hall were transferred to the museum. In the eighteenth century and first half of the nineteenth century there was access through the courtyard of the school of mines to a simulated mine, with four training mines and two roadways. In the vestibule of the main building of the university is a piece of coal, weighing 2,560 kilograms, which was brought from Newcastle, England, in the nineteenth century.

The early-Soviet period (from 1919) was marked by an increase in the museum's full-time research staff and the replenishment of its collections and funds, mainly through properties, such as mansions and companies which had been nationalised. The crystallographic, geognostic, mineralogical, numismatic and topographical collections suffered significant redistribution of their exhibits and educational materials; for example, to the geological, mineralogical and technical departments of the school of mines. On the other hand, the art collection of the conference hall was transferred to the museum and, at the same time, the museum acquired a significant number of new collections and educational materials.

During WWII the most valuable artefacts (diamonds, gold and platinum nuggets, *etc.*) were evacuated for storage in *Sverdlovsk* (now Ekaterinburg), on the border between Europe and Asia. They were returned to the museum in 1944. Part of the collection was stored in the cellars of the mining institute.

The museum was opened to the public for the first time in 1951. Almost until the mid-1960s most activity at the museum was concentrated on increasing the collections, especially in the field of the geological survey of the Union of Soviet Socialist Republics. Since then the main collection has increased, the systematisation of the exhibits has been carried out and the displays have been modernised.

In 1996 the mining university was included in the state register of the most valued examples of the cultural heritage of the peoples of the Russian Federation, and the collections of the museum became part of the Museum Fund of Russia. The museum continues to be a cultural, educational and scientific division of the university. The museum's collections have been the basis for the academic, educational and research activities of the school of mines for two centuries and the museum plays a vital role in the training of specialists in the fields of geology and mining.

The museum occupies a total area of 3,000 square metres. There are 30 members of staff, 19 of whom are curators, divided into the three departments of geology, mineralogy and the history of mining machinery. The museum attracts about 40,000 visitors each year. Tours of the three sections of the exhibitions, lasting 90 minutes, are offered, as well as hour-long themed tours.

The museum has an extensive collection of gems and mineralogical specimens from some 80 countries all over the world, amounting to more than 240,000 items. There are samples of anthracite, lignite and bituminous coals from all known coal deposits in Asian and European Russia, the Donbass coalfield, Kazakhstan, Poland, Uzbekistan and several other world deposits. There are also a scientific library, some documents about mines and examples of machines used in mining operations, such as several cutter-loaders dating to the middle of the twentieth century. The museum houses unique collections of mêlée weapons, metallurgical equipment, minerals, models and miniatures from the history of mining, jewellery, ores, paleontological fossils, rocks and stone-cutting products. There are also art works, such as statuettes of animals and vases carved from anthracite or jet.

There are extremely high-quality models relating to coal or coal mining, many of which were made in-house in the nineteenth century or the first half of the twentieth century. These include models from 1896 of a coke oven at the Berestov ore mine at a scale of 1:6 and of the Kizelugol Colliery in the Ural Mountains in the middle of the twentieth century at a scale of 1:100. Other models deal with winding and ventilation, including a number of headgears, steam-winding engines and centrifugal ventilation fans and their associated evasé-vent structures. One set of cased models depicts coal-cutting and tunnelling machinery and the surface and underground features of the Tsentralnaya Colliery, at a scale of 1:132, in the Donbass coalfield of eastern Ukraine, including the geology through which the shafts were driven. Behind these are models of an aerial ropeway, furnace structures and wall displays showing shotfiring tools and the development of miners' flame-safety lamps from the 1850s to the 1930s.

A model of a 1950s tracked roadway-tunnelling machine shows how, rather than using explosives to drive a roadway, machines like this used powerful cutting heads to cut a way through the rock, with the waste being moved through the middle by conveyor for disposal. The upper caterpillar track is necessary to keep the machine braced against the cutting surface rather than to support the roof. Machines such as this were developed from the late-nineteenth century onwards.

*References*

https://museum.spmi.ru/

http://www.russianmuseums.info/M105

personal communications, Elena Popova, director, 2016; Vladimir Litvinenko, rector, 2016; Mikhail Shabalov, director, 2021

### 312. Ugol'noy Mir Neryungri (The Coal Mining World of Neryungri)

The town of Neryungri has about 65,000 residents, many of whom are involved either in gold mining or in the large coal-mining industry that has developed here. There are five small opencast mines, one large opencast mine, about 10 kilometres outside the town, and an underground colliery. These mines are now the largest producers in Russia of coal for steel making. Neryungri has become a tourist destination as a result of tours offered of the opencast operation and the adjacent maintenance sheds in which the enormous mining machines are serviced and repaired. Tour guides and interpreters explain the technical details of the coal mine and outline its daily operation.

*References*

http://www.mechel.com/sector/mining/yakutugol/

http://www.nytimes.com/1982/01/26/world/challenge-of-siberia-conquering-riches-and-cold.html

### 313. Vorkutinskiy Muzeyno-Vystavochnyy Tsentr (Vorkuta Museum-Exhibition Centre)

The Vorkutinskiy Muzeyno-Vystavochnyy Tsentr is located just north of the Arctic Circle in Vorkuta (the easternmost town in European Russia) in the Komi Republic in the far north-west of the Russian Federation. The area of the Bolshezemelskaya tundra was the territory of the Arctic tribe of the Nenets, in whose language Vorkuta means 'a place with bears'. Vorkuta lies on the Vorkuta River just north of the Arctic Circle. During WWII Vorkuta developed into a town, achieving that status formally in 1943. The end of the *gulag* system of prison camps, which operated in the Soviet Union from 1930 until 1955, gave impetus to the rapid development of Vorkuta into a city, with mass construction of housing for thousands of young workers.

**Figure 313.1. Rudnik Mine drift entrance in 1930, with slogan exhorting miners to produce to the maximum (© Vorkutinskiy Muzeyno-Vystavochnyy Tsentr).**

**Figure 313.2. Model of roadway in sloping seam, with clothing, equipment, props and slogan exhorting miners to do the work of three men (© Vorkutinskiy Muzeyno-Vystavochnyy Tsentr).**

**Figure 313.3. Image from the collections of the Vorgashorskaja Mine in 1983 (© Vorkutinskiy Muzeyno-Vystavochnyy Tsentr).**

The Vorkutinskiy Muzeyno-Vystavochnyy Tsentr deals with the history of the region from the 1930s to the present day, including its development with the exploitation of the coal. The decision to open the museum was made in 1958 and more than 50 enterprises and institutions in the city participated in the work of the organising committee. The museum opened to visitors in 1960 with exhibitions about the leading industrial enterprises of the city in a building of 750 square metres. In 1968 the museum was granted regional status, giving it the right to conduct research activities throughout the Pechora coalfield. As a municipal institution of the urban district of Vorkuta, the funding for the museum comes from the city budget. The museum employs 18 staff with 20 volunteers and there are 17,000 visitors a year.

The memories section of the displays includes information about the pioneers of the Vorkuta coalfield and the coal-industry workers, geologists, scientists and ordinary residents of the city who participated in its development. One image on display shows the entrance to the drift of the Rudnik Mine in 1930, with a slogan above, 'Miners, we will achieve the half-year plan, we will produce 900 tons [915 tonnes] of coal [per day]' (*cf.* **354.1**). Memories of those who worked in the Vorkuta coal mines are also recorded. A reconstruction of an underground roadway in a sloping seam displays miners' clothing, hand drill, pick and wooden props. A slogan above says, 'To work on the home front as on the front, every man must do the work of two or three', exhorting the inmates of the 3823 *gulag* to greater effort during WWII. The camp was situated on the left bank of the river Kozhim and was popularly called *shpalorezka* (cutter of sleepers), presumably referring to the cutting of wooden sleepers (as are still used on railway tracks) for the tracks laid underground and on the surface for moving tubs of coal.

Maps and charts cover geological and topographical maps of the north of Russia, including geographical and geological institutions. Documents represent the period from 1930 to 2010, with certificates of individual employees of the coal industry and of miners' groups. Technical papers contain samples of the technical documentation, such as data sheets, diagrams, drawings, specifications and various types of reports relating to industrial production and coal mining. The photographic section has albums and individual photographs which illuminate different aspects of the working life of individuals, groups and companies and also important events associated with the coal-mining industry. The photographic collection, one of the museum's most extensive collections, covers the period from the 1930s to the 2010s. The art collection is of works by Vorkuta artists and sculptors depicting the work done by miners.

The museum stores cover a wide range of topics. There is material on the development of technology in the region, most of which is geological equipment, and accessories, components, tools and specialty products for the coal industry, machine parts and mechanisms. There are items used in the manufacture of coal products. A large collection of clothing from the 1940s to the 1970s includes work clothes and cotton jackets, gloves, helmets made from fibre, trousers and steel and rubber boots from earlier periods, as well as modern items of protective clothing. There is a small collection of miners' uniforms, including that of the Honoured Miner of the Russian Federation, the official title bestowed by the state on citizens of the Russian Federation for professional achievements. Finally the badges collection contains examples of the widely presented awards made to employees in the coal industry and a small number of personnel badges. In total the museum's collections contain more than 82,000 items.

*References*

http://encyclopedia2.thefreedictionary.com/Pechora+Coal+Basin

http://museumworkuta.ru/about/

https://en.wikipedia.org/wiki/Honorary_titles_of_Russia

personal communications, Galina Trukhina, director, 2016; Tatiana Zashikhina, chief curator, 2021

# Serbia

Serbia has four lignite coalfields: the largest is the Kolubara in the west, followed by the Kostolac in the north-east, the Kosovo in the south and the Metohija in the south-west (euracoal). The first coal mine in Serbia was Vrdnik in northern Serbia, north of the Kolubara coalfield. It was sunk in 1804 and belonged to the Ravanica monastery until 1894. The river Danube in Serbia was crossed in 1834 by the first steamship and in 1856 a French-Serbian company was established with four boats on the three main Serbian rivers: the Danube, Morava and Save. The first mine in the Kostolac coalfield was sunk in 1873; from 1881 the Stari Kostolac mine was owned by Georg Weifert (1850-1937), who also opened a brick and tile factory next to the mine in 1885.

In the second half of the nineteenth century, young, educated, engineering, geological and mining professionals returned to Serbia, leading to modernisation of mining legislation and technology, the establishment of a state administration for mining and the opening of new mines. In the early-twentieth century Senje mine (**314**) was the largest and most important colliery in Serbia. Kostolac coal became more important after WWI with the construction of new power stations and industrial plants in the major towns along the Danube. In 1943 the German occupiers intensified production in the Klenovnik and Kostolace mines and introduced armed control of the workforce of internees. After WWII all mines were nationalised under the communist state-directed economy. Kolubara Colliery operated from the end of the nineteenth century, but underwent considerable expansion in the mid-twentieth century with the development of the opencast mine, now the country's largest supplier of lignite. The opencast mine is now only 40 metres from the nearest village (see Vujić 2014 for the history of coal mining in Serbia).

Serbia became an independent state in 1992 when Yugoslavia broke into its constituent parts. In September 2000 massive protests against Slobodan Milosevic (1941-2006) were led by students and coal miners who refused to work. As 70 *per cent* of energy came from coal-fuelled power stations, Milosevic sent the army to try to force the miners at Kolubara Colliery back to work, but he was ousted from power on 5 October. In 2020 Serbia still mined 35-40 million tonnes of lignite a year, burned in three antiquated 1950s power plants, making Serbia one of the worst polluters in Europe. Serbia applied for EU membership in 2012 and is expected to become a member in 2025. To achieve membership, though, Serbia would have to move rapidly to phase out coal mining and move to environmentally friendly alternative forms of energy (https://www.euronews.com/2020/10/21/how-serbia-s-addiction-to-coal-could-cloud-its-future-in-europe). State-owned Chinese banks have, however, been investing in new power plants and the expansion of Kolubara Colliery from producing 9 million tonnes a year to 12 million tonnes in 2020 (https://chinadialogue.net/en/climate/serbia-forges-ahead-with-china-backed-coal-power-plant/).

### 314. Senjski Rudnik (Senje Coal Mine)

Senjski Rudnik is Serbia's oldest active coal mine and the country's oldest preserved industrial-heritage site. It is located in the east of Serbia on the southern slopes of the Carpathian Mountains, 150 kilometres south-east of Belgrade. The mine began in 1853 with the sinking of the Aleksandar shaft, after a pig herdsman put a 'stone' on his fire and it burst into flames. The shaft was subsequently named after Aleksandar Karađorđević (1888-1934), prince of Serbia from 1842 until his abdication in 1858. The shaft is still preserved, along with the 1860 administration building, 1922 workshop and 1930 stores building, which together form the mining museum. It is claimed that the Ubedljivo steam engine, made in 1872, installed in 1874 and used to take three shifts of miners underground each day, is the oldest working steam-powered winding engine in Europe using the traditional system of winding with two wheels (*cf.* **298**).

The mine has always been state-owned. In 1892 a narrow-gauge mineral railway was built linking the mine to Ćuprija. This was extended in 1908 to the river Flat, allowing rapid development of the mine. In 1893 the entire shift underground, somewhere between 28 and 55 men, were killed when a fire in the adjacent beech forest spread into the mine on 21 July. This is the name day of the Holy Great Martyr Saint Procopius, decapitated in Caesarea in CE 303, who is the protector of miners in the Orthodox Church. A church dedicated to Saint Procopius of Scythopolis was erected in 1900 in memory of the dead miners. It was officially decided that 21 July would be celebrated each year as the Day of the Miner. During WWII Senjski Rudnik was under Bulgarian occupation, but the management of the mine and its production were in the hands of the German invaders.

There are three sections to Senj village. First, Topolovac includes the railway. Second, Mine displays the miners' housing, the mining museum and the preserved entrance to the Aleksandar shaft, with the mine's administrative building constructed above it in 1860. This is one of the few entrances to a drift mine to be preserved in Serbia. Third, Stables, the smallest part, has the barn for housing the pit ponies.

The mining museum opened in the second section of the village in 1980, dealing with the social and technical history of Senjski Rudnik, and in 1983 was declared an important cultural monument of the Republic of Serbia. In 2009 a project started to develop the museum to modern standards with support from the Council of Europe and the EU. The main aim is to make Senj into a centre for industrial-heritage tourism along the lines being promoted by the city of Zabrze with the Muzeum Górnictwa w Węglowego (**297**). A second aim is to establish an ecomuseum, so that local people and small industries can use the natural resources of the area to develop a sustainable future here.

The museum tells the story of the historical and technical development of Serbian coal mines from the earliest periods up to the modern day. The exhibition area covers 550 square metres. Displays include simple mining implements from ancient and Roman times (when eastern Serbia was part of the province of *Dacia*), mechanised mining machinery and tools for excavation, several collections of mine lamps, and geological and mining maps, as well as a collection of mining symbols, photographs and archival material related to mining. The exhibition covers the development of the Senjski Rudnik Mine, the workforce, the miners' trade-union movement and the role of miners in WWII. It is completed with post-war mining in the period when Serbia was one of the six constitutional republics of the Socialist Federal Republic of Yugoslavia (1945-92). The blacksmiths', carpentry and machine workshops are also open for visitors.

*References*

ERIH, *s.n.* Senjski

https://en.wikipedia.org/wiki/Senjski_Rudnik

# Slovak Republic

The main lignite coalfields of the Slovak Republic are the Baňa Čáry in the west, the Baňa Dolina in the south and the Hornonitrianska Baňa Prieviedza in the central west, with reserves estimated at 1 billion tonnes. The seams are up to 20 metres thick at depths between 150 and 450 metres; 90 *per cent* of the coal is used for power generation (euracoal).

By the early-eighteenth century a shortage of wood had developed for roasting the gold ore in which the country is rich. Rewards were offered by the Viennese Court Chamber for the identification of coal deposits, in particular to supply the smelter at Banská Štiavnica (at that time in Hungary: **235**). The 1788 decree of the chamber, however, exempting landowners, who were mainly nobles, from needing consent from the mining courts to mine coal had the effect of stifling industrial development. So the first exploitation of coal, in Handlová in the Prieviedza coalfield, did not start until 1837. The 1854 General Mining Act declared coal to be a reserved mineral whose extraction did not require the consent of the landowner. Coal contributed little to the industrial development of Slovakia, however, because there was no quality coal and the deposits were remote from the major metallurgical centres.

Real development of the coal industry only began in the early-twentieth century. Industrial mining began in the Upper Nitra coalfield in 1909 and a thermal power plant was constructed in Handlová in 1910. This gradually increased its capacity so that by 1953 it was the largest such plant in Slovakia. The heart of the lignite industry was established 150 kilometres north-east of Bratislava around the town of Prievidza at the mines of Handlová and Nováky, sunk in 1940. After WWII the mines were nationalised under the communist state-directed economy, with the Nováky Colliery renamed as the Youth Mine in 1957. The Cígeľ mines (**315**) were split off in 1959 as an independent company. The area around the towns of Handlova, Nováky and Prievidza is considered to be the most polluted region of the country. The power plant at Nováky is now the only lignite power plant in Slovakia. The Upper Nitra/Prievidza coalfield is still the major source in Slovakia of lignite, which is the only domestic primary-energy source, but the future is limited for this low-quality coal (see Sombathy, Kúšik and Mižák 2018 for the history of coal mining in the Slovak Republic).

### 315. Hornonitriansky Banský Skanzen (Hornonitrianskeho Open Air Mining Museum)

The Hornonitriansky Banský Skanzen is located in Cígeľ, a village in the Trenčín region of central western Slovakia. The open-air museum is located on the site of Baňa Cígeľ (Cígeľ Mine). It is run by the Association for the Rescue of the Mining Heritage of the Upper Nitra, which is a not-for-profit charitable trust. The association was set up in 2003 and over the next 10 years about 35,000 visitors came to the museum from Czechia, the Russian Federation and various parts of Slovakia. On the surface there are old and contemporary mining machines, including an armoured flexible conveyor, used for clearing waste in a development heading. The headgear is made of reinforced concrete with the winding wheels roofed over, as is normal practice in continental Europe. The exhibition presents the history of lignite mining in Slovakia, especially in the Upper Nitra coalfield, including amusing tales and myths of the miners.

Visitors are taken on a tour around part of the underground of the mine, in the western part of the Handlová seam, which was worked by retreat mining (**382**). The site is open six days a week, but tours must be booked 24 hours in advance. The three-hour tour includes an introduction to security regulations and some background on the mining. Visitors wear hard hats and travel into and out of the mine by a monorail train hanging from the roof by a single iron rail at a maximum speed of 12 kilometres per hour. They spend a total of 30 minutes on the train and cover a distance of 2,312 metres. The 566 metres of the central section of the tour is undertaken on foot. The tour is unusual because it is organised in the operational area of the colliery, which is still actively mining lignite.

*References*

http://slovakia.travel/en/upper-nitra-open-air-mining-museum

http://www.banskyskanzen.sk/

http://www.hbp.sk/index.php/en/company/history

showcaves

# Slovenia

The main lignite coalfields of Slovenia are the Rudnik Trbovlje Hrastnik in the centre of the country, the Premogovnik Velenje in the north and the Goričko in the north-east, with total reserves of 1.256 million tonnes (euracoal). Foreign corporations began mining in the Rudnik Trbovlje Hrastnik coalfield, starting in 1804 at Bukova Gora. This was just south of Trbovlje, which in 1849 was connected to the Austrian southern railway, which linked Vienna in Austria with Trieste in Italy close to the Slovenian western border.

Industrial-scale coal mining developed in the nineteenth century with the introduction of the railways on which the coal could be moved. Exploration started at Velenje in 1875 and the 160-metre-deep Stari jašek (old shaft) in Velenje was sunk in 1889. Poor social conditions led to unrest amongst the miners of Trbovlje, which became a centre of communist agitation, and clashes and sitdown strikes in 1924 and 1934. During WWII the coal mines of Slovenia were annexed by Germany, whose repressive measures led to 90 *per cent* of the population opposing the Germans by 1944. In 1945 Slovenia became part of Yugoslavia with a state-directed economy under which the coal mines were nationalised.

There are huge layers of lignite just below the surface in Velenje. During the twentieth century, the topmost layers of lignite were extracted in opencast mines, but eventually the mines had to go underground to reach the lower seams. Over the years more than 2,000 people had to be resettled and more than 800 buildings were demolished to allow expansion of the mines. Low performance, strenuous physical work and high production, combined with some of the thickest lignite seams in the world (an average of 60 metres thick and in places up to 170 metres), led to the introduction of the first longwall faces in 1947, which were extensively developed from 1952 onwards. The Velenje longwall-mining method was developed with hydraulic-powered roof supports. The approach is to top-cave hanging seams by extending the extraction area above the protected space at the face. Velenje, the only mine still operating in Slovenia, is one of the largest and most modern mines in Europe. All the Velenje output of 3.2 million tonnes is used in the nearby Šoštanj power station. Improvements to the mechanisation are still continuing, as it is planned to continue operating until 2054. Slovenia became an independent state in 1991, when Yugoslavia broke into its constituent parts, and joined the EU in 2004. A plan for the gradual closure of the mine and a programme to phase out coal usage is to be prepared by 2022, as the EU moves to become carbon-neutral (see Vrankar 2019 and https://en.wikipedia.org/wiki/Trbovlje for the history of coal mining in Slovenia).

**316. Muzej Premogovnistva Slovenije (Coal Mining Museum of Slovenia)**

The Muzej Premogovništva Slovenije is located in the Šaleška Dolina Valley in the northern part of Velenje. The museum was founded in 1957, housed in Velenje Castle and run by the Ivan Napotnik Velenje Cultural Centre. It opened to the public in 1966. At the end of 1998 coal extraction moved from the Škale area to the new Preloge coalfield. So in 1999 the museum was relocated to the abandoned underground roadways of Škale Pit with its reinforced-concrete headgear, with roofed winding wheels above the upcast shaft, boxed-in for ventilation purposes.

The museum was established as a joint venture of the municipality of Velenje, Velenje Colliery and Velenje Museum. The museum was forced to close in about 2012. The site was purchased by the city council, which refurbished the underground and reopened the 90-minute tour, escorted by guides who are former coal miners. On the surface the museum occupies 9,000 square metres, situated in a further 1,000 square metres of parkland. The museum employs 10 members of staff and attracts 25,000 visitors a year. The aim of the activities of the museum is to preserve the old mining values of camaraderie, solidarity and mutual assistance..

There are 1,500 metres of roadways underground at a depth of 160 metres. These are reached by a cage which is 120 years old. The tour begins with a visit to a small chapel of Saint Barbara (**96**), and a display about a disaster in 1893 when 18 miners were killed in a methane explosion. After this there are refreshments in the original miners' underground canteen, followed by displays of modern mining methods. A modern mechanised longwall face is displayed, equipped with a single-ended ranging-drum shearer using chain haulage, an armoured face conveyor and immediate-face-support powered roof supports. The operation of the machinery is demonstrated. Sets with life-size models illustrate the activities of miners underground, using modern audio-visual techniques. At the end visitors ride on an underground train.

On the surface there are exhibitions about the everyday life of the miners, from the earliest times to the present day, with documents, lamps, photographs and tools. There is also an art collection on display depicting the working life of the mine. The exhibition of minerals does not deal specifically with lignite, but illustrates the geological structures of the Šaleška Valley. A miner's house shows a typical interior from the first part of the twentieth century. Other topics covered are the history of Slovenian coal mining, including the traditional initiation of apprentice miners of 'jumping the leather'. Originally novice miners had to jump across a mining shaft to be accepted into the trade. Miners used to wear a brown coat, together with a cap and a tanned calfskin apron hanging from their waist at their back on which they could slide down into the pit. The apron became a trademark and the honorary symbol of the trade. In the sixteenth century shafts became too wide to jump across, so the leap was replaced by a symbolic jump acrosss a calfskin apron. In 1735 Maria Theresa (1717-80), empress of Austria and queen of Hungary, had regulations produced for the whole of the Habsburg Empire for training in administrative, legal and technical skills for management in the mining and metallurgical industries. A school with a two-year course was established for this at *Selmecbánya* (now Banská Štiavnica in the Slovak Republic). In 1762 Maria Theresa approved the upgrading of the school into a new type of higher-education training (Labuda 2009, p. 92), making this the first technical university in the world (followed in Germany by Freiberg in 1765 and Berlin in 1770, and in Russia in Saint Petersburg in 1773: **311**). The traditional costume was adopted by the mining academy as the uniform for the students. In the nineteenth century the miners' costume changed to black (*cf.* **68.2**), but still with the apron. The ceremony was revived in about 1923 and again in 1961 after falling into abeyance during WWII. The ceremony is now held annually on Miners' Day, 3 July, which commemorates a five-day hunger strike by miners from the Zasavje region, which began on 3 July 1934.

*References*

ERIH, *s.n.* Velenje

http://muzej.rlv.si/en/

personal communications, Peter Op 't Veld, consultant, 2015; Stojan Špegel, director, 2016

showcaves

# South Africa

There are 19 coalfields in South Africa, concentrated especially in the north-eastern quarter of the country, 96 *per cent* of which is bituminous coal. This constitutes 7 *per cent* of international coal resources (Ratshomo and Nembahe [2016], pp.1-3). It has been claimed that charcoal, made from coal, was used to smelt iron and copper during the Iron Age (CE 300-1800), but without giving any references to archaeological excavations showing this (http://www.eskom.co.za/content/10637 - Draft EIR Bravo 5 Chapter 7d.pdf, p. 83).

Anthracite was first identified in northern Natal in the mid-1830s, but only became a commercial proposition with the construction of the railway from the coalfield to the port of Durban in 1888. Natal coal was expensive and difficult to mine because it was in thin deep seams, which were dangerous with methane gas and dusty. The seams had been broken by volcanic action, so had to be worked by pillar-and-stall (**47**). By 1950 80 *per cent* of Natal mines were longwall mining, but hand-filling of the broken coalfaces continued into the 1950s. Bituminous Transvaal coal was of lower quality, but was cheaper to mine, being in thick seams close to the surface and better situated in relation to labour and markets, while able to be worked by mechanised longwall mining. Many of the miners doing the heavy physical work have historically been black, who had often moved from rural areas to settle around the mines, where they provided cheap labour (Alexander 2008).

Mining was stimulated by the Anglo-Boer wars (1880-1, 1899-1902), being needed for the steam troop ships (McFadden 2016), and continued with exports of coal overseas. Production decreased with the advent of oil power, but revived with the energy crisis of the 1970s and the opening in 1976 of a new coal terminal at Richard's Bay and by 1982 23.9 *per cent* of coal production was exported. In 1983 sanctions were imposed by America and the UK on the apartheid regime (1948-94), resulting in the closure of many mines. The sanctions were lifted in 1994 and in 2015 South Africa was the fifth largest exporter of steam coal. South Africa has about 67 billion tonnes of coal reserves. In 2016 more than 90 *per cent* of South Africa's electricity was generated using coal, 51 *per cent* from underground mines and 49 *per cent* from opencasting (Ratshomo and Nembahe [2016], pp. 3-6).

### 317. MuseuMAfricA (formerly the Africana Museum)

MuseuMAfrica is a cultural-history museum located in Johannesburg in north-eastern South Africa. The museum occupies the 1913 building of the former fruit and vegetable market. The collection began in 1933 when the Johannesburg Public Library purchased a large collection of ethnological books and objects from John Gaspard Gubbins (1877-1935), who had been assembling his library of Africana since 1924. From the mid-1930s the Africana Museum expanded its collections policy to encompass all aspects of African history and material culture, especially Afrikaner, Tswana, Xhosa and Zulu. The museum has a wide remit, with displays and material about the struggle for racial equality of Mahātmā Gandhi (1869-1948), leader of the movement for independence for India, who worked as a lawyer in South Africa and led the community in non-violent civil disobedience between 1893 and 1914 (see also **318**), and of Nelson Mandela (1918-2013), who led the African National Congress and was president of South Africa (1994-9). The museum also deals with the economic, political and social history of Johannesburg, including the effects of coal mining.

After the release of Nelson Mandela and the end of apartheid in South Africa in 1994, the museum underwent a comprehensive refurbishment and redisplay in line with its new name of MuseuMAfrica (which distinguishes it from the American Museum Africa website: http://themuseumafrica.org/). Subsidiary museums of MuseuMAfricA are the Bamberg Museum of Fashion, the Bensusan Museum of Photography, the Geology Museum and the Workers' Museum.

The Geology Museum holds more than 21,000 mineral and rock samples from throughout southern Africa. The Geology Museum houses some items of coal-mining historical interest, mostly samples from mines. There are also current negotiations for a mining section to be opened in the museum that would cover the physical aspects of deep and opencast coal and gold mining. More than 20 different minerals, including coal, clay, diamond, gold, manganese and quartz, would be on display.

*References*

https://en.wikipedia.org/wiki/MuseuMAfricA

https://www.joburg.org.za/play_/Pages/Play%20in%20Joburg/Culture%20and%20Heritage/Links/Newtown%20Guide/guide/Museum-Africa-.aspx

personal communication, Katherine James, curator of geology, 2014

### 318. Talana Museum

Talana Museum is located on the site of the first formal battle of the Second Anglo-Boer War, fought on 20 October 1899. The museum is situated just outside the

**Figure 318.1. Exterior of Henderson Hall, which houses the archives, coal-mining exhibits and reception, with a Joy continuous miner in front of the building (© Talana Museum).**

**Figure 318.2. Simulated underground with wooden chocks and safety notices (© Talana Museum).**

town of Dundee in northern KwaZulu-Natal, a province on the eastern coast of South Africa. The Zulu name *Talana* means 'the shelf where precious items are stored', so is an appropriate name for a museum. The museum has only a small staff, with support from volunteers and a friends' organisation. It is managed by a board of trustees, who are responsible for the day-to-day management, development of the museum and fundraising. The museum is a department of the local authority, which provides funding for staffing and running costs.

The 180 hectares of the battlefield and the original 42 buildings constitute the Heritage Park, with a walking trail up Talana Hill in the footsteps of the British troops. The economic, military and social history of the region is depicted in 22 of the buildings. The first organised coal mining in South Africa was begun on the slopes of Talana Hill by the Dundee Coal Company, registered on the London Stock Exchange in 1889. The remains of local coal mines can still be seen in the area. Dundee became the heart of the richest coal-producing area in South Africa, with the result that the town expanded rapidly and was nicknamed coalopolis. The town square and some of the shops have been reconstructed to illustrate this period. The museum has four extensive exhibits illustrating the history and development of coal mining in northern KwaZulu-Natal. The Chamber of Mines exhibition deals with the early history of the mines, mining methods, uses of coal and support services within the mining industry, such as dust control, geology, surveying, life in the mines and the mines-rescue service.

The considerable geological collection on display reflects the coal geology of the region, as well as displaying other minerals from the wider region and the gold mines of northern Natal. Collected from the 1940s to the late 1960s by the inspector of mines, he was also responsible for the extensive collection of miners' lamps and safety lamps on display, some damaged in the many accidents in the mines over the years. Outside there are many machines, from both low-seam and thick-seam mining, as well as a Joy continuous miner from the early days of mechanisation.

A series of scenes simulate underground mining at the Durnacol and Hlobane Mines, portraying the different methods of coal extraction in low-seam and thick-seam mining. The first has models of kneeling miners, safety notices and wooden chocks; then machinery, such as flame-proof vehicles and a rubber-tyred free-steered vehicle; and miners wearing battery packs to power their cap lamps and self-rescuers. All machinery on display originally operated in the various mines.

On the centenary of the trial of Mahātmā Gandhi in the Dundee Court House in November 1913 (**317**), an exhibition opened dealing with the arrival from 1850 in Natal of indentured Indians and the conditions under which 9,221 of them worked in the Dundee area in the Natal coal mines, as well as on farms. The Miners' Rest restaurant is located in a pre-1914 corrugated-iron miner's home, which was moved from Ballengeich Mine to be re-erected at the museum. The building was destroyed in a fire in 2010, but has been reconstructed to reflect the original building.

The museum constitutes the national coal-mining museum. It is also the official archive repository for coal-mining documents and material relating to the KwaZulu-Natal coalfield, including the records of 92 mines in the region, many of the geological reports and thousands of photographs and mining documents. The museum also holds archival material for the rest of the country, as there is no other official place where such records can be deposited. The photographic mining collection has been digitised and is available to be consulted on request. The museum also has extensive archives on the military history of the area.

*References*

http://www.talana.co.za

personal communications, Pam McFadden, curator, 2012, 2021

# South Korea

Coal is found in four of the nine provinces of South Korea. The coalfields run in a line diagonally across the country from the north-east to the south-west, with the addition of one on the west coast and one on the south-eastern coast (CMMCP 2016, fig. 28.1). Large-scale coal mining began under Japanese rule (1910-45), although mainly in the north. Japan has no anthracite reserves, so the Korean high-quality anthracite was valuable to Japan. During WWII about 450,000 Korean men were sent as forced labour to Japan, including for service in Japanese coal mines. Following the end of the war between North and South Korea (1950-3), it is estimated that some 50,000 of the South Korean prisoners-of-war were not returned to the south, but were made North Korean citizens in 1956 and forced to work in the North Korean coal mines. They were permitted to marry, but their descendants, some as young as seven, have had to continue working in the mines, under extremely gruelling conditions (https://www.bbc.co.uk/news/world-asia-56178271). There are no mining museums in North Korea.

Coal mining in South Korea is overseen by the government-owned Korea Coal Corporation, a successor to the corporation set up by the Japanese occupiers (see Andrews and Hi 1956 for the history of mining in Korea in the first half of the twentieth century). Mining reached its peak in the late 1980s, since when it has declined by 91 *per cent*. The corporation operates three anthracite mines, producing 1.2 million tonnes in 2020, and has invested in overseas bituminous mines, such as in Mongolia. South Korea now depends on imported coal for 97 *per cent* of its fuel requirements.

### 319. Boryeong SeokTan BakMulGwan (Boryeong Coal Museum)

The Boryeong SeokTan BakMulGwan is situated in Boryeong-si on the coast of the Yellow Sea in the province of South Chungcheong in the west of South Korea. Coal mines were developed here during the first half of the twentieth century to provide fuel for heat. The coal mine in Seongju-myeon, Boryeong City, ceased production in 1989 and was completely closed in October 1994. The museum opened in May the following year, concentrating on the history of coal mining and all the machinery and processes involved in mining. The museum occupies an area of 1,700 square metres.

The first of the five parts of the exhibition deals with the formation, locations, types and uses of coal. The second covers the history of the coal industry and the distribution of coal deposits, focusing on the contribution of coal to modern history and the economy. The third comprises models of a mining village and Yeongbo Colliery and its environment. The fourth covers the processes involved in coal mining and the arduous labour carried out by miners. The final part describes the processes of digging, mining and transporting coal. Beyond the exhibition is a mine cage, which takes visitors down two floors, with lights and sounds simulating the feeling of going 400 metres down a shaft. The simulated underground has a roadway that is ventilated by cold air from an adjacent closed mine. Visitors are taken around by guides who are former coal miners with careers of more than 30 years, who also recount their own personal mining experiences. The tour takes between 30 and 60 minutes. Outdoors there is a wide range of machinery used to mine the coal and then to transport it. Coal briquettes were a universal method of heating in Korea until the 1980s, and there is provision for visitors to make their own briquettes to complete the visit experience.

*Reference*

http://english.visitkorea.or.kr/enu/ATR/SI_EN_3_1_1_1.jsp?cid=1064964

### 320. Mungyeong SeokTan BakMulGwan (Mungyeong Coal Museum)

The Mungyeong SeokTan ParkMuelGwan is located in Mungyeong-si, in the province of North Gyeongsang in the centre of South Korea. Mungyeong was one of the most productive coalfields in South Korea. The museum deals with the history of coal mining and its contribution to the rapid industrialisation of South Korea. The site is 50,000 square metres, of which the museum building occupies 1,805 square metres. The museum has 6,730 items in its collections, such as books, documents, fossils and machinery. The intention of the museum is that it should combine an educational experience with an enjoyable visit that will allow people to come as a family, so that the younger generation can understand the life of the older generation in an appropriate setting.

The central exhibition gallery has three floors. The first deals with the formation of coal, mineral resources and fossils, the coal industry and the life of miners. The second displays types of coal and coal mines, the history of the coal industry and the production of briquettes (**319**), and a video on the history of coal mining. The third floor shows disasters, together with recorded interviews, and a day in the life of the miners in the mining town.

There is an outdoor gallery, with large mining machinery displayed under a clear roof, supported by pillars. The machinery includes a battery locomotive and a diesel locomotive, both pulling minecars, a rocker-shovel, a single-strand armoured flexible conveyor and a rope-haulage drum and motor. There is a shrine in honour of miners who died of pneumoconiosis (**92**), where visitors can pay their respects to their dead relatives. There is an exhibition of what a mine from the 1960s was like, with an office, a collapse in a mine, how the mining was carried out and provisions for eating underground.

The life of coal-mining families is shown in the miners' village. There is a house from about 1970, when four families lived in one house. It reproduces the living conditions of two of those families, with adjacent photographs of actual homes. Also displayed are various facilities from about 1960 to 1970, such as a shop selling both raw and cooked meat and a barber, at a time when long hair was in fashion for men but the miners needed short hair to deal with conditions underground. There is a traditional shrine, with its own well from which to draw water, not just for purification before worshipping, but also for use by the general community, because water was not piped to all the homes.

The underground experience deals with all aspects of coal mining. This is based on a roadway of the Silver Star Gang Kaenaegi Mine, driven in 1963 and worked until it closed in 1994. The mine was 800 metres deep, with 400 kilometres of roadways. It employed 4,300 people and produced more than 295,000 tonnes of high-quality coal each year. There are simulated roadways with timber roof supports and modern steel pit props, together with set scenes. First there is the role of the deputy, or official, in checking that the underground is safe by testing for gas and inspecting the walls and roofs before deploying staff to different tasks. These included drilling a heading into the rock, while setting arched girders to support the roof, and a rocker-shovel in a bore-and-fire heading. Models are shown of miners eating their lunch underground, which was always done quickly because miners were only paid for what they actually produced. A drift-haulage paddy car is shown at a manriding station. Finally there is a scene of a mine locomotive pulling mine cars taking miners out of the pit at the end of the shift.

In order to generate income, the museum also offers three different film sets, which have been used for making many popular films and soap operas. One is a reproduction of brightly coloured houses and fortresses of the Goguryeo period (37 BCE to CE 668), based on research on historical documents. The other two are the Ainsi and the Liaotung fortresses, which were important in the seventh century CE, and the villages inside them, including government buildings and a market selling herbs and alcohol.

*References*

https://en.wikipedia.org/wiki/Mungyeong

https://www.tripadvisor.co.uk/Attraction_Review-g2024849-d4068015-Reviews-Mungyeong_Museum_of_Coal-Mungyeong_Gyeongsangbuk_do.html

personal communication, Jisangmyeong, member of staff, 2016

### 321. TaeBaek SeokTan BakMulGwan (Taebaek Coal Museum)

The TaeBaek SeokTan ParkMuelGwan is located in Taebaek-si in the province of Gangwon-do in the north-east of South Korea and deals with the history of the South Korean coal industry. The site occupies 23,811 square metres, with three floors below ground and one above ground. Outside there is a full-size mine headgear. The museum's collections include fossils, historical records, mine products, mining equipment and rocks.

The first floor depicts mining scenes, using equipment, sound effects, special lighting and life-size models, and has various galleries dealing with geology, the origins of coal deposits and the search for coalfields, mining methods and uses of coal, safety in the mine, the life of coal miners and the history of the TaeBaek region. Included in the displays are a full-scale replica of a double-acting rotative beam engine with Murdoch-Watt sun-and-planet gearing and Watt parallel motion, a simulated roof fall in a mine, models of mines rescuers wearing self-contained rebreathing apparatus and a replica of the interior of the quite-sparse room in a coal miner's house. One outside hall displays small digging equipment and rocks of minerals organised by type. The second outdoor gallery is a replica of a mining shaft, with four mine cars. Outside there are large pieces of mining machinery protected by covered arched girders, such as a four-wheel diesel locomotive, with either hydraulic or mechanical transmission, of the 100- to 150-horsepower class. The mine cars have independent taper-bearing wheels with a capacity in the region of 2.5 cubic metres.

*References*

coalmuseum.or.kr

http://english.visitkorea.or.kr/enu/ATR/SI_EN_3_1_1_1.jsp?cid=791768

**Figure 321.1. Display of mining equipment outside the museum, protected by covered arched girders and headgear in the background (© Geoff Preece).**

**Figure 321.2. Full-scale replica of double-acting rotative beam engine with Murdoch-Watt sun-and-planet gearing and Watt parallel motion (© David Preece).**

**Figure 321.3. Four-wheel, diesel locomotive of the 100-150 horsepower class; the mine cars have independent taper-bearing wheels with a capacity in the region of 2.5 cubic metres (© David Preece).**

# Spain

Spain has deposits of anthracite in the coalfields of Puertollano in the south (in the province of Ciudad Real in the south-western part of of Castile-La Mancha), Teruel in the mountainous region of Aragon in the north-east, and in the north-west in Nalón in Asturias and León-Palencia, south of Asturias. The main lignite coalfields of La Coruña and Meirama are in the extreme north-west (euracoal). Spain is rich in mineral ores, which were extracted in *Hispania* in the Roman period using deep mines, but coal appears to have been exploited only where it outcropped.

The first references to the coal mine in Arnao (**324**) date from 1591, making it the oldest identifiable coal mine known in the Iberian peninsula. The most important coal-mining areas were Catalonia (**332, 336**) and the Nalón coalfield of Asturias (**324, 327**). Exploitation of coal only really began in the mid-nineteenth century. The Royal Asturian Mining Company (RCAM) was established in 1833 and the Commission for the Geological Map of the Realm, staffed by mining engineers, was established in 1849 to identify deposits of industrial minerals (Ranestad and Sharp 2020, p. 5).

Women were not employed underground during the nineteenth century (**324**) and in 1900 no workers under the age of 16, male or female, were allowed to be employed underground (Verde, Soto and Gómez 2020). Women were, however, permitted by law to start working underground in 1992 and the first did so in 1996 (Rodríguez-Valencia 2011). In 1988 the author was told by mining engineers at the Geological and Mining Institute of Spain in Madrid that there were women working underground in Asturias at that time, but that was probably a misinterpretation of women applying for underground work with Hunosa (**335**) in 1984 and 1987 (*ibid.*).

The coal mines of Adrall were situated in Catalonia on the Spanish side of the border with Andorra. Exploitation of the Catalan mines began in the nineteenth century, but reached their peak in the period 1939 to 1975 (Llorens 2012), especially in the early years in power of Francisco Franco (1892-1975), who ruled Spain from 1936 until his death. Between 1940 and 1959 coal was regarded as a strategic commodity, leading to increased pressure on the workforce, who were forbidden to take industrial action (Martínez 2020). This led to many accidents; although disease was even more serious, silicosis was only seriously addressed in the 1960s (*ibid.*). From the 1960s coal began to be displaced by oil. Even in the 1970s, however, it was still compulsory to use coal for heating apartment blocks; for example, in Madrid. When Spain became a member of the EU in 1986, the Spanish were advised to close all mining facilities that were not profitable. The result was the gradual shutting down of mining operations. Spain has coal reserves of 4,550 million tonnes, but in 2018 coal contributed only 2.3 *per cent* to electricity generation. In 2018 the government committed to achieving carbon neutrality by 2050. In 2020 30 coal mines and 15 coal-fired power plants were closed.

### 322. Centro de Interpretación de la Minas de Aliaga (Centre for the Interpretation of the Mines of Aliaga)

The Centro de Interpretación de la Minas de Aliaga is based in Santa Barbara (**96**), a district of Aliaga, a municipality in the province of Teruel. The traditional economy of Aliaga was agricultural, until the coming of coal mining and the major development of the neighbourhood of Santa Barbara between the 1940s and 1960s. There were three major mines producing high-quality coal for the nearby thermal power plant, which was built in the 1940s with an annual output of 280 kilowatts per hour, one of the largest in Europe. When the plant was enlarged in 1958, the mines could not supply sufficient coal, leading to a decline in the mining area and the gradual abandonment of the collieries. The power plant itself closed in 1982.

The interpretation centre preserves the history of coal mining in Aliaga and the power plant and the effect of both on the area, with the construction of new settlements to house the workers. The centre was the initiative of the Santa Barbara Cultural and Recreational Association of former mineworkers and opened on 4 December 1999, the feast day of Saint Barbara (**96**). It is housed in the former working-men's club, which was built to provide recreation for the miners. The centre deals with mining activity in the municipality of Aliago, mainly from the opening of the power station and the consequent development of the mining environment. It has a small collection of fossils, minerals and rocks from different geological periods. This relates to the nearby geopark, which has 11 stopping points. There are models, panels and videos about the history of the mines and the power plant. A model demonstrates how the coal was transported by overhead wires.

*References*

http://es.globedia.com/centro-interpretacion-mineria-santa-barbara-aliaga

http://www.redaragon.com/turismo/que_hacer/default.asp?accion=pagina&CosaQueHacer_ID=1573

### 323. Centro de la Interpretación de la Minería (Centre for the Interpretation of Mining)

The Centro de la Interpretación de la Minería is located in Barruelo de Santullán in the province of Palencia, Castile and León. The deposits of the Rubagon coalfield were discovered in 1838 and exploitation by a company set up by the Collantes brothers, Saturnino (1799-1864) and Fernando (1811-90), began in 1845. The coal seams were readily accessible from the surface, so encouraging the development of the economy of the region, especially after it was linked to the national rail network. The first power station was built early in the twentieth century. Calero Mine in Barruelo operated from 1911 until 1972 and then again from 1994 until 2002. It was the most important mine in the coalfield, but also the most dangerous in Spain, with almost 100 miners dying during its history. The mines were acquired by Hunosa (**335**) in 1966. The last two deep mines in the area, in Velilla del Rio Carrion, finally closed in 2014.

The centre is managed and funded by the city council of Barruelo de Santullán. It opened in 1999 and in its first year of operation attracted some 22,000 visitors. The centre itself is composed of three elements: the mining museum, the simulated mine and the cultural centre. The mining museum is situated in a former national school. The museum occupies some 600 square metres, with nine galleries on three floors. The galleries cover the subjects of the origins of coal and its geological history in terms of how it is layered in seams below ground, together with mineralogy, the lithosphere, paleobotany and plate tectonics. Visitors are able to play an interactive game manipulating the various elements that influenced the formation and distribution of coal. Other aspects of coal mining are dealt with, such as the role of explosives, timbering of the roofs and walls of mines, the use of steel roof supports and ventilation systems. The medical and safety details are stressed, including accidents underground, firedamp explosions and diseases such as pneumoconiosis. Then there is the movement of coal from the mine, and its screening, washing and sorting according to the size of pieces and quality. Finally the role of coal itself is covered, such as for fuelling steam locomotives. On the top floor of the museum there are details of the headquarters of the coal industry in Barruelo and a model of Calero Mine. Guides are available who can add further explanations to the exhibitions.

The simulated mine is located at Alto del Tomillo, 1 kilometre from the museum. Above the mine is a headgear donated and moved by the town council of La Union in Murcia. Visitors are escorted in groups by tour guides, who explain the processes involved in the extraction of coal. The two underground roadways, each about 70 metres long, are set out as in a real mine, with steel roof supports, ventilation tubes and wooden pit props, along with mining machinery and tubs for transporting the coal.

*References*

https://es.wikipedia.org/wiki/Centro_de_Interpretación_de_la_Minería_(Barruelo_de_Santullán)

https://es.wikipedia.org/wiki/Cuenca_minera_palentina

https://es.wikipedia.org/wiki/Pozo_Calero

http://www.turismobarruelo.com/cim/cim-en.php

**324. Museo de la Mina de Arnao (Arnao Mining Museum)**

Arnao is located in the city of Castrillón in Asturias in northern Spain. In 1793 the government canalised the river Nalón and built a road to move the coal, and also constructed the first coke furnace in Spain. In 1792 all production by the Royal Mines of Langreo from the coal deposits on the two banks of the river Nalón was reserved for the navy to be used at the weapons factory established in 1784 in Turbia near Langreo. The region was secure as it was distant from the border with France, with whom Spain had frequently been at war. This continued to be the most important arms factory in Spain until only a few decades ago.

The coal mine in Arnao was the first in Spain to use explosives underground and to move coal by rails and was one of the driving forces behind the industrialisation of Spain. Its coal was not, however, suitable for the iron industry, so in 1853 RCAM built a zinc factory in the area to utilise the coal and a settlement for the workers, which included a company store, hospital and school. The workers' settlement still survives almost completely intact and guides take visits to the settlement. In 1855 the company constructed a railway line to connect the mining facilities with the dock at San Juan de Nieva. In 1858 Queen Elizabeth II of Spain (1830-1904) visited Arnao and became unusual as a woman in going underground at a Spanish colliery.

Arnao Mine was unusual in that its 1 kilometre of roadways went out under the sea, at a depth of 205 metres, as that was where the most profitable seams were. The derrick, which was the framework supporting the drilling machinery over the mine shaft, was constructed around 1858 in a brick building. It was later covered with zinc sheets in the modernist style devised in the early-twentieth century by the head of the workshops, Thomas Acha Zulaica. The mine operated from 1834 until 1915, when it was flooded by the sea and closed.

In 2007 Castrillón Council, which funds the running costs of the museum, began the extensive renovation work on the mine necessary to make the roadways safe for public access. The €5 million regeneration project mainly involved restoration of the mining galleries and associated facilities. The conservation of the headframe and underground galleries and renovation of the working men's club were completed in 2012, and the complex opened in 2013. The mining buildings and derrick occupy 1,178 square metres and the reinstated roadways measure 110 metres. The museum occupies the old mining buildings constructed between 1833 and 1915 by RCAM and is managed now by SADIM, a company belonging to the group. The museum attracted 6,000 visitors during the first six months after it opened. There are nine members of staff and three volunteers.

The complex is comprised of two different areas. First is the surface area, which includes the nineteenth-century mining building where a foundry, offices and sawmills were originally located, until it was converted into a casino for the workers' entertainment. The casino has been renovated and converted into the visitor centre. The tour begins with an audio-visual display in the centre, followed by exhibitions in the engine room on the geology of the area and the history of RCAM up to the closure of the mine. The original wooden derrick is a masterpiece of industrial architecture and preserves the original cage-and-pulley system. It is the oldest in Spain and has been declared a building of cultural interest, as has the wooden headframe.

The second area is formed by the nineteenth-century roadways, which are lined with brick masonry and are reached by visitors in a glass lift down the original shaft, the oldest in Asturias. They include galleries where the living and working conditions are explained by guides. Tours are also available in English and French. A steam locomotive, called Eleonore, is installed in this area. It was manufactured in 1880 and is believed to be the oldest narrow-gauge locomotive in Spain. The roadways are 100 metres long and only about 20 metres deep. The tour lasts about one hour for a maximum of 12 visitors per group. Windows have been installed in some parts of the floor of the gallery, for viewing other galleries, which are below sea level and flooded by seawater. The old, unflooded galleries also form a mining amphitheatre, or cultural space fitted out for concerts, exhibitions and recitals.

Staff of the museum are trying to recover photographic and written material of the history of Arnao Mine. They are also working on oral-history recordings and promoting cultural activities related to the mine, such as interviews and meetings with the descendants of miners. The collections include books from the school of the workers' settlement, a commemorative medal given by RCAM to special workers, musical instruments that belonged to the miners' orchestra, nineteenth-century miners' lamps, topographical plans of work and buildings, and wooden shields made by the miners to honour a visit about 1920 by Alfonso (1907-38), prince of Asturias (the title given to the heir to the throne of Spain).

*References*

personal communications, Iván Muñiz López, cultural director, and Guillermo Laine San Román, technical director, 2014, 2015; Eusebi Casanelles, former director, Museu de la Ciènca i de la Tècnica de Catalunya, 2021

www.museominadearnao.es

### 325. Museo de la Mina de Mequinenza (Mequinenza Mining Museum)

The Museo de la Mina de Mequinenza is located inside the Museo de Mequinenza situated in Mequinenza, a town and municipality of the province of Zaragoza, in Aragon. The first coal-mining operations started in the Mequinenza coalfield around the middle of the nineteenth century. The unique geology and its convenient location next to three rivers made the coalfield one of the most singular in Spain. The difficulties of land transport and the distance from the railway meant that the coalfield used the unique system of river transport called the *Camino de Sirga* (towpath). The lignite was shipped directly from the mines to the rivers

**Figure 325.1. Two *llauts* going downstream loaded with coal and each carrying a mule to pull the *llauts* back upstream, photographed about 1930 (© Museo de la Mina de Mequinenza).**

**Figure 325.2. Loading gate and mock powered roof supports on a longwall coal face (© Museo de la Mina de Mequinenza).**

**Figure 325.3. Underground roadway with safes to store explosives securely – one in the centre of the picture and one on the right (© Museo de la Mina de Mequinenza).**

and loaded onto wooden vessels called by the Catalan term *llaut* or *llagut*. The *llauts* were about 20 metres long and 2.5 metres wide and carried between 20 and 30 tonnes of coal on each trip. The *llauts* probably derived from the long-term navigation of the river Ebro, which is documented from Roman times. They are similar to the barges used in the rest of Europe for moving goods on canals and rivers, except that they also had a mast and sail. On the way downstream the *llauts* were moved by the current of the river or by using the sail. When going upstream against the current, the *llauts* were at first pulled by men on the towpath and later by mules. The traditional navigation of the *llauts* and the unique *Camino de Sirga* system disappeared with the construction of the Mequinenza and Ribarroja reservoirs (1957/69), while the old town of Mequinenza was flooded by the waters of the river Ebro.

The Mequinenza museums offer a journey through the historical and industrial heritage of the town located on the banks of the three rivers. The museum opened in 2010 and is owned and funded by the city council of Mequinenza. In 2019 it had four members of staff and attracted 7,000 visitors.

The visit begins at the entrance to the museum, which is laid out like a roadway. The museum has more than a kilometre of simulated underground roadways. The layout of 500 metres of the roadways includes two joining at a junction. The displays in the roadways cover all the operating systems that have been used in the Mequinenza coalfield over the last 170 years. At the beginning of mining, the method was hand-getting. The workers used picks to dislodge the coal and then short-handled shovels to load the mineral onto transport moved by mules. The miners wore baggy shorts and berets. The walls were held in place with wooden shoring.

The next section, about 100 metres long, deals with manual drilling to make holes in the rock, while the roof was supported by wooden chocks. So the coal could then be blasted down using explosives, before being loaded by the miners onto armoured flexible conveyors. The coal was then transferred to small-tonnage dump trucks, which moved it to the loading point, where a cast-steel prop-and-bar system held up the roof. This is the method used in this region in recent times. A more-advanced system used in higher seams is the longwall method, of which 200 metres are displayed, with a coal shearer carving off the coal onto the armoured conveyor.

Since 2018 the Mequinenza museums have been part of the Iberian Network of Geomining Spaces. They have also been members of the UNESCO Global Network of Water Museums since 2020.

*References*

http://www.museosdemequinenza.com

personal communications, Sandra Lacosta Coso, head of visitor services, 2016; Javier Rodes, director, 2021

### 326. Museo de la Minería de Puertollano (Puertollano Mining Museum)

Puertollano is situated between the Calatrava and Puertollano Ranges, in the foothills of the Sierra Morena. Coal was discovered here in 1873 and Puertollano became the largest industrial city in the region. It used to be a major coal-mining centre, but now there is just one opencast mine. National and EU grants have enabled the city to diversify into the manufacturing of fertilisers, petrochemicals and solar panels.

The Museo de la Minería de Puertollano opened in 2006. There are two floors of displays dealing with the geology of the region and the origins of coal mining in the area. It occupies an area of 450 square metres and attracts about 30,000 visitors a year. There are two members of staff and two volunteers.

The museum is located in the Parque del Pozo Norte (Pozo North Park) on the site of the Pozo Norte Mine. The design of its main building is based on the appearance of a coal-preparation plant. Rising from its centre between the two wings of the building is the Castillete North Pozo. This headgear is 30 metres high and was originally erected here in 1928 to serve the Pozo Norte Mine, which was one of the most productive mines in the coalfield until its closure in 1973.

A lift constructed to simulate a mine cage takes visitors down to the basement, where there are 250 metres of a simulated mine roadway. The simulated underground is equipped with mining machinery and tools covering the period from 1873 to 1975. The museum's collections are artistic, cultural and historical, including instruments, material from the mines, miners' clothing and photographs. There is a scale model of a coal mine.

There are other monuments to mining in the Puertollano region, such as eight preserved headframes. The largest of these is the Santa Maria headgear, 40 metres high, which operated between 1910 and 1974 at the Asdrúbal Mine. It was re-erected at a cost of €60,000 at a roundabout at the opposite end of the city from the museum and unveiled on 4 December 1999, the feast day of Saint Barbara (**96**). The legacy of mining in the city can be seen in the largest spoil heap of the coalfield, restored in 2010 to form an urban park. Finally on the hill of Saint Anna, overlooking Puertallano, there is a monument to the coal miners in the form of a sculpture of a miner, 9 metres high, by José Noja (born 1938).

*References*

https://es.wikipedia.org/wiki/Museo_de_la_Miner%C3%ADa_de_Puertollano

https://www.guiarepsol.com/es/fichas/museo/museo-de-la-mineria-16867/

personal communication, Jose Raúl Menasalvas Valderas, director, 2014

### 327. Museo de la Minería y de la Industrie de la Industrie de Asturias (Museum of the Mining and Industry of Asturias)

The Museo de la Mineria y de la Industrie de Asturias (generally known as MUMI) is located in the valley of the river Nalón in north-western Spain in the city of El Entrego in Asturias. It is 2.5 kilometres north-west of the Pozo Sotón (**335**); in 2016 an agreement was signed to amalgamate the two institutions and one ticket is now issued for admission to both. MUMI deals with the mining industry in the nineteenth and twentieth centuries. The museum occupies a total of 5,100 square metres and attracts about 70,000 visitors a year. There are eight members of staff. It is managed and funded by the Fundación de Nuevas Tecnologías y Cultura (Foundation for New Technologies and Culture) and opened to the public in 1994.

The museum was built on the levelled spoil heap of the Mina San Vicente. It is housed in an imposing brick structure, consisting of a large central cylindrical body, above which the steel headgear rises, visible from both outside and inside. On either side are two concrete wings, housing the first-aid room, together with the changing area, lockers and pithead baths. The lamproom displays a comprehensive collection of lamps, showing their evolution from primitive oil lamps through the first safety lamps to those used today.

The ground floor of the central hall details the technical history of mining in Europe, with both full-size and scaled models showing mining between the sixteenth and eighteenth centuries, while a mezzanine floor above deals with more recent developments, including the use of explosives in mining. Particular attention is paid to how various problems were solved. Different types of machines were used to remove water underground, originally driven by either human or animal power. The displays include a full-size treadwheel that can be turned by visitors walking inside the wheel to spin a gearwheel, a technology that is essentially Roman in origin. There are woodcuts of exactly this design in Agricola (**108**), used for powering rag-and-chain pumps for mine drainage (a medieval example still exists in Canterbury Cathedral, UK, for lifting stone). The invention of the Newcomen engine (**360, 371**) enabled water to be removed using steam power. All these methods of dealing with water underground are demonstrated, as well as different models of industrial steam engines.

**Figure 327.1. Exterior of the Museo de la Minería y de la Industrie de Asturias (© Museo de la Minería y de la Industrie de Asturias).**

**Figure 327.2. Human-powered treadwheel, with notices saying 'children must be supervised by an adult' and 'attention – no more than one person' (© Museo de la Minería y de la Industrie de Asturias).**

**Figure 327.3. German Deutz four-wheel diesel locomotive, with mine cars with independent taper-bearing wheels (© Museo de la Minería y de la Industrie de Asturias).**

Ventilation is another major problem underground and the displays show a mechanically operated bellows, originally driven by a waterwheel or using animal power. Finally there are ways of raising and lowering the heavy equipment needed underground. The displays include an original reciprocating winder/haulage; the gearing ratio shows that the winding drum would have revolved fairly slowly, so that this twin-cylinder steam engine would have been used as a low-speed capstan for raising and lowering heavy equipment, possibly during shaft sinking. Safety underground is demonstrated by the first mines-rescue service, smoke masks, ambulance brigade and a bird cage for resuscitating canaries after they had been used to detect carbon monoxide gas underground. Contrary to popular belief, canaries, or in some countries mice, pigeons or rabbits (Drexler 2011, p. 244), are not taken underground routinely to identify the presence of methane. Their rapid respiration rate means that they collapse before humans would in the presence of carbon monoxide, so are used only after an explosion underground.

Rapid and efficient transport of men, materials and the coal itself became increasingly important as collieries increased in size. Various methods of transport are shown including, for example, a German Deutz four-wheel diesel locomotive, probably of 50 horsepower. Its mine cars have independent taper-bearing wheels with a capacity in the region of 2.5 cubic metres.

The Spanish economy was transformed by the exploitation of coal, especially in the Asturias coalfield, with the development of explosives manufacturers, ironworks, power plants, railways and steelworks, as agriculture and craftsmanship were replaced by mining and industrial activity. The need for an abundant labour supply drove population growth in the coalfields and gave birth to the largest group of workers all dedicated to the same activity. This new class of workers headed the first trade-union movements in Asturias: the mining union of Asturias was founded in 1910, followed in 1916 in León and in Palencia in 1917 (Fernández 2015). Miners were active on the side of the republic in the Spanish Civil War (1936-9), followed by massive repression (*ibid.*).

Trade unionism is involved intimately in the history of the site. In 1926 the owners of the Mina San Vicente ran out of funds to pay the miners' wages. The miners' union offered to lend the owners 95,000 pesetas, which would have kept the mine operating for another three months. When the owners did not use the loan to pay their workforce, but instead disappeared with the money, the union took over ownership of the pit. So thereafter the workers had better conditions than in other mines. One display deals with the infirmaries of various pits, including Figaredo, La Camocha, Minas de Lieres and Sanatorium Adaro, with examples of the medical tools and models of other therapeutic provision. But at the Mina San Vicente, the medical provisions were available to the miners' families, not just to the miners themselves.

An hydraulic lift takes visitors from the machinery hall down to a cut-and-cover replica coal mine, which has examples of Belgian, British and German machinery used in Spanish mines. Former coal miners take 40-minute tours of the replica roadways, illustrating methods of roof support, mining methods, transport underground and ventilation systems. The roadway of this replica mine links into a roadway running right through to the bottom of a shaft at the other part of the Mina San Vincente. A gallery extension is currently underway to enable visitors to travel 350 metres by mine train to the San Vincente shaft and come up to see the mine buildings there.

*References*

ERIH, *s.n.* El Entrego

http://www.mumi.es/

personal communication, Susana Menéndez Martínez, head of administration, 2014

### 328. Museo de la Siderurgia y Minería de Castilla y León (Museum of Iron, Steel and Mining of Castile and León)

The Museo de la Siderurgia y Minería de Castilla y León is located in Sabero in the province of Castile and León. The museum is organised by the Fundación para el Turismo y las Artes de Castilla y León – Consejeria de Cultura y Turismo de la Junta de Castilla y León (Foundation for Tourism and Arts of Castilla and Leon – Ministry of Culture and Tourism of the Council of Castile and León). It occupies an area of 2,000 square metres. In 2019 the museum had two members of staff and attracted 37,118 visitors.

The first coal mining in the Sabero Valley was carried out by a British company in the 1830s, and between 1841 and 1846 there was mining in the vicinity of the hermitage of San Blas (Saint Blaize). In 1841 Miguel Iglesias Botias formed the Palentina Society of Mining, which acquired the rights to work the Sabero Numbers 1, 2 and 5 Mines, whose main period of mining was between 1850 and 1860. The construction began in 1846 of the ironworks of San Blas, the first major steel complex in Spain that worked with coke. It had housing for the workers, offices, two blast furnaces, workshops with line shafting to drive the machinery and an impressive forge in neogothic-style to house the large machines. Economic reasons precipitated its closure in 1866. In 1991 the San Blas ironworks were designated as a monument of cultural interest. Subsequent investment by the Council of Castile and León enabled the restoration of the building and its conversion to a museum, which opened in 2008.

The museum outlines the restoration, development and closure of the San Blas ironworks and also the history of the mining sector, one of the most important industries of Castile and León. The museum is divided into five main areas. The first is in the magnificent neogothic hall of the

Figure 328.1. Tub, cast-iron roof support, compressed-air receiver, surveyor's dumpy level and diesel flameproof underground locomotive in San Blas foundry neo-gothic hall (© Museo de la Siderurgia y Minería de Castilla y León).

Figure 328.2. Line shafting to drive machinery (© Museo de la Siderurgia y Minería de Castilla y León).

foundry of San Blas, with the forging and rolling mills. The second deals with the history of the Sabero Valley from prehistoric times up to the industrialisation of the area in the nineteenth century. The blast furnaces, boiler, gears, puddling furnace, sledgehammer and steam rolling mill of the foundry are housed in a gallery that resembles a basilica, so that it is known as 'the cathedral of iron'. Archaeological excavations here have uncovered the original floor of the ironworks, and exact replicas of the machinery reproduce the whole process of casting and rolling of iron as it was done 160 years ago. The replica workshop of the foundry houses the original drill and lathe, manufactured by James Nasmyth (1808-90) in England in 1848.

The fourth area deals with the development and operation of the Sabero coalfield, the first to be exploited in Spain and the first to suffer the consequences of restructuring. A large interactive model explains the characteristics and history of mining of the 50 square kilometres of the coalfield, which can be extrapolated to other Spanish coalfields. This covers the period from the first farms in 1839 to the end of mining in 1991. A selection of drawings, objects and photographs concentrates on both the technical aspects of work in the mine and the sociological aspects of the mining community. The basics of mining are outlined, from the startup of a mine, including use of a kibble (**355**) in shaft

**Figure 328.3. Medical room and pharmacy with stretcher in foreground and collection of instruments and pharmaceutical products (© Museo de la Siderurgia y Minería de Castilla y León).**

sinking, through coal cutting and lighting to movement of the coal, and showing how these tasks have evolved over time. The sociological side deals with how a mining community is created, the impact on the social fabric when the landscape is changed by new methods of exploiting its resources and the architectural changes. These include the creation of a dominant social class, including the roles of mine managers and technicians, and the evolution of mining housing. The industrialised area of San Blas had a small pharmacy, which supplied both workers and their families. The space occupied by the pharmacy has been preserved, with a complete collection of instruments and pharmaceutical products. The current space in the museum dedicated to mining is only an introduction to this topic, as a second phase, dedicated in its entirety to mining, is still to be developed.

*References*

ERIH, *s.n.* Museum of the iron and steel industry and mining of Castilla and León

https://www.museosiderurgiamineriacyl.es

https://www.spain.info/en/places-of-interest/museum-steel-mining/

personal communications, Roberto Fernández Fernández, director, 2016, 2021

### 329. Museo Minero de Escucha y CIM [Centro Interpretativo Minero] Pozo Pilar (Escucha Mining Museum CIM [Centre for the Interpretation of Mining] and Pilar Shaft)

Escucha is a municipality in the province of Teruel. There has been mining of the thick seams of the Oliete coalfield for about 150 years of both jet, which was used for jewellery, and lignite for fuel, mainly in heavy industry. The railway to Zaragoza was constructed in 1904 at the same time as the mechanisation of the mines. From the 1940s coal mining was the basis of the economy of Teruel. Production of lignite doubled between 1978 and 1985, 60 *per cent* of it from opencast mines. The last deep mine in the coalfield closed in 2003 and opencast coal is now used mainly to fuel the power plants.

The closure of the mines has caused real deprivation in this region, so mining tourism is now very important, both economically and socially. The Escucha project is being directed from the regional level with considerable funds that allow large-scale renovations. The Parque Turistico Minero de Escucha (Escucha Mining Tourist Park) leads to the Pilar Pozo, or mine interpretation centre, at the museum, on the premises of the former mining company Empresa Nacional Calvo Sotelo de Combustibles Líquidos y Lubricants (ENCASO). The museum occupies 2,500 square metres and opened in 2002, while the Pozo Pilar

**Figure 329.1. Exterior of the museum to the right and entrance to the drift to the left; headgear on the hill in the background (© Museo Minero de Escucha y CIM Pozo Pilar).**

**Figure 329.2. Internal view of the drift equipped with cable-haulage system; the tension, and the gloss on the rope and the rollers, show it is still in use (© Museo Minero de Escucha y CIM Pozo Pilar).**

**Figure 329.3. Roadway supported by wooden props and showing the dip of the strata (bedding planes) through which the roadway has been driven (© Museo Minero de Escucha y CIM Pozo Pilar).**

occupies 1,300 square metres and opened in 2014. The museum, which employs 11 staff and has six volunteers, attracts 23,000 visitors a year, and the Pilar Pozo, with three staff and three volunteers, attracts 7,500 visitors. The museum is managed and funded by the Escucha municipality.

The Museo Minero de Escucha is based on a genuine lignite mine, so the safety provisions are as stringent as for a modern operational mine. This means that visitors must wear appropriate personal-protective equipment, there must be an alternative means of egress, and adequate ventilation with back-up support is essential. There is a triple system of communication between the surface and the underground and continuous monitoring of air velocity, humidity, temperature and vibrations underground, all communicated to the surface. Visitors are taken underground in groups of up to 15 people by mining train on pre-booked guided tours, down an incline of about 200 metres with a vertical drop of 33 *per cent*. This is followed by a 400-metre walk, lasting about 75 minutes. Various underground galleries, with life-size models, display the different tools used in each period of mining. There is fixed lighting in some parts of the mine, but in the remainder visitors rely on cap lamps.

The shaft of CIM Pilar Pozo was sunk to a depth of 380 metres in the late 1960s. The museum is located in the premises of the former mining company, along with other facilities, such as offices, pithead baths, warehouse and workshop. The first section is a photographic exhibition that reveals what life was like for a miner, including cabinets where all types of tools are stored, as well as documentation that was used in these facilities. Secondly there is a series of re-creations of the settings of different buildings on the surface of the mine, including the first-aid room, laboratory, office and warehouse. Visitors are accompanied by a guide to explain about the various structures. The final section is at the foot of the headgear, where the miners entered and left the cage and where waggons filled with coal were brought to the surface and then returned underground empty. Underground there are connections between the Pilar Pozo and the Museo Minero.

Outside there is an exhibition of machinery from various mines in the area, with explanatory panels on how each operated and was used. The museum works with the Compañia General Minera, the Department of Electronic Engineering at the University of Valencia and the Department of Geotechnical Engineering at the Polytechnic University of Valencia.

*References*

http://www.museomineroescucha.es/

personal communications, Juan P. Cañizares Parrilla, manager, Parque Turistico Minero de Escucha, 2016, 2021

### 330. Museo Minero de la Comarca Andorra-Sierra de Arcos, MWINAS (Mining Museum of the Andorra-Sierra de Arcos región, MWINAS)

The Museo Minero de la Comarca Andorra-Sierra de Arcos is located in the Parque Minero (**333**) in Andorra, a municipality of the region of Andorra-Sierra de Arcos in the province of Teruel. The presence of coal in the Val de Ariño (Ariño Valley) was recognised in the nineteenth century, although its use then was confined to family farms and small industrial enterprises. Industrial-scale exploitation began in the 1920s and was developed further during the Spanish Civil War (1936-9) by ENCASO (**329**). Coal mining began to wind down in the Teruel región with the availability of petroleum, but ENCASO was replaced by Empresa Nacional de Electricidad (ENDESA) and in 1981 coal mining was given a boost by the construction of the Andorra power station.

The museum occupies the surface of the former underground mine, Pozo de San Juan (Saint John shaft), about 1.5 kilometres from the city of Andorra. The museum opened in 2005 with an exhibition paying tribute to mining. The opening took place the same year as the closure of the last underground colliery in Andorra, the Mina la Oportuna in the municipality of Alloza (an opencast mine continued on the same site for about five years afterwards). The museum is open each weekend and on public holidays with two guides, who take guided tours to other parts of the park. Five retired miners volunteer at the museum. They restore and maintain the mining machines and the site and sometimes take tours to describe their former working lives. The Comarca Andorra-Sierra de Arcos supports the museum by employing the staff and paying maintenance costs.

Much of the infrastructure of the mine has been preserved, including the first-aid room, carpenters' workshop, headframe, lamproom, warehouse and winding-engine house. The headgear and buildings have been conserved, along with the landscaped environment around the

**Figure 330.1. Twin cutting head of Alpine roadheading machine with winding-engine house and headframe in background (© Archivo de la Comarca Andorra-Sierra de Arcos).**

**Figure 330.2. Double-ended ranging-drum shearer on heavy-duty armoured face conveyor roof; support provided by two-leg immediate-face-support chocks (© Archivo de la Comarca Andorra-Sierra de Arcos).**

museum. A section of the former mining railway provides a rail journey of about 1,200 metres around the whole of the museum grounds. The museum currently occupies two of the mine buildings, totalling around 1,500 square metres. The old warehouse now houses an exhibition space, multi-purpose room, reception and storage area. The restored winding engine can be seen in operation. It was produced by Robey and Company of Lincoln, UK, a major manufacturer of agricultural and industrial machinery from the 1850s to the 1980s. The winding engine is powered by an English Electric 370 volt DC motor, supplied by a Ward-Leonard motor-generator set, and is equipped with disc braking.

As well as items on display, the museum has an extensive collection of drawings, mining equipment, models, photographs and tools. The collection is still increasing as volunteers are continuing to locate and collect items of interest. The mining company donated documentation relating to mining in the area since the late 1940s.

There is currently a large outdoor area of green land, in which some 30 large mining machines are exhibited, such as cutting and drilling machines. They range from the oldest to the latest technological examples, some of which were working until 2005. They include a skid-mounted, mobile, hydraulic-drilling rig with a built electro-hydraulic powerpack; rigs of this type may be used for geological exploration, methane drainage, drilling shotfiring holes in rock and other mining-related drilling operations. A second item on display is a double-ended ranging-drum shearer, used on longwall coalfaces from the late-twentieth century through to the present day, produced by Eickhoff of Bochum, Germany. It is displayed on a heavy-duty armoured face conveyor with roof support provided by two-leg immediate-face-support chocks; a typical installation would see a line of these supports up to 200 metres in length. In 2020 a mining railway was installed to run through the site, with a replica of one of the locomotives that served on the Andorra-Escatrón line.

The slender steel headgear (**333.1**), which stands 44 metres high, is the emblem of the park. It was built with hot rivets, instead of screws, on the same principle as the Eiffel Tower (**106**). Visitors can ascend the headgear to view the mining town and the landscape of the Val de Ariño.

*References*


https://en.wikipedia.org/wiki/Endesa_(Spain)

https://www.museomineroandorra.com

http://www.turismoandorrasierradearcos.com/mwinas.php

http://www.100patrimonioindustrial.com/Ficha.aspx?id=28

personal communications, Maria Angeles Tomás Obón, culture and tourism officer, Comarca Andorra-Sierra de Arcos, 2016, 2021


### 331. Museo Minero de Utrillas (Utrillas Mining Museum)

The Museo Minero de Utrillas is located in the Parque Temático de la Minería (Mining Theme Park) in Utrillas, a municipality of the city of Teruel. The theme park is based on the site of the former Santa Barbara (**96**) Mine. This was 160 metres deep, with roadways at depths of 60, 105 and 150 metres. A railway linked the mines of Utrillas with Zaragoza between 1904 and 1966 and terminated in 7 kilometres of narrow-gauge track from the mine shafts to the coal washery. Now 1,300 metres of this narrow-gauge railway have been restored, along with a newly built station occupying 70 square metres, located next to the Saint Barbara shaft and headgear.

The museum has one member of staff and 35 volunteers, with support from the Asociación de Amigos del Museo Minero de Utrillas (Association of the Friends of the Utrillas Mining Museum). The museum occupies a building of 200 square metres, which had been used between 1910 and 1915 as a school for the miners' children. It was restored between 2003 and 2004 for use by the museum, including the creation of a replica of a school classroom of the 1940s and 1950s. The headgear was rebuilt in 2007. The museum opened in 2015 at a cost of €225,000 and is publicly funded.

The exhibition covers a number of topics related to coal mining. There is an exhibition on the development of mining lamps from the oil lamps of Roman *Hispania* through the electric carbide lamp to the modern cap lamp, including acetylene, battery and oil lamps, with 1,500 examples on display. Similarly the way different tools and large machines were used in mining throughout the period is outlined with explanatory panels. There is a gallery devoted to geology, mining and palaeontology in the former miners' hospital. This includes a model of an *Aragosaurus ischiaticus*, the first dinosaur found in Spain, whose femurs were discovered in Utrillas in 1987, as well as a collection of marine fossils from the area. There are displays in the boiler house, where two boilers manufactured by the British company, Babcock and Wilcox, powered the cages in the shaft. Locomotives are displayed in the machine shop. A photographic exhibition is presented in the miners' locker room about the last mine in the central coalfield, Trinity Pit in Escucha, which closed in 2002.

The most important item in the museum's collection is the 2-2-0 steam locomotive Hulla (Spanish for 'coal'), works number 1166, built in 1903 by the German company, Orenstein and Koppel. Conservation of the locomotive was carried out in 1987. It was restored to full working order in 2000 with the repair of the boiler and now runs on the railway through the theme park. The museum has also recently restored a 1958 German Deutz underground locomotive (*cf.* **327.3**).

**Figure 331.1. Display of some of 1,500 examples of the development of mining lamps from Roman oil lamps through electric carbide to modern cap lamps (© Museo Minero de Utrillas).**

**Figure 331.2. Flameproof diesel underground locomotive running on a narrow-gauge track from the newly built boarding station located next to the Saint Barbara shaft and headgear (© Museo Minero de Utrillas).**

*References*

http://treneando.com/2015/12/07/nuevo-parque-tematico-de-la-mineria-y-del-ferrocarril-de-utrillas/

http://www.elperiodicodearagon.com/noticias/escenarios/parque-tematico-mineria-utrillas-abrira-19-marzo_1093855.html

parquemineroutrillas.com

personal communication, Yolanda Luengo Polo, councillor for tourism, Municipio de Utrillas, 2016

### 332. Museu de les Mines de Cercs (Museum of the Mines of Cercs)

The Museu de les Mines de Cercs is located high in the mountains not far from Colonia Sedó in the Berguedà district of Catalonia. The mines of Figols and the area around the upper reaches of the river Llobregat were the main sources of Catalan coal. Mining first began to be modernised in this area when the mines were purchased in 1893 by Jose Enrique de Olano Loizaga (1858-1934). He developed the settlement of Sant Corneli (Saint Cornelius) and in 1904 the Sant Joseph railway to the mine. In 1900 the company occupied the tenth position in the state ranking of the sector. In 1908 King Alfonso XIII (1886-1941) paid a visit to the mining complex and granted Olano the title of count of Fígols. In 1911 Olano set up the Carbones de Berga, which became the largest corporation in the mining sector in Catalonia.

The museum deals with the history of mining in the Berguedà district and the relationship between coal, geology, the landscape, and social and economic history. It occupies an area of 5,317 square metres and employs four members of staff. The museum, which opened in 1999, is organised and funded by the Cercs local authority.

The permanent exhibition is located in a building that was originally a convent and is divided into two areas. The first is coal exploitation, the importance of mining infrastructure, transport and coal as both a domestic and an industrial fuel as a primary source for generating electricity. The second is the daily life of miners, including the 1940s mining colony of Sant Corneli, dispensary, leisure, school, women's work, the employer and the company, workers, and social and labour demands. Life in the mining settlements was difficult due to their isolation and the fact that their inhabitants were totally dependent on the mine and the mining company. Until the Spanish Civil War (1936-9) and, despite their isolation, mining colonies lived intensely political and social lives. The constant confrontations with the company, the struggle for better working and social conditions, anticlericalism and the strength of the anarchist union CNT-FAI are an important part of the history of coal mining in this region. The museum has a 16-minute audio-visual dealing with the anarchist revolt and strike of miners and textile workers at Alt Llobregat in 1932.

Finally there is an audio-visual presentation on the mining of the High Berguedá and a display of the 1944 home of a miner. The miners and their families lived in housing rented from the company. A few homes, measuring 56 or 70 square metres, were reserved for foremen and administrative staff, but most of the homes had only 35 square metres. The most important place in the house was the kitchen.

A mine train takes visitors inside the roadways of the Sant Romà Mine where they can experience what work was like inside the mine, in the original setting. Working conditions underground were also extremely harsh, with very low wages, poor ventilation, the dangers of explosions, the risk to health of poor lighting and dust, and long working hours from 6 a.m. to 6 p.m. Back on the surface visitors are also able to see the ancillary functions, such as the washery and the workshops, and also the mining settlement and an exhibition of mining machinery. The most important item in the museum's collection is a Bucyrus steam bulldozer dating to 1922 (*cf.* **409, 422**).

A new building is faced with a reflective surface, the lower half of which is covered with alpine plants. An exhibition, 'L 'Espectacle Mine' (the mine show), uses material from the Carbones Pedraforca Company, which is active in Berguedà. with both underground and opencast mining. This is illustrated using holograms, image and sound effects, and visual projections. It is based on the idea of archaeologists of the future entering the storeroom of the mine and finding tools used in mining, such as lighting and safety systems, tools and, in particular, the old computer for the mine, which gives much detail about how mining used to work.

*References*

ERIH, *s.n.* Cercs

personal communication, Susana Menéndez Martínez, director, 2016

showcaves

www.mmcercs.cat

### 333. Parque Minero (Mining Park)

The Parque Minero has its headquarters at the Pozo de San Juan (**330**). An initiative to develop tourism and outreach through the mining history of the region came from a group of volunteer former mineworkers, who persuaded the local and regional authorities to carry out restoration of the Pozo de San Juan. The steel headgear is the emblem of the park (**330**).

Tourism is very important, both economically and socially in this region, where the closure of the mines has caused real deprivation; in a period of 20 years the town of Estercuel lost 50 *per cent* of its inhabitants. The Parque Minero is aimed at achieving sustainable development

**Figure 332.1. View of the museum, high in the mountains; entrance to underground on left; main museum building in centre; new museum building (with red roof) to lower left (© Museu de les Mines de Cercs).**

**Figure 332.2. Underground locomotive pulling mine tubs (© Museu de les Mines de Cercs).**

**Figure 332.3. New red-roofed building, faced with a reflective surface, the lower half of which is planted with alpine plants (© Eusebi Casenelles).**

through historical memories and the innovative use of old mining areas. Throughout the year the Parque Minero is the base for a varied cultural offering, with events such as competitions, conferences, exhibitions and music, but always related to the world of the mine. For example, for several years the winners of the Lámpara Minera (Mining Lamp), the major award of the Festival Internacional del Cante de las Minas de La Unión (International Choral Festival of the Mines of La Union), did their first performance in the Parque Minero.

The park is an open-air ecomuseum. Mining in the Andorra-Sierra de Arcos region was developed in the second half of the twentieth century. The mining company ENDESA (**330**) worked four opencast mines – Corta Alloza, Corta Barrabasa, Corta Gargallo and, the last to close in 2012, Corta Gargallo Oeste – covering a total of 60 hectares. The first restoration work began in the 1980s, with the planting of 14,000 fruit trees and 2.7 hectares of vineyard. Corta Alloza in the Val de Ariño was the first opencast site that ENDESA restored.

The entire region is seen as a mining landscape. There are reference points throughout the territory, giving a complete overview of the history of this sector of the Andorran economy, including methods of environmental restoration and subsequent land use. The landscape created by opencast mines has been restored and the void of the opencast filled by a lake covering 6 hectares. For its work on transforming the mining landscape into an area of ecological value, the ENDESA Company was awarded the Aragon Environment Prize in 2005. In 2010 the Corta Alloza wetland was added by government decree to the register of unique protected wetlands of Aragon. Currently, there are two main lakes in the park, but there are also other smaller ponds that can regulate the water flow and facilitate an increase in wildlife. The lakes provide an ideal habitat for a wide variety of amphibians, animals, birds and plants.

Three vantage points, located strategically along the valley, provide views over the landscape and an understanding of the relationship between mining and the environment. The opencast process of the opening up of the mines, the creation and then restoration of tailings areas and the final restoration of the landscape are also explained. Free access to this area is prohibited and visitors are taken in groups by bus between the three points. Interpretation is by guides accompanying the groups, rather than by explanatory panels.

**Figure 333.1. Headgear of the Pozo de San Juan, which is used for the emblem of the Parque Minero (© Archivo de la Comarca Andorra-Sierra de Arcos).**

*References*

http://mineriaypaisaje.com/val-de-arino-it/

http://www.turismoandorrasierradearcos.com/mwinas.php

http://www.turismoandorrasierradearcos.com/noticias_ext.php?id=57

personal communications, Maria Angeles Tomás Obón (**330**), 2016, 2021

### 334. Pozo Julia de Fabero (Julia Shaft of Fabero)

The Pozo Julia de Fabero is located in Fabero in the province of Castile and León. The Pozo Julia was sunk to a depth of 275 metres in the early 1950s, with 3,600 miners employed between 1955 and 1960. The mine was one of those worked by the Fabero Anthracite Company, which pioneered mechanisation of mining in Spain. In 1962 longwall advance mining (**382**) and a mechanised cutting system were introduced for the first time in Spain, which meant that fewer miners were required to work the colliery. Representatives from the most important mines in the country, including Hunosa (**335**), came to Fabero to learn about the new forms of mechanisation.

The Fabero Anthracite Company was the most important and extensive employer in the municipality and made several investments, such as a bakery, commissary, housing and workshops. At the national level it was a very prominent company, being granted the title of model enterprise and in 1952 won first prize in a national exhibition of mining and metallurgy held in Madrid. Production at the mine always exceeded 100,000 tonnes a year. In 1958 production reached 387,570 tonnes and in 1962 246,000 tonnes. Between 1955 and 1960 the colliery employed 3,600 workers.

Of the five major mining companies operating in the Fabero coalfield in 1999, only Fabero Anthracite survived. It was absorbed into the Union Minera del Norte S.A (UMINSA) and continued in production. A business decision was made in 2002 to end underground mining, based on the difficulties of extraction and the high maintenance costs of the facilities. The remaining workforce of fewer than 100 workers was relocated to other jobs with UMINSA in the area.

In 2007 UMINSA gave the sites of Pozo Julia and Pozo Viejo (old shaft) free of charge to the city of Fabero. The donation included the hospital, miners' houses, warehouses and workshops, together with the commissariat located at the Pozo Viejo. The museum site occupies 60,000 square metres and opened in 2007. It employs two members of staff, with between 14 and 20 volunteers assisting each year. The museum attracts about 8,000 visitors a year and is managed and funded by the city of Fabero, which works actively with the Asociación de Mineros Cuenca de Fabero (Fabero Coalfield Miners' Association).

The site preserves all aspects of mining in the past in Fabero. The mining and other facilities are preserved as they were in the 1950s. Inside there are, for example, the first-aid room and the room where the miners' clothes were stored, suspended from the ceiling by ropes, with a padlock at the bottom of each to which only the individual miner had the key. Elsewhere on the site there are the conveyor belts, running under covered gantries, one taking coal into the screens building and a second taking the coal sorted from the dirt to drop from a hopper into waggons below. In various buildings photographs and prints illustrate mining work, highlighting the arduous conditions in which the miners operated. Members of the Asociación de Mineros Cuenca de Fabero provide pre-booked hour-long tours of the site.

*References*

http://www.aytofabero.com/historia_mina.php

personal communication, Paz Martínez Ramón, mayor, Ayuntamiento de Ayto (Ayto City Council), 2016

www.aytofabero.com/pozo_julia/

**335. Pozo Sotón (Woodland Shaft)**

The Pozo Sotón is located in north-western Spain in the city of El Entrego in Asturias. It is 2.5 kilometres south-east of MUMI (**327**), with which it has been amalgamated since 2016. Coal mining began in the valley of the river Nalón in the nineteenth century. The Pozo Sotón was sunk in 1914 and deepened in 1922 by the Duro Felguera Company, established in 1858 and named from its founder, Pedro Duro (1810-86), and his foundry. The mine was sunk to provide fuel for the company's foundries in La Felguera and Langreo, with the work being completed in 1921. The name Sotón is the adjectival form of *soto*, which means 'place of forest' or 'wooded land' and is derived from the Latin *saltus* (forest). The name refers to the heavily wooded slopes of the valley of the river Nalón in which the mine is situated. The mine was extended in the late 1950s. Between April and June 1962 the miners of Asturias came out on strike in demand of higher wages and improved working conditions. The regime of the fascist dictator, Francisco Franco (1936-75), responded by using the force of the civil guard and dismissing many miners from their employment in the mines and evicting them from Asturias. During 1962 the duke of Parma, Carlos Hugo de Borbón (1920-2000), worked underground here under the pseudonym Javier Ipiña. This was both to see the miners' working conditions and to try to enlist their support for his claim to the Spanish throne, which was ultimately unsuccessful. His identity was discovered after two months and he was dismissed, although the Carlist faction thereafter referred to him as 'the mining prince'.

In 1967 the mine was acquired by Hulleras del Norte S.A. (Northern Coal Company), the company founded that year to mine coal in Asturias and usually known as Hunosa; the *norte* (north) of the name refers to the north of Spain. The colliery has two headframes, powered by two identical 850-horsepower winding engines. These are now the only example of twin headgears surviving in the Nalón coalfield. They stand above two independent shafts, one 690 metres deep and the other, which is used for the visits, 595 metres deep. It is believed to be the deepest visitor coal mine in Europe. The site ceased active production of coal in December 2014, having been declared a monument of cultural interest in March 2014.

The staff consists of six miners and about 30 visitors a day go underground. The funding of the site is underwritten by Hunosa, with the staff of MUMI providing advice on displays and presentation. The site opened to the public in June 2015 and the amount open on the surface was extended in March 2016. The Pozo Sotón has 140 kilometres of roadway, 600 metres deep, of which 12 kilometres are used for the underground tour. The tours last between four hours 30 minutes and five hours, in groups of up to 30 visitors. Visitors are able to experience using some equipment, such as a Holman Silver 303 airleg-fitted rock drill used for boring shotfiring holes. Companies like Holman Brothers Limited, based in Camborne, UK, supplied this sort of equipment around the world. Guides demonstrate the operation of some of the equipment, such as a pneumatically powered rocker-shovel.

Above ground the site still has all the other buildings normally found at a colliery, such as pithead baths. The buildings are constructed in the same style as those of the

**Figure 334.1. Brochure of the Pozo Julia, with text saying 'El Bierzo [an administrative area of Castile and León] from gold to coal' (© Pozo Julia de Fabero).**

**Figure 334.2. Left to right, conveyor taking coal to drop into waggons and screens building (© Pozo Julia de Fabero).**

Felguera foundry and include communal housing for the miners. In 2016 Hunosa constructed a memorial with the names of all those who had been killed underground since the company was established in 1967, amounting to some 500 names.

*References*

https://es.wikipedia.org/wiki/Carlos_Hugo_de_Borbón

https://es.wikipedia.org/wiki/Huelga_minera_de_Asturias_de_1962

https://es.wikipedia.org/wiki/Pozo_Sotón

personal communication, José Enrique Mencía, press officer, 2016

www.cemmineria.es

www.memorialminero.es

www.pozosoton.es

### 336. Surroca Museo de la Minería del Carbón (Surroca Coal Mining Museum)

Palau-Surroca is located in the province of Girona, in Catalonia. The Surroca-Ogassa coalfield, known locally as Sant Joan de las Abadesses (Saint John of the Abbesses), began to be worked in the late-eighteenth century and continued until the mines closed in the 1967. In 1880 a railway was opened to Ripollès to enable the coal to be moved to Barcelona more quickly. In the period between 1860 and 1928 more than a million tonnes of coal were produced. This was lower than the production levels of other Spanish coalfields, as the geological faulting of the rugged terrain and high mountains posed considerable difficulties for the mine owners. In the general area of Palau-Surroca there are the remains of inclined planes and headframes.

The history of the coalfield was meant to be preserved in the Surroca Museo de la Minería del Carbón**.** A building was restored to house the museum and a small gallery was opened. The financial crisis since 2008 means it has now closed, as has happened with many Spanish sites. There is no money to operate or develop the museum and funding is unlikely to be available for many years. There are now plans to open a museum about the Surroca mines in Ogassa, where the coal was turned into charcoal and the miners built a church dedicated to Saint Barbara (**96**). Visits can be made to the Ca l'Ermengol overhead cable station.

*References*

https://150elements.mnactec.cat/es/index.php/company/minas-de-surroca/

http://www.turismeruralcancamps.com/en/environment/

http://www.turismeruralcancamps.com/mines-dogassa/

personal communication, Eusebi Casanelles (**324**), 2014

# Sweden

Sweden is famous in the mining world mainly for its metal mining, especially of copper, of which it was the largest producer in Europe in the seventeenth century, when its mineral wealth made Sweden a great military power. It was eminent in mining technology and mining education in the eighteenth century. The museum at Falun in Dalarna, adjacent to the great Falun copper mine (which dates back to the eleventh century), still has the wooden working models of mining machinery used to train mining engineers, who travelled there from throughout Europe. The Geological Survey of Sweden was established in 1858 (Ranestad and Sharp 2020, p. 5). Women who worked underground were usually the wives of male miners, regularly paid half the male wage (Henriksson 1996). This was mainly in the smaller ironstone mines when there was a labour shortage, and women were no longer required for underground work when the population increased in Sweden in the mid-nineteenth century.

Bituminous and sub-bituminous coal and lignite are found only in north-west Skåne in southern Sweden, with 1 million tonnes of reserves. Coal mining started in this region in the middle of the seventeenth century, although it ceased in 1680. The coal was got by hand from very low seams, until mechanised coal cutters were introduced in the 1920s. The clay industry also developed in Skåne, offsetting the fact that the coal on Skåne had a relatively low calorific value and could not be transported without creating a lot of fragmentary, or small, coal. Production gradually decreased during the 1950s and 1960s, until the last mine closed in 1979 (personal communication, Magnus Ericsson, adjunct professor of mineral economics, Luleå University of Technology, 2021, for the history of coal mining on Skåne). In 2018 Sweden set a target of 100 *per cent* electricity generation from renewables by 2040 (euracoal; but see also **230**).

*References*


https://www.forskning.se/2019/11/07/machokultur-i-gruvor-pa-vag-att-utmanas/

http://www.falugruva.se

personal communication, Johanna Nybelius, education officer, Falu Gruva, 2021


### 337. Gruvmuseet Bjuv (Bjuv Mining Museum)

So important was coal mining in Bjuv that the municipal coat of arms is an image of a flaming coal torch, as used underground in the eighteenth and nineteenth centuries. Shafts 1 and 2 of the Bjuv Colliery were sunk in 1872 and 1873. The museum is on the site of shaft 3, which opened in 1896. Production by the company Kropp AB finished

**Figure 337.1. Headgear on site (© Gruvmuseet Bjuv).**

in 1979, the last coal mine in operation in Sweden. The Gruvmuseet Bjuv, which opened in 1985 and is run by Bjuvs Kommun (the municipality), occupies the winding-engine hall, an area of about 200 square metres. The museum only opens in summer, when there is one member of staff on duty. It attracts approximately 750 visitors a year.

The museum deals with how Bjuv changed from a farming community into an industrial one with the coming of coal mining, along with the popular movements that took over the role of the previous peasant society – religious revivalism, sport, and the temperance and labour movements – which fought for a better society for the community in this difficult transitional period around the end of the nineteenth century. Full-scale models show work in the mine and the social lives of miners and their community. The museum also covers the geological history of the formation of coal in the Carboniferous era.

*References*

https://www.bjuv.se/kultur-och-fritid/kulturarv-och-museer/gruvmuseet.html

personal communications, Helena Sjölander, culture and leisure assistant, Bjuvs Kommun, 2012; Catarina Månsson, unit manager, Library and Cultural Heritage, Bjuvs Kommun, 2021

showcaves

### 338. Nyvångs Kolgruva Museum (Nyvång Colliery Museum)

Nyvångs Kolgruva was opened in 1907 by the Skåne Kolbrytningsaktiebolaget Company, with the Carl Cervin shaft, which was 100 metres deep. It was named after Carl Cervin (1850-1919), the major investor in the company. The first coal did not reach the surface until 1911, so there was no income generated for the first four years and his

**Figure 338.2. Winding-engine house over the Carl Cervin shaft with colliery spoil heap, the Bowl, in the background (© Nyvångs Kolgruva).**

**Figure 338.1. Exterior of the Nyvångs Kolgruva, with two winding wheels to the left; the slogan says 'welcoming visitors to the Nyvångs Kolgruva' (© Nyvångs Kolgruva).**

investment kept the company going until the mine became profitable. The mine was bought in 1916 by the Hőganäs Coal Corporation, which developed it into the largest coal mine in Sweden, as well as establishing the largest school in the country for training coal miners. At one time the mine employed 500 people. The company sponsored choirs and brass bands and provided houses with low rents and free medical care for the miners and their families. The mine closed in 1966 and the shaft was capped. Employees of the mine were able to buy their rented homes at subsidised rates and were also offered other jobs within the Hőganäs Coal Corporation.

The Nyvångs Kolgruva Museum was started in 1985 by former employees of the mine and occupies the former machine hall. This used to house three turbines in a large power-generating station, with seven steam boilers producing between 10 and 20 megawatts of power. The mine supplied coal to the nearby power station, which was not closed until 1970 and occupies about 160 square metres. There are no members of staff and the museum is operated by about 15 volunteers, most of whom are local residents or relatives of deceased coal miners. The adjacent water tower supplied fresh minewater, pumped from the underground, for the local community and for steam for the compressors. The winding-engine house is now the gymnasium of Nyvång school.

The not-for-profit society, Föreningen Nyvångs Gruva (Nyvång Coal Mine Association). has about 250 members, all paying an annual contribution towards the running costs of the museum. There is a small fee for about 700 visitors every year, and charges for light refreshments ordered in advance for pre-booked groups of visitors also contribute income. The building is owned by the community and the society pays no rent. The museum deals with the history of the local coal miners. The visit also includes a film about mining in the 1930s. The nearby spoil heap, known as the Bowl, is approximately 60 metres high.

*References*

http://www.geocaching.com/geocache/GC1B6FC_nyvangs-gruvmuseum?guid=d7c93f65-2ddc-4f6d-af7d-0a216147e5a5

http://www.lundarallarna.se/weckoblad/num23/index.html

http://www.nyvangsgruva.se/

personal communications, Per Ivar Svensson, board member, 2013; Jan Petersson, chair, 2021

showcaves

**339. Tekniska Museet (Technical Museum)**

The Tekniska Museet, officially known as the National Museum of Science and Technology, opened in 1936 and is Sweden´s largest museum of technology. The museum has a national charter to preserve Sweden's technological and industrial history as part of its cultural heritage. The museum is located on the outskirts of Stockholm, the capital of Sweden, on the south-east coast of the country. The galleries cover around 10,000 square metres and the museum attracts more than 300,000 visitors every year. The museum has a staff of around 50, together with a number of attendants.

The mining collection consists of around 1,050 objects. There is a replica mine in the basement of the museum,

**Figure 339.1. Models of miners in simulated underground showing typical nineteenth-century clothing (© Tekniska Museet).**

covering approximately 400 square metres. This uses models to show scenes of the miners' daily life. It does not represent any particular type of mine or a particular time, but aims to give a cross-section through history and different types of mines, including coal.

The mine is a permanent exhibit, entered through a simulated mine shaft by going down a wooden staircase. There is also a mine lift which can be used, but only on request and operated by museum staff. The shaft has hoist, ladders and water pumps, and is boarded with timber logs. The first section shows early mining, with open-light sources and mining by lighting a fire. Mining with explosives started with gunpowder in the seventeenth century. In the nineteenth century nitro-glycerine and later dynamite were used. The main work for the miner was the drilling of holes, first by hand and later, from about 1910, by pneumatic drills. Compressed-air and electrical power began to change mining methods from the early-twentieth century. The 'Swedish method' of using lightweight pneumatic-boring machines equipped with carbide drill bits was introduced in the 1940s.

Recently a new area was added featuring a modern mining environment up to the present day, with simulators for explosives, interactive games and machinery. Visitors are able to ride a vehicle underground, and to operate a drill and a simulator of a loading machine. The mine ends in the modern control room of a twentieth-century LKAB Minerals Company mine, showing the remote control of heavy mining machinery and fully automatic loading machines, trains and drilling machines and the major use of computers.

*References*

https://www.tekniskamuseet.se/en/discover/exhibitions/the-mine/

personal communications, Eva Derlow, curator, 2012; Lotta Oudhuis, archivist, 2021

showcaves

# Switzerland

Coal was discovered in the canton of Zurich in the north of Switzerland during roadworks. There were a number of unsuccessful attempts at mining, in 1548, 1663 (for fuel to fire bricks), 1708 and 1763. Since 1915 the Natural History Museum in Basel has held samples of Swiss bituminous coal, mainly from the Swiss Plateau along the east-west axis of the country and the Valais region of southern Switzerland (https://www.nmbs.ch/home/en/museum/sammlungen/geowissenschaftlich/mineralogie/bodenschaetze-der-Schweiz.html). The Käpfnach and Riedhof Pits (**340, 341**) were the only two economically important mines. The seams are extremely thin, so coal mining in Switzerland was never profitable, except during wartime when coal supplies were cut off from outside the country.

### 340. Bergbaumuseum und Besucherbergwerk (Mining Museum and Visitor Mine)

The Bergbaumuseum und Besucherbergwerk is located at Horgen. The Zurich Government established mining here in 1784, which continued until 1911. Käpfnach Pit was the largest coal mine in Switzerland, with 80 kilometres of roadways. The coal seam was, however, only 48 centimetres thick, and the coal contained a lot of water and other impurities such as sulphur. Some 80 miners were employed from 1917 during WWI, but the pit closed in 1921. Mining was revived in 1941 during WWII, with 250 miners employed. They produced some 54,600 tonnes of coal, approximately 4 *per cent* of the total used in Switzerland. Mining ceased again in 1947 after the war had ended. All the machinery was sold off and the roadways were backfilled.

The Bergwerksverein Käpfnach (the Käpfnach Mining Society) was founded in 1982 as a not-for-profit organisation. The aim was to ensure the survival of a nationally important monument of the industrial culture of the region, and to open and manage the show mine. By law the society was required to acquire the rights to the remaining coal in the mine, estimated at about 300,000 tonnes. The society has about 600 members, 35 of whom are active members and work as volunteers, conducting tours or maintaining the mine and the museum. By 2012 there had been more than 215,000 visitors. In 2015 there was a fire in the museum which caused a considerable amount of damage, causing the museum to close for reconstruction, but it is now open again.

The museum is housed in the former coal magazine and has items related to coal mining. Groups of up to 20 visitors are taken into the mine by an electrically powered railway. The train travels for 1.4 kilometres. The roadways in the mine were very low, because of the thin seams. The roadways have not been heightened, unlike some tourist mines elsewhere.

*References*

http://www.bergwerk-kaepfnach.ch

http://www.forums9.ch/ausflug/kaepfnach/

personal communication, Brigit Ruoff, information officer, 2016

showcaves

### 341. Bergwerk Riedhof (Riedhof Mine)

Mining began at the Bergwerk Riedhof in 1787 under government control. It ceased in 1814, because of the lack of transport links to move the coal. The pit reopened in 1917 during WWI, but closed in 1921. Mining was revived in 1942 during WWII, with 30 miners employed and 7 kilometres of roadways. They produced only about 1,180 tonnes of coal, although in 1946 production was higher than at Käpfnach Pit (**340**). Mining ceased again in 1946 after the war had ended. All the machinery was sold off and the roadways were backfilled.

The Verein Bergwerk Riedhof (Bergwerk Riedhof Society), a not-for-profit association, was formed in 2002 to preserve the industrial history of the area and the museum opened in 2003. The museum has exhibits covering all aspects of the history of the mine, including photographs and documents from the phase of mining between 1942 and 1946.

The underground of the mine could not be reopened. So a simulated underground was created in an underground bunker constructed during the Cold War (1947-91) to provide protection for local people in the event of a nuclear war. The bunker was abandoned after the Cold War ended. The simulation is reached by a concrete stair accessed from the middle of a grassed area. The bunker was divided into rooms, which have been used to display mining conditions at different periods; one, for example, showing wooden roof supports. As it is run by volunteers, the museum is open only occasionally during the year. As well as the general tour, a special four-hour tour is available for those who are physically fit. This involves crawling on the wet floors to simulate working in a low seam.

*References*

http://www.bergwerk-riedhof.ch/

showcaves

# Taiwan

There are small and insignificant amounts of coal in the central region of Taiwan, but the main concentration of coal, mostly bituminous, is in the north, covering an area of 3,000 square kilometres; all three mining museums are located in this area. There had been mining of coal in Taiwan for hundreds of years, but industrial production only began in 1876 at Badouzi, Keelung, the first government-managed coal mine (Xue 2016, p. 218). Eventually there were some 500 mines in Taiwan, 18 of them in the immediate vicinity of the Xīnpíngxī Méikuàng Bówùyuánqū (**344**). The average coal seams are only around 30 to 60 centimetres thick.

Japan colonised Taiwan in 1895, following the Sino-Japanese War (1894-5). The army immediately took over Keelung Mine and Japanese companies acquired the other mines, followed by intensive development. Production increased from over 100,000 tonnes in 1906 as new mechanised mines were sunk, until production reached 2.85 million tonnes in 1941 (*ibid.*, p. 218).

Air raids during WWII decimated the mining infrastructure, so that production fell below nineteenth-century levels. Taiwan has only a very small amount of natural gas and the sole major indigenous fuel available is coal. Since 1949 Taiwan has been administered separately from mainland China and had to rely entirely on coal to provide the electricity essential for the development of industry in the 1970s and 1980s. The district of Ruifang, in the eastern part of New Taipei City (**342**), was the most productive coalfield in Taiwan after WWII. The total production of the coalfield was more than 10 million tonnes from 1911. The coal-mining industry here closed down in 1990, because of a decrease in mining resources and the introduction of costly new policies to promote mining safety. After WWII the Pingxī district of New Taipei City (**343, 344**) was the third most productive coalfield in Taiwan, with a total production of 3,721,315 tonnes during this period. Production finished here in 1997, as a result of competition from foreign coal, and in the whole of Taiwan in 2000.

### 342. Hóudòng Méikuàng Bówùyuánqū (Houtong Coal-Mine Ecological Park)

The Hóudòng Méikuàng Bówùyuánqū is situated by the bank of the river Jilong in the district of Ruifang, in the eastern part of New Taipei City. The mining facilities that remain in this area after the closure of the mining industry include the mines, bridges and railway systems for transporting coal, factories and offices of the mining industry, related community-activity centres, the welfare shop, canteen and dormitories for mining workers, and the mining villages themselves. This industrial heritage made the site especially suitable for the development as an ecomuseum (Zhen-Hui and Yung-Jaan 2015). Under pressure, however, from the mining communities and other bodies, such as associations and government, in the late 1990s the local authority decided instead to build a coal-mining museum. This failed in the early 2000s, and was then redesignated as an ecomuseum that used the tangible and intangible cultural resources and the built environment to promote an integrated redevelopment policy.

The master plan of the Hóudòng Méikuàng Bówùyuánqū focused not only on the mining heritage and the antiquities inside the properties, but also on the integration of the natural environment, the lifestyle of the local population and the social processes. The framework proposed was for one core zone, two satellite regions, six community-living circles, one fieldwork school camp and various discovery trails. The decaying mining landscape and environment were landscaped, and some mining-industry artefacts have been restored for exhibition or tourism facilities. Examples include the arched bridge across the river Jilong, the warehouse and office close to the coal-preparation plant, the room for drying sand and some of the communal baths for workers and miners. Many of the buildings date to the Japanese colonial era (1895-1945).

Three galleries and three tourist information centres were established to create educational and visitor facilities, and to preserve the coal-mining heritage. Also preserved were the remains of three of the mines, the coal-preparation plant, coal bridge, garage, mine building, railway station, transportation system and facilities for the mine workers. The boundary of the Hóudòng Méikuàng Bówùyuánqū takes in the mining villages of Gongqiao, Guangfu and Hóudòng. Visitors can meet retired miners and their families, who share their memories and provide a local perspective on mining history and related stories and technologies. Visitors can also experience the whole of the mining operational process.

The park opened in 2010 and occupies 11 hectares of land. It was founded and is managed by the Tourism and Travel Department of the New Taipei City Government. Some local organisations hold related activities; for example, a guided tour in the flowering season of the rare *Bretschneidera sinensis Hemsl*, a tree standing 10 to 20 metres tall with large inflorescences, found in China and Vietnam, but only recently identified in Taiwan and Thailand. Visitors to the park reached a peak of 128,600 a month.

**Figure 342.1. Coal-mine transport bridge (© Zhen-Hui Liu).**

**Figure 342.2. Display model showing a reconstruction of the coal-preparation plant and the coal-mine transport bridge on the left (© Zhen-Hui Liu).**

*References*

http://tour.ntpc.gov.tw/page.aspx?wtp=3&wnd=336&id=E1_38 2000000A_011585

http://www.travelking.com.tw/eng/tourguide/taipei/houtong.html

personal communications, Zhen-Hui Liu, assistant professor, Providence University, Taiwan, 2016, 2021

### 343. Jīngtóngkuàngyè Shēnghuóguǎn (Jingtong Mining-Industry Life Pavilion)

The Shidi Coal Mine was adjacent to the Jīngtóngkuàngyè Shēnghuóguǎn in the Pingxī district of New Taipei City. Starting in 1917, First Pit, Second Pit, Third Pit, Fourth Pit, Fifth Pit and Sixth Pit were excavated in this coalfield. The Pingxī railway, completed in 1921, was constructed by the mining company to transport the coal.

The region gradually became a well-known tourist destination after production ceased in 1995, because of its tangible and intangible cultural heritage. The integrated mining area was identified as a suitable site for implementing the ecomuseum concept in the future (Zhen-Hui and Yung-Jaan 2015). The Jīngtóngkuàngyè Shēnghuóguǎn opened in 2005. It is managed by the Pingxī District Office of the New Taipei City Government. The pavilion has 139 square metres on the first floor and 147 square metres on the second floor. There is one member of staff running the pavilion and three selling refreshments and souvenirs, supported by 22 volunteers. It attracts about 100,000 visitors a year.

The Jīngtóngkuàngyè Shēnghuóguǎn building was originally built to house railway workers. It is located next to Jīngtóng railway station, which was constructed by the Japanese in the 1930s. The building has been extensively renovated to house the mining museum, located near the Jīngtóng Méikuàng Jìniàn Gōngyuán (Jingtong Coal Mine Memorial Park). It has a simple wooden exterior, partly covered with vines, and inside there are exhibitions on the development of mining and the railway in Jīngtóng and on the cultural life of Pingxī, including the sky-lantern festival. Free guided tours are provided of the exhibitions to enable visitors to ask questions.

*References*

https://tour.ntpc.gov.tw/zh-tw/Attraction/Detail?wnd_id=60&id=110044

http://tour.ntpc.gov.tw/page.aspx?wtp=3&wnd=336&id=E1_382000000A_011610

http://www.mdpi.com/2071-1050/7/10/13249

personal communications, Zhen-Hui Liu (**342**), 2016, 2021

### 344. Xīnpíngxī Méikuàng Bówùyuánqū (New Pingxī Coal-Mine Museum)

Also known as the Táiwān Méikuàng Bówùguǎn (Taiwan Coal-Mine Museum), the Xīnpíngxī Méikuàng Bówùyuánqū is located in the Pingxī district of New Taipei City. Until 2019 the museum had active mining rights in its underground. The museum is of both ecological and historical importance: in the summer there are tung flowers, mountain lilies and aquatic plants.

The current site of the Xīnpíngxī Colliery produced coal from 1965 until 1997. The first narrow-gauge electric railway was laid in the Taipei Songshan district for moving coal, as well as iron, salt, sugar and timber. Other narrow-gauge railways still continue in operation as tourist attractions elsewhere in Taiwan, but this is the only colliery-related railway to survive.

The museum opened in 2002 and employs 10 members of staff, with six volunteers. It is run as a limited cooperative partnership and is partially funded by the Cultural Affairs Department of the Taiwan Government, and is also recognised as a local cultural museum. The museum occupies around 45,000 square metres, with the tunnel measuring 400 metres. It attracts 50,000 visitors a year. Some notices are in Japanese characters and the English language as well as Chinese characters, although not in the modified/simplified Hànyǔ Pīnyīn system, but in traditional script, which is still sometimes used in Taiwan.

The drift and facilities all still survive, such as the coal-preparation plant, with hoppers to allow the coal to drop into waggons beneath it, pithead baths, railway system and storage bunkers. There are also many traces of the historical lifestyle of the miners and the original miners' village. The simulated underground reproduces mining operations, from cutting the coal to moving it out of the mine. The drift roadways are supported with concrete in less stable ground near or at the surface and then with steel arching further in. The only light in the pit is from the headlamps, and fresh air is drawn in from the outside by large fans. The tunnel is also used for training mines-rescue teams.

The exhibitions on the surface include data and historical information about early mining, mining objects and photographs. There is a display of helmets, with the explanation that foremen wore white helmets, mechanical and electrical engineers blue, and miners yellow. Women worked underground, usually with their husband, until government legislation in 1964 permitted only men to be employed underground. This was to avoid the danger of both husband and wife being killed, leaving children without any parent to raise them.

Temperatures were high in the Taiwan mines, so miners tended to work stripped to the waist. They often had to return home dirty, although some mines did provide bathing facilities in large tubs. These were similar to the Japanese communal bath known as an *ofuro*, which also served for relaxation and comradeship.

The most important items in the museum's collection are two locomotives, originally imported from the Nichiyu Company in Japan in 1942. They were powered by two direct-current motors of 220 volts and were fitted with a pantograph (*cf.* **254**) to take their power from an overhead line, much like a tramway. For safety reasons they have been changed to battery-operated, still using the same motor but driven by direct current. One is on display in the exhibition and the other is used for taking visitors around

**Figure 344.1. 1942 mine locomotive taking visitors to see firefly display at night (© Xinpingxi Meikuang Bowuyuanqu).**

**Figure 344.2. Coal-preparation plant on site and plant to store clean coal, with hoppers to allow coal to drop into waggons underneath (© Xinpingxi Meikuang Bowuyuanqu).**

the site, and in April and May to observe the firefly display at night. Having a circular window at the front, they were nicknamed the *dú yǎn xiǎosēng*. This translates literally as 'the little one-eyed monk', which is a monster from Japanese folklore, known as Hitotsume Kozō in Japan and now also a creature in geek sub-culture.

### *References*

https://en.wikipedia.org/wiki/Taiwan_Coal_Mine_Museum

https://ja.wikipedia.org/wiki/%E4%B8%80%E3%81%A4%E7%9B%AE%E5%B0%8F%E5%83%A7

https://www.museum.ntpc.gov.tw/xmdoc/cont?xsmsid=0G275735901206394658&sid=0G297834803293950118

personal communications, Jiunn-yih Gong, president, 2016, 2021

# Thailand

Reserves of lignite were first identified in Thailand in 1917. In 1927 King Prajadhipok (King Rama VII, 1893-1941) issued a royal command that the coal resources of the district of Mae Moh should be preserved for the benefit of the kingdom. Production initially was small-scale, at less than 1 million tonnes a year, until 1979 when production was increased, quickly reaching 20 million tonnes and peaking in 1997 at 22 million tonnes (Ratanasthien 2011).

Approximately 3,200 hectares of mining land are owned by the Royal Forest Department, and there are sufficient coal reserves for a further 50 years. It was estimated that Thailand had 1,063 million tonnes of reserves in 2016, when it consumed 38.7 million tonnes, 21.7 million tonnes of which were imported (https://www.worldometers.info/coal/thailand-coal/).

The lignite Mae Moh coalfield, in the eastern part of the province of Lampang in northern Thailand, was identified in 1955. Many fossils have been discovered here, showing that the mountains in this region were once a large lake. Mae Moh Mine and the adjacent 2,455-megawatt Mae Moh power station were established in 1978 to utilise the coal, with the latest upgrading being in 1995. The power station produces 6.5 *per cent* of the electricity used in Thailand. At the end of 2020 the workable reserve of the mine was 313 million tonnes, sufficient to fuel the power station until 2049. Mae Moh opencast mine covers about 28 square kilometres and produces about 50 million tonnes a year, representing 20 *per cent* of all coal produced in Thailand. In 2019 it was announced that the mine was preparing for closure.

Coal-fired power stations have been constructed in the various coal-mining areas and the Electricity Generating Authority of Thailand (EGAT) has established small parks with native vegetation and viewing points overlooking the opencast mines. In 2021 EGAT announced that it was replacing two planned coal-fired power plants with a gas-fired complex (https://www.pinsentmasons.com/out-law/news/thailand-to-replace-coal-fired-plants-with-gas-fired-complex).

### 345. Pipitaphan soon thanhin lignite suksa (Muang Mae Moh) Chalermprakiat Prabaat Somdet Prapokklao Chaoyuuhua (Mae Moh Mine Museum – Lignite Study Centre, in commemoration of His Majesty King Prajadhipok)

The Pipitaphan soon thanhin lignite suksa was founded in 2003 next to Mae Moh Mine, in honour of King Prajadhipok (1893-41). It was established by EGAT as a learning centre and is fully supported financially by EGAT. There are 11 members of staff and about four volunteers. Each year the museum holds a training programme for 100 to 120 student volunteers, who are rotated after training, as junior guides at weekends and on some special occasions. The museum is one of the largest in Thailand, with a total size of 1,600 square metres. It attracted 56,271 visitors in 2006, the year it opened, and there were 63,003 visitors in 2013.

The museum begins with a short animated movie of the history of mining and then has seven zones. The first deals with the history of King Prajadhipok (also known as King Rama VII), his benevolent works and his role as a pioneer of the energy sector in Thailand and reserving it for national benefit. His handwriting is engraved onto the brass plate about the importance of sustainability in relation to globalisation and a project for the preservation of plant-genetic diversity. The second section covers the history of the geological sciences and of the formation of the four main types of coal, including the origins of the Earth, palaeontology and the evolution of life. There are displays of plant and animal fossils found in the opencast mine, positioned in the layers of a stratigraphic model of the mine. The next section deals with how lignite is acquired and used for electricity generation, the technology involved in power generation, the environmental controls on the power station and the restoration of the landscape after opencasting. This leads on to the discovery of the basic science involved in electricity and then to other forms of power, such as nuclear and renewable energy. Finally the displays deal with how the museum participates in community activities aimed at developing sustainable local communities in the Mae Moh district. The museum also presents a four-dimensional animated film about lignite mining and electricity generation.

The museum presents the standardised-management systems being used at Mae Moh Mine to preserve the environment and benefit the community. The soil overburden removed prior to exposing the coal seams of the mine has been transferred to east of the mine and planted with Mexican sunflowers. The Mae Moh botanical garden, in which the viewing area and the museum are located, has a number of large mining machines on display, from the earliest period up to the modern day. These are positioned as they would be when operating.

*References*


https://www.gem.wiki/Mae_Moh_power_station

http://maemohmine.egat.co.th/museum/index.html

http://www.tripadvisor.co.uk/ShowUserReviews-g2237316-d6957558-r236684747-Maemoh_Mine_Musuem-Mae_Mo_Lampang_Province.html#

personal communications, Pakai Chundakowsol, head of public relations, Royal Thai Embassy, 2017, 2021

**Figure 345.1 Entrance to the museum (© Pipitaphan Soon Thanhin Lignite Suksa).**

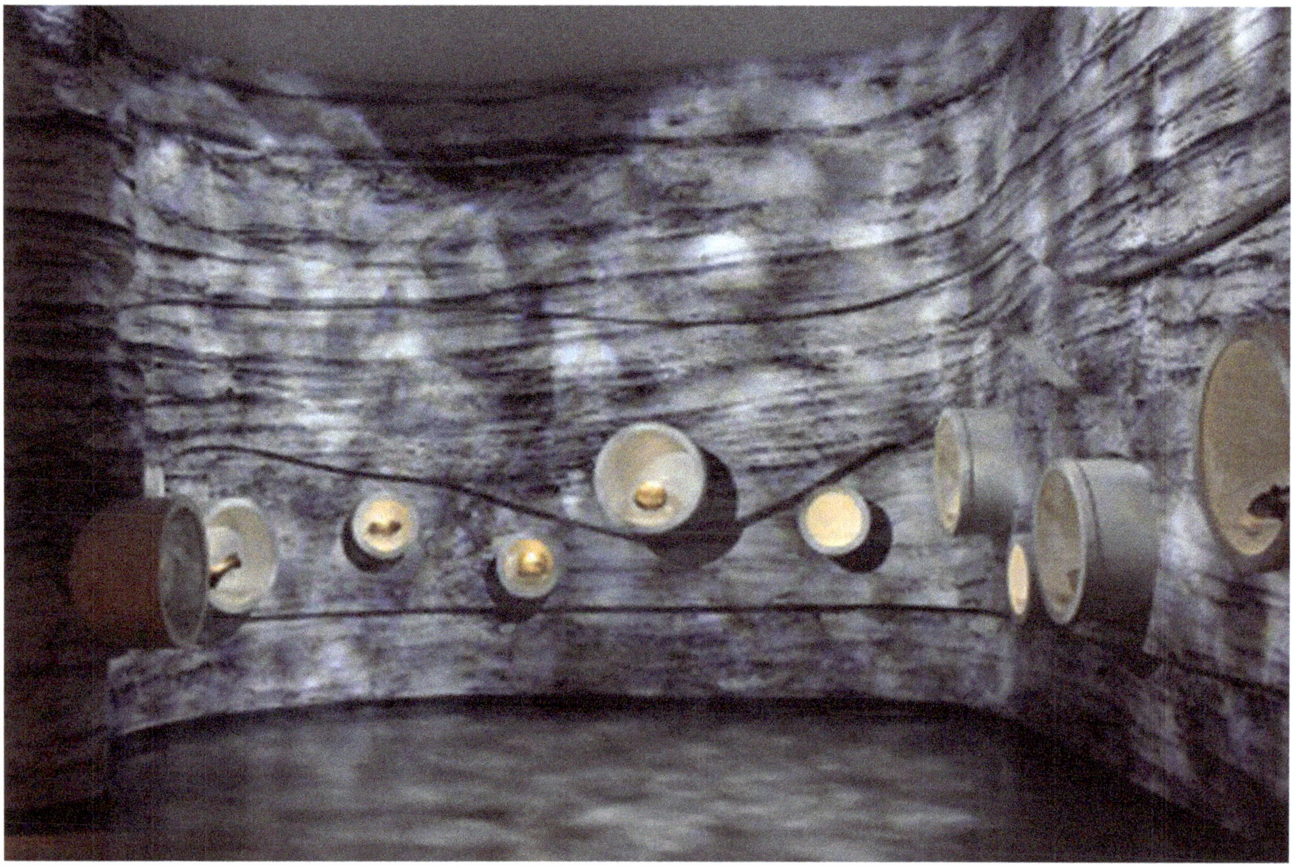

**Figure 345.2 Displays of plant and animal fossils found in the opencast mine, positioned in the layers of the stratigraphy (© Pipitaphan Soon Thanhin Lignite Suksa).**

# Turkey

Deposits of lignite and sub-bituminous coal are widespread throughout Turkey, while reserves of bituminous coal are concentrated in the Zonguldak coalfield on the Black Sea in the mid-north-west of the country (euracoal; CMMCP 2020, fig. 33.1). The industrial history of Zonguldak started with the discovery of coal in the Üzülmez Valley of the Ereğli region in 1829 during the reign of the Ottoman emperor, Sultan Mahmud II (1808-39). Correspondence began between Istanbul and the Turkish embassy in Vienna, the capital of Austria, which is the European country geographically closest to Turkey. As a result the first miners came from Montenegro. Coal production began in the 1840s, accelerating after 1850 when the coalfield was developed with an inflow of funding from Turkish and foreign capital. There was economic and social development as new settlements were established around each group of mines.

Despite the increased need for coal, the economic crisis in the Ottoman Empire stopped the necessary investment. In 1875 the number of foreign companies increased and after 1890 companies from countries such as France, Germany and Italy began to operate in the Zonguldak coalfield. Zonguldak is located at the mouth of the river Kiliçsu, from which timber was shipped for constructing vessels of the Ottoman navy. Coal became an essential material for the navy with the introduction of steam vessels. After the War of National Liberation (1919-23), foreign companies left the province of Zonguldak in 1924. Coal was an essential fuel, so Zonguldak increased in importance to the republic of Turkey. The region began to develop in the early years of the republic, with the Zonguldak coalfield, the Kardemir iron and steel factory in the city of Karabük and the Çates thermal-coal plant in the town of Çatalağzı. The city of Zonguldak grew from a fishing village with only a few hundred peasants to become the leading industrial centre of Turkey with hundreds of thousands of workers. The place-names in Zonguldak, coined by miners during its development, were adopted as the addresses and house-numbering systems by the authorities. In 1936 the author, İsmail Habib Sevük (1892-1954), who had visited Zonguldak, wrote:

> There is also a coal train passing through the middle of the muffled bazaar, with one-ton waggons lined up behind it, full when coming, empty when leaving; a dusty, smoky, raucous coal train, unable to accelerate due to the crowd, blowing a whistle on its neck. Should this train be removed for the sake of the health and aesthetics of the city? Well, when the city is gone, there is this train and this line. The miners resisted. The main thing here is the city, which is the mine. Take the city, not the train. In fact, that charming Zonguldak came out of these waggons like a huge tree comes out of a pinecone. There were 18 houses here 40 years ago. Even Zonguldak, which I saw 15 years ago, is a miracle of industrial cities, having developed suddenly. The land said there cannot be a city here, but the mine said that it had to become a city quickly (Sevük 1942).

In 2011 it was estimated that Turkey had the most dangerous mines in the world in proportion to the amount of coal produced (Drexler 2011, p. 244). Turkey is the eleventh largest producer of coal in the world and has reserves of 17.46 billion tonnes of lignite and 1.3 billion tonnes of bituminous coal (CMMCP 2020, 33). In 2018 85.2 million tonnes of lignite were mined which, together with imports, met 29.7 *per cent* of energy supply (euracoal).

Although her relics were originally kept in Constantinople (Istanbul) and Turkey was her legendary homeland, there was no tradition of veneration of Saint Barbara (**96**) in Turkey, which had become a Muslim country in the fourteenth century. In the 1950s, however, Turkish mining graduates of German universities introduced the annual celebration of her feast day (http://www.bergbaumuseum-grube-anna2.de/index.php/276-2/).

### 346. Karaelmas Maden Şehitleri Müzesi (Karaelmas Mine Martyrs Museum)

The Karaelmas Maden Şehitleri Müzesi is located in Zonguldak. The building occupies an area of 300 square metres in an enclosure of 3,000 square metres. The museum opened in 2020, with two personnel and two security guards, and attracts 20 people a day. The museum is operated by the Türkiye Taşkömürü Kurumu Genel Müdürlüğü (General Directorate of Turkey Hard Coal Enterprises). It is intended that the museum will be a focal point for industrial-heritage tourism in Zonguldak. The museum deals with the historical impact of coal on the region, the miners who worked in the local mines and other local memories of coal mining.

On display are mining machines produced in the central workshop of the Türkiye Taşkömürü Kurumu factory, maps of the coalfield and documentaries about the importance of coal in how the city of Zonguldak developed. There is information about coal mining underground and the above-ground facilities. The display includes a number of objects related to coal mining, a large-scale interpretation of the relationship between underground and surface structures, and an aerial view of the colliery site.

**Figure 346.1. Aerial view of site (© Türkiye Taşkömürü Kurumu).**

**Figure 346.2. Exterior of museum building (© Türkiye Taşkömürü Kurumu).**

**Figure 346.3. Image of a miner (© Türkiye Taşkömürü Kurumu).**

*References*


http://karaelmasmadensehitlerimuzesi.com/index

personal communications, Ekrem Murat Zaman, mining engineer consultant, 2021; Kazim Eroglu, general manager, Türkiye Taşkömürü Kurumu, 2021


## 347. Zonguldak Maden Müzesi (Zonguldak Mining Museum)

The Zonguldak Maden Müzesi is located in the centre of Zonguldak. Zonguldak Colliery first opened in 1880, but closed in 1910. It was nationalised in 1940, and was used as an air-raid shelter during WWII. In 1964 it was owned by the Ereğli coal plant and used as a training centre for mineworkers, as it still is today. It opened to the public in 1997 and some 13,000 tourists visited the drift mine between 1997 and 2007. There had been attempts to create a permanent coal-mining museum since 1930, but it was not until 2008 that the planning and preparatory works for the construction were started. Construction and landscaping were completed in 2012, contracts were let in 2014 for the work on the exhibitions, and the museum opened to the public in 2016. The museum has 15 members of staff and is organised and financed by the Ministry of Culture and Tourism. It is hoped that the museum will attract at least 10,000 visitors a year.

The museum occupies about 7,000 square metres, with 1,000 square metres of indoor exhibition space, 700 metres of roadway, located 50 metres underground and accessed by a drift, and the training facilities, which still belong to Türkiye Taşkömürü Kurumu (Turkey Hard Coal Enterprises). At the entrance a multifunctional hall shows movies and other audio-visual shows about coal mining, and about the relationship between the city and the colliery. This is followed by historical and technical information about mining in Zonguldak.and on the second floor material about health and safety, palaeontology, social care, the development of the mine and the life of the coal miners. Fossils, the formation of coal and coal derivatives are explained on the third floor. Altogether there are 300 works on display.

The underground still has its ventilation tubes and is lined with concrete. There are displays underground of equipment used for mining, with models such as a ripper, the miner who removes the rock above the coal seam and sets arched girders to raise the height of the roadway as the coalface advances. He is shown using a percussive rock drill with an airleg to drill shotholes to take explosives.

Above ground there is a large headgear whose rope configuration shows it is based on the *Koepe* system (**127**), but with the winding engine on the ground. The most important feature of the museum is the presence of an active coal mine next to the exhibition building and the educational activities for miners that continue in it.

*References*


https://www.google.co.uk/search?q=coal+mining+museum+in+Zonguldak+Turkey&rlz=1C2SFXN_enGB499GB519&tbm=isch&tbo=u&source=univ&sa=X&ei=z0obUvGoEOby0gXi04GgDQ&ved=0CF0QsAQ&biw=1256&bih=885

https://zonguldak-ktb-gov-tr.translate.goog/TR-186810/zonguldak-maden-muzesi.html?_x_tr_sl=tr&_x_tr_tl=en&_x_tr_hl=en&_x_tr_pto=sc

personal communication, Ekrem Murat Zaman (**346**), 2016

# Ukraine

Ukraine ranks seventh in the world after America, China, India, Russia and Australia in coal reserves, about 70 *per cent* of which is bituminous and 30 *per cent* coking-bituminous coal (euracoal). There are reserves of 34.4 billion tonnes in the Donetsk coalfield in the eastern Donbass in the south-east, which occupies 60,000 square kilometres, has 90 *per cent* of the country's reserves and continues into western Russia (**307, 311**). The smaller L'vo-Volyn coalfield in the western Donbass, of only 7,500 square kilometres, extends into Poland (*ibid.*).

During the reign of Tsar Alexis I (1629-76), a serf, Grigoriy Kapustin, was sent in 1666 to lead an expedition to search for coal in Russia and Ukraine. Coal mining began in 1668 on the bank of the river Siverskyi Donets, samples from which were shown to Tsar Peter the Great (1672-1725) during the two Azov campaigns (1695-6) of the Russo-Turkish War (1686-1700). Coal deposits were identified at Lisichansk in the oblast of Luhansk in the eastern Donbass coalfield in 1710, and in 1721 Grigoriy Kapustin opened up the coalfield close to where the river Kurducha joins the Severskiy Donets. Major mining developed from 1870 and by 1913 the eastern Donbass produced 87 *per cent* of the coal in the Russian empire. Under the communist state-directed economy, there were three five-year plans and extraction increased from 128 million tonnes in 1937 to 166 million in 1940. Germany invaded Ukraine in 1941, with the retreating Russians destroying part of the mining infrastructure. In 1943 the retreating Germans destroyed the rest of the infrastructure (Ponty 2016, p. 224).

The western Donbass coalfield in the Dnepropetrovsk region of south-eastern Ukraine has major deposits of iron ore and anthracite, bituminous and lignite coal, although no lignite is exploited in Ukraine. The first geological investigations were only carried out in this coalfield in 1952 and the first mine, the Pershotravenskaya Colliery, began production in 1963 and the Ternovskaya Colliery in 1964. Altogether 11 mines were sunk in the western Donbass coalfield, 10 of which are still operational. The coal mines are some of the deepest in Europe, being up to 1,798 metres deep. Coal accounts for 29.5 *per cent* of Ukrainian electricity generation, all from the L'vo-Volyn mines. The self-proclaimed people's republics of Donetsk and Luhansk mean that in 2017 control was lost of about a quarter of the eastern Donbass anthracite. The unrest meant that the government ceased to maintain the mining infrastructure in the remaining area and discontinued $600 million of mining subsidies. This forced the government to import bituminous coal from America, Russia and South Africa (euracoal). The government's policy predictions are a reduction in dependence on coal to 16.1 *per cent* in 2025 and 14.3 *per cent* in 2030, with renewables constituting 20 *per cent* by 2035 (*ibid.*). In 2016 Ukraine joined the Deep and Comprehensive Free Trade Area with the EU, with a view to applying formally for EU membership in 2024 and becoming a member in the 2030s, something bitterly opposed by the Russian Federation. Membership of the EU would require Ukraine to phase out use of coal.

### 348. Muzey Istoriyi i Trudovoy Slavy Chao DTÉK Pavlohraduhol′ (Museum of History and Labour Glory of Joint Stock Company DTÉK Pavlogradcoal)

The Pavlograd Historical Museum began collecting coal-mining material about 40 years ago, but transferred most of its collection to the Muzey Istoriyi i Trudovoy Slavy Chao DTÉK Pavlohraduhol′ and the Shakhterskaya Slava Muzey (**349**) when they opened, retaining only enough to illustrate the role of coal mining in the Pavlograd region. The Pavlograd Historical Museum has two cases of coal-mining objects, as well as awards, photographs and a model of the Heroes of Space Coal Mine made by secondary-school pupils.

The Muzey Istorii i Trudovoy Slavy is located in Pavlograd in the Dnepropetrovsk region. The museum was founded in 2009 and opened to the public in 2011. It was officially registered as number 66 by the Ministry of Culture of Ukraine in 2012. The museum occupies 172 square metres and has one member of staff. The museum has between 1,100 and 2,200 visitors a year, especially school pupils and students from technical colleges and universities. The museum receives museological advice from the Historical Museum of Dnepropetrovsk named after D.I. Yavornytsky. Its running costs are funded by the DTÉK Pavlohraduhol′ Company.

The aims of the museum are the preservation of the history of the West Donbass mining industry, its formation and development, and the origins and preservation of miners' traditions and their contribution to the history and culture of the region. It also presents the role of the mining company in the regional economy and in the development of the infrastructure of towns and communities in Ukraine. The museum operates in a social partnership with the company and concentrates especially on educational work and promotion of information through mass media and television.

The first area of the museum is a room for a preliminary introduction to the history of the mining company, its administrative structure and activities. Secondly an office where items such as worksheets are prepared. Finally a

**Figure 348.1. Cups awarded to the DTÉK Pavlohraduhol' Company for high performance standards (© Muzey Istoriyi i Trudovoy Slavy Chao DTÉK Pavlohraduhol').**

hall of displays deals with the 58 years of history of the development of the West Donbass coal deposits and natural resources, demonstrating the milestones achieved in this development and the people whose work made possible the development of the coalfield. This includes photographs of famous miners who achieved high levels of production. There are cups awarded to the DTÉK Pavlohraduhol' Company for high performance standards, as well as diplomas and certificates earned by colliery employees, including for outstanding achievements in sport.

*References*

personal communications, Viktor Sumchenko, procurement manager, URS Federal Services International Incorporated (AECOM Corporation), 2016; Victor Ivanovich Pil'guy, museum director, 2017

### 349. Shakhterskaya Slava Muzey (People's Museum of Mining Glory)

The Shakhterskaya Slava Muzey is located in the town of Ternovka, in the Dnepropetrovsk region. The name of the museum, НАРОДНЫЙ МУЗЕЙ ШАХТЕРСКОЙ СЛАВЫ, can be transliterated as either the Shakhterskaya Slava Muzey (Russian version) or the Shakhtars'ka Slava Muzey (Ukrainian version), but the museum prefers to use the Russian form. On the outskirts of Ternovka Donbasska, there are four mines close together: Dnieprovska, Samarskaya, Ternovskaya and Zapadno.

To mark the twenty-fifth anniversary of the Western Donbass Colliery, the Ternovskaya Mine Joint Stock Company DTEK Pavlograd Coal opened the Shakhtars'ka Slava Muzey in 2005 with material from the Pavlograd Historical Museum (**248**). The museum is organised and

**Figure 349.1. Reconstructed underground roadway with arched girders supporting the roof and model of miner; displays to the right including map of coal areas (© Shakhterskaya Slava Muzey).**

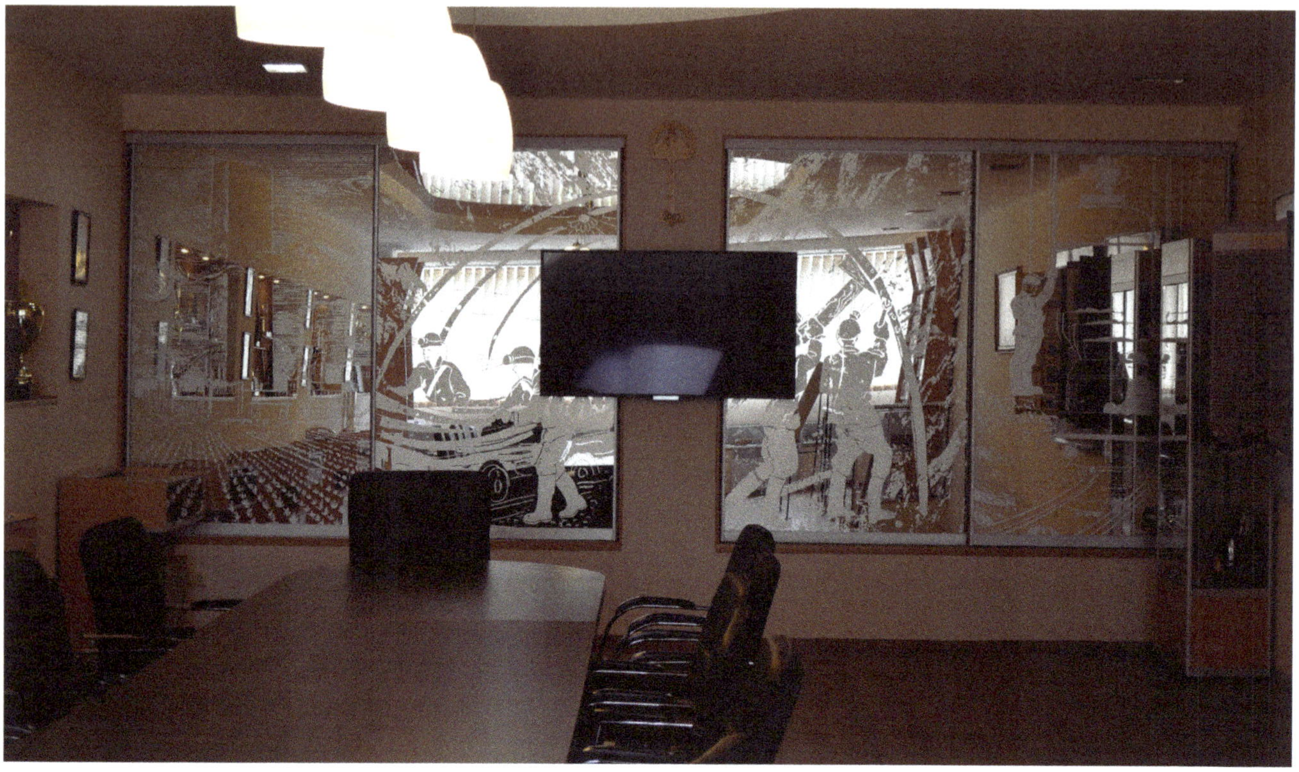

**Figure 349.2. Meeting room with mining scenes etched on glass (© Shakhterskaya Slava Muzey).**

**Figure 349.3. Memorial to miners, showing Christ with his right hand raised in blessing and holding a Bible in his left hand and Saint Barbara below him (© Shakhterskaya Slava Muzey).**

funded by sponsors and the company and is overseen by a board of 10 company employees. The museum attracts about 1,400 visitors each year. There is one member of staff, supported by about five volunteers. In 2007 the museum was awarded the title of People's Museum by the Culture and Tourism Board of the Dnepropetrovsk Region Administration.

The premises were repaired and furnished prior to opening. The organising committee was chaired by the director of the company between 2003 and 2005, A. Koval, and collected documents, machinery, photographs and other exhibits. These date back to as early as WWII and were donated by individuals and by the Trade Union of the Coal-Mining Industry (known as PRUP) and the Independent Union of Miners.

In 2013 a complete redisplay was carried out. A new area was created for storage of more than 1,500 items, the conference room was redecorated with mining scenes etched on glass and the display area was doubled to 160 square metres. There are 30 different displays covering the history of coal mining in Ukraine, starting with the period before Peter the Great, with a model of a pony pulling a tub of coal with the roof supported by wooden props and a model of man pulling a sledge loaded with coal. The displays go through to the modern day, with a reconstructed underground roadway with arched girders supporting the roof and a model of a miner. One of the aims of the museum is to popularise mining as a profession, as mining has currently lost prestige as an occupation amongst young people.

**Figure 349.4. Images of miners killed in fighting in eastern Ukraine from 2014 onwards (© Margaret L. Faull)**

The exhibits on display include fossilised plants and a fossilised tree trunk, about 250 million years old, found underground. The most important item is a gasoline mine lamp dating to the 1840s. There are the first piece of coal produced during the sinking of the Western Donbass Colliery in 1979, the first miner's hard hat worn underground at the mine and the album of a famous foreman, Vasiliy Bubnov, of Mine-Construction Administration Number 3. The conference room has a small library with leaflets about coal mining and publications by the museum.

The formal displays of the museum are accessed through a large conference hall (not the same as the conference room/library), in which the museum holds regular theoretical and practical conferences, lectures and seminars, with participation by mining engineers and other specialists from the DTEK Pavlograd Coal Company. Around the walls of this conference hall are works of art and traditional crafts made by employees of the DTEK Company and works produced by school children related to coal mining.

The museum also celebrates various other groups, such as the participants in WWII and various foreign wars and the work of miners after the 1986 explosion of the Chernobyl nuclear power station in the very north of Ukraine. A tunnel, 168 metres long, was dug by 400 coal miners, mainly from the Ural Mountains of Russia, underneath the power station to create a capsule which was filled with concrete to form a plate 2 metres thick. All the miners were exposed to high doses of radiation, especially on entering the shaft, and 25 *per cent* died. There are also photographs of miners killed fighting against pro-Russian separatists in eastern Ukraine from 2014 onwards. Finally a formal memorial with images of Christ and Saint Barbara **(96)** lists the name of miners who have died in the course of their work.

As well as the formal displays, the museum has a number of large mining machines. These are located physically in the grounds of the adjacent technical college, where they are available for the public to see and are used in the training of mining students.

*References*

https://upload.wikimedia.org/wikipedia/commons/6/63/Samarskaya_Mine_Ternivka.JPG

http://trade-union.dp.ua/profsouzi-mira/45-profsouzi-mira/408-international-coal-mining-museums-questionnaire

personal communications, Valentin Kardapolov, museum director, 2014; Anna Vasylyevna Sergeeva, museum director, 2017

# United Kingdom

Coal deposits are found in all four countries of the UK (Trueman 1954). In Northern Ireland there are widespread deposits of lignite, but most mining in the eighteenth and nineteenth centuries has been of the anthracite coalfields of Ballycastle in the north-east and Coalisland in the north (euracoal). The main coalfields of England are of anthracite and bituminous coal in Cumbria, County Durham and Northumberland, in the north-east; in the centre of England in the Midlands, the Yorkshire and the South Lancashire, stretching the entire length of the border with Wales; and of lignite in the Bristol, Gloucestershire and Somerset coalfield in the south-west, including in the Forest of Dean. There is a small deposit of bituminous and semi-anthracitic coal in Kent in the south-east. North Wales has an extension of the anthracite South Lancashire coalfield and the South Wales coalfield is also of anthracite. In Scotland the lignite coalfields are the Ayrshire in the south-west and the Clackmannan in the south-east (CMMCP 2016, fig. 35.1; euracoal; Goodman 2020, p. 69). The highest-quality coal is anthracite from Carmarthenshire and Pembrokeshire in Wales (Goodman 2020., pp. 2, 58).

Mining of bituminous coal in the UK started in the Roman period (Travis 2008, pp. 159-73; **352**), although use of coal dates back to the Late Iron Age in the first century BCE (*ibid.*, p. 10). There was a break until the twelfth century and throughout the Middle Ages mining was on a very small scale, with short leases and just a few miners allowed to ensure that the reserves were not exhausted (Goodchild 2000, p. 13; Moorhouse 1981, p. 778). Coal only became acceptable for household heating in the Tudor period (1485-1603) with acceptance of the chimney (Goodman 2020, pp. 72-5), suitable for the smaller residences of the middle and working classes. The dissolution of the monasteries (1536-41) released land resources and capital for investment. Tudor landowners developed coal resources on their land themselves and this continued into the seventeenth and eighteenth centuries (**371**), as they could afford the heavy cost of sinking shafts. Sixteenth-century coal mining in England was developed with advice from the German mining engineer, Daniel Hechstetter (1544-81) from Augsberg, who was invited to England in 1564 by Queen Elizabeth I (1533-1603), and introduced the techniques described by Agricola (**108**). Indeed the term kibble (**355**) for a sinking bucket came from the German noun *Kübel* (bucket), whereas the modern German term is *Abteufkübel* (sinking bucket).

Britain had large amounts of high-quality coal, abundant water power and extensive iron-ore deposits in close proximity, enabling it to become the first industrialised country in the world (**379**). Important developments, such as the smelting of iron with coke (**379**) and the Newcomen engine which allowed water to be pumped from pits (**371**), were followed by deeper mining and increased need for safety inventions, such as the Davy lamp (**355**). In the nineteenth century, as seams close to the surface began to be worked out, the cost of sinking deeper shafts and installing expensive machinery meant that capital had to be obtained from entrepreneurs, which caused major social changes (**382**). Throughout the nineteenth and twentieth centuries Britain continued to be a leader, along with Germany, in innovations in mining technology, in provisions for safety (**80**) and in improved conditions for the workforce (McIvor and Johnston 2007). The first geological survey in the world was established in Britain in 1835 (Ranestad and Sharp 2020, p. 5). The reliance on coal led to the development of railways to deliver it to distant markets, so that by the 1850s coal had become the dominant fuel (Goodman 2020, p. 128) and 95 *per cent* of consumers relied on coal (*ibid.*, p. 213). As a result the running of households changed radically (*ibid.*, pp. 129-296).

There are coal-mining museums in all the coalfields of Great Britain, but none in Northern Ireland, although the Coalisland & District Development Association has recently opened a Coalisland Heritage Centre and Museum with displays on local mining (https://coalisland-lnp.com/project/coalisland-museum-heritage-centre/). The three national coal-mining museums of England, Scotland and Wales complement each other, as a small nineteenth-century colliery (**386**), a large late-nineteenth-/early-twentieth-century pit (**387**) and a small twentieth-century mine (**388**).

In the twentieth century coal began to be displaced by oil (**404**), then by gas and cheaper coal imported from places like Colombia (**61**). Despite the activities of the miners' union, conditions both for miners underground and their families on the surface continued to be poor, although alleviated by the work of the Miners' Welfare Fund, established in 1922 under the 1920 Mining Industry Act (Turner 2021). The conditions involved in the operation of a mine in the 1930s and the appalling standards of housing are graphically described by George Orwell (1937, pp. 32-68). Tall chimneys were constructed to comply with the 1956 Clean Air Act, following thousands of deaths in the coal-smoke smog in London in 1952. But this resulted in acid rain killing water life in northern Europe, forest dieback and erosion of stone buildings. The necessity of finding cleaner forms of energy to combat climate change spelled the effective end of coal mining in the UK. So the demise of coal that began in 1913 (**404**) culminated in the closure of Kellingley Colliery (**386**) in 2015 (the history of the British coal industry from before 1700 up to 1982 has been detailed in publications sponsored by the NCB/ British Coal Corporation: Ashworth 1986; Church 1986; Flinn and Stoker 1984; Hatcher 1993; Supple 1987).

### 350. Aber Valley Heritage Museum

The Aber Valley Heritage Museum is located in the Senghenydd community centre, in Glamorganshire in south-east Wales. The museum is operated by volunteers of the Aber Valley Heritage Group, registered as a charity in 2015. Its objects are to highlight the history and heritage of the Aber Valley and surrounding areas, in particular the history and heritage of mining and its effect on people's lives. The two main communities in the Aber Valley were Abertridwr and Senghenydd, which developed in the late-nineteenth and early-twentieth centuries and whose collieries were the main source of employment in the valley. Now that all the pits have closed, the area has struggled with unemployment and economic decline. Members of the group have acquired a collection of artefacts and photographs to display in the small museum, along with two interactive touch screens.

Senghenydd is infamous as the site of the worst disaster in British mining history. There had already been one explosion at the Universal Colliery in 1901, in which 81 miners were killed. An investigation afterwards showed that the cause of the 1901 disaster had been a coal-dust explosion (**80**). Universal Colliery was a dry and hot mine, with even more dust than usual produced by the way in which the coal tubs were filled and because the tubs had an open framework at their ends. The report on the 1901 disaster strongly recommended that in future roadways and all dusty areas should be watered. The 1911 Coal Mines Regulation Act, which covered a number of safety issues, included the provision that fans should be able to be reversed to pump in fresh air if necessary. The owners of the Universal Colliery had been given an extension to the requirements of the Act, which was meant to expire on 16 September 1913, but had still not been implemented. So, at the time of the explosion on 14 October 1913, the fans could still not be reversed, greatly increasing the death toll of 439 miners and one rescuer. The report on this disaster found that the initial gas explosion that set off the coal-dust explosion may have been caused by a spark from a signalling joint box. The enquiry led to negligence charges under the 1911 Act. The colliery manager was fined only £24 (£2,900 at today's prices) and the colliery company £10 (£1,210).

**Figure 350.1. Memorial to the dead of the 1913 Senghenydd Colliery disaster (© Amgueddfa Cymru-National Museum Wales).**

Until this explosion the worst mining disaster in Britain had been the 1866 explosion at the Oaks Colliery (**355**), although the numbers killed are similar. The Big Pit National Coal Museum (**388**) has a number of items connected with the Senghenydd disaster, including a pocket watch presented to Charles Adam by Llanhillieth Colliery for bravery during the rescue attempts and another watch used to identify the body of Evan Weston. The Winding House has a memorial card for the disaster (**403.3**).

The colliery closed in 1928, but the buildings were not demolished until 1963, with one shaft being kept open for pumping and to support the ventilation of nearby Windsor Colliery. The shafts were capped in 1979. It was only in 1981 that the NCB erected a monument to the men and boys killed in 1901 and 1913. This copy of the original headgear of the mine stands 6 metres high on the road into Nant-y-Parc school. About 400 metres away from the Aber Valley Heritage Museum is the Universal Colliery Memorial Garden and a new national Welsh mining memorial erected on the site of the Universal Colliery, 100 years after the disaster. The bronze sculpture, 'The rescue', by Leslie Godschall Johnson (born 1941), shows a rescuer holding a miner's lamp aloft and supporting an injured miner. Ceramic tiles with the names and ages of the miners killed in the disasters are set into a memorial wall and each mining disaster in Wales is commemorated in the ceramic paviers. It is hoped eventually to replace the museum in the community centre with a purpose-built centre on the same site as the monument.

*References*

ERIH, *s.n.* Caerphilly

http://your.caerphilly.gov.uk/abervalleyheritage/

personal communication, Ceri Thompson (**388**), 2021

### 351. Aberfan Disaster Memorial and Graves

The 1866 Oaks Colliery disaster was the worst in English coal-mining history (**355**) and the 1913 Senghenydd disaster was the worst in the UK as a whole (**352**). Yet the 1966 Aberfan disaster, in South Wales, is the catastrophe that has aroused the greatest horror and is most remembered as embodying the dangers and tragedies of coal mining, even though no actual miners were killed.

For 50 years prior to the disaster, the NCB had been dumping millions of tonnes of mining waste from Merthyr Vale Colliery (which closed in 1989) on an enormous spoil heap looming over the little village of Aberfan (Orwell 1937, pl. [12], for an example in Swansea, Wales). There had been warnings that several springs of water underneath the spoil heap rendered it unstable, but these warnings had been ignored by the NCB, which claimed that the tip was stable. In 1965 the headteacher of the primary school submitted a petition to the local authority about the danger posed by the tip. At 9.15 a.m. on the morning of 21 October 1966 the spoil heap began to move down the hillside. Directly in its path was the Pantglas primary school, for 240 children under the age of 11, most of them with collier fathers. More than 40,000 cubic metres of the advancing spoil heap engulfed first two cottages, then the school and continued on to wipe out 18 houses of the terrace on the opposite side of the road.

Almost half of the children in the school were either crushed to death or suffocated and the occupants of the houses were also all killed. Parents and miners dug in a frantic attempt to reach the children, some with their bare hands. In the first hour after the slide began only 10 survivors were found and no more were found alive after 11 a.m. An exhibition of the most up-to-date mining machinery was taking place in North Wales. The manufacturers immediately put all the machines on display onto low loaders and rushed with them to the scene, as did miners from all the collieries in the surrounding area. The army was brought in to provide a ring of soldiers around the village to prevent anyone coming in who might distract from the rescue effort being carried out by more than 2,000 emergency-service workers and volunteers. Queen Elizabeth II (born 1926) visited eight days later, one of the few occasions when she has cried in public. She was given flowers by a child with the message 'from the remaining children of Aberfan'. In total 116 children between the ages of seven and 10, together with five of their teachers including the headteacher, and a further 23 adults lost their lives. The victims ranged from three months to 82 years of age. An entire generation of village children was wiped out. Some families lost more than one child and parents who helped at the school were also amongst the dead.

As the enormity of the disaster unfolded, the NCB sought to divert blame away from the organisation in their evidence to the tribunal of enquiry, which sat for 76 days The report of the tribunal, published in August 1967, laid the blame, however, on the ignorance, blundering ineptitude and failure of communications by the NCB. One father demanded that his child's death certificate give the cause of death as 'buried alive by the National Coal Board'. A study found that half the survivors suffered from post-traumatic-stress disorder. An appeal for funds raised £1.607 million (the equivalent of £27 million today), of which the government took £150,000 towards the cost of removing the remaining spoil heap. This was not repaid by the government until 1997, without any account for inflation, which would have meant an additional amount of £1,486,000, nor any interest.

Almost all the dead were buried together in two long rows in the Bryntal cemetery in Aberfan on 27 October 1966. The graves have white granite arches above them and granite cherubs hold open granite books with inscriptions and poems chosen by each family. One mother, killed in her home, was buried with her two sons. As family members die, their names are added at the foot of the relevant grave. The remains of the school were totally removed and a memorial garden was created on the site. Stone walls within the garden show the location of the walls of the classrooms. The Aberfan and Merthyr Vale community centre, paid for from the disaster fund, was erected in place of the houses that were destroyed. The fiftieth anniversary of the disaster was commemorated in 2016 and, even today, small tributes of flowers and toys are left at the foot of one of the trees in the garden.

The disaster led firstly to the passing of the Mines and Quarries (Tips) Act in 1969, which laid down requirements on the owners and managers of mines and quarries, as well as local authorities, to ensure the safety of tips and to prevent disused tips constituting a danger. The Quarries Regulations 1999, 30 years later, specified the exact geotechnical and other measures that had to be taken to ensure that tips were designed, constructed, operated and maintained so that instability or movement were avoided.

*References*

https://en.wikipedia.org/wiki/Aberfan

http://www.bbc.co.uk/news/resources/idt-150d11df-c541-44a9-9332-560a19828c47

personal communication, Ceri Thompson (**388**), 2017

### 352. Apedale Heritage Centre

The Apedale Heritage Centre is located in Newcastle-under-Lyme, Staffordshire, in the West Midlands. A civilian settlement at Holditch, near the Chesterton Roman fort, 2.5 kilometres from Apedale, has evidence for the use of local coal in the early-second century CE for the manufacture of pottery and iron goods. There is then a break in evidence for coal usage until the fifteenth century, followed by major expansion of the coal-mining industry in the nineteenth century. The mines ranged from deep collieries, such as the Burley Pit, which opened in 1850

**Figure 351.1. Memorials and graves of victims and their families in lower row (© Margaret L. Faull).**

**Figure 351.2. Memorial garden on site of Pantglas primary school, with walls showing the outline of the classrooms (© Margaret L. Faull).**

Figure 352.1. Exterior of Apedale Heritage Centre (© Andy Bennett, Apedale Heritage Centre).

Figure 352.2. Underground showing arched girders, mine tubs and rails (© Andy Bennett, Apedale Heritage Centre).

and where an explosion in 1878 killed 10 workers, to the numerous small drift mines, or footrills (*cf.* **371**). One such footrill opened in 1941, and at its height employed between 100 and 140 men. It finally shut in 1998, when pumping stopped at the adjacent Silverdale Colliery. Some of the workings were saved as the Apedale Heritage Centre, which opened in 2001. The heritage centre occupies some 1.01 hectares, adjacent to the 184-hectare Apedale Valley Country Park, in what was formerly an industrial valley. Some of the original colliery buildings still survive on site, including the winding-engine house.

The centre is a charity run by a board of five trustees and operated by some 30 volunteers, with a management team made up of volunteers responsible for the various activities of the centre. Income comes mostly from donations, sales in the café and charges for mine tours. Grants have been obtained for some specific projects. Tours of about an hour are taken in groups of up to around 12 visitors, always escorted by an experienced miner. The tours go about 50 metres underground through some of the original mine workings, which are accessed by foot along a drift 200 metres long. The roadways vary in height between 1.5 and 2 metres and are supported by a number of different methods, including metal rings and traditional wooden posts. Approximately 30,000 visitors come to the site every year, of whom 3,000 go underground.

At the entrance is a stone plaque, which was previously located in the local miners' hall. The displays on the surface indoors deal with mining in the area, as well as local history and industries. Items on display include mine lamps, miners' personal belongings, pit tools, maps and photographs. There are also a mines-rescue van, with a dummy on a stretcher for first-aid training, and a model of a typical drift mine. Three local mine disasters are described, including the worst in the area's history, the 1918 Minnie Pit disaster, in which 155 men were killed; many of the bereaved families had already lost men in WWI. The colliery was badly affected by methane gas (**80**), but no definite cause was ever identified as the cause of the explosion. There is a reproduction of the interior of a miner's cottage, with parlour, kitchen and scullery, plus yard with outside toilet and coal shed. Other industries covered are brickmaking and the ironworks that were on the site before the mine was sunk. Some specimens from the geological and fossil collection are on display, as is most of the rest of the collection, although there is a small store of items, as well as a small library.

Outside there are the auxiliary winding engine from Hem Heath Colliery and an Eimco rocker-shovel, which are operated during special events, together with a two-deck mining cage. Around the site are various items, such as electric winders, hydraulic chocks, mine cars, mine notices and tubs. The Moseley Railway Trust runs a railway on the same site and items from their collection are on display in the museum, including a Hunslet 28-horsepower flameproof diesel mine locomotive from Calverton Colliery near Nottingham. The railway trust has a large collection of its own and is currently constructing a dedicated museum building on the site.

*References*

ERIH, *s.n.* Chesterton

http://www.apedale.co.uk/

personal communications, Andy Bennett, trustee, 2012, 2021

showcaves

### 353. Ashington Colliery Trail

Ashington Colliery in Northumberland opened in 1867. The small settlement grew gradually into a town, known locally as the 'largest mining village in the world'. At its height the pit employed up to 5,500 coal miners. When the colliery closed in 1986, the site was cleared of buildings and developed into a business park. A small lake and a covered seating area for relaxation in the form of a decorative pavilion were provided for the local community.

The Ashington Colliery Trail was created with funding from the Heritage Lottery Fund to commemorate 119 years of mining. There are 11 panels along the paths of the business park, dealing with the history of the colliery and the coal-mining culture and heritage of the town. The order of the panels is first the colliery workshops, then the railway lines, the marshalling yards, pit shafts, coal-preparation plant, coal depot, area training centre, the human story, the workers' entrance and colliery housing.

*References*

http://www.ashingtoncollieryheritage.org/colliery-heritage-trail/

personal communication, Barry Mead, former director, Woodhorn Museum, 2016

### 354. Aylesham Heritage Centre

The village of Aylesham, in Kent, was only established in 1926. It was built to house miners brought in from other coalfields (the north-east, Scotland, South Wales and Yorkshire) to open up the Kent coalfield, especially at the nearby Snowdown Colliery. These miners brought with them brass-band performances, rugby league and the traditional Welsh male-voice choir, all of which still survive in the area. Aylesham was also intended to house miners working at Adisham (a colliery planned, but never sunk) and Wingham (where work was abandoned in 1914 after no coal was found), so was designed to accommodate 30,000 residents. As these pits never became operational, however, only about 1,000 houses were actually built.

Aylesham Heritage Centre opened in 2001 in one of these houses and attracts about 300 visitors each year. It is run by eight volunteers and is funded by donations and grants. It has a collection of lamps, machinery and mining

**Figure 353.1. View of lake and pavilion in centre of site of former Ashington Colliery (© National Coal Mining Museum for England).**

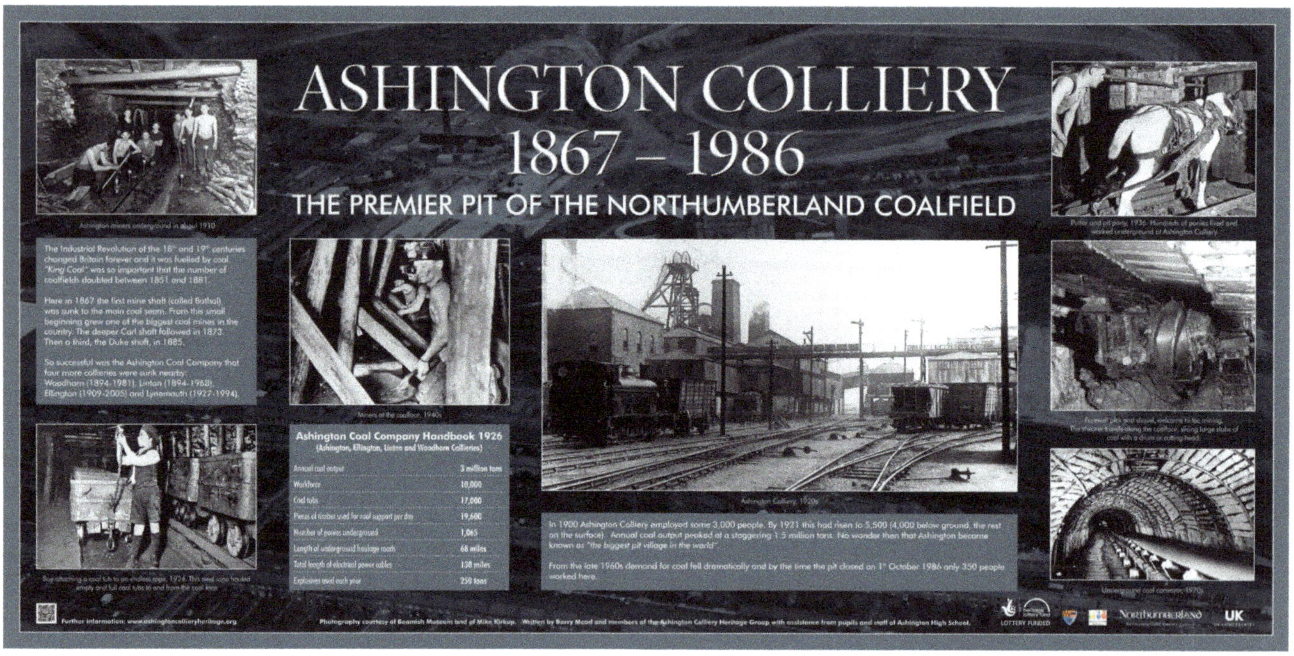


**Figure 353.2. Panel with history of Ashington Colliery between opening in 1867 and closure in 1986 (© Ashington Colliery Heritage Group).**

equipment, including the National Union of Mineworkers' banner from Snowdown Colliery (**396**) and a poster exhorting miners to produce as much coal as possible during WWII (*cf.* **313.1**). Also still standing in the village is the Kent Mines Rescue Station, which served the south-eastern Kent coalfield.

*References*

https://www.facebook.com/groups/77615626368/

http://www.healeyhero.co.uk/rescue/Station/Aylesham.htm

personal communication, Kerrylyn James, volunteer, 2016

Figure 354.1. Former Kent Mining Rescue Station, which is still standing in Aylesham (© Kent Mining Heritage Foundation).

Figure 354.2. Poster exhorting miners to produce as much coal as possible during WWII (© Colin Varrall).

### 355. Barnsley Main Colliery (formerly the Oaks Colliery)

Barnsley Main Colliery, in South Yorkshire, occupied the same site as the Oaks Colliery, sunk in 1838. It worked the Barnsley seam, known locally as the Barnsley eight foot, which was especially gassy. There had been concern for many years about the dangers of using candles for lighting underground, as the flame was likely to ignite any methane present. After two major explosions in 1812, William Reid Clanny (1776-1850) invented a lamp with a candle with a glass surround. In 1815 Sir Humphry Davy (1778-1829) devised a much less cumbersome flame-safety lamp, in which the air could reach the flame through tiny holes in a metal surround, but the holes were too small for the flame to pass in the opposite direction. The Barnsley seam was not able to be worked until this invention. after which owners were prepared to take the risk of working the deeper and more gassy seams. Paradoxically, the invention that Davy had hoped would save lives in fact led to many more explosions and deaths. The Barnsley seam was considered so dangerous that some men refused to work there. In 1847 73 men were killed in an explosion and in 1866 the mine had to be closed down after a succession of explosions. The first explosion, on 12 December, shortly before the change of shift, threw the cage up into the headgear of number 2 shaft, thereby putting it out of action. The rope of the cage at number 1 shaft was broken, but was able to be repaired, enabling rescuers to bring up 50 bodies and 20 badly burned men, only six of whom survived. After treatment by local doctors at the scene, all the men were sent home to die, as there was no hospital for the area.

The rescuers were led by Parkin Jeffcock (1829-66), a prominent mining engineer, who had travelled from his office in Derby as soon as he heard of the disaster, and Tewart, the underviewer (modern undermanager). The following day there were two further violent explosions, killing 27 rescuers who were underground and throwing the cage of number 2 shaft into the headgear. It was believed that all the rescuers were dead, as well as all the miners. Early on the morning of 14 December, however, the signal bell of number 1 shaft was rung. A bottle of brandy was lowered on a rope and was removed, showing that there was still someone alive underground. An engine was brought from a nearby factory and attached by a rope to a kibble, a very large bucket in which men could stand for sinking and inspecting shafts (*cf.* **33.2**). It was intended to send down Thomas William Embleton (1810-93), mine manager (modern production manager), of the 13 J. and J. Charlesworth Collieries (although not including the Oaks), and a boy, John Edward Mammat, rather than two men, in order to reduce the weight. The winding-engine man was very reluctant to do the wind, because the shaft was filled with debris and he had no depth indicator. Mammat's father said that he would stand beside the winding-engine man, who then agreed. They brought up Samuel Brown, the only survivor of the group of rescuers.

On 15 December there were a further four explosions. The underground fire was only finally stopped by capping the three shafts. In all 361 men and boys, some as young as 10 years of age, and 27 rescuers were believed to have died, but it is now suggested that the toll could have been as high as 427. They left at least 658 dependents. Commemorative pieces of crockery were produced and sold to raise funds for the families left destitute by the disaster (**386**). This was the worst disaster in English mining history and the second worst in the UK, being exceeded only by that at Senghenydd (**350**). The disaster was widely reported, including in France in *Le Journal Illustré*, 151 (30 December 1866 to 6 January 1867), which claimed that 'Saint Barbara appeared in the midst of the flames'. As no report in the UK mentions Saint Barbara (**96**) and as this was not a Catholic area, the claim appears to be a journalistic invention.

A meeting of notables shortly afterwards expressed:

> Its cordial sympathy with the proprietors of the Oaks Colliery, not simply for the great loss of property entailed but by the crushing sorrow endured by the terrible loss of lives of their workmen, and the bereavement of those connected with them.

The order of priority of these sympathies may say something about the general attitude at the time to coal miners.

After the colliery had been reopened with a new shaft nearby, there were a further 17 explosions in the years after 1866. At least 80 bodies were never recovered after the 1866 explosion and in later years miners occasionally came across human bones. When the surface of the site of the Oaks Colliery was being cleared by Amalgamated Construction Limited (AMCO) for a new canteen at Barnsley Main Colliery in the early 1980s, the top of the Oaks number 1 shaft was uncovered. Embedded in it were the old bank-top winding-pulley gear and the kibble used in the final rescue. The kibble came close to being discarded and sold as scrap metal, but is now on display at the NCMME (**386**).

Barnsley Main Colliery closed in 1991. The ownership of the pithead and associated electric-winding engine and winding-engine house were transferred to Barnsley Metropolitan Borough Council to remain as a memorial to those who lost their lives underground at the colliery. These structures date from about 1900, but were modernised in 1956 with the construction of an upper floor to the winding-engine house and an upper part to the heapstead building. The pithead is not open to the public at present, but stands in a grassed area and can be viewed from the roadside (**255.2**). The windows and doors have been bricked up to protect against vandalism. It was given grade II listed status by English Heritage in 2013 as a nationally rare survival of a nineteenth-century arrangement of winding-shaft structures.

The miners who died in 1866 were buried in the cemetery at Christ Church, Ardsley, together with a memorial in the form of a short obelisk, installed in 1879. The inscription says that it was erected by public subscription:

> TO COMMEMORATE THE DEATHS OF 354 MEN AND BOYS WHO LOST THEIR LIVES IN THE EXPLOSION AT THE OAKS COLLIERY, ON DEC$^{R}$ 12$^{TH}$ AND 13$^{TH}$ 1866 OF WHOM 35 ARE INTERRED HERE.

There are now only three gravestones, commemorating four miners, aged between 13 and 25. A few other gravestones are now illegible, but there are wide stretches of grass in the churchyard, representing the graves of miners whose families could not afford gravestones.

A memorial to the rescuers in the form of a tall ashlar obelisk was erected in 1913 by Samuel Joshua Cooper (1830-1913), a local businessman and benefactor. A bronze group shows the winged Greek goddess of wisdom and war, Pallas Athena, carrying a wounded warrior. The inscription on the memorial says that it was erected:

> AS A TRIBUTE TO THE BRAVERY OF PARKIN JEFFCOCK AND OTHER HEROES OF THE RESCUE PARTIES WHO LOST THEIR LIVES OWING TO FURTHER EXPLOSIONS ON DECEMBER 13$^{TH}$ 1866. ALSO TO COMMEMORATE THE SIGNAL BRAVERY OF JOHN EDWARD MAMMAT AND THOMAS WILLIAM EMBLETON IN DESCENDING THE PIT AND RESCUING THE SOLE SURVIVOR ON DECEMBER 14$^{TH}$ 1866.

**Figure 355.2. Kibble found in original Oaks shaft, used to rescue the last survivor, Samuel Brown, on display at the NCMME (© National Coal Mining Museum for England).**

**Figure 355.1. Commemorative teapot showing the surface of the Oaks Colliery after the second explosion (© National Coal Mining Museum for England).**

**Figure 355.3. Memorial to the victims of the Oaks disaster erected in Ardsley churchyard in 1879 in front of gravestones and burials of some of the victims (© Margaret L. Faull).**

A further bronze memorial shows a miner's wife carrying her child and rushing to the colliery on hearing the explosion. Behind the woman is a white circle, which stands for both a mine winding wheel and the circle of life. Coal flows symbolically down her back, representing the backbone of the coal industry and the strength that it gave to the UK. The coal lands just above the head of a miner working on his side in the thin seam that would become his grave, with a Stephenson's lamp next to him.

The work was produced free of charge by Graham Ibbeson (born 1951), a local sculptor and son of a miner. One of his relatives was killed in the disaster. This memorial was the first to commemorate all the victims of the disaster. It was unveiled in the centre of Barnsley in May 2017 on the one hundred and fiftieth anniversary of the report on the explosion being presented to Parliament and also on the seventieth anniversary of the 1947 Barnsley Main explosion, in which nine miners died.

*References*

http://discoverdearne.org.uk/research-reveals-higher-death-toll-1866-oaks-colliery-disaster/

https://historicengland.org.uk/listing/the-list/list-entry/1413541

https://www.gracesguide.co.uk/Thomas_William_Embleton

http://www.bbc.co.uk/news/uk-england-south-yorkshire-22986039

http://www.dmm.org.uk/uknames/u1866-02.htm

http://www.oaks1866.com/

### 356. Beamish, the Living Museum of the North

Beamish, the Living Museum of the North, which opened to the public in 1970, occupies a site of about 142 hectares in County Durham. There are 250 employees of the museum, with about 300 volunteers. The museum is a charitable trust, supported by nine local authorities and administered by a joint committee of city, county and district councils. It is funded with grant support from the Heritage Lottery Fund and the Arts Council England, together with income raised from catering, retail and admission fees from the 589,000 visitors each year.

Beamish forms the regional museum for the area and has a large number of buildings moved from elsewhere in north-eastern England. It concentrates mainly on the region in about 1825 and 1913. The various parts of the site are joined by a tramway. Although the buildings cover many aspects of life in the area, the site on which Beamish stands was once at the heart of the Durham coalfield. The early 1900s, which are portrayed at Beamish Colliery, was a prosperous period for north-eastern mines. The Mahogany drift mine opened about 1855 and worked sporadically until the 1950s. It has been partially reopened so that visitors can see first-hand how coal was worked in the early 1900s. The guided tour goes approximately 18 metres along a main roadway, then turns left into a gallery where two coalfaces are exposed. The first is a machine-cut face, where the coal was extracted using compressed-air tools. Some 14 metres further is a hand-cut coalface where miners worked using only picks and shovels. The height of the roadway and gallery varies from 4 metres to 6 metres.

In the colliery yard is a replica of a typical colliery lamproom, known locally as a lamp cabin. Safety lamps are housed on wooden racks, with numbers identifying the individual miner to whom each lamp belonged. Costumed staff in the lamproom introduce the guided tour of the drift mine, which takes about 15 minutes, and also explain the development of safety lamps and demonstrate the cleaning and maintenance of lamps.

Across the colliery yard is the 1855 winding-engine house, a tall stone building which was moved in 1974 from Chophill Colliery. This houses a steam-winding engine built in 1855 by J. and G. Joicey of Newcastle-upon-Tyne. The engine is probably one of the smallest of its type, but was capable of lifting 0.91 tonnes of coal a time and in a 24-hour period would raise 636 tonnes of coal. Coal was lifted at 7.32 metres per second and men at 3.66 metres per second. The engine was in use until Chophill Colliery closed in 1962. It is the sole survivor of a type once common throughout the northern coalfield and is operated every day throughout the summer season.

The wooden heapstead adjacent to the winding-engine house came from Ravensworth Park Mine in Gateshead. The heapstead demonstrates how men and ponies were raised and lowered in the shaft and how coal was brought to the surface, weighed and sorted before being loaded into chaldron waggons below for transportation by rail. A chaldron was a measure of volume, especially for coal, in use from the thirteenth century until 1835, and defined as weighing 2,690 kilograms from 1694. A standard railway chaldron waggon measured about 3.05 metres in length and stood around 1.91 metres high.

The colliery railway, centred on the pit yard, allows occasional demonstrations of shunting operations, using the restored collection of some 20 Londonderry chaldron waggons. The main line is standard gauge, 4 feet 8.5 inches (14.35 centimetres), but a narrow-gauge line has also been constructed to serve the stockyards and surface activities. Operational locomotives include an 1871 Head Wrightson locomotive number 1 and an 1877 Stephen Lewin locomotive number 18. The associated pit village shows typical cottages and gardens, board school and Methodist chapel.

The museum has a collection of some 200 miners' lamps, ranging from pre-safety lamps through to modern electric lamps, and a disaster-glass collection of some 60 items. Engraved glassware, commemorating deaths and disasters in north-eastern collieries, was sold to raise money to help support the widows and families of those who were killed; for example, a goblet in memory of Matthew Lee, killed at

Figure 356.1. Drift mine underground (© Beamish Museum).

Figure 356.2. Head Wrightson number 1 locomotive and chaldron waggon (© Beamish Museum).

Beamish Mary Pit by being crushed while erecting a new screens building in 1889, aged 23, and a milk jug inscribed 'West Stanley Disaster 168 lives, Feb 16th 1909' (see also **368**).

*References*

https://historicengland.org.uk/listing/the-list/list-entry/1120965

personal communication, Jacki Winstanley, marketing and public relations officer, 2014

showcaves

www.beamish.org.uk

### 357. Bedwellty Park

The mining community at Tredegar in the Sirhowy Valley in South Wales, 10 kilometres west of Blaenafon, was established about 1800. Plas Bedwellty, in Tredegar, was occupied in the mid-nineteenth century by the manager of the Tredegar Iron and Coal Company. The owners, the Morgans of Tredegar House, gave the house and grounds to the local community in 1901 and it was converted into offices for the urban district council. In 1851 a block of coal, weighing 20.32 tonnes, was cut by an expert coal miner, John Jones, at the Yard Level Pit for display at the Great Exhibition at the Crystal Palace in London. It is believed to be the largest single block of coal ever cut in the UK. While being transported a piece weighing 5.08 tonnes broke off, so it was decided that it would not be possible to move the block intact 258 kilometres to London. A smaller block from the same colliery was exhibited at the Festival of Britain in London in 1951, 100 years later, although production at the Yard Level Pit had ceased in 1873.

The large lump of coal measures 1.52 metres by 1.22 metres by 3.96 metres. It is protected from disintegration by a wrought-iron band and stands on an eight-wheel dram on a specially laid short length of plateway in the grounds of Plas Bedwellty. The 1951 block is also bound with iron bands. A shelter of steel and perspex was erected over them in 1992 to protect against the weather. The same year both blocks were listed grade II as a unique monument of the coal industry of South Wales.

**Figure 357.1. Block of coal weighing 15.24 tonnes and measuring 1.52 by 1.22 by 3.96 metres, cut for display at the 1851 Great Exhibition (© Amgueddfa Cymru-National Museum Wales).**

*References*

ERIH, *s.n.* Tredegar

http://www.bedwelltyhouseandpark.co.uk/

http://www.britishlistedbuildings.co.uk/wa-1879

### 358. Bellingham Heritage Centre

The Bellingham Heritage Centre in Northumberland was originally opened in the village in 1994, but moved to its present site in 2000 and expanded in 2008. It comprises a single L-shaped building, with the museum on the ground floor and office space, storage area and work room on the first floor. Two former railway carriages were added in 2011, one forming the tea room and the other an education space and exhibition area. There is also a former railway parcel shed, now used as a store. The museum preserves the mining heritage of the area of North Tyne and Redewater. Among the displays are a blacksmith's anvil and a coal-fuelled forge, such as would be found at a mine, and tools used at the three local collieries of Hareshaw, Lewisburn and Plashetts.

There are no full-time staff. The museum has one part-time cleaner and the remainder of the staffing is carried out by about 30 volunteers. The museum is funded by admission fees, catering and sales of items such as books. Grants from organisations such as the Heritage Lottery Fund pay for special projects. The museum attracts about 5,000 visitors a year, although the tea room caters for about 20,000 people annually.

*References*

http://www.bellingham-heritage.org.uk/

personal communications, Ian Roberts, chair, 2014; Stephen Gibbon, chair, 2021

showcaves

### 359. Bersham Colliery Mining Museum

Bersham Colliery Mining Museum near Wrexham possesses the last headgear still standing in the North Wales coalfield. The steel-lattice headgear was moved from nearby Gatewen Colliery for the number 2 shaft in 1930, and the earlier timber headframe was destroyed by fire in 1937. The surviving headgear is a scheduled ancient monument and the other colliery buildings are listed grade II. The mine opened in 1868, with the coal being used especially by the local iron industry and the brick, tile and terracotta works in nearby Abenbury and Ruabon. The colliery closed in 1986.

The colliery canteen, pithead baths and offices were built in 1954. They have been converted into offices and

Figure 358.1. Blacksmith's anvil and forge fuelled by coal (© Bellingham Heritage Centre).

Figure 359.1. Pit-wheel memorial to the dead of the Gresford Colliery disaster (© Ceri Thompson).

enterprise workshops, although many original features survive, including tiled dados and the central wall dividing the showers from the lockers. The museum exhibition is located in the red-brick former winding-engine house, 230 square metres in area, with a barrel-vaulted concrete roof. This still houses the electric-winding engine. The building and the headgear are owned by Wrexham County Borough Council, and some money is about to be spent on the winding-engine house to arrest further deterioration. A group of six to eight volunteers open the site by appointment and for one weekend a month from Easter to September, attracting some 30 to 100 visitors per open day.

The display shows pictures, lamps and other mining gear, including a steam-winding engine from the former Point of Ayr Colliery and the top of a headgear from the former Gresford Colliery. The Gresford disaster on 22 September 1934 was the worst disaster in the North Wales coalfield, resulting in the deaths of 266 miners. Only 11 bodies were able to be recovered. At the official enquiry the miners were represented by Sir Stafford Cripps (1889-1952), a Labour member of parliament, who subsequently pressed for nationalisation of the coal industry. The Nationalisation of the Coal Mines Act 1946 was passed while Sir Stafford was president of the Board of Trade (1945-7). The disaster is commemorated in the hymn 'Gresford', more usually known as the 'Miners' Hymn', which is one of only three British songs for which the audience normally stands. It was written by Robert Sant (1905-50), a miner from north-eastern England.

*References*

https://en.wikipedia.org/wiki/Gresford_disaster

http://sniff.numachi.com/pages/tiGRESFORD;ttGRESFORD.html

http://www.britishlistedbuildings.co.uk/wa-16567

http://www.wrexham.gov.uk/english/heritage/bersham_colliery.htm

personal communications, Steve Grenter, heritage services manager, Wrexham County Borough Museum and Archives, 2015; Margaret Jones, secretary, North Wales Miners Association Trust, 2021

showcaves

### 360. Black Country Living Museum

The Black Country Living Museum is located in Dudley. The area of the West Midlands located north and west of Birmingham has generally been known as the Black Country since at least the 1830s. In the Industrial Revolution (**379**) it became one of the most industrialised parts of England with coal mines, coking, iron foundries and steel mills, all producing a high level of air pollution, leading to the description of the region as black.

The open-air museum occupies an area of some 10 hectares and opened in 1978. It is run by an independent not-for-profit trust, and raises all its funds itself from admission charges, donations, and shop and café sales. The museum employs 210 staff and has approximately 160 volunteers. It attracts between 250,000 and 300,000 visitors a year.

The collections constitute a comprehensive record of industrial activity and society in the Black Country, chiefly spanning the period 1800-1950. They include four original industrial features on site, comprising limekilns, two mine shafts and a canal arm, together representing a microcosm of the classic Black Country industrial landscape. About 60 buildings, engineering structures and items of street and canal furniture, all with regional provenance, have been relocated to this core area. There are over 80,000 items in the collections, of which 1,650 relate to coal mining, ranging from photographs to tools and clothing.

The Black Country was famous for its Thick Coal, a seam 9 metres thick. Coal extraction peaked in 1872 when the coalfield accounted for 10 *per cent* of national production. Seams of ironstone and fireclay are present in the Coal Measures and were often mined from the same shafts as the coal. There are over 50 known workings on the museum site. Two of these – the mid-nineteenth-century shaft of Racecourse Colliery and the Brook shaft – have been excavated, and original plant and equipment of Black Country provenance have been added around them to recreate an authentic colliery area. Racecourse Colliery contains a single-cylinder colliery winding engine of about 1860 used at Amblecote Colliery from about 1910 to 1926. It is operated regularly and steam is raised in a cross-tube vertical boiler made locally by the Cradley Boiler Company of Cradley Heath in 1954. In total the museum has 10 boilers, including three examples of the earlier egg-end boiler and a rare haystack boiler of about 1770, one of the earliest surviving steam boilers anywhere. There is also a steam donkey engine of about 1870 to 1910 and a Cappell-type extractor fan from Birmingham University. The site at the Brook shaft has an electrically operated cage and sheds that contain a comprehensive range of colliery-surface equipment. The two shafts are linked by a tub tramway, with a gauge of 2 feet (61 centimetres) and an early tipping waggon of riveted construction from Rowley Regis quarry. The collection includes detonators, miners' lamps and candles, protective gear for both miners and pit ponies, scientific instruments and other equipment from mines in the region. There is also some of the distinctive rattle chain, made of three separate chains, which was used to lift the coal to the surface (an improved type was patented by H.P. Parkes of Tipton in 1836). The museum holds a comprehensive collection of bricks, tiles and fireclay products, representing the former importance of the related brick, coal and fireclay industries.

Of especial interest are firstly the re-erected Black Country's workers' cottages, with authentic furnishings, typical of those occupied by coal miners. Secondly the replica of a Newcomen atmospheric engine, opened at the museum in 1986 after 10 years of research. A print of 1719 shows the first known working atmospheric steam engine.

**Figure 360.1. Mid-nineteenth-century Racecourse Colliery preserved on its original site at the Black Country Living Museum, with vintage bus in foreground (© National Coal Mining Museum for England).**

**Figure 360.2. Rocking great-balance beam, with the pump side projecting outside the engine house (© National Coal Mining Museum for England).**

**Figure 360.3. Typical twentieth-century Black Country miner's house (© National Coal Mining Museum for England).**

This was installed in 1712, probably at the Coneygree Coal Works, just to the east of Dudley Castle, to pump water from the coal mines on the estates of Lord Dudley (1704-31). The only surviving *in-situ* Newcomen engine is at Elsecar (**371**) and the Black Country Living Museum replica is the only existing full-size working Newcomen engine. The fire engine, as it was known, is installed in a brick building from which a wooden beam projects through one wall. Rods hang from the outer end of the beam and operate pumps, which are located at the bottom of the mine shaft and raise the water to the surface. The engine itself is simple, with only a boiler, a cylinder and piston-and-operating valves. A coal fire heats the water in the boiler, which is little more than a covered pan, and the steam generated then passes through a valve into the brass cylinder above the boiler. The cylinder is more than 2 metres long and 52 centimetres in diameter. The steam in the cylinder is condensed by injecting cold water and the vacuum beneath the piston pulls the inner end of the beam down and causes the pump to move. The engine is operated on a number of occasions during the year.

*References*


ERIH, *s.n.* Dudley

personal communication, Sara Lambert, executive assistant, 2014

showcaves

www.bclm.com


### 361. Brandy Bottom Colliery

The Bristol and Gloucestershire coalfield was one of the two main English coalfields (**368**) from at least the later-thirteenth century, because coal could be moved from both areas by sea, especially to London. The site of Brandy Bottom Colliery is located to the west of Pucklechurch and is one of the few surviving intact colliery sites of the south Gloucestershire coalfield. The colliery is first referred to as Brandy Bottom in 1856, having previously been called Lord Radnor's Pit as it was owned by the third earl of Radnor (1779-1869). The first, northern, shaft of the mine had been sunk by 1837, situated on the line of a horse-drawn railway, known as the dramway (*cf.* **393**). This took coal until 1850 from the Coalpit Heath area to either Bristol or the river Avon near Keynsham. In 1871 Handel Cossham (1824-90), a local entrepreneur who in 1851 had started to develop Parkfield Colliery 1 kilometre to the north-east, took over the lease and sank another shaft, 206 metres deep. Cossham then connected the two pits underground and worked them as a single unit, which may have been an important factor in the preservation of the original layout. When he died, the colliery was put up for sale by auction. The layout of the site and its buildings are shown on a surviving plan of 1900 prepared for the sale. When worked out sometime before WWI, Brandy Bottom Colliery, which was sometimes referred to as Parkfield South Colliery, continued to ventilate Parkfield Colliery and serve as an emergency exit. Parkfield closed in 1936, removing any need to maintain Brandy Bottom.

Figure 361.1. Remains of Cornish engine house on right; 1837 heapstead in left foreground; chimney in background (© Robin H. Whitworth).

Figure 361.2. Entrance to 1871 horizontal-engine winding house (© Robin H. Whitworth).

The site was scheduled as an ancient monument of national importance in 2001. This was because all of the most important buildings shown on the 1901 prospectus still survive, albeit in a ruinous state and overgrown with vegetation, and so preserve almost completely the surface layout of a nineteenth-century colliery. Historic England notes that it is very unusual for a site of this date to survive in such a complete form and that the undisturbed buried remains of engine bases, boiler settings and additional features will be present and represent considerable potential for the future study of this area's coal-mining industry.

The two groups of buildings, of stone and brick, occupy an irregularly shaped area, with maximum measurements of approximately 100 metres in length and 35 metres in width. The northern group dates to the first stage of the pit in the early-nineteenth century and the southern to the last 30 years of the nineteenth century. Ancillary features include two large spoil heaps and a pond, 20 metres by 10 metres and several metres deep, that possibly provided water for the boilers or for firefighting. The northern group is comprised of the heapstead; the vertical-engine house, containing the bed for the 61-centimetre-diameter cylinder, 1.4-metre-stroke engine; and the Cornish engine house, believed to retain its original internal engine settings, including the bed for the 1.5-metre-diameter cylinder, 2.4-metre-stroke Boulton-and-Watt-design engine used for pumping water out of adjacent 1837 shaft.

Both these engine houses are in poor condition without roofs and with some walls having collapsed. The square chimney still stands 23 metres high, the lower 15 metres being in stone and the remainder in red brick, with an octagonal top capped with stone. The southern group includes the heapstead and the associated 1871 horizontal-engine house, which has an arched roof that still survives, although none of the engine settings remain. The horizontal-engine house is the only roofed structure on site. A Sirocco-type fan was installed sometime between 1915 and 1936 for ventilating Parkfield Colliery. The brick ducts, which were part of this system, still survive and are thought by Historic England to represent a rare example of its type and age in south-western England. In 2016 the northern duct was covered with a cast-concrete roof after being photographed and part of the brick base of the fan housing was found.

In 2008 the Avon Industrial Buildings Trust acquired a lease of the site from its owners, Ibstock Brick. By 2021 the trust had almost finished stabilising the fabric of the buildings through a combination of work by contract masons and between 10 and 12 volunteers. The trust received grants to cover the cost of most of this work. Tree clearance around the perimeter fence has improved views of the site for passers-by. At present the site is closed to the general public for safety reasons, apart from days when the volunteers are working, but the long-term intention is to provide public access.

*References*

https://historicengland.org.uk/listing/the-list/list-entry/1019400

http://www.aibt.org/brandy-bottom.html

personal communications, Robin H. Whitworth, committee member, Avon Industrial Buildings Trust, 2016, 2021

### 362. Cefn Coed Colliery Museum

Cefn Coed Colliery is located in the Dulais Valley in Neath Port Talbot in the south of Wales, quite close to the coast. It was believed to be the deepest anthracite coal mine in the world when it was sunk by the Amalgamated Anthracite Company in 1926/7, with two shafts each more than 732 metres deep. Five coal seams were worked at different levels, the deepest being the Peacock seam, named because the coal had a blue/green sheen like a peacock's feathers. The colliery acquired the nickname of the slaughterhouse because it was one of the most dangerous coal mines in Wales, with many roof falls and high levels of methane gas. Production began in 1930, with over 140 pit ponies working underground. By 1945 there were more than 900 men working at the pit (and over 5,500 in all the collieries in the Dulais Valley). In the 1950s and 1960s touring rugby-union teams were brought to Cefn Coed to be shown a Welsh coal mine before international matches at Cardiff Arms Park. When they went underground, the winding engineman would leave off the slow-banker brake and allow the cage to drop by gravity, in the way miners were lowered underground (Orwell 1937, p. 21). He would then apply the brake at the very last moment, with the aim of putting the visiting players off their game when they met the Welsh players. Although it closed in 1968, the colliery continued to be used for ventilation and as an emergency exit for the Blaenant drift mine, which was developed by the NCB in 1963 on the same site and which worked until 1990.

The Cefn Coed Colliery Museum opened in 1978 in the surface buildings of the mine. These were all listed in 1991 as grade II*, being a unique survival of a complete 1920s colliery complex. The museum has one seasonal worker, together with 20 volunteers. The museum is run and financed by the county borough council of Neath Port Talbot and attracts about 10,000 visitors a year.

The underground roadways are not accessible, but there is a simulated underground gallery. The main buildings on the site are the chimney, compressor house, electricity-generating house, two headframes and the number 2 shaft winding-engine house, which still contains its original engines. The 1927 Worsley Mesnes horizontal duplex-cylinder steam-winding engine is the most important item in the museum's collection, although now powered by electricity. Boilers situated in the centre of the site provided steam to these engines, as well as to the engine house of the number 1 shaft, which no longer survives. There are six Lancashire boilers, which were installed in 1926/7, although one is dated to 1910 and was brought from another site. Four of the boilers were converted in

**Figure 362.1. Hunslet Austerity 0-6-0ST, no. 2758, on display at the entrance to the museum (© National Coal Mining Museum for England).**

**Figure 362.2. Lattice-girder headframes, standing 18 metres high (© National Coal Mining Museum for England).**

1958 to run on methane gas piped up from underground. A Hunslet Austerity 0-6-0ST locomotive number 2758 is displayed at the entrance to the museum.

The site includes a number of items believed to be unique now in Wales. The boilers, along with the associated steam-capstan engine, are the only examples surviving in Wales. The headframes stand 18 metres high and date back to 1926/7, the only surviving headframes in this coalfield dating from before WWII. They are also one of only two sets of the once-widespread lattice-girder construction (**204**) left in South Wales. In 2015 it was discovered that both headframes were in very poor condition, as a result of which the site was closed to the public for one season while dismantling both headgears entirely and then rebuilding them as before. Once the headgear of number 2 shaft had been refurbished, one end of the museum reopened to the public in 2017.

*References*

http://www.britishlistedbuildings.co.uk/wa-11858-63

https://en.wikipedia.org/wiki/Cefn_Coed_Colliery_Museum

museum leaflet

personal communications, Keith Davies, manager, 2016; Simon Chipps, library assistant, 2021

showcaves

www.npt.gov.uk/1655/

### 363. Chatterley Whitfield Mining Museum

The first shaft was sunk in 1838 at Chatterley Whitfield Colliery, one of 22 collieries of the Staffordshire coalfield, in the West Midlands, at the time of nationalisation. This area had been well-known for its coal since the fourteenth century and mining developed strongly after the completion of the railway in the nineteenth century. In the 1930s Chatterley Whitfield was the major mine of the coalfield, surrounded by other pits and employing more than 4,000 men. Its deepest shaft went down 610 metres and altogether there were about 80 kilometres of roadways underground, working various seams at different levels. In 1937, and again in 1938, it became the first colliery in the UK to mine a million saleable tons (984,000 tonnes) in one year. The water make at Chatterley Whitfield was about average at 2.46 million litres a day (*cf.* **386**), which was pumped to a pond and then to the boiler and coal-preparation plant. There were five shafts, close together, each with its own winding engine. and an impressive spoil heap.

The site shut in 1977, but reopened in 1979 as the Chatterley Whitfield Mining Museum. As well as an extensive range of colliery buildings with their fittings intact and several shafts with winding engines and winding-engine houses, the museum also acquired the British Coal Collection. This consisted of some 5,500 items deemed to be important examples of mining heritage collected by the NCB/British Coal Corporation (**384**). The British Coal Collection had been built up from 1971 in Nottinghamshire at Lound Hall, which in 1976 was renamed as the National Mining Museum. Lound Hall closed in 1989 and the collection was moved to Chatterley Whitfield. These included not only a large amount of mining machinery, but also clothing and personal items; fine art; photographs; matters relating to trade-union and industrial relations, training and welfare; and items such as 15 stoves from continental Europe.

The last mine left working in the northern area of the coalfield was Wolstanton Colliery, which was connected to Chatterley Whitfield underground by a roadway 6 kilometres long. As the deepest colliery in the region, with shafts 914 metres deep, Wolstanton was responsible for the pumping that kept Chatterley Whitfield dry. At first the museum offered genuine underground tours using the Winstanley shaft, which was 215 metres deep, but in 1985 Wolstanton ceased coaling and the pumps were finally turned off in May 1986. The underground at Chatterley Whitfield would flood without this pumping and the museum could not afford the cost of taking on the pumping itself. The rising water also meant that methane began to build up in the workings and monitoring showed that it was reaching dangerous levels in the museum level, so the shafts were capped. British Coal funded the construction of an artificial underground, using a railway cutting and some shallow old workings. This was accessed by capping the two shafts at a lower level than normal, to allow the top parts of the shafts to be used to give visitors the experience of riding in a genuine cage.

Although genuinely representative of underground mining from about the 1850s to the modern day, visitors crave authenticity in a visit to a mining museum. The result was a sharp drop in visitor numbers. The museum took out a loan of £30,000, having an understanding that the local authority would meet the repayments. This understanding was not, however, backed by a legally binding agreement. When Stoke-on-Trent Council decided not to continue making payments, the museum was declared bankrupt in August 1993 (*cf.* **277**). It thus became the first museum in the UK to go into liquidation. In order to protect it, English Heritage, having already listed all the buildings in 1994, designated the entire site as a scheduled ancient monument, but could not prevent the liquidators, Touche Ross, from selling the museum's collections at auction. The Charity Commission deemed that the British Coal Collection, however, constituted a separate charitable trust, which had to be kept intact. Although a number of items in the collection had already been sold at the auction, the Charity Commission arranged for ownership of the rest of the collection to be transferred in 1995 to the NCMME, which published a catalogue of what had survived of the collection (Salway 1999).

The site belongs to the local authority and is now effectively derelict. When ownership was transferred from British Coal to the council, there was a covenant that the site had always to be used as a mining museum or ownership would revert to British Coal. British Coal offered to remove the covenant from the parts of the site without buildings, so that the vacant areas could either be sold or be used for income-generating activities, such as workshops, that could have funded the operation of the museum. The council did not, however, accept the offer and the covenant has now expired. Various plans have been proposed for

**Figure 363.1. Headgears at Chatterley Whitfield (© National Coal Mining Museum for England; photographer John Cornwell).**

**Figure 363.2. View of the three headgears and boiler-house chimney (© Margaret L. Faull).**

reopening the site; for example, with the buildings used as craft workshops, but none so far have come to fruition. The Friends of Chatterley Whitfield organisation was set up in 2001 to campaign for the site to be reopened. They open it to the public on a heritage open day in September each year. The underground experience has been dismantled and the annual tours access only selected surface buildings of the 34 on site. Some buildings on the periphery of the site have been converted into small industrial units, as well as the Friends' Heritage Centre. Otherwise the rest of the buildings can only be seen at a distance through the wire-mesh fencing. Historic England has placed the 15 scheduled structures and six listed buildings on the register of buildings considered to be at risk on the grounds that they are 'in need of major repair and regeneration'.

*References*

http://chatterleywhitfieldfriends.org.uk/

http://historicengland.org.uk/listing/the-list/list-entry/1015947

https://historicengland.org.uk/advice/heritage-at-risk/search-register/results/?searchType=HAR&search=Chatterley+Whitfield

personal communication, Nigel Bowers, chair, Friends of Chatterley Whitfield, 2021

https://www.europeanheritagedays.com/story/124d6/The-Hidden-remains-of-an-abandoned-Mining-Museum-in-the-UK

showcaves

### 364. Chislet Colliery

Coal was not found in Kent until 1890, when coal seams were identified during boring for a possible tunnel under the British Channel. Sinking of Chislet Colliery began in 1914 and the first coal was brought to the surface in 1918. Coal was moved by the railway line from Ashford to Minster. When British Rail switched to oil for running its locomotives, the main market for Chislet coal disappeared and the colliery closed in 1969. The only surviving buildings from the colliery are the loading hoppers, which are derelict, and the pump house, which is situated on the marshes. The former Chislet Welfare Club, which opened in 1937, is now the clubhouse of the Chislet Colliery Bowls Club. A locomotive, Yorkshire Engine Company 0-6-OST 2498/1951, worked at the colliery from 1959 until 1969, when it was acquired by the Buckinghamshire Railway Centre. It was rechristened as Chislet in honour of its provenance and is kept at Quainton in south-eastern England.

*References*

https://en.wikipedia.org/wiki/Kent_Coalfield

http://www.brc-stockbook.co.uk/chislet.htm

personal communication, Gary Cox (**380**), 2016

### 365. Church Farm Colliery

The Church Farm Cornish winding-engine house is preserved at Mangotsfield, Emersons Green, 11 kilometres north-east of Bristol. The date on which the mine was sunk is not known, but it was being worked during the 1860s. The Mangotsfield Collieries Company Limited was formed in 1881 and a Cornish engine was installed at what was described as the Deep shaft. The engine worked at seven strokes a minute, showing that this was an engine for drainage rather than for winding. Two seams were worked: the Great vein, which was 84 centimetres thick at a depth of 78 metres, and the Little vein, 58 centimetres thick at a depth of 82 metres. The colliery was abandoned in 1891 because of a lack of capital for investment.

Today only the Cornish winding-engine house survives, conserved in a green area in the centre of the Church Farm housing estate; the site has been landscaped and bars fitted to openings to prevent entry into the building. The winding-engine house was listed as grade II in 1987 as a very rare survival above ground of what had been an extensive coalfield. It is built of pennant sandstone and has two storeys, with an arched doorway on the ground floor, which has dressed keystones and quoins. No internal structures or machinery survive.

*References*

https://historicengland.org.uk/listing/the-list/list-entry/1116365

personal communication, Ken Kemp, secretary, South Gloucestershire Mines Research, Group, 2021

### 366. Clipstone Colliery Headstocks

Clipstone is a small village in north Nottinghamshire in the northern part of the East Midlands. Work at Clipstone Colliery on developing the Top Hard seam (known in Yorkshire as the Barnsley seam), 675 metres deep, began in the early-twentieth century, but was delayed by WWI. It was operated by the Bolsover Colliery Company from 1922 and was producing about 3,600 tonnes a day by the 1940s. In 1953 the shafts were deepened by 245 metres to the Low Main seam. Two new winding engines, with two new headstocks, were installed over the coaling shaft and the manriding shaft. The headstocks were designed by the Manchester firm of architects, Young and Purves. They were linked to a central powerhouse and were made by Head Wrightson Colliery Engineering of Thornaby-on-Tees and Sheffield. Friction winders (**127**) were expensive, but more effective for deeper mines, and the Clipstone shafts were amongst the deepest in the country.

Clipstone was producing close to a million tons (984,000 tonnes) a year and employed more than 1,300 men at its peak in 1986. British Coal (**384**) closed the colliery in 1993. RJB Mining (subsequently renamed as UK Coal) started mining in the Yard seam in 1994, at a depth of 870 metres, but ceased production in 2003. In 2000 the headstocks, winders and powerhouse, which were then fully operational, were listed as grade II. When erected the headstocks were believed to be the tallest in Europe and are still the highest in the UK. The friction winders are amongst the few surviving examples of friction winders in the UK and the buildings are good examples of modernist design. The headstocks and associated building form an architecturally cohesive scheme designed by professional architects. Everything else on the colliery surface has been demolished and the site cleared.

Since 2003 there has been serious vandalism on the site and theft of metals for their scrap value, so that the machinery can no longer be operated and the building is becoming derelict. In 2003 the Welbeck Estates, which owned the site, submitted applications to English Heritage to delist the building and the headstocks and to the local authority for permission to demolish them.

The Clipstone Colliery Regeneration Group was set up in 2013 to campaign for retention of the listed complex, as a major feature of the landscape in Clipstone. In 2021 Adco Properties Limited, Mansfield, which had purchased the site, announced plans to renovate the buildings and restore the site as a multi-purpose leisure facility, low-rent facilities for small businesses and a museum of the history of mining at Clipstone and the area, scheduled to open in mid-2022.

*References*

https://historicengland.org.uk/listing/the-list/list-entry/1380235

https://www.chad.co.uk/news/people/leisure-plans-old-mine-site-clipstone-headstocks-following-purchase-3087445

https://www.facebook.com/Save-Clipstone-Colliery-Headstocks-106659108993/

http://www.engineering-timelines.com/scripts/engineeringItem.asp?id=1406

### 367. Dean Heritage Centre

The Dean Heritage Centre is located in the Forest of Dean at Cinderford in the Soudley Valley of Gloucestershire. It aims to record and preserve the social and industrial history of the Forest and its people. The centre opened in 1983 and is an independent registered charity and a company limited by guarantee, overseen by a board of trustees. The museum employs 10 members of staff and has approximately 40 volunteers. The centre raises all its own funding through admission charges, corporate sponsorship, donations and sales in the shop and cafe.

The museum occupies a 2-hectare site with a reconstructed Victorian cottage and a charcoal burner's camp. Five exhibition galleries explore the history of the Forest from the Ice Age to the present day. A significant part of the galleries is dedicated to the coal-mining industry of the Forest, from mining disasters and rescues to everyday mining tools and objects. Gallery five focuses on the large-scale collieries of the Forest and the deeper shafts that were

**Figure 365.1. Church Farm Colliery Cornish winding-engine house (© Ken Kemp, South Gloucestershire Mines Research Group).**

**Figure 366.1. Clipstone headstocks (© Neil Cossons).**

**Figure 367.1. Beam engine from Lightmoor Colliery, Gloucestershire (© Dean Heritage Centre).**

**Figure 367.2. Reconstruction of a typical Forest of Dean Free Mine with tunnels dug into the hillside to reach the coal seam rather than by vertical mine shafts (© Dean Heritage Centre).**

made possible by the advent of steam power. A working beam engine from Lightmoor Colliery, Gloucestershire, was originally made by Hewlett's iron foundry (1823-76), at Camp Mill, the site of the present museum. The museum has a reconstruction of a typical Forest Free Mine with adits dug into the hillside. Forest mines are gas-free, so originally miners were able to use candles and later carbide lamps (for an example see **391.3**), which had a stronger flame than candles and did not go out in draughty conditions.

The exhibition includes a three-dimensional geological model of the Forest of Dean, showing the mineral deposits and underground workings. It was made in 1837 by Thomas Sopwith (1803-79), an English mining engineer, geologist and surveyor. The model can be dismantled to reveal the main coal seams and collieries, whether accessed by shaft or drift. The museum also holds the sheets produced by Sopwith and the index map which shows the collieries, enclosures and railways.

Coal and ironstone mining in the Forest are organised quite differently from in the rest of the UK, being regulated through a system known as Free Mining, with mines being known as gales. Anyone born and living in the hundred of Saint Briavels (an ancient administrative area approximating to the Forest of Dean), who has worked for a year and a day underground and has reached the age of 21, can claim their right to be a Free Miner. This means they are entitled to mine for coal, iron or stone in the Forest of Dean, subject to certain conditions laid down by the controller of mines, the Gaveller. In 1244 it was recorded that Free Miners had the exclusive right to mine ore in the Forest. The earliest written description of this system is in a document dated to 1612 and entitled 'The Dean Miners Laws and Privileges', which refers back to earlier documents that do not survive. It claims that the 'customes and franchises' that had existed since 'tyme out of mynde' were confirmed by Edward I (1239-1307) in recognition of the support given by the miners at the sieges of Berwick-upon-Tweed in 1296 and 1305, where it is assumed that they worked on tunnelling under the walls. Newland church in the Forest of Dean has the only image of a medieval miner surviving in the UK. Dating from the fifteenth century, the brass Greyndour memorial depicts a helmet and crest, and a miner with a pick in his hand, a candle on a stick held in his mouth and a hod (an open trough on a pole for carrying coal) over his back (Goodman 2020, p. 125).

Following a report from a Royal Commission set up in 1831, the Dean Forest (Mines) Act 1838 still regulates Free Mining today. The Forest of Dean mines were specifically exempted from the Coal Industry Nationalisation Act 1946, under which all mines employing more than 30 men were taken into public ownership. The 1838 Act permitted Free Miners to sell their gales to non-Free Miners and the NCB used this provision to buy some of the larger mines in the Forest; the last NCB mine in the Forest closed in 1965. The 1838 Act also restricted the right to mine to men, but this was amended in the 2010 Equal Opportunities Act to include women.

Since at least the late-fifteenth century, Free Mining has been administered by the Gaveller, an officer of the British Crown, who manages the mines and quarries in the Forest. The Gaveller is responsible for the administration of Free Mining customs and the collection of mineral royalties. The Gaveller also allots gale boundaries (*i.e.* mining plots), settles disputes over land, accepts surrenders of gales and keeps the registers required by the 1838 Act. There are probably about 150 Free Miners today, although only a few coal mines are still being operated (**377**). In 2006 the only maternity unit in the hundred of Saint Briavels, at the Dilke Memorial Hospital in Cinderford, was transferred to Gloucester, outside the Forest, despite public opposition. This meant that any babies born there would not be eligible to become Free Miners. It has been noted that home births in the hundred are much higher than the national average, and it is assumed that this is in order to keep the tradition alive.

*References*

https://www.publications.parliament.uk/pa/cm200809/cmpublic/equality/090707/am/90707s01.htm#end

http://www.wyedeantourism.co.uk/freeminers

personal communications, Nathalie Hall, manager, 2014; Nicola Wynn, collections officer, 2021, 2022

www.deanheritagecentre.com/

### 368. Durham Cathedral Miners' Memorial

There has long been a close connection between coal mining and the Church in north-eastern England. The Durham and Northumberland coalfield was one of the two main English coalfields from at least the later-thirteenth century (**361**). Amongst the main owners of the medieval coalfield were the prince bishops of Durham, who ruled the diocese of Durham and Durham County Palatine from 1080 until the county palatine was abolished in 1836. This was virtually a separate state from the rest of England, in which the prince bishops could appoint sheriffs, hold their own parliaments, levy taxes, mint coins and raise armies.

Durham Cathedral was built in 1093 to serve as the cathedral for a community of Benedictine monks. The 1183 Boldon Book, documenting the holdings of the bishop of Durham, refers twice to the *faber . . . pro carbone quem invenit*, which the editor translates as 'the smith . . . for the coal which he provides' (Austin 1982, pp. 17, 25), and to *quidam carbonarius*, translated as 'a certain coal-miner' (*ibid.*, p. 71). At first sight, these would appear to be very early references to coal mining, but Latin *carbo* and Middle English *col* usually referred to charcoal. Coal was generally called 'stones to burn' or 'sea coal', referring initially to the fact that coal was often collected from the beaches where it had been washed up by the sea and later to the coal that went by sea to London. Also charcoal, not coal, was used by smiths in the production of iron until the early-eighteenth century, when it was replaced by coke (**379**). So the Boldon Book references are almost certainly to charcoal and a charcoal-burner.

As a result of the dissolution of the monasteries under Henry VIII (1491-1547), the monastery passed into the hands of the Crown in December 1539, but the cathedral was refounded in May 1541, with the last prior as dean and 12 former monks becoming the first canons. Shute Barrington (1734-1826; bishop 1791-1826) broke up a miners' strike in 1812 using his own troops. Edward Maltby (1770-1859; bishop 1836-56) spoke in the House of Lords about how the pit owners did not attend to the moral and religious improvement of their workforce. Brooke Foss Westcott (1825-1901; bishop 1890-1901) was known as the Miners' Bishop for the interest that he showed in the welfare of the coal miners. In 1892 he negotiated a peaceful end to a long miners' strike. His successor, Handley Moule (1841-1920; bishop 1901-20) came back from London immediately after the 1909 West Stanley Pit disaster (**356**). In gratitude for this, he was presented with a miner's lamp from West Stanley Pit, which is still kept at Auckland Castle, the residence of the bishops of Durham. He visited the mines and invited retired miners and their wives to Auckland Castle. Hensley Henson (1863-1947; bishop 1920-39), on the other hand, considered that strikes were morally wrong and disliked trade unions so strongly that mounted police had to be deployed outside Auckland Castle to protect it from stone throwing. Towards the end of his life he did, however, organise a charitable campaign to buy boots for the children of miners. David Jenkins (1925-2016; bishop 1984-94) was appointed at the beginning of the Miners' Strike (**404**). When he was enthroned he enjoined the government to heal rather than divide society and declared that, "the miners must not be defeated".

The Miners' Memorial in the cathedral was created in 1947 under Alwyn Williams (1888-1968; bishop 1939-52). John Cosin (1594-1672; bishop 1660-72) was responsible for much of the unique woodwork in the cathedral, a fusion of gothic and contemporary jacobean styles. Part of his organ screen was given to Ramside Hall in the early-nineteenth century, where it was used as a fireplace. The Pemberton family returned the fireplace so it could be used to create the Miners' Memorial. The inscription in gold letters in the centre of the memorial states:

**Figure 368.1. Early-thirteenth-century Norman towers of Durham Cathedral from the cloisters, laid out in 1093, with the lavabo, or stone basin, in which the monks washed, in foreground (© Margaret L. Faull).**

**Figure 368.2. Miners' Memorial in Durham Cathedral with book to left of memorial listing miners killed in colliery accidents and miner's lamp above (© Durham Cathedral).**

> Remember before God the Durham miners who have given their lives in the pits of this county and those who work in darkness and danger in those pits today.

At the bottom of the memorial is a quotation from the Old Testament Book of Job (chapter 28, verse 4):

> He breaketh open a shaft away from where men sojourn
> They are forgotten of the foot that passeth by.

A book to the left of the memorial lists miners killed in colliery accidents, with a miner's lamp above in reference to Christ's words in the New Testament (John, chapter 8, verse 12), 'I am the light of the world'.

The Durham Miners' Gala, organised by the National Union of Mineworkers, has been held annually since 1871 on the second Saturday in July, except between 1915 and 1918 during WWI, in 1921, 1922 and 1926 during strikes and between 1940 and 1945 during WWII. The gala is still held, even though the last coal mine in the north-east, Wearmouth Colliery, closed in 1993 (**399**). The gala does not have a specific starting or finishing time, as colliery bands accompanying the pit banners each start out from the village of their own coal mine to meet in the marketplace in Durham. The bands stop en route in front of the County Hotel, where union leaders and important guests assemble on the balcony, before marching to the

**Figure 368.3. Banner of South Hetton Lodge in procession at the Durham Miners' Gala, 2012, with slogan taken from the *Communist Manifesto* (© Geoff Preece).**

racecourse, where the crowds are addressed by invited speakers. After the speeches four or five bands, with their banners, march to the cathedral for the annual Miners' Gala Service, which has been held annually since 1896. In the past new banners would be brought to be blessed. The words of the inscription on the Miners' Memorial have been set to music and were sung for the first time at the service in 2007.

*References*

http://www.chroniclelive.co.uk/whats-on/arts-culture-news/miners-bishops-durham-ups-downs-11410427

http://www.durhamminers.org/gala

http://www.englandsnortheast.co.uk/CoalMiningandRailways.html

http://www.englandsnortheast.co.uk/PrinceBishopsDurham.html

http://www.sunnisidelocalhistorysociety.co.uk/cathedral.html

### 369. Durham Mining Museum

The Durham Mining Museum deals with mining in northern England, comprising the counties of Cumberland, Durham, Northumberland and Westmorland. The Friends of Durham Mining Museum is a company limited by guarantee and a charity, established in 2000. The charity was set up to run the museum and to preserve the knowledge and history that helped drive the Industrial Revolution (**379**) in the region. There are 20 members, of whom nine volunteer regularly. The charity meets one evening a month at the town hall in Spennymoor in County Durham, where the museum is located.

The museum focuses on the mining history, heritage and culture of the Durham coalfield in particular, but does include mining in general as well. The museum was established in 1999. Its small-exhibits section consists of an artefact display of approximately 186 square metres, which traces the history of coal and coal mining plus displaying mineral samples illustrating the formation of coal. Items include helmets, surveyor's equipment and a colliery time clock. It also has a screen playing mining-history digital video discs. There are two computers loaded with the museum's website, so that visitors without computers can access the website to allow them to research their family history through the information about mining in the region.

The second part of the museum is downstairs, underneath the exhibition section. This is approximately 93 square metres of a life-size reconstruction of an underground roadway. There is a simulated coal seam, 610 centimetres thick, with traditional wooden pit props supporting the roof and a life-size model of a pit pony drawing a tub.

The Friends also run a website with very detailed data about miners who were killed during accidents and other information about mining in the north of England. The website contains over 90,000 pages of information covering all forms of mining, both coal and metalliferous, in the four old northern counties of England, plus the Isle of Man and parts of North Yorkshire and North Lancashire. It has 2,219 individual pages of mines known to have existed in these areas since the early 1700s, plus an alphabetical list of over 23,000 known mining deaths. The website attracts on average 3 million hits *per annum*.

*References*

personal communication, Roy Lambeth, chair, 2012

showcaves

www.dmm.org.uk

### 370. Easington Colliery Memorials

Easington Colliery was located on the coast of County Durham. So dominant was the mine in the community that the village in which it was situated is actually called Easington Colliery. Sinking of the mine began in 1899 with three shafts, respectively 143, 344 and 436 metres deep, but production did not begin until 1910 because of problems involved in sinking through water-bearing strata. The adjacent village was constructed with a series of terrace houses, together with clubs, library, shops and other amenities, such as the infant and junior schools, built in 1911. Tubs of colliery waste were carried to the beach by an aerial ropeway with the pylon standing actually in the North Sea, a common feature of mines in the north-east that were close to the sea. The conveyors and the pylons were removed once the collieries closed and tonnes of coal waste were cleaned from the beaches of eastern County Durham, at a cost of about £10 million. The 1971 crime film 'Get Carter' gives an idea of what they looked like before this remedial action. The scene on the blackened beach at the climactic end of the film was filmed at Blyth staithes and Blackhall beach near Blackhall Colliery.

On 29 May 1951 the miners' picks threw off sparks when they hit a patch of pyrites (iron disulphide), which set off a firedamp explosion. This threw the coal dust into the air, where it exploded. Coal-dust explosions (**80**) were usually prevented by stone dusting or water spraying, so it is surprising that such a major disaster should occur more than 40 years after their cause had been identified. The subsequent enquiry found that the General Regulations (Precautions against Coal Dust) 1939 had been contravened by poor management permitting bad practice, such as deputies not being instructed as to how many shots they could fire underground and especially in allowing the build up of firedamp in spaces left as the mine was worked by retreat mining (**382**).

The coal-dust explosion killed 81 miners, as well as two of the mines-rescue team. The Easington Colliery disaster memorial dates to 1953/4 and is located at the Easington

**Figure 370.1. Social-realist bas-relief of a miner, inscription and colliery tub in Easington Colliery cemetery marking communal grave of those who died in 1951 Easington Colliery disaster (© Margaret L. Faull).**

**Figure 370.2. Wrought-iron gates at entrance to the memorial gardens (© Geoff Preece).**

Colliery cemetery. It marks the communal grave with a bas-relief of a miner in social-realist style, flanked by a colliery tub on each side and the inscription:

> REMEMBER BEFORE GOD THOSE WHO GAVE THEIR LIVES IN THE EASINGTON COLLIERY DISASTER IN MAY NINETEEN HUNDRED AND FIFTY ONE TO WHOM THIS GARDEN OF REMEMBRANCE IS DEDICATED

In front are four benches on either side of an arc-wall chain-type coal cutter, driven by compressed air and supplied to the Easington Coal Company in 1932 by Anderson Boyes of Motherwell, Scotland, which closed in 2013. The whole monument is listed as grade II, both for its artistic merit and for its historic interest as a witness to the impact of the disaster on the community and for its expression of the mining heritage of the region.

A memorial garden was created on the site of the former pit and opened on 29 May 2001, the fiftieth anniversary of the disaster. The elaborate wrought-iron gates at the entrance bears the names of the National Union of Mineworkers, Easington Lodge and the Durham area. Below are those of Lawrence Daly (1924-2009), who was general secretary of the NUM from 1968 to 1984 and steered the union through

**Figure 370.3. Abstract representation of pit shaft in the memorial garden (© Geoff Preece).**

**Figure 370.4. Pit pones Sparky and Carl from Easington Colliery in retirement at the National Coal Mining Museum for England (© National Coal Mining Museum for England).**

major strikes in 1972 and 1974, and Tommy Hepburn (1795-1864), who was leader of the Northumberland and Durham Pitmen's Union, the first mining trade union, formed by Hepburn in 1825. The union was broken by the coal-mine owners following a strike in 1832 and in order to gain employment Hepburn had to promise not to undertake any trade-union activities. He then dedicated his life to gaining shorter hours of work for miners and their education.

The roadways of Easington Colliery went up to 6.4 kilometres out under the North Sea. It was difficult to get machinery into these narrow undersea roadways. So Easington Colliery was the last deep mine in the UK to use pit ponies, which could carry out salvage work in the confined spaces. The ponies at Easington were stabled underground, and only came out for a short summer break. From 1987 pit ponies at Easington Colliery at the end of their working lives were retired to live out the rest of their lives at the NCMME (**386**); the first of these disliked working underground so much that he refused to go even close to the entrance to the drift, because he could smell the underground (*cf.* **18**). The last four Easington ponies left the pit in 1994. Sparky, a strawberry roan, died at the museum in 2007. The last surviving pony, Tony, which did not go to the museum, died at the age of 40 in 2011.

The mine closed in 1993 and the headgear was demolished in 1994, along with virtually all the other surface buildings. The Easington Colliery band, founded in 1915, has to raise its funding itself, having lost the colliery's financial support. The loss of the 1,400 well-paid colliery jobs was devastating for the local economy. A further blow was the creation of a new unitary local authority, Durham County Council, which meant that the employees of Easington District Council were moved to work in Seaham, 10 kilometres to the north, and the local-authority premises in Easington Colliery village were demolished. In 2000 the village was used in the film 'Billy Elliot' to represent the fictitious town of Everington. The 1911 schools are now derelict. They were sold in 2003 to a development company. Planning permission was refused for their demolition and the building of residential units. Durham Constabulary has had to introduce patrols around the cemetery and memorial gardens following anti-social behaviour there by unemployed youths.

*References*

http://dmm.org.uk/pitwork/html/easington1.htm

https://en.wikipedia.org/wiki/Easington_Colliery

https://en.wikipedia.org/wiki/Get_Carter

https://historicengland.org.uk/listing/the-list/list-entry/1437889

http://www.geograph.org.uk/photo/1578485

### 371. Elsecar Heritage Centre

Elsecar is located south of Barnsley in South Yorkshire (see also **372, 376**). The Elsecar Heritage Centre occupies the 2 hectares of the former colliery workshops and ironworks, first developed from the 1790s by the earls Fitzwilliam of Wentworth Woodhouse (**402**). The centre is within the conservation area of the village of Elsecar, initially built for housing the workers, as the owner wanted to be seen as a good employer. It is now recognised as one of the first model villages and a unique example of the aristocracy embracing the Industrial Revolution (**379**) by creating a new estate village.

A canal was cut into the village in the 1790s and a railway with extensive sidings was created in the 1850s. The site of Elsecar New Colliery still exists, including the original winding shaft, pumping shaft and 1795 engine house. The colliery was sunk by the fourth earl Fitzwilliam (1748-1833) in 1794/5 to coincide with the 1798 opening of the canal and of two major ironworks locally. A small footrill (*cf.* **352**) was created at the same time, around 500 metres to the south-west, through which distinguished guests could enter the colliery without using the steam-wound cage.

The Elsecar workshops of dressed sandstone, built in 1850, were taken over by the NCB at nationalisation in 1947 and continued in use until their closure in the 1980s. Elsecar New Colliery worked until the 1850s, when Simonwood Colliery was sunk a short way up the canal between the railway and the canal. Elsecar Main Colliery was sunk in 1905 on the south-east side of the railway and closed in 1983.

Barnsley Council began a programme of conservation after purchasing the site in 1988. In the early 1990s the site was developed as a heritage centre, with small craft workshops in the converted workshops, a play area in the joiners' workshop and a large event space in a former rolling mill. Almost all the buildings are listed grade II* and a large part was scheduled as an ancient monument in 2020. The railway now operates on a 1.6-kilometre section of the branch line, using historic steam and diesel locomotives.

The most important element of the site is the Newcomen beam engine (Clayton 1962), of the type designed by Thomas Newcomen (1664-1729) in 1712 as the first steam engine in which the mechanical work was carried out by a piston moving in a cylinder. It did not use the pressure of steam, but instead the weight of the atmosphere pushing a piston in a cylinder against a partial vacuum, created by condensing steam with cold water (**360**). Elsecar New Colliery was much deeper than those previously sunk in the area and so a Newcomen beam engine, with a 107-centimetre diameter, was needed to extract water from the mine, allowing the exploration of the deeper coal seams.

**Figure 371.1. Exterior of the Newcomen engine house (Barnsley MBC, Arts, Museums & Archives).**

**Figure 371.2. Shaft of the Newcomen engine (© Barnsley MBC, Arts, Museums & Archives).**

As originally built in 1794/5, 1,755 litres of water were pumped each minute. After the installation of a 122-centimetre cylinder in 1801 and new pumps in 1812 the rate increased to 2.73 litres per minute. The last major modification to the engine itself was the replacement of the wooden beam with a cast-iron beam in 1836. The engine worked from 1795 until 1923, when it was replaced by electric pumps. The engine last worked in the 1950s but, despite having lost its original boiler, was preserved as a static engine by the NCB in view of its historic status.

This scheduled ancient monument is believed to be the oldest steam engine in the world still in its original location and the only Newcomen engine still *in situ* anywhere in the world. The engine house was derelict, but has now been restored by Barnsley Council.

*References*


ERIH, *s.n.* Barnsley

http://historicengland.org.uk/listing/the-list/list-entry/1004790

https://historicengland.org.uk/listing/the-list/list-entry/1151096

http://historicengland.org.uk/listing/the-list/list-entry/1151097

https://research.historicengland.org.uk/Report.aspx?i=16223

http://www.elsecar-heritage.com/history

personal communications, John Tanner, project development officer, Arts, Museums and Archives, Barnsley Metropolitan Borough Council, 2014, 2021


**372. Elsecar Old Colliery**

The complex of eight Elsecar collieries (**371, 376**), south of Barnsley in South Yorkshire, was begun about 1750 and taken over by the second marquis of Rockingham (1730-82) in 1752. The shafts were sunk about 15 metres deep to work the Barnsley Bed seam with a horse gin. In about 1762 the horse artist, George Stubbs (1724-1806), painted a life-size study of the marquis of Rockingham's racehorse, Whistlejacket, now in the National Gallery in London. At about the same time, the marquis commissioned Stubbs to paint Elsecar Old Colliery with the colliery overseer, Thomas Smith, in the foreground. He is shown supervising the horse gin, while seated on a bay horse. This was the only industrial painting that Stubbs ever produced. A portrait by Stubbs of Smith, painted several years later on his retirement at the age of 70, hung in the marquis' antechamber until the death of the marquis in 1782. Both paintings are still in private hands. The fourth earl Fitzwilliam (1748-1833) inherited the pits on the death of the marquis in 1782. The earl installed steam-winding engines at Elsecar Old Colliery in 1796.

One of the prime functions of the Elsecar collieries was to supply coal to heat the marquis' country house of Wentworth Woodhouse (**402**). Elsecar Old Colliery was the focus of royal visits, including the future king, William IV (1765-1837), who entered by the nearby footrill (**371**) in 1828. Elsecar Old Colliery appears to have been a

**Figure 372.1. 1793 image and cross-section of Elsecar Old Colliery, surveyed by the Earl of Fitzwilliam's steward, Joshua Biram (© Sheffield Libraries Archives and Information, WWM-MP-55).**

**Figure 372.2. Detail of 1793 image showing horse gin (© Sheffield Libraries Archives and Information).**

Figure 372.3. Detail of 1793 image showing horse gin (© Sheffield Libraries Archives and Information).

Figure 372.4. Possible eighteenth-century pithead building (© Barnsley MBC, Arts, Museums & Archives).

trophy pit for the marquis of Rockingham, and later the earl Fitzwilliam, as a symbol of their industrial enterprise and enlightenment. This would have been consolidated when the new ironworks were built about 91 metres behind the colliery. The ironworks were constructed and leased by Walker Brothers between the 1790s and 1820, including being a leading supplier to the Royal Navy during the Napoleonic Wars (1803-15).

Elsecar Old Colliery appears to be on the line of the 'deep level intended to be pursued' on a 1757 map of the Elsecar collieries (Barnsley Archives and Local Studies, A/3491/Z). A plan dating to 1793, drawn by the Earl's steward, Joshua Biram (**376**), shows two colliery buildings, each with a horse gin alongside it (Sheffield Libraries Archives and Information, WWM-MP-55). A hand-written daybook, dating to 1839 and labelled 'Elsecar Old Colliery', gives details of individual labourers at the pit, as well as banksmen, horse boys and numerous other roles. There are about 20 separate lists, with about seven labourers in each list, paid between 2 and 3 shillings a day (Barnsley Archives and Local Studies, A/3299/B).

In 1848 the pit was renamed as Elsecar High Colliery, by which time its active shafts had been moved onto the Milton ironworks site, when it was recorded as employing 87 workers and was mainly supplying the ironworks. The colliery was eventually worked out and closed in 1888.

Field research by Barnsley Council and Historic England has identified the possible site of the original Elsecar Old Colliery. The 1750 shaft may be intact, albeit under some paving, but has not been archaeologically explored. By the 1840s the site had become the house of the head engineer of the Milton ironworks and is now occupied by a cluster of private residences. There are certainly similarities between one of the buildings and the headgear buildings shown on the 1793 plan, so it is probable that parts of other buildings survive from the original colliery, although this still remains to be proved.

*References*

http://news.bbc.co.uk/local/sheffield/hi/people_and_places/history/newsid_9093000/9093809.stm

https://books.google.co.uk/books?id=mFrO5o2X2EcC&pg=PA218&lpg=PA218&dq=George+Stubbs+painting+of+Elsecar+Old+Colliery&source=bl&ots=PIBLhpr1pH&sig=JZSOlia9Dwilb4RKz2VIdeAr4I&hl=en&sa=X&ved=0ahUKEwiLnvmNw8TKAhUFcw8KHSM5AyAQ6AEIHzAA#v=onepage&q=George%20Stubbs%20painting%20of%20Elsecar%20Old%20Colliery&f=false, numbers 65, 67.

https://en.wikipedia.org/wiki/Elsecar_Collieries

personal communications, John Tanner (**371**), 2015, 2021; Paul Stebbing, archives and local studies officer, Arts, Museums and Archives, Barnsley Metropolitan Borough Council, 2022

### 373. Elvington and Eythorne Heritage Centre

The Elvington and Eythorne Heritage Centre is located in Kent. The centre was opened in 2013, to coincide with the centenary of the first coal being raised at the nearby Tilmanstone Colliery on 12 March 1913. The centre occupies a large room in the community centre. The aims of the centre are to research and archive the heritage of the mining community of Elvington and also of Tilmanstone Colliery, which was located close to the village of Tilmanstone, 4 kilometres north-east of Elvington. The colliery closed in 1986 and all of the buildings were demolished, although the foundations and basements of a couple of the original engineering workshops were exposed during clearance for a new development on the colliery site. The three shafts are sealed with metal covers. The divide station for the 12-kilometre aerial ropeway from Tilmanstone Colliery to Dover harbour survives in a derelict state. This was created by the colliery manager and owner, Richard Tilden Smith (1866-1930), who believed that the East Kent Railway (**396**) was overcharging for coal transport.

The committee of the Elvington and Eythorne Heritage Group has 13 members. The group was set up with funding from the Kent Coalfield Regeneration Trust and Dover District Council and its running costs are funded by local grants. The group has a Facebook page but no website. Since 2011 the centre has held a heritage event every year, which attracts about 100 to 200 people. The centre is also open for four hours most Tuesdays or by prior arrangement, with free entry and two or three volunteers on duty. About 50 to 100 people visit the centre throughout the year, mainly people visiting from elsewhere, rather than local residents.

The group has collected numerous items of Kent mining heritage, mostly small pieces of colliery equipment, such as helmets, lamps and picks, and maps, paperwork and photographs. The most important items are the original Tilmanstone Colliery banners, which had been kept in an attic since the colliery closed and were donated to the group in 2013. Both banners are plain on one side, unlike the normal practice of images on both sides. The older banner bears images of one of the colliery's headgears, a miner's lamp and the white cliffs of Dover, with the slogan 'unity is strength'. It was decided that this banner was too old and the images and text too damaged for the banner to be used in parades, so a second banner was created. The original banner will be conserved and displayed at the Kent Mining Museum (**380**).

The later banner was designed by Malcolm Pitt (1944-2010), who was a branch secretary of the Kent Area National Union of Mineworkers and who worked at Tilmanstone Colliery for a number of years. The second banner is inscribed 'Miners' Strike February 1981'. It shows both of the colliery's headgears. There is a miner in modern orange overalls and wielding a pneumatic drill on one side of the headframes, with the word Beresford (the top seam worked at Tilmanstone) underneath. On the other

**Figure 373.1. Original National Union of Mineworkers' banner of Tilmanstone Colliery (© Colin Varrall).**

**Figure 373.2. National Union of Mineworkers banner (1981/6) of Tilmanstone Colliery (© Colin Varrall).**

side is a medieval miner, based on the miner's brass in the Forest of Dean (**367**), with the word Milyard (the bottom seam worked at Tilmanstone) underneath. It is believed that this banner, which appears to be made of yacht-sail fabric, was painted by Malcolm Pitt's widow, Elizabeth, along with another smaller banner that was created for the Support the Tilmanstone Women's Group for use during the 1984/5 Miners' Strike (**404**).

*References*

https://en-gb.facebook.com/groups/531974076829457/

https://www.youtube.com/watch?v=OxvdKprSsS0 for the working of the Tilmanstone aerial ropeway

personal communication, Colin Varrall, volunteer heritage representative, Tilmanstone Colliery and Elvington Mining Community, 2015

**374. Haig Colliery Mining Museum**

Whitehaven harbour, in Cumbria on the north-western coast, was built in the eighteenth century for shipping coal from the coastal Cumbrian mines. The shafts at Haig Colliery, in Whitehaven, were sunk in 1914/18 to a depth of 366 metres. The pit was named after General Sir Douglas Haig (1861-1928), commander-in-chief of the British forces in WWI. There were serious accidents in 1922, 1927, 1928 and 1931, with up to 39 miners killed.

**Figure 374.1. Entrance to Haig Colliery Mining Museum (© Helen Davies).**

In 1939 one of the fastest manriding conveyors, of 350 horsepower, was installed. In 1980 there were 938 miners and the pit was producing 372,800 tonnes of coal a year. Haig was the last deep mine in the Cumbrian coalfield when it closed in 1986. Most of the colliery buildings were demolished, apart from the headgear and winding-engine house dating back to 1917, together with the two steam winders by Bever and Dorling, a company in Bradford, West Yorkshire, which closed in 1932. One of the winders has been fully restored and one is still derelict. The buildings and headgear were listed as grade II in 1987 and the entire remaining complex was scheduled as an ancient monument in 1998. on the grounds that the engines are particularly rare survivals of twin-cylinder horizontal single parallel-drum winding engines and that the engine in number 4 winding-engine house is the only *in-situ* working example in the world.

A group of volunteers set up the Haig Colliery Restoration Group, which in 2015 had 137 members. The group obtained ownership of the site and opened the complex as the Haig Colliery Mining Museum in 1999. In 2015 the museum was being run as a registered charity and company limited by guarantee with a board of nine directors. It employed eight members of staff and had 28 volunteers. The museum had a large collection of mining artefacts and memorabilia, but the majority of delicate pieces were

**Figure 374.2. One of Haig Colliery's two Bever and Dorling steam-winding engines (© Helen Davies).**

kept in storage because of the condition of the buildings. Important items in the collection include the letters from 1803 to 1830 of the manager, John Bateman, to the owner, Lord Lowther (1787-1872), the letters from the manager, Robert Steele, following the 1910 Wellington Pit disaster (in which 136 miners were killed), sketches by the North Durham artist Ron Gribbons (1928-98) and oil paintings by Paul Schofield (born 1938).

The museum was closed for 18 months for restoration with a grant of £2.4 million from the Heritage Lottery Fund. It reopened in February 2015, and exceeded its target of 15,000 visitors annually. In December 2015, however, the museum closed again, because of lack of external financial support. West Cumbria Mining then purchased the site from the liquidators, who also transferred ownership of the collection to West Cumbria Mining. The company has catalogued and packaged the collections, which are currently in storage on site. The company has held public update events at which the museum catalogue has been available to view and will allow access to the collections if given prior notice. The visitor centre is used as an office, while there are plans to convert the old powerhouse building, which is currently unoccupied, into a training centre.

*References*

ERIH, *s.n.* Whitehaven

https://historicengland.org.uk/listing/the-list/list-entry/1017644

https://historicengland.org.uk/listing/the-list/list-entry/1263938

http://www.museumsassociation.org/museums-journal/news/13012016-cumbrian-mining-museum-faces-closure

personal communications, Pamela Telford, Haig Colliery Mining Museum, 2015; Helen Davies, head of communications, West Cumbria Mining Limited, 2021

showcaves

### 375. Hartley Colliery Disaster Memorials

Hartley Pit, on the north-eastern coast of England, was first established sometime before 1291. The roadways went out under the North Sea and the consequent flooding led to the closure of the mine in 1844. New Hartley Colliery, 1.6 kilometres inland, began production in 1846. In line with the 1842 Act, women, girls and boys younger than 10 years of age were not employed underground (**378**). The Husker shaft, 3.7 metres in diameter, was the only shaft and was used for ventilation, for moving coal, men and materials, and to accommodate the pumps, which were worked by a beam engine. An airtight timber brattice down the entire length of the shaft divided the downcast side from the upcast, with the upcast air created by a furnace at the pit bottom (*cf.* **386**).

On 16 January 1862 the beam of the pumping engine broke and fell down the shaft. The 40-tonne beam demolished the brattice, which created a pile of debris at the base of the shaft, thereby blocking it to a depth of 27 metres. As it was time for shifts to change, both the night and the morning shifts were underground, amounting to 204 men. The first rescue attempt was carried out by the owner, Charles Carr, and the viewer (mine manager), Joseph Humble, together with three other men, including Her Majesty's Inspector of Mines (the first Inspector of Mines was appointed officially in December 1843, followed by legislation leading to four more in 1850). Two men were brought up alive from the jammed cage, as well as Thomas Watson, a local preacher, who had been lowered down the shaft to comfort and pray before they died with the four men who had fallen from the cage. Despite desperate attempts from the surface and also by the trapped men below, no further miners were rescued.

The trapped men tried to keep themselves alive by eating the corn stored underground to feed the pit ponies. When rescuers eventually broke through six days after the accident, they found that some of the miners had only just died of gas poisoning. Dead miners were found lying with their arms around their sons, brothers with brothers and one man had his stepson on his knee. On the body of James Amour, an overmen, was found a note written in pencil on a piece of newspaper:

> Friday afternoon, at half-past two.
> Edward Armstrong, Thomas Gledson, John Hardy, Thomas Bell, and others, took extremely ill. We also had a prayer-meeting at a quarter to two, when Tibbs, Henry Sharp, J. Campbell, Henry Gibson and William Palmer. Tibbs exhorted us again, and Sharp also.

On 26 January the bodies of the 204 dead were taken in procession to Earsdon church, 6.4 kilometres from New Hartley village. The procession was so long that the first hearse had arrived at the church before the last one had left the colliery. The cemetery was not large enough for the number of graves required and the duke of Northumberland donated additional land to extend the cemetery. The list of names on the memorial erected above the graves gives the same surname again and again, as do the inscriptions on gravestones clustered around the memorial, showing how many of the miners were from the same family. One woman lost her husband and six of her seven sons and another lost her three sons and her husband, who was 72 years old. Two of the victims were only 10 years of age. The inscription on the memorial states:

> Erected
> To the memory of the 204 miners, who lost their lives in Hartley Pit, by the fatal catastrophe of the engine beam breaking, 16th January 1862

The top of the Hartley shaft is now covered by a large block of stonework between two rows of terraced houses in the village of New Hartley. There are two badly eroded memorials on the wall of the stonework.

Queen Victoria (1819-1901), who had lost her own husband, Prince Albert (1819-61), a month earlier, telegrammed twice, expressing her concern and requesting

Figure 375.1. Memorial erected in Earsdon churchyard with the names of those who had died inscribed on all four sides, flanked by individual gravestones of some of the miners who died (© Margaret L. Faull).

Figure 375.2. 'Reading the Queen's letter', following the Hartley Colliery disaster; oil painting by Henry Hetherington Emmerson, 1862 (© National Coal Mining Museum for England).

Figure 375.3. Silver medal obverse with two miners working with pick and shovel to free two miners trapped under rockfall, overseen by winged figure with sword (© National Coal Mining Museum for England).

information, and subsequently sent a personal letter of condolence. The message from the queen said:

> The Queen, in the midst of her own overwhelming grief, has taken the deepest interest in the mournful accident at Hartley, and up to the last had hoped at least a considerable number of the poor people might have been recovered alive. The appalling news since received has afflicted the Queen very much. Her Majesty commands me to say that her tenderest sympathy is with the poor widows and mothers, and that her own misery only makes her feel the more for them.

This was read out to the families by the local clergy. The reading of the queen's letter by an elderly collier and his wife (presumably parents of a dead miner) was depicted in an 1862 oil painting by Henry Hetherington Emmerson (1831-95), now in the NCMME (**386**). In her journal the queen recorded 'the accounts of the colliery disaster are terrible, – such awful misery'. The mine never reopened, and a new colliery was sunk nearby in 1874.

Public meetings were held in Blyth, Hexham, London, Manchester, Morpeth, Tynemouth and elsewhere to raise funds for the bereaved families. The City of London gave 100 guineas and the London Coal Exchange 1,000 guineas. In total about £75,000 (£8,368,500 in present values) was raised. The NCMME has one of the glasses sold to raise funds after the disaster, which was presented to the museum by the daughter of a woman whose relatives were miners in the Nottinghamshire coalfield. The inscription says:

> Hartley Colliery
> Disaster Jany 16 1862
> 204 Lives Lost

Money was allocated from the relief fund to reward the rescue team. On 20 May 1862 a gold medal was presented at a civic reception at Newcastle Town Hall to William Coulson, who had led the rescue operation, and silver medals were given to the 37 men in his team of shaft sinkers. Each medal had the name of the recipient engraved on its edge. An unknown number of bronze medals were also given to others who had assisted. After paying for the medals, the amount left from the allocated £1,587 was distributed amongst the rescuers. The sums varied between £4 and £30 (equivalent to £3,350), distributed in proportion to the amount of time each had spent in the shaft.

Several days after the disaster a large meeting was held at Newcastle of miners from collieries throughout the region. The meeting drew up a petition for there to be legislation for the provision of two shafts at every pit. At the subsequent inquest into the deaths, the jury exonerated Joseph Humble, and expressed their strong opinion that it was essential that all collieries should have a second means of escape. From 1855 Her Majesty's Inspector of Mines had been required to make a report to Parliament every year, which in this year included the views of the inspectorate on the disaster. In August 1862, only six months after the inquest and less than seven months after the disaster, the British Government passed an act requiring all new mines to have two shafts. Owners of existing mines were given until the end of 1864 to provide a second shaft.

*References*

https://en.wikipedia.org/wiki/Hartley_Colliery_disaster

http://www.dmm.org.uk/archives/a_hart01.htm

http://www.dmm.org.uk/individ0/i02738.htm

http://www.mining-memorabilia.co.uk/hartleydisastermedal.htm

http://www.ndfhs.org/Articles/HartleyDis.html

### 376. Hemingfield (formerly Elsecar Low) Colliery

Hemingfield Colliery is located south of Barnsley in South Yorkshire (see also **371, 372**). Sinking of Elsecar Low Colliery by Benjamin Biram (1804-57), the steward of the fifth earl Fitzwilliam (1786-1857) of Wentworth Woodhouse (**371, 402**) began in 1840, but production did not start until 1848, because of problems with flooding. In the 1840s Biram was embarking on various technical innovations, including mechanical ventilation and new types of miners' lamps. Hemingfield was the site of Biram's experiments with ventilation where some of the earliest mechanical fans were installed. Biram also invented an anemometer (**391.2**) to measure airflow around collieries and check whether the ventilation was working. By 1850 production had reached some 454 tonnes per day. The colliery experienced problems with water and firedamp, and in 1852 the worst explosion killed 10 miners, although it was noted that the death toll would have been much higher had it not been for Biram's innovations. In 1856 he invented machinery for washing coal in screens buildings. Although he opposed trade unions, hundreds of miners voluntarily attended his funeral.

The mine closed in 1920, because the Barnsley Bed seam was exhausted. Both shafts were kept open to pump water, using first a Cornish beam engine and then, from at the latest 1934, electric pumps until into the 1990s. The headgear above the pumping shaft was rebuilt in concrete around 1934. Its distinctive main headgear had the wheels mounted one above the other (*cf.* **231, 277, 386**), which gave it the local name of the bicycle wheel pit. The headframe was demolished in 1940 and replaced with the surviving concrete structures. A small two-man riding cage, used to access the underground pumps at Elsecar Low Colliery, is now on display at the NCMME (**386**).

The pit was designated as low, because it was on the valley side and to distinguish it from the high and mid-Elsecar collieries. The colliery was known locally as both Elsecar Low Colliery and Hemingfield Pit. It is not known when the first name ceased to be used. After suffering significant neglect, theft and vandalism, the site was in danger of being lost when the owners, UK Coal, went into liquidation in 2013 and the pit passed into the hands of the receivers. The

**Figure 376.1. Headgear and associated buildings now (© Friends of Hemingfield Colliery).**

same year the Friends of Hemingfield Colliery was set up as a company limited by guarantee. The Friends managed to raise about £30,000 in grants and loans and were able to purchase the site in 2014.

The Friends are now restoring the site, with about 20 active volunteers. The immediate aim is to clear the land of the vegetation that has grown up over many years. The ultimate goal is the renovation of the original 1840s colliery buildings and the two headframes, and the conversion of the site into a museum and visitor centre, as well as a multipurpose community space. To this end the Friends are building up a small collection of, for example, bricks from various local Yorkshire collieries. At present the site attracts up to 200 visitors when it is opened to the public once a month. Work at the site is funded by donations, grants and loans from members of the Friends.

The site is about 5,000 square metres and includes the 1845 vertical-winding-engine house, its 1890 annexe, the electrical-equipment house, which has been through multiple rebuilds, and two small concrete headgears. A row of pit houses survives opposite the site, together with the Cornish pumping-engine house, which has been converted into a dwelling. In 2020 the site was scheduled and the vertical-winding-engine house was listed as grade II*. Alongside the site there is a canal basin with coal-drops where barges were loaded with coal, which was listed as grade II in 1987.

**Figure 376.2. Small two-man riding cage formerly used to access the underground pumps at Hemingfield Colliery (© National Coal Mining Museum for England).**

*References*


https://barnsleymuseums.art.blog/2020/08/06/fans-lamps-and-whirligigs/

https://hemingfieldcolliery.wordpress.com

http://www.aditnow.co.uk/Mines/Hemingfield-Colliery-Coal-Mine_14683/

personal communications, John Tanner (**371**), 2014, 2015, 2021; Steve Grudgings, chair, Friends of Hemingfield Colliery, 2015; Alf Evans, local historian, 2016

**377. Hopewell Colliery Museum**

The Hopewell Colliery Museum is located 2 kilometres to the east of Coleford in the Forest of Dean. The Old Furnace level here is a drift leading to the coal seam. It was built of dressed stone between 1810 and 1820 by David Mushet (1772-1847), who was both a colliery owner and ironmaster, and also a well-known metallurgist and geologist. The tunnel is elliptically vaulted and there are stone revetments to the banks of the cutting at the entrance. The tunnel originally had a plateway through its centre, with a drainage channel next to it, but this has been removed and the floor covered with gravel. It was listed as grade II in 1998 as a rare surviving example in the Forest.

This is one of only seven Free Mines still to exist, although this is an increase of five on five years ago, and there are currently plans for additional Free Mines in order to maintain mining rights in the Forest of Dean (**367**). The topside of the mine is run as a commercial colliery during the winter and the owner also works Phoenix Mine, dating back to 1821, on the opposite side of the road. The worked-out part of the mine is open as a tourist attraction during the summer. It first opened to the public in 1997 and is self-funding; none of the other Free Mines are open to the public. There are two staff plus two seasonal workers.

Visitors, wearing cap lamps, are taken down the one-in-four New Road adit, 90 metres long, with parallel rails on the floor to allow tubs to be brought down when necessary. At the bottom is a display of an already-worked sloping seam, 46 centimetres high, presented as it would have been when the pit was working 200 years ago. There are 300 metres of roadway lined with dry-stone walling. The tour takes 45 minutes and visitors return to the surface by the Old Furnace roadway, dating to 1820, and along a short forest path, which follows the course of an early tramway. The total route is about 530 metres long and the colliery covers about 1.5 hectares.

The winch for the mine is homemade, a converted hand winch with a petrol engine. The mine also has tippling gear and a screens to sort the coal from the spoil. On the surface there are an underground dram, hand tools and jigger picks on display. There is also a small display in the entrance building of mining equipment, photographs and tools.

*References*

https://historicengland.org.uk/listing/the-list/list-entry/1376766

http://www.hopewellcolliery.com/

personal communications, Richard Daniels, manager, 2016, 2021

showcaves

**Figure 377.1. Entrance to the underground of Hopewell Colliery Museum (© National Coal Mining Museum for England).**

**Figure 377.2. Underground of Hopewell Colliery Museum (© Hopewell Colliery Museum).**

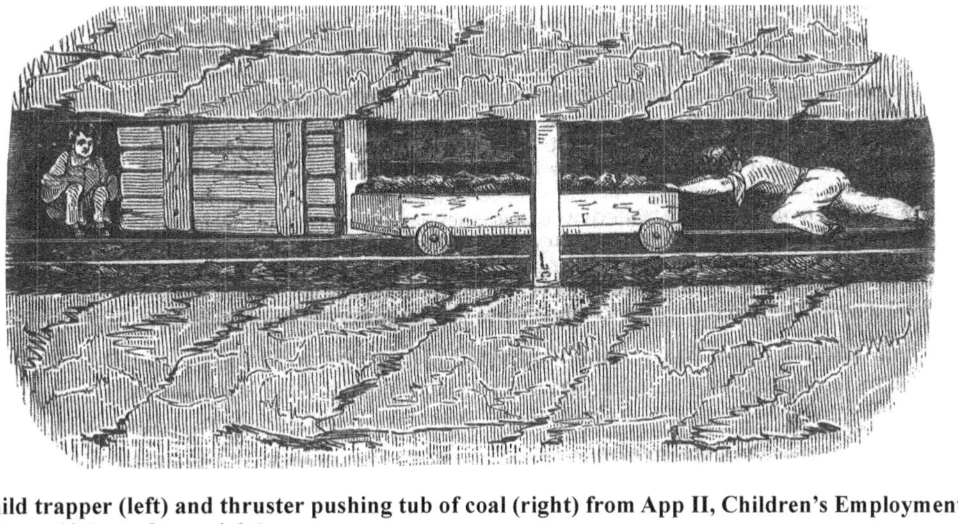

**Figure 378.1. Child trapper (left) and thruster pushing tub of coal (right) from App II, Children's Employment Commission report, p. 166, Figure 10 (out of copyright).**

**Figure 378.2. Child working as hurrier, from the Children's Employment Commission report, p. 78, Figure 2 (out of copyright).**

### 378. Huskar Pit Disaster Memorials

Huskar Pit was located at Silkstone, a village between Barnsley and Penistone in the southern part of Yorkshire. On 4 July 1838 heavy rain and extremely large hailstones fell for two hours in the afternoon. An observer returning after the weather had cleared said that the whole area was so white with hailstones that he thought that it must have snowed, despite being the middle of summer. The storm put out the fire of the boiler that powered the winding engine of the main vertical shaft. The workers stranded underground were instructed to remain where they were until the boiler could be restarted. A group of 40 children had, however, already worked a nine-hour shift and were so tired that they did not want to wait. They knew that there was a second route out of the pit up an adit, known locally as a dayhole, and decided to walk up that drift to the surface.

The children were not aware that the hailstones had filled a small stream near the pit, causing it to overflow into the drift. The water was held back by a pair of air doors that opened inwards. The children went through the lower pair of air doors, which opened outwards. When they opened the higher pair of air doors, the water being held back by the doors swept them down the adit and trapped them against the lower pair of doors. Huskar Pit was connected to the nearby Moorend Colliery for ventilation by a roadway running off the drift between the two sets of air doors. Some of the older children managed to escape along this roadway to Moorend Colliery, where they raised the alarm, but the onrush of water drowned 26 children between the ages of seven and 17. One was aged seven, seven were eight, four were nine and four were 10.

The bodies of the children were recovered and taken on carts to their respective homes. It was said that every home in the village had lost at least one child and three had lost two children. An inquest jury sitting at the Red Lion Inn, Silkstone, returned a verdict of accidental death. The children were buried together three days later in the churchyard of All Saints church in Silkstone, the 15 boys in four graves and the 11 girls in three graves placed at the feet of those of the boys. One girl, aged 11, was buried at the foot of the grave of her brother, aged 16. A memorial was erected at the head of the communal grave with the names and ages of the 26 children and a long inscription.

In 1988 150 years later, a service was held in the churchyard and the local community erected a second memorial in Nabs Wood, Silkstone Common, close to where the drift came to the surface. This shows a boy and a girl crawling on their hands and knees, as they would have done pulling tubs of coal. Local children, whose ages and genders matched those of the victims, also planted 26 hazel trees in the wood, one for each child killed.

In 2010 a stained-glass window, designed by local artist, Julie Tyler, and created by a stained-glass artist, Rachel Poole, was placed in the north-eastern chapel of the church. In the forefront of the window are blue ribbons and blue drops, representing the stream that overflowed. Behind this are the brown, gold and green fields and hills of the area around Silkstone. Suspended above the landscape are 26 four-pointed stars, representing the souls of the victims. The stars are of varying sizes, indicating the different ages of the children. Three panels at the bottom of the window list the names of all 26 children.

The report in the *Times*, the most influential newspaper of the time, of the deaths of so many children shocked the nation, which had not realised that such young children were working such long hours underground and in such dangerous and onerous conditions. Queen Victoria (1819-1901), who was aged only 19 herself and had come to the throne a year earlier, made it known publicly that she had sent condolences to the bereaved families.

The seventh earl of Shaftesbury (1801-85) and his wife were devout Christians. He had been so horrified by the working conditions in the northern textile factories that he and his wife had agreed that he had a moral obligation to sponsor a bill in the House of Commons to attempt to ameliorate conditions. Shaftesbury had intended to return to normal politics after the 1833 Factory Act had been passed, which would not have been possible if he continued promoting social legislation. Once he was aware of the even worse conditions in the underground mines, however, he felt obliged to forego his political career, although he had just been offered a lucrative position in Ireland and the family was not financially secure. In 1840 the Children's Employment Commission (Mines) was set up, with Shaftesbury as its chair. Its work included taking evidence directly from women and children who actually worked in the pits, which was quoted verbatim in the commission's report, and the members went underground to see conditions for themselves. The sub-commissioners who gathered evidence were horrified by what they found. This was the first report of a royal commission whose text was accompanied by drawings, showing how women and children laboured underground for up to 12 hours a day. The mine owners fought back, claiming that the cramped conditions underground in thin-seam mines meant that workers had to become accustomed to them by working as children, and denied that many children started work before they were nine years old.

The report was published in 1842. It revealed that children as young as five (and in one case only three) regularly worked as trappers, opening and shutting ventilation doors to allow tubs of coal to pass through and to direct air around the underground. The children were allowed a candle for light on their first day, after which they had to work in the complete darkness, giving rise to the saying that someone is 'not worth a candle'. The report stated that 'this occupation is one of the most pitiable in a coal-pit'. After a few years as a trapper, the children went on to working as hurriers, moving tubs of coal weighing up to 2.5 hundredweight (127 kilograms) between 4 and 6

**Figure 378.3. Lancashire pit-brow lasses with their equipment (photograph taken in 1887 by Herbert Wragg; image courtesy of National Coal Mining Museum for England).**

miles (6.4 and 9.7 kilometres) a day. The report illustrated how women, who were naked to the waist, pulled tubs full of coal on their hands and knees by chains passing through their legs and attached to a belt around their waist (*cf.* **253**), even when pregnant (Orwell 1937, p. 30). The report stated that this made 'girls unsuitable for marriage and unfit to be mothers' and justified the inclusion of a drawing of such a woman, which could be considered obscene, claiming that:

> This illustration of the circumstances of this degrading labour is so much more forcible than any verbal description, that we must claim permission to subjoin it.

As well as the working conditions, the immorality was stressed of completely naked male miners working alongside half-naked women. The report stated that:

> I have often been shocked in contemplating the hideous, and anything but human, appearance of these men, who are generally found in a state of bestial nakedness . . . black and filthy are they are in their low, dark, heated and dismal chambers, they look like a race fallen from the common stock.

Although a number of the members of parliament in the House of Commons and peers in the House of Lords were

**Figure 378.4. Fountain in Piccadilly Circus, popularly known as Eros, commemorating work of the Earl of Shaftesbury (© Clare de Sainte Croix).**

themselves mine owners, the effect of the publicity about the Huskar Pit disaster, the publicly expressed sympathy of the queen for the Huskar families, the immoral situation of naked and nearly naked men and women working side by side, and the appalling working conditions underground outlined in the report meant that there was sufficient impetus to get the bill passed.

Shaftesbury had to accept that he could not get everything that he wanted. He had to choose between getting all female workers or all children out of the mines. He

> expressed his poignant regret that . . . rather than lose the Bill . . . he was obliged to sacrifice the children to save the women.

One aspect in the background, but never explicitly stated, was that of sexual assault and rape (**386**), and this was probably why Shaftesbury prioritised protection of women and girls over that of the boys. The miners did not work in family groups, unlike in Japan (**250, 253, 273, 277**) or Taiwan (**344**), which might have given some protection to the women.

The Mines and Collieries Act passed on 4 August 1842 made it illegal from 1 March 1843 to employ any women or girls in work underground, as well as any boys younger than the age of 10. If the Act had been in force in 1838, it would have saved the lives of 12 of the children killed at the Huskar Pit. Finally in 1872 the Mines Regulation Act was passed forbidding the full-time employment of boys under the age of 12 and collieries were obliged to provide half a day's schooling in every working day for their child labour force, making it no longer economical to employ children. The regulations forbidding women working in any job that involved spending a substantial amount of time underground, which especially affected women engineers working on tunnels, or of being on the surface of a colliery between midnight on a Saturday and 6 a.m. on a Monday were not repealed until the 1989 Employment Act repealed sections of the 1842 Coals Mines Act and the 1954 Mines and the Quarries Act, which also prohibited this type of work for women; when the author was appointed as director of the-then Yorkshire Mining Museum (**386**) in 1986, this regulation was still in place and her appointment was technically illegal, but 'overlooked' by the Mines Inspectorate.

After 1842 women continued to work on the surface of mines in the screens buildings, sorting the coal from the dirt, especially in Lancashire in north-western England. They were known as pit-brow lasses and a number of images still survive of them in the trousers that they wore. There were fewer than 1,000 women working in the screens by 1953, when an agreement was reached between the NCB (the employers) and the National Union of Mineworkers (the trade union) that elderly or injured male miners should be given priority for surface work. The last pit-brow lass in the Wigan coalfield of Lancashire finished work at Golborne Colliery in 1966 and the very last woman colliery-surface worker in the UK retired in 1972 from Whitehaven (**374**).

Shaftesbury never achieved government office, but was revered for the work he had done, becoming known as the 'poor man's earl'. When he died in 1885, aged 84, the streets of London were lined with ordinary working people. They were all wearing black or something black, such as a ribbon, in order to pay tribute to his endeavours on their behalf as his coffin passed by. Shaftesbury had been offered burial in Westminster Abbey, an honour accorded only to genuinely national figures, but chose instead to be buried in the church of Saint Giles-in-the-Fields. In 1893 a fountain, now a listed structure, was erected in Piccadilly Circus in London to commemorate Shaftesbury's philanthropic work. The inscription around the base says:

> Erected by public subscription to Anthony Ashley Cooper KC Seventh Earl of Shaftesbury. During a public life of half a century he devoted the influence of his station, the strong sympathies of his heart and the great powers of his mind to honouring God by serving his fellow men. An example to his order, a blessing to his people and a name to be by them ever gratefully remembered.

The fountain is surmounted by a naked winged figure, often confused with Eros, the Greek god of erotic love, but actually representing his brother, Anteros, the Greek god of selfless love, and subsequently renamed as the angel of Christian charity. Shaftesbury's biographer claimed that:

> No man has in fact ever done more to lessen the extent of human misery or to add to the sum total of human happiness (Battiscombe 1974, p. 334).

*References*

https://en.wikipedia.org/wiki/Mines_and_Collieries_Act_1842

https://en.wikipedia.org/wiki/Shaftesbury_Memorial_Fountain

http://www.dmm.org.uk/pitwork/html/daz.htm

http://www.healeyhero.co.uk/rescue/pits/Huskar/Huskar1.htm

personal communication, Jill Clapham, librarian, NCMME, 2017

### 379. Ironbridge Gorge Museum

As the name implies, the 1781 iron bridge at Ironbridge was the first arched bridge in the world to be made of cast iron. The bridge crosses the river Severn 1.6 kilometres from Coalbrookdale in western Shropshire in the West Midlands.

The Coalbrookdale area was already an important industrial region, because of its plentiful coal, as once again this name implies, as well as iron ore and easily accessible water transport. In 1622 a technique developed in Dudley, West Midlands, by Dud Dudley (1600-84), was patented for a method of smelting iron ore with coke made from coal, instead of with charcoal produced from wood. The iron manufactured by this method was not, however, acceptable to the ironmasters, who continued to use charcoal. In 1709 Abraham Darby (1678-1717), the great-grandson of Dudley's sister, developed a commercially viable way of smelting iron with coke (Goodman 2020, pp. 131-5). Darby was working at Coalbrookdale, 36.5 kilometres north-west of Dudley. This was just in time, because in the 1720s Daniel Defoe in his *A tour thro' the whole island of Great Britain* had recorded the depletion of the woodland in the early years of the Industrial Revolution.

As well as the bridge, the first iron railway locomotive, wheels and tracks were made at Coalbrookdale. The plentiful supply of coal in the UK meant that the ironmasters no longer had to rely on the limited amounts of timber available and were freed to produce the large amounts of cast iron demanded by the machines of the Industrial Revolution (mid-eighteenth to mid-nineteenth centuries). In 1986 Ironbridge was designated by UNESCO as a World Heritage Site, covering approximately 6 square kilometres.

Coal from the Coalbrookdale coalfield, which is 11 kilometres from north to south and 5 kilometres from

**Figure 379.1. Bridge made of cast iron from which Ironbridge takes its name (© Ironbridge Gorge Museum Trust Ltd).**

**Figure 379.2. Pit headgear and outbuildings at Blists Hill, a typical Shropshire coal-mining town (© Ironbridge Gorge Museum Trust Ltd).**

east to west, was used in the Roman period. The longwall system of mining is believed to have been developed here in the first quarter of the seventeenth century, and in 1635 Benthall Mine was able to produce 27,000 tonnes of coal. The coalfield has 21 named seams and peak production was reached in 1871 with 907,000 tonnes.

The Ironbridge Gorge Museum Trust was set up as a registered charity in 1967, with the aims of education and heritage conservation. This included the Ironbridge Institute as a centre for industrial-archaeology research and teaching, a movement that spread to America in the 1970s, then to Europe. The International Committee for the Conservation of the Industrial Heritage (TICCIH) was founded in 1973 after the first TICCIH conference (see also **228**), had been held in Ironbridge.

The Ironbridge Gorge Museum is one of the largest independent museums in the world, being responsible for 36 scheduled ancient monuments and listed buildings within the Ironbridge Gorge World Heritage Site. The trust operates 10 museums that collectively tell the story of the birthplace of the Industrial Revolution. There are around 170 members of staff and the museum has more than 350 volunteers. The museum has extensive collections, much of which are on permanent display.

One of the museums, Blists Hill Victorian Town, is a 23-hectare re-creation of a typical Shropshire Victorian coal-mining town. The first shaft was sunk at Blists Hill in the 1780s, where the mine worked five different seams and operated until 1916. The present headgear is thought to be on the site of the 1780s shaft. As it was so essential to so many industries, the theme of coal runs through many of the other museums and exhibitions; for example, the former Coalport China Works coal-fired bottle ovens are now used to house china displays at the Coalfport China Museum. The Coalbrookdale Museum of Iron displays the remains of the water-powered blast furnace used by Abraham Darby to perfect the commercial smelting of iron with coke. Demonstrators and skilled craftsmen throughout the 10 museums show how iron was worked, as well as how china and glass were produced.

*References*

https://en.wikipedia.org/wiki/Abraham_Darby_I

https://en.wikipedia.org/wiki/Dud_Dudley

http://www.ironbridge.org.uk/

personal communications, Matt Thompson, senior curator, 2012; Georgina Grant, senior curator, 2021

showcaves

### 380. Kent Mining Museum

The Kent Mining Museum will be situated in the 81-hectare Betteshanger Country Park, which is also the site of the further-education Hadlow College. A consortium of Betteshanger Sustainable Parks, Hadlow College and the Kent Mining Heritage Foundation were awarded £1.3 million by the Heritage Lottery Fund towards setting up the Kent Mining Museum. Hadlow College went into receivership in 2019 and the museum site was sold to a local housing developer, Quinn Estates, but work is continuing on the museum with the building 85 *per cent* completed. Archives for the museum are being collected by the Kent Mining Museum Foundation staffed by former Kent miners.

The Betteshanger Country Park visitor centre previously had display panels dealing with the history of coal mining in Kent and with Betteshanger Colliery in particular, now removed in preparation for the new museum. The museum will be housed within a dedicated new-build visitor centre in the park, which is located on the site of the spoil heap of the former Betteshanger Colliery, which closed in 1989. Having been provoked over pay and work rates, in 1942 the miners here went on strike, the only such strike during WWII. The colliery was the last to return to work at the end of the Miners' Strike (**404**). The museum will link to the pithead site through a mining-heritage trail. The only building surviving from the colliery is the former office block, Almond House, named after Geoff Almond, the last miner to have died underground at the mine. A statue, the 'Waiting miner', was sculptured by H.R. Phillips in 1966 and was originally sited at the entrance to Richborough power station. The statue has now been re-erected on the site of the former colliery, with an inscription on the side of the plinth listing all the men killed between 1896 and 1989, the period during which the Kent coalfield was working.

The aim is to open the museum sometime in 2023, with a team of 57 volunteers, a museum-and-heritage manager, a learning-and-engagement officer and a community curator. They will work with community-heritage centres, supporting them on collections management, collections care, engaging new audiences and developing changing exhibitions at the museum. Visitor figures are projected at 40,000, from amongst the 100,000 to 150,000 to the Betteshanger Country Park annually. The budget for developing the Betteshanger Sustainable Parks project is estimated at £40 million.

The museum will preserve the coal-mining heritage of Kent, with stories of the coal-mining communities formed by migration to Kent from other mining areas (**354, 396**), the impact of coal mining on Kent and the demise of the industry. The collection includes oral and written personal stories and memories, photographs, records and documents, and small and large-scale artefacts. The Kent mining collection is currently dispersed across two heritage centres (**354, 373**), several personal collections and two museums. These organisations will work together to preserve the collections and present them in the new museum and at the two heritage centres.

**Figure 380.1. Almond House in Betteshanger Country Park, the former offices of Betteshanger Colliery (© Colin Varrall).**

**Figure 380.2. 'Waiting Miner' statue on site of the former Betteshanger Colliery (© Colin Varrall).**

*References*

http://www.dovermuseum.co.uk/Exhibitions/Coal-Mining-in-Kent/History/Betteshanger-Colliery.aspx

personal communications, Sarah Brown, museum and heritage manager, Betteshanger Sustainable Parks, 2015; Gary Cox, chair, Kent Miners' Festival, 2015, 2016; Colin Varrall, volunteer heritage representative, Tilmanstone Colliery and Elvington Mining Community, 2021

### 381. Lancashire Mining Museum (formerly Astley Green Colliery Museum)

The Lancashire Mining Museum is located on the 6-hectare site of Astley Green Colliery, near Tyldesley in Lancashire, which opened in 1908 and closed in 1970. The shaft of the pit was pioneering engineering for its time, having been sunk through layers of unstable strata and sands to reach the coal seams. The original Lancashire Mining Museum opened in 1957 in an 1827 grade II listed building at Buile Hill in Salford. The museum was closed in 2000 by Salford Council on the grounds of financial savings, in the same year that the council opened the Lowry Centre at a cost of £106 million. Wigan Borough Council offered to take over the collection, but the offer was not accepted. The collection, of 1,611 items, was donated to the Museum of Science and Industry in Manchester in 2002; only three items are on display, the remainder being in storage. The

**Figure 381.1. Entrance to Lancashire Mining Museum showing headgear (© Lancashire Mining Museum).**

**Figure 381.2. 3,300 horsepower twin-tandem compound steam-winding engine (© Lancashire Mining Museum).**

Buile Hill building was to have housed the local-authority education department, but is now derelict.

Astley Green Colliery Museum opened in 1985. It is an accredited museum run by the Red Rose Steam Society Limited, a registered charity and a company limited by guarantee. There is no direct funding from outside sources, so the museum relies on donations, grants and site-usage fees. It has been known as the Lancashire Mining Museum since 2018. There are no paid staff, as the museum is run by about 20 volunteers under the direction of a management committee. The charity has a total of 120 members and the site attracts some 6,000 to 7,000 visitors every year.

The main exhibits are the colliery buildings, which are listed grade II. The capstan engine is housed between the legs of the latticed headgear. Built by John Wood and Sons in the 1950s for Agecroft Colliery in Salford, the engine was used for reroping the friction winder (**127**) and for shaft inspection. The drum of the capstan engine alone weighs 59 tonnes. The emergency winder displayed on the site was built by Walker Brothers of Wigan in 1940. It came from Boothstown Mines Rescue Station and is made up of a trailer housing a winding drum and an eight-cylinder Gardner diesel engine, which would be towed to collieries in the area when their main winding engine was out of commission.

The steam line includes a vertical boiler, vertical weir pump used to feed water to the header tanks above the boiler or directly to the boiler if required, and a single-cylinder horizontal engine. The engine was made by Marshall, Sons and Company of Gainsborough, Lincolnshire, and drives the circular saw used to cut wood for the boiler. The pit bank was originally under a large building housing the tub-track circuit, from which tubs that had been brought up the shaft went into the screens, then returned to the shaft. Some of the original track is still visible, as are a number of tubs and a cage from Bold Colliery.

The colliery's rescue room, with garage facilities below, now houses the museum's offices and workshops. The surrounding landscape is also important, as the Bridgewater Canal (opened in 1761) to the south provided the means of transporting coal by barge to a number of destinations, including Barton power station. The offices of the pit management were housed in the lodge, where there are now a number of smaller exhibits, including a reconstructed lamproom, replica underground-roadway display and a small winch engine from Agecroft Colliery. Due to the limited covered accommodation for exhibits on site, there is some equipment which is not on display at present.

Of especial note are the restored 3,300-horsepower twin-tandem compound steam-winding engine. The engine is 102 years old and could bring 7 tonnes of coal to the surface from 805 metres deep in two minutes at 93 kilometres per hour. It was the largest steam-winding engine ever used in the Lancashire coalfield and one of only 10 of this type that survive in the UK. The museum's lattice-steel headgear is a prominent feature, standing 30 metres high above the low-lying landscape and can be seen from many kilometres away. The winding engine, winding-engine house and headgear are the only ones surviving in Lancashire. The site is a scheduled ancient monument, including the top 10 metres of the innovatory shaft.

The charity is currently involved in two major projects: the conservation of the headgear of the number 1 shaft, at an estimated cost of £500,000; and the construction of a narrow-gauge railway. This surface railway will carry visitors in authentic mine rolling stock hauled by one of the museum's 28 colliery locomotives.

*References*

https://collection.sciencemuseumgroup.org.uk/search?q=Y2002.19

https://historicengland.org.uk/listing/the-list/list-entry/1017061

https://lancashireminingmuseum.org/2018/10/11/the-former-lancashire-mining-museum-buile-hill-salford/

personal communications, Geoffrey Jones, chair, Red Rose Steam Society Limited, 2014; Trevor Barton, chair, 2021; Chiara Fiaccavento, documentation and enquiries officer, Science and Industry Museum, 2021

showcaves

www.agcm.org.uk

### 382. Ledston Luck Colliery

Ledston Luck Colliery is located in the east of West Yorkshire. It was owned by the Micklefield Coal and Lime Company Limited and sunk in 1911. Peckfield Colliery, also owned by the Micklefield Company, was sunk in the 1870s. Peckfield seems to have been closely associated with Ledston Luck, so that the sinking dates appear to have become confused in some sources. The two winding-engine houses at Ledston were also built in 1911, and were among the first to use electric winders, at a time when most other mines were still using steam. The shafts were sunk 195 metres deep to work the Flockton seam and then, from the early 1980s, the Middleton Little seam was worked by retreat mining at a depth of 178 metres. In this method two roadways were driven on either side to the end of the section of coal to be worked and then the face worked backwards towards the shaft, with the area already worked and the associated roadways left to collapse, the collapsed area being known as the goaf. In advance mining the coal is removed close to the shaft and roadways driven at either end of the coalface as it advances, creating a coalface close to the shaft bottom, with the roadways running back to the shaft at either end. As the face advanced into the coal seam, the roadways had to be extended and roof supports set to keep them stable (Orwell 1937, pp. 27-8). Retreat mining is much more efficient and cost effective, once the roadways have been driven, provided that there is sufficient capital available to support the cost of the development before any coal begins to be cut and income generated. Peak production at Ledston appears to have been reached in 1982 with 520 tonnes of coal in that year.

Ledston Luck belonged to what are known as the 'eastside collieries' in Yorkshire. Westside pits, such as Caphouse Colliery (**386**), worked the relatively shallow seams close to where they outcropped in the western part of the coalfield. They were sunk mainly by local landowners who lived in the local area and were on much better terms with their workforce. The owners also had to allow relatively reasonable conditions and pay, because the workforce always had the option of going to work instead in the textile mills of western Yorkshire or as farm labourers. The eastside collieries worked the deeper seams of the eastern coalfield, which required a greater investment of capital and so tended to be financed by companies or entrepreneurs based away from the area, such as in London. They deliberately sited the pits as far away from other villages as possible, to remove alternative sources of employment. The miners and their families were housed in purpose-built villages owned by the mining companies, so they could evict the families if the miners were to go on strike, as indeed happened at Denaby Main Colliery in 1877, 1885 and, most famously, in 1903. The bitterness engendered lasted for generations and led to worse relations during the 1984/5 Miners' Strike (**404**) at the eastside than the westside collieries; for example, at the westside Denby Grange Colliery the management allowed a colliery building to be used for sorting donations of food for the families of striking miners. This bitterness was demonstrated at Ledston Luck. When the miners returned to work after the strike, it is claimed that the mine manager chained the gates shut and said that he would decide when they could return to work, not them.

The brochure for 1986 says that the colliery had been 'highly profitable', but that 'production from a single seam will continue until late 1986 when the remaining reserves will be exhausted and the men will transfer mainly to the Selby mines'. English Heritage listed the two winding-engine houses as grade II when the mine closed in 1987. The buildings are constructed of red brick with sandstone dressings and the roof concealed by a parapet. One is square in plan and the other rectangular, both built in the free-classical style.

The colliery buildings were developed into a small enterprise park under the auspices of Leeds City Council. Being close to the A1 motorway and just north of the town of Castleford, this should have been successful. A number

**Figure 382.1. Front elevation of winding-engine house number 1 (© National Coal Mining Museum for England).**

**Figure 382.2. Winding-engine houses numbers 1 (to left) and 2 (to right) (© National Coal Mining Museum for England).**

of the units are, however, now empty. The winding-engine houses, despite their listed status, are in an extremely bad state of disrepair, but surprisingly are not on the register of listed buildings at risk.

*References*

https://historicengland.org.uk/listing/the-list/results?q=Ledston+Luck+Colliery&searchtype=nhle

http://www.derelictplaces.co.uk/main/industrial-sites/13282-ledston-luck-colliery-castleford-oct-09-a.html#.VsSPT_mLTcs

http://www.28dayslater.co.uk/ledston-luck-colliery-castleford.t61908

personal communication, David Cross, library assistant, NCMME, 2016

### 383. Lofthouse Colliery Disaster Memorials

Lofthouse is located 6 kilometres north of Wakefield in West Yorkshire. The two shafts of Lofthouse Colliery were sunk in 1873/7. It was originally called Lofthouse Station Colliery, because it was so close to the 1858 Lofthouse and Outwood station on the Great Northern railway. By 1895 the mine was producing over 900 tonnes of coal a day. In 1945 a method of using equipment to handle minecars on the surface was invented at Lofthouse Colliery by Lodge, the colliery engineer, and Cordon, the mine manager, and named the Lofco system from a combination of the first part of their surnames. The system, which was then developed by Qualter Hall and Company Limited of Barnsley, Yorkshire, meant that only one man was required for coal-winding. Loaded cars were brought up on alternate winds and propelled by remotely controlled oil-hydraulic rams to the same side of the shaft. After emptying the minecar did not go through the tippler, but was returned to the cage in time to discharge the next loaded car onto the opposite side of the shaft. The system was adopted in collieries throughout the UK and was similar to a system devised independently in Czechia (**69**).

In the early hours of 21 March 1973 30 men were working on the night shift in the South 9B seam, at a depth of 229 metres. They were cutting a new face when they broke into an abandoned nineteenth-century flooded shaft. It is estimated that more than 13 million litres of water swept into the colliery. Although 23 men were able to outrun the flood, seven men were trapped. NCB surveyors worked on the tower of the church of Saint Paul in Alverthorpe, as it was located almost exactly above the flooded roadway. The frantic rescue operation lasted for six days, but managed to recover only one body. The bodies of the other six men are still entombed underground. The dead men left six widows and 22 children. A memorial service was held on 6 May 1973 at Wakefield Cathedral and memorial services have since been held regularly for the survivors and the relatives of the dead men. A memorial in the form of an obelisk with seven faces, each one representing one of the lost miners, was erected immediately above the spot where the six miners are still buried. The NCMME (**386**) has items relating to the disaster, such as a type N1834 Ericsson portable mining magneto telephone with generator used in the rescue attempts.

The NCB surveyors were aware of the existence of the shaft, but it had been believed that it was not as deep as the modern workings, unlike the records of the British Geological Survey, which showed that the shaft was as deep. The BGS records had not been checked by the NCB and the centralised records held by the NCB did not show workings abandoned before legislation in 1872 (**384**). At the enquiry into the disaster, Arthur Scargill (born 1938), then compensation agent for the Yorkshire National Union of Mineworkers and later president of the union (1982-2002), argued that the NCB should have been aware of these records. The report of the enquiry, published in September 1973, stated that ‘the disaster could have been foreseen’.

The most important result of the disaster was the Mines (Precautions Against Inrushes) Regulations, which came into force in April 1979. This required all owners and managers of mines to take precautions against inrushes of gas, water or any materials which flow or are likely to flow when wet. Owners and managers were also required to examine all records and sources of information such as archives, geological records, libraries and people with knowledge of the area and its history. As a result, NCB staff were sent to all archive repositories that had collections relating to coal mining to check records of possible other abandoned shafts and workings. The West Yorkshire Archaeology Unit had carried out a detailed study of all tithe-award and estate maps for West Yorkshire, recording all field names of archaeological significance. These included names relating to coal mining, such as ‘pit hill’ or ‘cinder field’, and many maps showing the actual location of coal mines (**383.2**). The unit offered to provide all this material to the NCB, but the offer was not accepted.

Lofthouse Colliery closed in 1981 when its reserves were worked out and the miners were transferred to work in the Selby coalfield. The site was landscaped in the late 1980s by Wakefield Council to form a country park, where there are now many bird species and mature deciduous woodland. The paths in the park are named after the various coal seams that were worked at Lofthouse Colliery. To commemorate the fortieth anniversary of the disaster, a new steel memorial, resembling a shaft of light, was erected in the park. This stands where the mine shaft used by the miners to descend into the pit was located.

*References*

https://en.wikipedia.org/wiki/Lofthouse_Colliery_disaster

http://www.bbc.co.uk/news/uk-england-leeds-21863448

### 384. Mining Heritage Centre

Under the 1850 Mines Act managers were obliged to keep a working plan of their colliery and to make it available to government inspectors, but it was not until 1872 that

**Figure 383.1. Steel memorial, resembling a shaft of light, in Lofthouse Colliery Country Park (© Margaret L. Faull).**

it became compulsory to deposit plans of workings when they were abandoned. In 1925 38 miners at the Montagu View Pit drowned when they extended the roadways underground by breaking through into Paradise Pit. The owners had not consulted the plans of the flooded mine, even though these were kept at a nearby colliery. The result was the establishment of a central repository for plans. This had already been proposed as early as 1796, after six men at East Denton Colliery were drowned when they broke into flooded workings of an old mine. William Thomas, the manager, suggested to a meeting of the Literary and Philosophical Society in Newcastle that plans of abandoned workings should be lodged with the Justices of the Peace. It was next suggested by John Buddle, the manager of Heaton Colliery, following the 1815 disaster there in which 75 miners were killed when they broke into flooded abandoned workings. Rescuers could not break through until eight months later and found that some of the dead miners had kept themselves alive for a time by eating one of their pit ponies (*cf.* **80**). The centralised records did not, however, show workings abandoned before the 1872 legislation, resulting in the 1973 Lofthouse Colliery disaster (**383**).

After nationalisation of the coal industry in 1947, records were at first kept regionally by the NCB at area headquarters. The British Coal Corporation (the 1987 successor company in title to the NCB) devoted two years to establishing a national mining-records office in 1992 at Bretby in the East Midlands. When the coal industry was privatised in 1994, the Coal Authority was set up to deal with matters of public interest, such as mining permits,

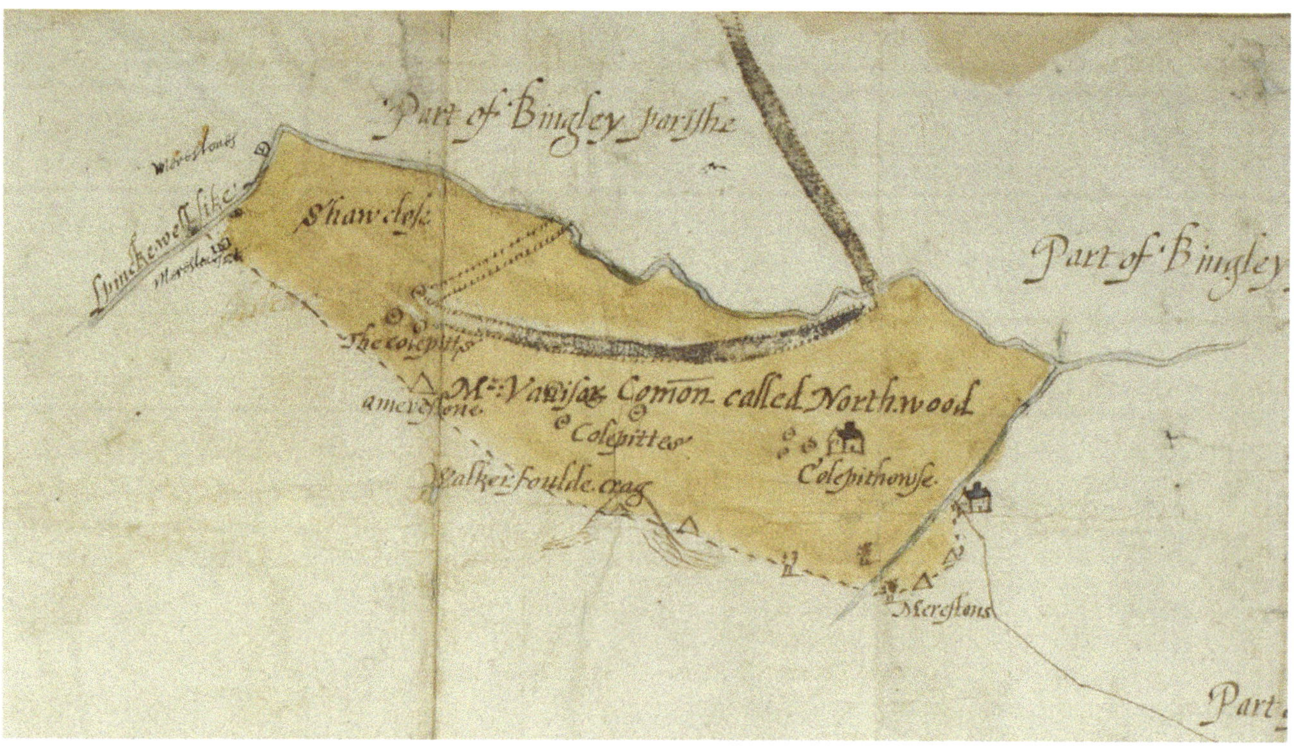


**Figure 383.2. Part of 1610 map of Baildon Common, Yorkshire, by Robert Saxton, marking seven coalpits and *Colepithowse* (reprd permission of Special Collections, Leeds University Library, YAS DD146-11-1-13).**

**Figure 384.1. Entrance to the Mining Heritage Centre (© The Coal Authority).**

**Figure 384.2. Rolled plans (© The Coal Authority).**

mining records and the effects of past mining operations; for example, subsidence or surface flooding from closed mines (of which it is estimated there are some 70,000). The records were moved in 2001 into a purpose-built store at the headquarters of the Coal Authority in Mansfield in Nottinghamshire in the East Midlands. The Mining Heritage Centre is managed and funded by the Coal Authority, has three members of staff and occupies 600 square metres.

Abandonment plans for all closed collieries pass to the Mines Inspectorate, who then transfer them to the Mining Heritage Centre. The centre houses more than 120,000 colliery-abandonment plans that have been deposited since the passing of the 1872 Coal Mines Regulation Act. These plans depict both the extent of coaling operations and the location of any associated entrances into coal mines (*cf.* the much wider focus of **243**). The national collection covers England, Scotland and Wales, but not Northern Ireland, where the mines were privately owned; responsibility for mineral rights, licensing and most abandoned mines there was vested, not in the Coal Authority. but in the Ministry of Commerce under the Mineral Development Act (Northern Ireland) 1969. The Geological Survey of Northern Ireland has an abandoned-mines database with information on over 2,000 shafts and adits and, in some cases, plans of underground workings.

The entire collection at Masfield has been scanned and visitors to the centre consult the electronic images to ensure preservation of the original documents. Since 1992 the numbers using the centre have reduced drastically; for a fee the Coal Authority undertakes searches and supplies copies to meet requests for information, thus preventing unnecessary travel. Current visitor numbers are about 1,000 each year.

In addition to the principal collection of mine-abandonment plans, the Mining Heritage Centre also holds a number of other substantial coal-related documents. There are 3,800 boxes of the surveyors' notebooks that were used underground to record observations and the measurements from which the plans were prepared. The coal-holdings register and associated claim files relate to the nationalisation of the coal industry on 1 January 1947. Under the Coal (Registration of Ownership) Act 1937, anyone owning coal in England, Scotland or Wales had to register their interest, for which they were subsequently compensated when the mines were nationalised. There are 26,500 files dealing with the minerals owned, the rights associated with that ownership and the compensation payable. There are also some 47,000 electronic copies of the photographic collection built up by the NCB/British Coal, which are now available on-line. The originals of these photographs are now housed in the National Archives in London.

Unfortunately the records do not include the detailed papers of individual mines, only plans relating to their abandonment. Papers relating to mines that had closed before 1947 were usually kept with the records of the families or companies that owned them and were deposited in local archive repositories along with their other papers. Following nationalisation coal-mining records came under the Public Records Acts, first passed in 1838. This meant that all records had to be deposited with the Public Records Office (PRO), now the National Archives, or else be destroyed. The PRO generally wished to keep only strategic records, which effectively meant documents of the national body and its regional offices, not those of individual collieries. Unless it was agreed that these records could be deposited with one of the designated regional depositories (for example, Leeds and Sheffield in Yorkshire), by law they had to be destroyed. The NCB was often criticised for throwing historical records down the shaft when a colliery closed, but the NCB had no choice in the matter as legally these documents had to be destroyed. On one occasion a by-passer rescued large-scale hand-drawn nineteenth-century plans from a skip in Castleford, West Yorkshire.

*References*

https://www.nidirect.gov.uk/articles/abandoned-mines

https://underthefieldsofheaton.com/an-1815-diary/

https://www2.groundstability.com/

personal communications, Dave Clarke, mining information manager, 2016; Helen Simpson, records manager, 2021, 2022

www.gov.uk/government/organisations/the-coal-authority

### 385. Museum of Cannock Chase

The Museum of Cannock Chase, which opened in 1989, is located in Hednesford in Staffordshire in the West Midlands. The museum deals with the wildlife and the domestic, industrial, local and social history of Cannock Chase, a former royal Forest, since 1958 designated as an area of outstanding natural beauty. The museum and collection are owned by Cannock Chase Council.

The museum has one temporary-exhibition gallery and five permanent galleries giving a detailed overview of the history of the Chase, especially the industrial origins of the area. It attracts about 35,000 visitors a year and has 22 members of staff, with five volunteers. The importance of coal mining here is emphasised by the winding wheel from Lea Hall Colliery, which opened in 1960. This was the first modern colliery sunk by the NCB, with its coal taken directly by conveyor belt to power Rugeley A power station.

In 2014 the museum opened a gallery specifically dedicated to mining in the Cannock Chase coalfield. The gallery cost £150,000 and allows visitors to experience a day in the life of a miner, including providing a crawlway to demonstrate the cramped spaces in which miners had to work (*cf.* **295**). Other aspects dealt with are the measures taken to ensure mine safety, such as mines-rescue awards and proto-breathing apparatus, as well as

Figure 385.1. Exterior of the Museum of Cannock Chase, with winding wheel at entrance (© Museum of Cannock Chase).

Figure 385.2. Model of headgear (© Museum of Cannock Chase).

first-aid competitions and training. Models and objects are used to demonstrate certain aspects, such as a pit pony with its full underground protective equipment, a miner dressed in his working clothes and a typical coal-mining headgear with winding wheels. The role of the National Union of Mineworkers is illustrated; for example, with a banner from the Midlands Area of the union. Banners were usually made of silk which, as a protein, deteriorates over time, especially with exposure to light, so this banner is displayed for only part of the year.

*References*

http://www.wlct.org/cannock/museum/

personal communications, Susan Dalloe, museum services manager, 2015; Yvonne Cooper, collections officer, 2021

### 386. National Coal Mining Museum for England (formerly Yorkshire Mining Museum)

The NCMME originally opened in 1988 as the Yorkshire Mining Museum and achieved national status in 1995. It is located at Overton, in West Yorkshire. The museum was originally set up and subsidised by the local authorities of the area, to commemorate the mining heritage of Yorkshire, at a time when nobody expected the closure of the mines. Indeed when there was local pressure to provide a tourist attraction by preserving and displaying the extensive excavated remains of the Roman fort of *Lagentium* at Castleford, 4.5 kilometres south of Ledston **(382)**, this was rejected by the leader of Wakefield Council (and chair of the NCMME), with the comment, "there'll always be pits in Castleford".

Since 1998 the museum has been funded by the government, with the funds being channelled through the Science Museum Group. It is a registered charity and company limited by guarantee with a board of trustees. The museum employs about 100 staff and has a varying number of volunteers. It attracts about 120,000 visitors a year. The museum is responsible for preserving the history of coal mining in England, where the last deep mine, Kellingley Colliery, closed in December 2015. Its in-house publication series has produced 16 books about coal mining in England.

Licences for the working of coal here were let in the 1780s. The shaft now in use to take visitors underground

**Figure 386.1. Caphouse Colliery, from left, red headgear, blue screens, conveyor belt, winding-engine house, green store and display buildings, store, chimney (© National Coal Mining Museum for England).**

**Figure 386.2. Hope Pit, from left, compressor house, engine house, headgear, heapstead, auxiliary fan house (front of headgear), powerhouse (© National Coal Mining Museum for England).**

**Figure 386.3. Nineteenth-century wooden horse-drawn waggon from the Little Eaton tramway, Derbyshire (© National Coal Mining Museum for England).**

**Figure 386.4. Late-nineteenth-century woven hazel corfe for moving coal underground from William Pit, West Cumberland (© National Coal Mining Museum for England).**

**Figure 386.5. Jib and gear box of AB Meco Moore cutter-loader shearer, believed to be the only surviving remains of a Meco Moore in the UK (© National Coal Mining Museum for England).**

is shown on a map of 1791, so it appears to be the oldest still in use, certainly in the UK. The museum occupies about 17 hectares of land, comprising the surface buildings of two mines, Caphouse Colliery and Hope Pit, together with land to the north on which there are the remains of Green Lane Pit. Here there is a wooden headgear and a horse-gin (**372.1-3**) above a small shaft down to an underground-drainage roadway. These and the associated spoil heap are scheduled. Also to the north is a water-treatment plant consisting of seven large settling tanks. About 4.5 million litres of water are pumped daily into the tanks to keep the underground of the museum dry and to prevent water coming out elsewhere in the area and causing local flooding by water from all the surrounding closed collieries. The water contains orange iron oxide and, after going through the tanks, is finally cleaned completely by an extensive reed-bed system. While the colliery was operating, the local farmer was dependent on the pumped water for his cattle and complained when pumping stopped and he believed that his water had been cut off. Earlier it had been pointed out to him that the iron in the water, about which he had complained, was actually beneficial to his cattle. Work is underway on trying to use the warmth in the water to provide heating on the surface (Faull 2011, *cf.* **113**).

At the eastern end of the site there was a miner's cottage, whose foundation were excavated in 2008, revealing two rooms (Preece, Roberts and Rose 2009). In the 1881 census it was occupied by Patricia Earnshaw and her family; the Earnshaw family still lives in the area and when one of her descendants visited the museum, he disclosed that she had worked underground before 1842, where she was raped several times.

The buildings on the Caphouse site include the original 1876 stone winding-engine house, containing the steam winder, and the heapstead building supporting the headgear. The winding wheels being one above the other (*cf.* **231, 277, 376**) indicates that when the winding-engine house was rebuilt in 1876 it was placed at a right angle to the previous building (*cf.* **231**). This was to allow production to continue using the previous winding engine during construction of the new house. When the new winding-engine house was commissioned and the old one demolished, the wheels had to continue to have this unusual arrangement. Also preserved on the site are the colliery canteen, control room, lamproom, medical room, offices, pithead baths and workshop. The equipment in the screens building is all able to operate and was used in the 1989 film of the novel 'The rainbow', by D.H. Lawrence (1885-1930); its distinctive blue colour was simply because that was the cheapest paint available when it needed painting when the pit was operating. The main shaft is quite narrow at 3.4 metres in width, so only two very small single-decker cages holding between either eight to 10 men or two small tubs holding 180 kilograms could be accommodated. So in 1973/4 an adit was driven 640 metres to the underground to allow a conveyor belt to take the output from underground to the screens building.

At Hope Pit there are a number of buildings, including the headgear with wheels more conventionally next to each other, compressor house, fan drift and pumping shaft, 206 metres deep and 3.7 metres wide. The Hope Pit colliery workshops have the line shafting and belt drives transferred from the nearby Park Mill Colliery workshops at Clayton West, where they continued to be used until it closed in 1980. The Royal Commission on the Historical Monuments of England considered that Caphouse and Hope Pit together form the only surviving intact nineteenth-century-colliery complex in the UK. The central core of buildings at Caphouse are listed as grade II*, while Hope Pit is a conservation area.

New visitor facilities have been erected on the Caphouse site, designed as they would have been had it still been a working colliery, but Hope Pit has been preserved as it was in the 1950s without any new structures on the site. It is the only colliery site in the UK that has not had any changes since it ceased to function as a mining site. So land was purchased outside the Hope Pit curtilage for erecting the museum's open-access collections-storage building in 2002. The 1995 collections-storage building on the Caphouse Colliery site contains the British Coal Collection, which was originally at Chatterley Whitfield (**363**).

The Caphouse and Hope Pit complex is an example of two collieries that were geographically separate from each other, but joined underground, with the upcast shaft at one and the downcast shaft at the other (as also **388**). The Caphouse site also has a nineteenth-century furnace shaft, 129 metres deep and only 3.4 metres wide (*cf.* **284**). Visitors go underground with former miners in groups of up to 19 people to the New Hards seam at a depth of 134 metres. There are 900 metres of roadway underground, 500 metres of which are open to the public. As this is a genuine underground and mining legislation applies to anywhere where coal 'is or has been extracted', the roadways still come under mining legislation. The underground interprets coal mining from the early-nineteenth century up to the modern day and includes an underground classroom. A mobile emergency winder is kept on the surface in case of failure of the winding engine while the cage is in the shaft,

Large machines are displayed outside. Various buildings detail the history of the two collieries and of coal mining in England. The museum has about 47,000 objects in its collections. These include 39 union banners, about 800 drawings, paintings and prints and around 100 three-dimensional artworks, such as coal or wood carvings and other sculptures, and something like 190,000 photographs, including glass slides, negatives, *etc.* The library of 18,728 individual titles has been judged by the British Library to be of national importance in its own right.

The collections include a number of unique items, including the kibble used in the rescue at Barnsley Main Colliery (**355**); a late-nineteenth-century woven hazel corfe from William Pit, West Cumberland (Salway 1999, 10.1.17, pls XX, XXI) for moving coal underground; a nineteenth-century wooden horse-drawn waggon from Derbyshire (*ibid.*, 10.8.3, pl. XXXI); and memorial glass for the Hartley Colliery disaster (**375**). More modern items include a British Jeffrey Diamond coal-cutting machine dating to 1900/10 (*ibid.*, 8.3.8, pl. XIV; *cf.* **8**) and the jib and gear box of a mid-twentieth-century AB Meco Moore cutter-loader shearer (*ibid.*, 8.3.42-

4), believed to be the only surviving remains of a Meco Moore in the UK. The museum also holds works by some important artists, such as some of the Spennymoor Settlement Painters, often known as the Pitmen Painters (**404**), including Tom McGuinness (1926-2006) and Norman Cornish (1919-2014), drawings of miners produced by Andrew H. Freeth (1912-86) for publication in *Coal News*, sculptures by Betty Miller (born 1926) based on her childhood in the mining village of Royston in South Yorkshire, and the complete collections of several photographers, including Martyn Pitt, Albert Walker and Harold White (1902-83).

*References*

ERIH, *s.n.* Wakefield

https://historicengland.org.uk/listing/the-list/list-entry/1013407

http://www.ncm.org.uk

personal communications, Andrew Smith, deputy director and mine manager, 2017; Jill Clapham, librarian 2017, 2021; Mark Carlyle, curator of industry, 2017

showcaves

### 387. National Mining Museum Scotland (formerly Scottish Mining Museum)

Lady Victoria Colliery was located at Newtongrange, 15 kilometres south of Edinburgh, in Midlothian in southern Scotland. From its very opening in the 1890s, Lady Victoria was the leading mine of the Scottish coalfields. The pit had a steam-winding engine of 2,400 horsepower, the largest ever installed in Scotland. It worked from 1894 until 1981 and is still operational, although converted from steam to electricity. The surface buildings at the colliery were located on both sides of the main road. The two parts of the site are joined by a covered gantry across the road, 7.3 metres above the ground. The colliery closed in 1981 and, with the shafts filled, the underground became inaccessible. The Scottish Mining Museum opened in 1984 and underwent a major renovation and redevelopment in 1999. In 2009 the Scottish Government funded a major programme of conservation of the brick-vaulted undercroft, elevator shed, picking tables and tippler floor. In 2012 its name was changed to National Mining Museum Scotland, echoing the names of the English and Welsh national museums (**386, 389**).

The museum is an independent registered charity and company limited by guarantee, governed by a board of trustees. The museum raises funds from various sources including admission charges, commercial activities and grant aid from the Scottish Government. There are over 20 members of staff and more than 20 volunteers who work with the curatorial and facilities teams. The museum is responsible for preserving the history of coal mining in Scotland, where the last deep mine, Longannet, closed in 2002.

The museum occupies a 5-hectare site, which has statutory protection. Its historical buildings encompass all the structures that would be found at a typical colliery: chimney, engine house, power station and heapstead building with tub circuit, tipplers and picking tables (similar to **99.3**), all built between 1890 and 1894. A coal-washery and hopper were added sometime between 1906 and 1914; the boiler house and power station were extended about 1924; the picking tables were extended in the 1930s; and the gantry to provide access to the baths was added about 1954. Most of the colliery plant, with various modernisations, is still *in situ*. The structures are steel-framed with brick cladding and roofs covered with sheet metal.

The museum includes a three-storey visitor centre, with exhibitions with interactive displays, reconstructions, sound and visuals, dealing with the history and technology of coal mining and the story of Scottish mining communities. There is a replica underground roadway and coalface with machinery. Altogether there are some 3,800 square metres of formal displays, including a short account of serfs (and their children who followed them in serfdom) in the Scottish mines for nearly 200 years from 1606 (Swinbank 2011).

The collections at the museum comprise over 60,000 items, including more than 10,000 examples of mining tools and equipment, plus art, clothing, geological specimens, lamps, models, safety equipment and trophies. There is large-scale equipment, such as cages, coal cutters, locomotives and winding engines. Highlights of the collections include fossilised trees 400 million years old and a large collection of trade-union banners from across the Scottish coalfields.

The library houses some 40,000 maps and plans, 20,000 books, journals, and trade catalogues and periodicals covering the history of the Scottish coal-mining industry. The 35,000 photographs include aerial views of collieries, gala days, miners at work and recreation, surface images, and underground and village scenes. The archives hold many original documents relating to Scottish collieries, including those of the Lothian Coal Company, which operated Lady Victoria Colliery from 1890 until nationalisation in 1946.

*References*

ERIH, *s.n.* Newtongrange

https://magazine.cim.org/en/in-search/liberating-scotlands-slaves-of-the-soil/

https://nationalminingmuseum.com

https://portal.historicenvironment.scot/designation/LB14604

http://www.scottishmining.co.uk/429.html

personal communications, Rowan Brown, director, 2012; Nicola Moss, curator, 2021

showcaves

**Figure 387.1. Entrance to the museum (© Trustees of National Mining Museum Scotland).**

**Figure 387.2. Winding engine built in 1894 by Grant Ritchie & Co. and in operation until 1981, now operated electrically (© Trustees of National Mining Museum Scotland).**

**Figure 387.3. Headgear and buildings of the museum (© Trustees of National Mining Museum Scotland).**

**Figure 387.4. Interior of heapstead building, with tub circuit, tipplers and entrance to cage (© Trustees of National Mining Museum Scotland).**

**388.** ***National Museum Wales: Big Pit National Coal Museum (formerly Big Pit Mining Museum)***

Big Pit in Blaenafon in south-eastern Wales was an amalgamation of several separate mines. Mining started on the site about 1812 with the Forge level, which supplied both coal and ironstone to the Blaenafon ironworks. The shaft of Kearsley's Pit was sunk in 1860, apparently mainly for ventilation. In 1880 the diameter and depth of the shaft were increased and it was renamed Big Pit. It was the main coal-winding shaft until 1973, when a drift was driven to the Garw seam and the colliery was renamed Blaenavon New Mine. At its peak the colliery employed about 1,200 men and produced some 22,680 tonnes of coal a year. The last face finished in 1979 and the colliery closed in 1980 owing to lack of workable reserves. It was by then the oldest working colliery in Wales.

Although some of the surrounding area has been landscaped, most of the colliery buildings have been preserved. The surface structures date from between about 1900 and the 1970s, with the surface plant having been modernised in 1939. The buildings were listed in 1995 as grade 2 or grade 2* as the most functionally complete colliery complex remaining in Wales. Eighteenth-century ironworks are related to the coal mines. In 2000 UNESCO designated the chapels, church, colliery, ironmasters' houses, ironworks, schools and workmen's institute as a World Heritage Site.

The colliery site reopened to visitors in 1983 as the Big Pit Mining Museum, to interpret the history of the coal-mining industry in Wales, where the last deep mine, Tower Colliery, closed in 2008. It was initially run by a trust that was a registered charity, funded entirely by self-earned income, such as admission charges. Following financial pressures, it was incorporated into Amgueddfa Cymru-National Museum Wales in 2001 as the Big Pit National Coal Museum. As part of the national museum, its funding is underwritten by the Welsh Government. The museum occupies a 15-hectare site, has over 60 members of staff and attracts around 150,000 visitors a year

Being on the north-eastern edge of the South Wales coalfield means that the mine is relatively shallow, self-draining and mainly gas free, the shaft is only 90 metres deep and the second means of egress is a walkable drift. Big Pit is an example of two collieries that were geographically separate from each other, but joined underground, with the upcast shaft at one and the downcast shaft at the other (*cf.* **386**). Ventilation is controlled by auxiliary fans and there is no need for mechanical pumps as the minewater drains away by gravity into old workings below, which extend all the way down the valley. A emergency winch is kept on the surface in case of failure of the winding engine while the cage is in the shaft

The colliery buildings include a blacksmiths' workshop (which at one time had nine forges), explosives magazine, fan house, pithead baths, sawmill and winding-engine house, which are all interpreted for their roles in the operation of the colliery. There are displays in the pithead baths with over 1,000 items about the history of coal mining in Wales and the mining communities.

Underground there are 2.4 kilometres of roadway showing earlier mining methods, such as a 40-horsepower electric direct-rope haulage, which moved empty tubs on the west side of the pit bottom. This is typical of haulages built by the Blaenafon Company. Underground tours are led by former miners, although a new scheme trains mining apprentices to become future miner guides. Visitors are taken down the 90-metre-deep shaft and through several preserved features, including haulage houses, roadways and stables. Pillar-and-stall working (**47**) and two longwall faces – one hand-cut and the other machine-cut – have been reproduced. The roadways still come under mining legislation.

Simulated underground workings using original large machinery have been created on the surface for visitors who do not wish to go underground. Modern coal mining can be explained here in more detail; for example, the use of compressed air and the movement of coal underground, such as with a 615 Eimco bucket loader dating to about 1980. A film theatre sets the scene and the interpretation covers roadway development and repairs, a machine-undercut conventional coalface, shotfiring and a 1970s power-loading coalface.

Big Pit houses the Welsh national coal-mining collection of around 5,000 objects, although a water-balance pithead gear from Brynpwllog old shaft, Blaencarno, Rhymney, Monmouthshire, dating to the early-nineteenth century and erected by the Rhymney Iron Company, is displayed in Cardiff over a simulated shaft. The museum holds a number of important objects, including items connected with disasters, such as five of the 25 Albert Medals awarded to rescuers for their bravery in saving life during the 1877 Tynewydd Colliery inundation, which was the first time this medal had been awarded for saving life on land (see also **394**). There is a collection connected with the 1984/5 Miners' Strike (**404**), as well as the 1910/11 Cambrian Combine Dispute (**394**) and the 1926 Miners' Strike and Lockout. The museum does not collect banners, as National Union of Mineworkers banners are held by the South Wales Miners' Library at the University of Swansea.

The coal-mining library is open to the public in the National Museum's Waterfront Museum in Swansea. The national coal-mining collection of around 15,000 images and paper objects are kept off-site in the collection centre of the National Museum in Nantgarw, which is open by appointment. The collection includes the last-surviving Mynydd Newydd Mine bible. An explosion at the colliery in 1844 killed five men. So to protect themselves the miners decided to construct a chapel underground, where they met for prayers every Monday morning before starting work. Bibles had to be replaced at intervals because of the

**Figure 388.1. Entrance to the Big Pit site (© Amgueddfa Cymru-National Museum Wales).**

**Figure 388.2. Simulated underground on the surface at Big Pit, showing a 615 Eimco bucket loader about 1980 and a compressed-air drill to the right (© Amgueddfa Cymru-National Museum Wales).**

Figure 388.3. Gold Albert Medal awarded to Isaac Pride for bravery in saving life during the 1877 Tynewydd Colliery inundation (© Amgueddfa Cymru-National Museum Wales).

damp conditions underground. The last bible was acquired in 1904 and was brought out when the colliery closed in 1955. Prayers underground are also known to have been held in north-eastern England; for example as late as the 1940s by John Henry Yole, warden of Lynemouth Primitive Methodist Chapel, at the meeting point for the miners at Ellington Colliery, and after their shift to give thanks that they had come through their work safely. His son-in-law, Robert Farrier, a circuit preacher, did the same underground at Lynemouth Colliery.

One particularly striking object on display at the National Museum in Cardiff is a silver regency vase presented by the county of Denbigh to the fifth baronet, Sir Watkin Williams-Wynn (1772-1840), to commemorate his service during the 1798 Irish Rebellion and in southern France in 1814 during the Napoleonic Wars (1803-15). From 1948 the vase was used as the Mitchell-Hedges trophy for the NCB's first-aid competition. It came into the ownership of the National Museum in 1994 on the privatisation of the coal industry.

*References*

ERIH, *s.n.* Blaenavon

https://museum.wales/articles/2088/Underground-Bible

http://www.britishlistedbuildings.co.uk/wa-15282-91

Figure 388.4. Early water-balance pithead gear from Brynpwllog Old shaft, Blaen-carno, Rhymney, Monmouthshire, over simulated shaft (© Amgueddfa Cymru-National Museum Wales).

http://www.museumwales.ac.uk/bigpit

personal communications, Peter Walker, head of site, 2012; 2014; Dai Price, head of museum: Big Pit National Coal Museum and National Roman Legion Museum, and Ceri Thompson, curator, Big Pit: National Coal Museum, 2021; Neil Taylor, grandson of John Henry Yole, 2021

showcaves

### 389. *National Museum Wales: Saint Fagans National Museum of History*

Saint Fagans National Museum of History is part of the National Museum Wales and is located on the outskirts of Cardiff, the capital of Wales, in southern Wales. As part of a national museum, its funding is underwritten by the Welsh Government. More than 40 historic buildings, representing those typical of various parts of Wales, have been moved here and re-erected in 4 hectares of natural woodland. The museum opened in 1948 and has 215 staff including seasonal and part-time staff. It attracts more than half a million visitors annually.

Three of the buildings relate to the coal-mining industry, which was extremely important to the Welsh economy in the nineteenth and twentieth centuries. Among the oldest is a row of fully furnished workers' cottages from the early 1800s. The Rhyd y Car Terrace in Merthyr was a small terraced row of six cottages built by Richard Crawshay (1739-1810) as housing for his ironworkers. The cottages have been fitted out with furnishings of different periods from about 1800 up to when the building was moved to Saint Fagans in 1985, prior to being re-erected in 1991. Gardens that would have provided the miners with vegetables have been laid out behind the cottages, with hedges or woven wicker fences between the plots.

The Workmen's Institute from Oakdale, which was a model village built in 1907 by the Tredegar Iron and Coal Company, dates to 1917. The company lent the funding for the institute and the miners then repaid the cost to the company. The institute was always the centre of village life in mining communities throughout the UK. At Oakdale the institute was an imposing building, housing a committee room, concert hall, library and reading room. Entertainment was provided in the concert hall on Friday and Saturday evenings; the brass band and the miners' choir practised regularly one night a week, before doing performances in the concert hall; while the library and reading room enabled miners and their families to improve their education and knowledge. The local branch of the South Wales Miners' Federation would meet in the committee room, where young miners learned how organisations and committees operated and how a formal meeting was chaired and conducted (see Arnot 1949-61 on the miners' union). Members of the union branch would go on to become elected local-authority councillors and eventually, for the most talented, members of parliament. Sometimes the union would pay for further education for those who showed most promise. For example, the

**Figure 389.1. Aerial view of St Fagans: left to right, at back Workmen's Institute, Gwalia Stores (behind trees) and Rhyd y Car Terrace (© Amgueddfa Cymru-National Museum Wales).**

**Figure 389.2. Rhyd y Car Terrace of coal miners' housing (© Amgueddfa Cymru-National Museum Wales).**

**Figure 389.3. Oakdale Workmen's Institute (© Amgueddfa Cymru-National Museum Wales).**

Labour politician, Aneurin Bevan (1897-1960), started work underground at the age of 13. He became head of the local lodge at 19, was sponsored by the federation to study economics, history and politics in London at 23, was elected to Monmouthshire County Council at 31, became a member of parliament at 32 and finally, at 48 as the minister for health, founded the National Health Service in 1948. The hall houses the banner of the Oakdale Colliery National Union of Mineworkers lodge and the library with its collection of books.

The Gwalia Stores, from Ogmore Vale, Bridgen, is furnished as in the 1920s. The stores encompassed an animal-feed section, bakery, gentlemen's outfitters, ironmongers and a pharmacy, as well as tinned and bottled food. This meant that the stores provided virtually everything needed for everyday life without having to go outside the village. The stores served the mining community from 1880 until they closed in 1973.

New galleries opened in 2018. One explores the history of Wales, including the coal industry at its height and the 1984/5 Miners' Strike (**404**). The display deals with the effects the industry had on its communities and how it has shaped Welsh history. A second new gallery includes the effects of the coal industry on the daily routines of men, women and children. Saint Fagans is the national repository for thousands of oral-history recordings from across Wales. There are over 300 recordings relating to coal and coal-mining communities, including recordings collected for *Geirfa'r Glowyr*, a dictionary of Welsh coal-mining terms. Recordings linked to the Rhyd-y-Car terrace and the Oakdale Workmen's Institute were used to inform the interpretation of those buildings.

*References*

ERIH, *s.n.* Cardiff

http://www.museumwales.ac.uk/stfagans/living-museum/

personal communications, Bethan Lewis, head of Saint Fagans, 2015, 2021

### 390. Pleasley Colliery

Pleasley Colliery in Derbyshire in the East Midlands opened in 1873 and closed in 1983. All the colliery buildings were removed, except for the two sets of headstocks and the two back-to-back winding-engine houses. The museum occupies approximately 500 square metres and attracts about 10,000 visitors a year.

The buildings were listed as grade II in 1986, following pressure from a local action group, but the site was then abandoned and the engine houses were left without their roofs. The site was scheduled in 1997. The Friends of Pleasley Pit organisation was formed in 1995 and opened the site to the public in 1999, funding their operations from donations and fundraising events. The Friends of Pleasley Pit are now part of the Pleasley Pit Trust, which registered as a charity in 1999. They run the site under licence from a government body, the Land Trust, who are the owners. The site has been open daily since 2011. The museum is run entirely by volunteers, about 50 of whom turn out each week. A second group of volunteers are working with Derbyshire County Council on managing the adjacent spoil heaps, which now form a country park.

The North Engine House dates from 1873 and houses the Lilleshall 1,500-horsepower horizontal-steam engine installed in 1902. Adjacent to the building are the headstocks built in 1901 of riveted-steel construction, which stand over the North shaft, which is now filled and capped. The Lancashire steam boilers were removed in 1983, although the boiler-house chimney still stands 40 metres high. At first the engine was run electrically, but the engine has now been rebuilt to a standard to enable it to run on steam.

The 1922 South Engine House stands on the same site as the 1873 engine house. The shafts were sunk to a depth of approximately 500 metres to the 1.6-metre Top Hard seam. In the 1920s the Top Hard seam was becoming exhausted, so the South shaft was deepened to a depth of approximately 850 metres to gain access to other seams. The deeper shaft meant that the Markham Company of Chesterfield installed a more powerful 3,000-horsepower double-horizontal piston steam engine, whose drum had a diameter of 6.5 metres. This engine has been restored to run electrically, and to a standard where it could run on steam. Both engines are run once a month from March to October.

The South shaft, which was the upcast shaft, was capped, but 2 metres were left at the top with a cage in position as when the pit was running. Adjacent to this is the banksman's cabin, containing the original pneumatic controls, rams and tub stops, while 10 metres away is the electrically powered turntable, which reversed the empty tubs to direct them back to the shaft side. This area exhibits mining machinery, including an AB coal cutter, an original battery-charging rack, a battery pony locomotive, a Dosco tunnelling machine, hand-boring machines with drill rods and cutting bits, mine cars, oil lamps, pit-pony equipment, including a straw-chopping machine, shotfiring equipment, a Siskol boring machine, steel and hydraulic roof supports, tubs and wooden pit props. Inside the engine houses there are working shaft signals and underground magneto mining telephones, as well as electrical and gas-testing equipment and surveying instruments.

*References*

https://historicengland.org.uk/listing/the-list/list-entry/1015641

https://historicengland.org.uk/listing/the-list/list-entry/1045855

http://www.pleasleypittrust.org.uk/

personal communications, Neville Buckle, secretary, Pleasley Pit Trust, 2016, 2021

**Figure 390.1. View of colliery showing the two headframes and back-to-back winding-engine houses (© Pleasley Pit Trust).**

**Figure 390.2. Winding engine in South Engine House (© Pleasley Pit Trust).**

### 391. Pooley Hall Colliery

Pooley Hall Colliery, near Polesworth in Warwickshire in the West Midlands, was sunk in 1897 next to a wharf on the 1789 Coventry Canal and a branch line connected the colliery to the Trent Valley railway line. So the coal could be moved by either canal or rail. Pooley Hall Colliery is believed to have been the first mine in the UK to generate its own electricity, using excess steam, in the early 1920s. The pithead baths were opened in 1928. There are reports by government-departmental committees and comments in royal-commission reports on pithead-bath provision dating from before WWI. The 1926 annual report of the Miners' Welfare Fund (Turner 2021) pointed out that none of these existing baths met the recommendations of the 1913 committee, in particular with regard to the minimum of 4 square feet (37 square centimetres) per man. The report stated that the Fund intended to finance construction of pithead baths to the new design at four collieries, in County Durham, Scotland, Wales and Warwickshire. So although there had been pithead baths provided by the coal owner or by the men themselves at other collieries well before then, the baths at Pooley Hall Colliery were the first to be constructed to the recommended standard. A simply worded booklet was issued or else an illustrated lecture was given to describe to the miners how the baths should be used, including the requirement to wear a modesty 'slip'. Colliery documents record fines of miners for 'indecency in the baths', referring to not wearing the slip.

In 1951 Pooley Hall Colliery amalgamated with Alvecote and Amington Collieries, when the roadways of the three mines met underground, to become the North Warwick Colliery, which closed in 1965. The colliery buildings, as well as part of the nearby 1509 hall, suffered from mining subsidence and so were demolished. There is no trace left of the historic baths and there are now industrial units where they used to stand. All that remains of the colliery today is a winding wheel and the spoil heap, now covered with scrub and woodland, immediately next to the canal. The colliery site has become part of the 62.5-hectare Pooley Country Park, run by Warwickshire County Council. The areas of mining subsidence have been transformed into ponds and one-third of the park is designated as a site of special scientific interest. In 1847 a coal mine was sunk on the Pooley Hall Estate, close to the hall; a footpath that was used by the miners to get to work from 1847 onwards is preserved as the Miners' Walk in the park.

The park's visitor centre houses a shop, with a tea room and a digital video disc on sale of former miners describing their work underground. There is a small display of mining equipment donated by former North Warwick Colliery miners, about 90 *per cent* of which is on display. There is an open-flame acetylene miner's lamp of a type developed in the 1930s; this has a container of calcium carbide which produces acetylene gas when wetted by a constant drip from the water container in the upper half of the body. The gas was burned at the stalk, the light of the flame being focused by the reflector behind it. There is also a shotfirer's combined pricker-key for unlocking explosives pouches, inserting detonators into explosives and unlocking and firing the exploder, dating from the 1960s onwards. Dating from the late 1970s onwards are a Davis of Derby Biram-type anemometer (**376**) for measuring the air flow in colliery roadways to ensure adequate ventilation and an Oldham T-series battery and cap lamp converted for use as a handlamp.

*References*

http://countryparks.warwickshire.gov.uk/country-parks-greenways/pooley-country-park-1

https://en.wikipedia.org/wiki/Pooley_Hall

personal communications, Paula Cheesman, country parks manager, Warwickshire County Council, 2015; Stuart Ikeringill, rural services manager, 2015

showcaves

### 392. Prestongrange Museum (formerly Scottish Mining Museum; then Prestongrange Industrial Heritage Museum)

Prestongrange Museum is located in Prestongrange in East Lothian on the coast in south-eastern Scotland. The site was of major importance in Scotland's Industrial Revolution (**379**), because it had a rich supply of natural resources, including clay, coal, fireclay, sand and seawater. Over time it has been a sixteenth-century harbour, seventeenth-century glass works, potteries in the eighteenth and nineteenth centuries, and a brickworks and colliery in the nineteenth and twentieth centuries. Evidence for all these industries can still be seen on the site. There had been coal mining here since the twelfth century, initially carried out by Newbattle Abbey (1140-1587). Flooding underground closed the mine in 1749. In 1830 a new shaft, the first deep shaft in Scotland, was sunk 128 metres deep to the Great seam by Matthias Dunn (1788-1869), with pumps to deal with the water.

The colliery closed in 1962, leaving a derelict site of 11.5 hectares. It was intended to clear the site, but a steering committee was formed in 1968 to preserve it and to collect exhibits. The museum opened as the Scottish Mining Museum in 1984, jointly with the site at Lady Victoria Colliery. In 1992 the management of the two museums was separated, with this site becoming the Prestongrange Industrial Heritage Museum, run by East Lothian Council, and Lady Victoria Colliery continuing as the Scottish Mining Museum (**387**). The museum is now funded by East Lothian Council and managed by the council's museums service, which employs eight members of staff, whose role includes working at the museum. There are 18 volunteers. The museum attracts about 4,545 visitors a year. The collections include those relating to the coal industry gathered by David Spence, a retired mining engineer, who was instrumental in ensuring the preservation of the site as a museum.

**Figure 391.1. Spoil heap of Pooley Hall Colliery immediately next to the Coventry Canal (© Warwickshire County Council).**

**Figure 391.2. Davis of Derby 'Biram' type anemometer for measuring the air flow in colliery roadways to ensure adequate ventilation, dating from the late 1970s onwards (© Warwickshire County Council).**

**Figure 391.3. Acetylene open-flame miner's lamp, developed in the 1930s (© Warwickshire County Council).**

**Figure 392.1. Exterior of 1874 beam-engine house (© East Lothian Council).**

**Figure 392.2. Colliery generating house with steam crane (not used at this colliery) to left (© National Coal Mining Museum for England).**

The main remains on the site from the colliery are the beam-engine house, generating house and pithead baths. The three-storey sandstone beam-engine pump house was erected in 1874 specifically to house the Cornish beam-engine pump. Its front wall is 2.1 metres thick, to enable it to support the main pivot-bearing of the huge beam. The beam engine was built in 1853 by J.E. Mare and Company at the Plymouth foundry in Devon, England, for a silver/lead mine in south Devon. In 1873 Harvey and Company of Hayle purchased the engine, installed a new cast-iron beam and lengthened the Oregon-pine piston rod. The beam engine was then sold to the Prestongrange Coal and Iron Company in 1874 and moved north to the colliery. The beam engine was able to pump 2,995 litres of water per minute and continued in operation until 1954, when it was superseded by electric pumps. It survived, however, to become the last Cornish beam engine remaining *in situ* in Scotland.

The generating house was built to produce electricity for Prestongrange Colliery, and is now used to display the museum's collection of mining artefacts. The red-brick building, which dates to the mid-nineteenth century, is of two storeys with a rectangular plan. It has contrasting brickwork banding and round-arched openings at the first-floor level. The pithead baths are of modern design. Several items have been moved from elsewhere to the site, such as a steam crane and a small headgear brought from Little Mill Colliery in Ayrshire.

The beam-engine house is a scheduled ancient monument, while the old generating house and the Hoffman kiln, dating to the later-nineteenth century, are listed buildings. The kiln is all that survives of the substantial brick and fireclay works, which were sited near the colliery.

*References*

http://portal.historic-scotland.gov.uk/designation/LB47020

http://portal.historic-scotland.gov.uk/designation/SM2813

https://en.wikipedia.org/wiki/Prestongrange_Industrial_Heritage_Museum

https://www.eastlothian.gov.uk/info/210593/museums/11878/museums_in_east_lothian/4

personal communications, Katherine Weldon, museums officer (exhibitions), 2016, 2021

### 393. Ram Hill Colliery

Ram Hill Colliery is located in south Gloucestershire. Coal has been extracted here since the later-eighteenth century. Ram Hill was one of a number of coal mines in the region, but is the only mining site on Coalpit Heath to survive without being transferred to private ownership. The mine closed in 1860 and ownership of the site passed in 1900 to the Great Western Railway, who left it untouched. Most of the site is now owned by South Gloucestershire Council, which permits open access.

The colliery and its associated dramway have been partially excavated. The dramway, or horse-drawn tramway (*cf.* **361**), still has its sleeper blocks *in situ*. It is preserved as part of an industrial-archaeological route, linking together a number of buildings and sites along its course. The Friends of Ram Hill Colliery organisation was set up in 2004 to raise funding to protect the remains and improve access and interpretation of the site. In 2005 the Ram Hill site was scheduled as an ancient monument, on the grounds that its features are very rare and of national importance, as it illustrates the technological development of horse winding and then steam winding on the same site. The site is currently managed by the Friends under licence from South Gloucestershire Council.

*References*

https://www.southglos.gov.uk/i-am/i-am-a-volunteer/friends-groups/ram-hill-colliery-2/

http://www.historicengland.org.uk/listing/the-list/list-entry/1021386

http://www.ramhillcolliery.org.uk/

http://www.winterbourne.freeuk.com/ramhill.html

personal communications, Roger Gosling, chair, South Gloucestershire Mines Research Group, 2016, 2021

### 394. Rhondda Heritage Park

The Rhondda Heritage Park is located on the site of the former Lewis Merthyr Colliery in Trehafod in South Wales. The colliery was established between 1870 and 1880 by W.T. Lewis, later Lord Merthyr (1837-1914), by the amalgamation of the Coedcae and Hafod shafts, both sunk in 1850. The present buildings are grouped around three shafts sunk in about 1880 and around 1890. By 1900 the Lewis Merthyr Consolidated Collieries Company Limited was raising almost a million tonnes *per annum*. In about 1929 the company was acquired by Powell Duffryn Collieries Limited which, in 1935, became the largest coal-producing unit in the UK. The Lewis Merthyr Colliery itself was, however, in decline and by 1957 output had dropped to some 200,000 tonnes. In 1958 Lewis Merthyr merged with the nearby Tymawr Colliery with all coal wound up the Tymawr shaft. By 1983 only the Four Feet seam was still being worked and the colliery closed that year.

The site occupies about 30,000 square metres and was opened in 1989 as a tourist attraction. It was later developed to include industrial and social-history collections telling the story of the Rhondda Valley from the peak of the Industrial Revolution (**379**) through to the modern day. It is owned and funded by the Rhondda Cynon Taf District Council. It has 12 members of staff and attracts about 30,000 visitors a year.

The site consists of various pithead buildings, including tram circuit, tippler, 1891 fan house with a rare surviving

Figure 393.1. Top of shaft (left-hand side) with horse gin (© Roger Gosling).

Figure 393.2. Arch at the dramway terminus (© Roger Gosling).

Figure 394.1. Sunrise view of Rhondda Heritage Park (© Rhondda Cynon Taf County Borough Council).

Figure 394.2. Headframes and winding-engine houses of Trefor engine at front and Bertie at the rear (© Rhondda Cynon Taf County Borough Council).

Sirocco fan, lamproom, two headframes and two winding-engine houses, one with a steam engine (Trefor) and the other an electrical engine (Bertie). These were named after Lord Merthyr's two sons. The site was listed as grade II* as a rare surviving pit top complete with surface buildings and a particularly rare tippler, for having the oldest headframe remaining in Wales, and for the group value of this architecturally important colliery complex.

The visitor centre has exhibitions, interpretation and a space for temporary exhibitions, mainly dealing with topics about the Rhondda, such as the 1919 Rhondda teachers' strike. There are reconstructed period shops dealing with shopping habits and rationing during WWII. The audio-visual presentation, 'Black gold – the story of coal', deals with the rise and fall of coal in the Rhondda; children who worked in the mine; Rhondda coal being shipped across the world; the Aberfan disaster (**351**); and the 1984/5 Miners' Strike (**404**) and the involvement of women. It ends with new sources of energy, such as wind turbines in the Rhondda. Another gallery, 'Coal society', focuses on the lives of those outside the mine, such as Methodist chapels, sports and women in the Rhondda.

There are also displays opened in 1991 in the two winding-engine houses and the fan house, depicting the story of three generations of a local mining family. First it deals with the 1877 Tynewydd Colliery disaster in which five miners were drowned and nine survived after being trapped underground for 10 days (see also **388**). Then the 1910/11 Cambrian Combine Dispute (usually known as the Tonypandy Riots), when the home secretary, Winston Churchill (1874-1965), sent troops into the valleys to restore order (see also **404**). Finally the 1920s fight for a minimum wage, led by A.J. Cook (1883-1931), general secretary of the Miners' Federation of Great Britain, who had started work as a colliers' boy in Lewis Merthyr in 1899.

In 1994 the park opened a simulated underground tour on the surface, called 'A shift in time', with former miners as guides. This replicates the underground roadways of Lewis Merthyr Colliery, with the darkness, environmental changes, heat, humidity and smells of a typical underground, as well as a simulated explosion.

There are also a number of larger pieces of mining machinery around the site, such as an underground locomotive and a section of power-loading equipment, left *in situ* at the closure of Lewis Merthyr Colliery, as well as chocks and drams that were brought up from the underground when the mine closed. The park has an archive, with material representing aspects of life in the Rhondda Valley, encompassing education, religion, sport, tourism and the original reports relating to mining disasters. These include the 1956 Lewis Merthyr disaster, when a rock fall ignited a pocket of gas, killing nine men. There is also a large collection of prints by the graphic artist and printmaker Peter Piech (1920-96), who moved to Wales in the 1980s. On display in the fan house are the National Union of Mineworkers' lodge banner from Maerdy Colliery, the 1984/5 Maerdy Women's Support Group banner and a Soviet flag from the Maerdy NUM lodge offices.

*References*

ERIH, *s.n.* Rhondda

https://britishheritage.com/the-tynewydd-colliery-disaster/

https://www.rctcbc.gov.uk/EN/Tourism/RhonddaHeritagePark/Home.aspx

http://www.britishlistedbuildings.co.uk/wa-13119-26

personal communications, Corinne Havard, 2016; Ceri Thompson (**388**), 2021; Esta Lewis, heritage and outreach services operations supervisor, 2021

showcaves

### 395. Snibston Colliery

Snibston Colliery opened in 1832, adjacent to a small settlement called Long Lane. This was soon renamed as Coalville, owing to the number of collieries founded in this area of north-west Leicestershire, in the East Midlands. The railway engineer, George Stephenson (1781-1848), was invited to build a railway in the area. Having seen its potential, Stephenson bought the Snibston estate in 1831 and opened the colliery and an adjacent railway in 1832. A third shaft, the Stephenson, was sunk in 1915. In 1942 the wooden headgear of number 1 pit was replaced by a steel-tandem structure, and a drift was driven in 1961/3. The mine closed in 1983 and the site was bought by Leicestershire County Council in 1985.

The local authority constructed a large, warehouse-style building adjacent to the colliery buildings to form the Snibston Discovery Museum, which opened in 1985. About 40 staff were employed and the museum attracted about 100,000 visitors a year. The museum concentrated especially on the social history and industry of Leicestershire, such as the hearse used in the funerals of miners killed in the Whitwick Colliery disaster of 1898 in which 33 men died and only nine bodies were recovered. It focused on how design, science and technology affected everyday life and had a mining-themed children's playground. The most important items on display were the remains of medieval coal mining uncovered during opencast mining (Goodman 2020, pp. 110-13; Hartley 1994). These included oak pit props dating to around 1459, used in a reconstruction of a medieval mine shaft, and a woollen tunic worn by a medieval coal miner about 1500/80. These discoveries were especially important, because it had often been thought that medieval coal mining was carried out only by bell pits. A shaft would be sunk down to coal near the surface and the coal removed all around the shaft, leaving the shaft and mined area together forming the shape of a bell underground. Once it appeared that the roof might be about to collapse, the bell pit would be abandoned, and a new bell pit sunk nearby;

**Figure 395.1. Winding-engine house and tandem headgear (© Leicestershire County Council).**

**Figure 395.2. Medieval woollen coal miner's tunic (© Leicestershire County Council).**

sometimes the miners did not get out in time and records of inquests on such medieval fatalities can be used to identify the location of medieval coal mines.

To prevent accidents the shaft of the bell pit was usually backfilled (Moorhouse 1981, pp. 777-8) and a tree planted in the middle of the filled shaft to indicate that it was unstable ground; the tree disappearing down the shaft would show that the shaft filling had collapsed. This was common practice both in the UK and on the Continent, such as in Germany. Mining with bell pits continued surprisingly late, with examples being known from the nineteenth century on Baildon Moor in Yorkshire, also shown on a seventeenth-century map (**383.3**). The discoveries in the opencast mine proved that medieval coal mining was also already being carried out using deeper shafts and then roadways running away from the shaft, as had already been suspected for the early-fourteenth century in West Yorkshire (*ibid.*, p. 778).

**Figure 395.3. Three bell pits at Bentley Grange, Emley, near Wakefield, West Yorkshire; third at rear to right still with tree growing from shaft (© Margaret L. Faull).**

From 1992 tours were led by former miners of the colliery buildings and machinery, including the lamproom and the control and first-aid rooms. In 2011 a simulated mine roadway was opened, showing a coalface and stables for pit ponies. The old mineral railway line was reconstructed and reopened in 2000, with steam locomotives from the museum's collection transporting visitors into the centre of Coalville and around the old pit bank.

In January 2015 Leicestershire County Council announced that the museum would be closed, the main museum building demolished and the land sold for housing, thereby it was claimed saving about £900,000 a year. Despite a strong local campaign against this, the museum closed on 31 July 2015. The mine buildings cannot, however, be demolished, because in 1999 the whole colliery site was designated as a scheduled ancient monument. This was on the grounds that it was one of the best surviving examples of a mining complex modernised by the NCB, dating from the period of major production of the British coal industry from the 1950s to the 1970s. At Snibston rare structures, such as a creeper system, are preserved *in situ*, along with machinery, including two electric-winding engines, a fan and locomotive engines. The tandem headgear is believed to be one of only two surviving in England. The shaft has been capped, but the shaft-top equipment remains, including rare double-decker cages. Initially the county council said that it intended eventually to open a small mining museum in the colliery, which it was estimated would cost £450,000 a year to run. The council has now decided that the mining museum will not go ahead, although the council will have to continue to maintain the scheduled ancient monument. The site is open to the public, with interpretation boards, set in a country park developed from reformed spoil heaps. The mining artefacts are kept in storage in the collections of the local authority.

*References*

https://historicengland.org.uk/listing/the-list/list-entry/1018472

https://historicengland.org.uk/listing/the-list/list-entry/1005786

https://www.nmrs.org.uk/mines-map/coal-mining-in-the-british-isles/leicestershire/snibston-colliery/

https://www.wyorksgeologytrust.org/misc/Baildon%20Hill%20geology.pdf

personal communications, Heather Broughton, assistant director, Communities and Wellbeing, Leicestershire County Council, 2012; Franne Wills, head, Communities and Wellbeing Service, 2021

showcaves

### 396. Snowdown Colliery and East Kent Railway

Sinking of Snowdown Colliery in Kent began in 1907. Water was always a problem in sinking shafts in Kent and the Snowdown shaft flooded at a depth of 79 metres and two shaft-sinkers were drowned (a typological error has led to this being reported as 22). The first coal was brought to the surface in 1912. At 915 metres Snowdown was always the deepest, hottest and most humid pit in Kent, known colloquially as Dante's Inferno, referring to the allegorical poem describing the journey through hell of the poet Dante Alighieri (1265-1321). It was considered to be the worst colliery in England in which to work. The pit shut in 1922 as a result of a strike, but in 1924 was bought by the Pearson Dorman Long Company. The company replaced the steam-winding engine with an electrical one and built Aylesham village close to the pit to house 650 mining families brought from Belgium, France, northern England and Scotland. The mine had been sited close to the mainline railway to the port of Dover to enable its coal to be shipped out very easily. Once the colliery had closed in 1987, colliery waste from Snowdown was moved to level the land at the terminus of the Channel tunnel, linking England to France and excavated between 1988 and 1994.

The colliery site is now in private hands and not open to the public. The machinery has all been removed, but most of the buildings survive, although in a derelict condition. In 2007 the fan house and the attached 1950s number 2 winding-engine house were listed as grade II, as being the last remains of the Kent coal industry and also as high-quality buildings, the winding-engine house being a good example of modernist architecture in an industrial context. The red-brick fan house belongs to either the original 1907 phase of the pit or its modernisation in 1924. The winding-engine house is of concrete faced with red brick to match the fan house and housed a friction winder (**127**).

A project is currently underway in relation to the East Kent Light Railway, built in 1911, which ran from a junction with the Kent coastal line at Shepherdswell through another junction to two termini, Wingham and Port Richborough, and served the Guilford, Hammill, Tilmanstone and Wingham Collieries (see also **373**). The 4 kilometres of the line reopened as a preserved heritage railway in 1995, run by about 45 volunteers. A signal box at Eythorne station is open to the public. It has details of the importance of the railway to the development of the Kent coalfield and its later decline to serve just the one profitable pit at Tilmanstone in the early 1950s, when it also ceased to carry passengers. The most important item is the restored Saint Dunstan Avonside 0-6-9 steam locomotive, which previously worked at Snowdown Colliery until it closed in 1987. The locomotive was built in 1927 with a gauge of 14.35 centimetres. The East Kent Mining Heritage Railway took ownership of the locomotive in 1992. The Heritage Railway also owns an operational diesel locomotive from Nottinghamshire Colliery, coal-transport bucket waggons from Betteshanger Colliery and a rare coal-tipping bucket waggon.

*References*

https://en.wikipedia.org/wiki/List_of_preserved_Avonside_locomotives

https://historicengland.org.uk/listing/the-list/list-entry/1391968

**Figure 396.1. Listed buildings today; fan house on the left and winding house on the right (© Colin Varrall).**

**Figure 396.2. Saint Dunstan steam locomotive from Snowdown Colliery after restoration (© East Kent Railway Trust).**

http://www.britishlistedbuildings.co.uk/en-502720-fan-house-and-winder-house-KOEPE-no-2-at#.VnG7h4_XLIU

http://www.culture24.org.uk/history-and-heritage/historic-buildings/art527823

http://www.derelictplaces.co.uk/main/industrial-sites/4346-snowdown-colliery-kent-october-2007-a.html#.VqnzW_mLTcs

http://www.eastkentrailway.co.uk

personal communications, Gary Cox (**380**), 2014, 2015; Matthew Plews, general manager, East Kent Railway Trust, 2015; Simon Potter, fundraising and development manager, 2016; Colin Varrall, volunteer, 2021; David Harris, chair, 2021

### 397. South Wales Miners' Museum

The South Wales Miners' Museum is located in Cynonville in the Afan forest park of south-west Wales. Initially it was proposed to establish this museum on the site of the former Glyncorrwg Colliery, which closed in 1970. The cost of the conversion would, however, have been considerable and access to the site was poor. The museum was founded in 1972 and opened in 1975 in temporary accommodation, the oldest of the Welsh coal-mining museums. By 2008 the charity had raised sufficient funds from grants to erect a purpose-built museum, with a third of the interior work done by volunteers. The museum is a registered charity governed by a board of trustees and run by volunteers, many of whom are former coal miners. There are one full-time and one part-time member of staff, who coordinate the activities of about 35 active volunteers at the museum. The running and maintenance of the museum is funded by admission charges and donations.

The museum depicts living conditions of miners and their families and the working life of miners in the Afan Valley, where the mines all closed in the 1970s. Inside the building are displays of photographs and various mining tools, together with a replica mining roadway. There is an exhibition about how children worked underground until 1842 (**378**) and a timeline of mining disasters in the Afan and Llynfi Valleys. Part of the collection relates to the Welsh Miners' Eisteddfod, established in 1947 and strongly supported by David ('Dai') Francis (1911-81), the general secretary (1963-76) of the South Wales Area of the National Union of Mineworkers. Outside there are a blacksmiths' workshop, lamproom and winding-engine house, as well as displays of large mining machinery, such as winding and transport equipment.

*References*

ERIH, *s.n.* Cynonville

personal communications, Sarah Chapman, volunteer coordinator, 2012; Ceri Thompson (**388**), 2021; Charlene Rogers, manager, 2021

showcaves

www.swminers.co.uk

### 398. Summerlee Museum of Scottish Industrial Life

The Summerlee Museum of Scottish Industrial Life is located at Coatbridge, North Lanarkshire, in the western lowlands of Scotland. The aim of the museum is to interpret the social and industrial history of central Scotland, especially the Monklands area, and 'to preserve and interpret the history of the local coal, iron, steel and engineering industries and of the communities that depended upon them'. The museum is operated by North Lanarkshire Council and has 30 staff. The museum occupies an 8-hectare site, which includes the original coal drops for transferring coal from the railway to the canal.

The main exhibition hall was created in the late 1980s from the former Hydrocon crane works, providing ample space in which to display large working machines, which are operated daily. Coal-mining machinery includes the 1926 steam-winding engine from Cardowan Colliery, made by Murray and Paterson of Coatbridge, and a chain coal cutter by the Anderson Boyes company of Motherwell dating to the early-twentieth century.

In 2008 the museum underwent a £10 million refurbishment. The exhibition hall reopened with displays, including about coal mining, such as a bronze bust commemorating 47 miners who died in the 1959 Auchengeich Mine disaster. There are also a Spirelmo smoke helmet and a primitive breathing apparatus used in mines rescues, with its associated hose-and-air pump made by Siebe, Gorman and Company Limited.

As part of the refurbishment a simulated coal-mine area has been created with a reconstructed adit, mine and miners' homes. A working tramway uses historic trams to take visitors to the reconstructed mine and a re-erected row of miners' cottages. The mine shows the tough conditions in which Lanarkshire miners had to work and illustrates different methods of mining at different eras, but mainly pillar-and-stall (**47**), with a reconstruction of the 1810 Farme Colliery engine. When the headgear was re-erected there were no known images; since then a picture has come to light showing that the reconstruction is not completely accurate.

The surface structures of the drift mine are based on a small independent mine of the 1930s which was near Motherwell and from which most of the haulage, hutches, rail and other equipment were collected. This area also includes the actual 1810 Farme Colliery engine, which is on loan to the museum from Glasgow Museums. This is the only surviving Newcomen engine (**371**) that was operated fully manually and is the only rotative Newcomen engine in Europe (there are two others, one of 1791 from the West Midlands and the other of 1821 from Leicestershire, in the Henry Ford Museum in Dearborn, Michigan, America). The cottages depict the living conditions of miners and their families from the 1840s to the 1980s, with reconstructed interiors of, for example, a home in the 1840s, with an iron

Figure 397.1. Exterior of the South Wales Miners' Museum and wooden tub (© South Wales Miners' Museum).

Figure 397.2. Exterior of museum with winding wheel (© South Wales Miners' Museum).

**Figure 398.1. Headgear and replica winding drum of Farme Colliery, Rutherglen, erected 1810 and worked into the twentieth century (© North Lanarkshire Council Museums and Heritage).**

**Figure 398.2. Interior of reconstructed adit coal mine showing pillar-and-stall working and wooden pit props (© North Lanarkshire Council Museums and Heritage).**

**Figure 398.3. Reconstructed interior of 1840s home with iron pot hanging from a sway above an iron grate for a coal fire (© North Lanarkshire Council Museums and Heritage).**

pot hanging from a sway above an iron grate that would have contained a coal fire (*cf.* Goodman 2020, p. 155).

The museum also has coal-mining collections that are not on display, including books, ephemera and models of coal-cutting machinery from the Motherwell firm of Anderson Boyes. The archive of Anderson Boyes is held by the archives of the local authority at the North Lanarkshire Heritage Centre, Motherwell, where the museum's photographic collection is also kept.

*References*

ERIH, *s.n.* Coatbridge

https://culturenl.co.uk/museums/visiting-us/summerlee/

http://www.healeyhero.co.uk/rescue/Station/breath.htm

personal communications, Justin Parkes, industrial history curator, 2012, 2021

### 399. Sunderland Museum

Sunderland is located in the north-east of England, where mining began at least by the later-thirteenth century (**368**). The Sunderland seams were very deep, however, and the first shafts were not sunk there until the early-nineteenth century. Wearmouth Colliery, also known as Monkwearmouth Colliery, was located on the north bank of the river Wear. This shut in 1993, having been the last mine operating both in Sunderland and the north-east of England. The Stadium of Light football ground, home of Sunderland football club, opened on the site in 1997. The stadium has a large replica miner's lamp outside to commemorate its past.

The history of coal mining in the north-east is preserved at Woodhorn Museum (**404**), but some of the museums in the area also have extensive displays about coal mining. The most important of these is probably the exhibition devoted

**Figure 399.1. Stadium of Light football ground, Sunderland, on the site of Wearmouth Colliery, with large miner's lamp outside (© Margaret L. Faull).**

to the history of coal mining in the region at Sunderland Museum. The museum has a considerable collection of coal-mining artefacts, including rosettes for pigeon racing (a popular hobby of coal miners) and posters exhorting people not to shoot racing pigeons, which are protected by law, and offering a £20 reward for information about people doing so. A small replica underground gives visitors an impression of what it was like to be in the confined dark environment of a coal mine.

*References*

https://sunderlandculture.org.uk/our-venues/sunderland-museum-winter-gardens/

http://www.englandsnortheast.co.uk/CoalMiningandRailways.html

### 400. Thornhill Collieries Disaster Memorials

The victims of two disasters at collieries in Thornhill, 54 years apart, are buried in the churchyard of Thornhill parish church. That at Combs Pit on 4 July 1893, resulted in the deaths of 139 men and boys, while that at Inghams Pit on 9 September 1947 killed 12 men. Combs Colliery was owned by the Ingham Collieries Company Limited and was considered to be a safe mine. The downcast shaft was known as the Drawing Pit, while the upcast shaft was known as the Water Pit, both sunk through the New Hards seam to the Wheatley Lime. This was the seam being worked in 1893 at a depth of 128 metres when there were 146 miners underground. A new manager, William Scott, had taken up his post in May 1893 and was not aware of the existence of a fault that produced gas. Because it was very wet, naked-flame lamps were in use, which ignited gas from the fault. By the afternoon of the disaster a crowd of about 20,000 people had gathered outside the colliery. In 1988 an original photograph was donated to the NCMME, showing the crowds on the bank outside the colliery and hearses lined up to collect the bodies. The donor said that he had been offered a lot of money for it, but preferred it to be preserved at the museum, which at that time was the Yorkshire Mining Museum (**386**).

The rescuers who went underground were at first beaten back by a fierce fire, but after the ventilation had been restored, rescuers found nine men alive. Seven of these survived, although one died subsequently, after having given evidence to the inquest. It was found that all those who died were suffocated by the gas. They were found in groups, looking as if they had simply fallen asleep. The men who died had had some warning of the gas gradually overwhelming them. One had a piece of board in his hand

**Figure 400.1. Original photograph of surface of Combs Colliery, Thornhill, on day of the 1893 explosion (© National Coal Mining Museum for England).**

**Figure 400.2. The seven survivors of the disaster at Combs Colliery, one of whom died subsequently (© National Coal Mining Museum for England).**

on which he had written 'look for my watch', a watch being a very valuable item for a working man. The body of one man was found with a piece of chalk in his hand with which he had had time to write a message to his wife on a corfe, saying, 'Good-bye Betsy, love; you must do the best you can. From your loving Tom'. On another corfe was written 'God bless us all'. The body of George Wilkinson, one of the deputies, was found kneeling in prayer. The youngest victims were aged 12 and the oldest 70 years. One woman lost her husband and her two sons, aged 13 and 15. Queen Victoria (1819-1901) sent a message of sympathy and enquiry, which was displayed publicly for all to see.

The bodies were laid out in the hall of the parish church for identification, and the funerals took place on 6, 7, 8 and 9 July. Many of the families could not afford gravestones, so a large number of those who died were buried in unmarked graves. On the list of dead the same surnames occur frequently. For example, the list includes James Scargill, aged 50, and Lot Scargill, aged 28, while identifications at the inquest included evidence from George Scargill, a gamekeeper, who identified the body of his brother, Robert, and that of Robert's stepson, Rufus Summerscales. Their gravestone in the graveyard of Thornhill parish church describes them as having fallen asleep. George stated that his brother had not been insured, so there would be no money for his widow and the six children that he had left. Reports at the time said that virtually every house in the village had lost at least one member, who was usually the sole breadwinner in the household. The two insurance companies paid out on all claims immediately without any discussion or dispute about what was claimed. A fund was set up by the Batley Co-operative Society to provide for the families, especially those who had not been insured. It was estimated that £21,000 would be needed, and in fact £34,546 (£4,540,000 at present values) was raised. The local gentry all contributed, including Lord Savile, who gave £250 (now £33,000), and a large fund that had been raised to celebrate the wedding on 6 July of the Duke of York, later King George V (1865-1936), was diverted to the disaster fund. The Yorkshire Miners' Association distributed £366 amongst its 64 members.

Inghams Pit began production in about 1860. It was joined underground to Combs Pit, so that coal from Combs came to the surface screens at Inghams. Part of the ventilation system was common to both mines, but they were worked as two separate collieries. The explosion in 1947 was of methane coming from abandoned workings, forced into the working area by the fixing and closing of a door. The enquiry concluded that the gas was ignited by someone opening and using the relighting device in a lamp that had gone out, something strictly prohibited underground, especially in the return airway. The culprit could not be identified. The mine manager, E.E. Cleaver, was asked by the rector of Thornhill parish church to read one of the lessons at the funeral service, which he said was one of the hardest things that he had ever had to do.

The enquiry into the 1947 disaster recommended that the lead rivets in lamps should be replaced with more-efficient magnetic locks and the Pyropho relighter used in the lamps should be replaced with a better relighter. But it was the major 1893 disaster that had a more important effect, as it was responsible for reconsideration of the practice of allowing naked-flame lamps in mines, even if the underground of the colliery were very wet, as at Combs Pit.

Unlike in most other major disasters, no memorial was erected for the 1893 disaster until 1980, when a small memorial in the churchyard was paid for by a local building company. This takes the form of an open book, carved from granite, inscribed to the memory of the miners who died in 1893 and 1947. In 1993 the local newspaper, the *Dewsbury Reporter*, printed a special edition to launch an appeal to provide a larger memorial, which would list the names individually of all the miners who died. The appeal committee has half a pit wheel, but has not yet been able to raise sufficient funds to re-erect the wheel and carve the names.

*References*

http://www.dmm.org.uk/ukreport/7434-03.htm

http://www.healeyhero.co.uk/rescue/pits/Thornhill/Ingham-JS-P-1A.html

personal communication, Tony Ridley, historian, Thornhill parish church, 2017

### 401. Washington 'F' Pit Industrial Museum

Washington is located in the city of Sunderland in north-eastern England, 8 kilometres south-east of Newcastle-upon-Tyne. The mines at Washington were sunk in the 1770s by William Russell, a coal-shipping agent. He sank eight shafts, known by the letters of the alphabet, J Pit not being sunk until the 1920s. Nearby F Pit had been sunk in 1777 and was the most productive, with four seams at depths of between 146 metres and 206 metres. In 1828 15 miners were killed in a methane explosion, five of them under the age of 12 and one only seven years old. In 1927 F Pit was modernised with the introduction of electricity and pneumatic picks at the coalface. At its peak in 1964 F Pit employed over 1,500 workers and produced 478,000 tonnes of coal, but closed in 1968. The last colliery in Washington ceased production in 1974. F Pit was presented to the people of Washington by the NCB for conversion into a museum, which opened in 1976.

Most of the buildings of J Pit have been demolished. The remainder of the site is original, apart from the entrance building, which was constructed by the Washington Development Corporation. The most important elements of the museum are the F Pit headgear, 1903 winding-engine house and its Victorian steam-winding engine, all listed as grade II. The headgear is a rare late example of steel latticework more typically used in the third quarter of the nineteenth century (**204**). The red-brick winding-engine house has a rectangular plan, with a hipped Welsh-slate roof. Inside the building is divided into two floors by a cast-iron mezzanine floor to give access to the drum and the engine.

The twin-cylinder steam-winding engine of 500 horsepower was built in 1888 by the Grange Iron Company of Durham City, for an unidentified coal mine. When it was moved to the new winding-engine house at F Pit in 1903, it was fitted with the Grange Iron Company's patented expansion-valve gear to make it more efficient and so to save fuel and money. The engine could raise a cage of miners or full coal tubs 305 metres to the surface in only one minute. The demolition of the boiler house in 1968 means that there is now no opportunity to run it with steam. The boiler located beneath the engine has been removed as superfluous and the area is now used for interpretive displays. The winder is operated using an electric motor, which means that its movement is not nearly as impressive as it was when it was a steam-powered engine. Nor does it have the characteristic aroma of a steam-winding engine. Few winding engines survive *in situ*, and even fewer in full working order. This is believed to be the last surviving winding engine of this kind *in situ* in the UK.

*References*

https://historicengland.org.uk/listing/the-list/list-entry/1018224

http://www.theguardian.com/uk/the-northerner/2012/aug/02/mining-museums

personal communications, Martin Routledge, keeper of history, Sunderland Museums, 2015; Jo Cunningham, exhibitions, collections and archives manager, 2021

showcaves

### 402. Wentworth Woodhouse

Wentworth Woodhouse near the village of Woodhouse near Rotherham, in South Yorkshire, is one of the finest Georgian houses in England and one of the most expensive ever built. It is believed to be the largest private house in the UK. The east front is 187 metres long, its buildings cover 1.2 hectares and it stands in a park of 105 hectares. It has a seventeenth-century core, but was remodelled between about 1724 and 1734. The fourth earl Fitzwilliam (1748-1833) inherited the house from his uncle, the second marquis of Rockingham (1730-82). The earl made money from mining and industry, and was responsible for sinking the Elsecar collieries (**371, 372, 376**).

In the immediate period after WWII, there was a pressing need for coal supplies, especially for the railways. The

**Figure 401.1. Exterior of Washington F Pit engine house from the south-east with spare winding wheels propped up outside (© Sunderland Museums).**

**Figure 401.2. Steam-winding-engine piston and associated valves (© Sunderland Museums).**

**Figure 402.1. Front elevation of Wentworth Woodhouse (© National Coal Mining Museum for England).**

richest and highest grade of coal in Yorkshire is the Barnsley seam. In the area of Wentworth Woodhouse, this seam is within 30 metres of the surface and so able to be extracted by opencasting. In 1946 the Labour Party minister of fuel and power, Manny Shinwell (1884-1986), ordered that opencasting should be carried out almost up to the house. A total of 40 hectares of the formal gardens, lawns, woods and pink-shale driveway were mined, in what was the largest opencast mine in the UK of that time, followed by mining of the adjacent fields into the early 1950s. The overburden was stocked in a pile 15 metres high in front of the family's living quarters at the rear of the house, and the formal gardens and wood were not reinstated after having been opencast. Wentworth Woodhouse was protected by being listed as grade I in 1952, but it is claimed that the house suffered from subsidence as a result of the opencasting.

The Fitzwilliam title became extinct on the death of the tenth earl in 1979 and Wentworth Woodhouse was sold in 1989. It was owned privately until its sale in 2016 to the Wentworth Woodhouse Preservation Trust, a registered charity. The adjacent village, built to house the earl's workers, is now owned by the Fitzwilliam Wentworth Amenity Trust, which aims to preserve the original character of the village. The house needs extensive restoration work, now estimated at about £40 million. In 2016 the government gave £7.6 million towards the cost of repairs. The first phase was completed in 2020 and work is continuing on phases two and three.

*References*

ERIH, *s.n.* Wentworth Woodhouse

https://en.wikipedia.org/wiki/Earl_Fitzwilliam

https://en.wikipedia.org/wiki/Wentworth_Woodhouse

https://historicengland.org.uk/listing/the-list/list-entry/1132769

http://www.wentworthwoodhouse.co.uk/

### 403. Winding House (formerly Elliot Colliery Engine House)

Deep mining began at the New Tredegar Colliery, at New Tredegar in the Rhymney Valley in South Wales, during the 1880s with the sinking there of the East and West Elliot Pits. The colliery was well known for the quality of its steam coal. The present Elliot Colliery winding-engine house was originally known as the East Pit winder and served the upcast shaft of the East Elliot Pit, now filled and capped to the north of the building. The winding engine was installed in 1891 with two 1.4-metre horizontal cylinders of 1.82-metre stroke and a parallel drum. The pit developed water problems soon after production began and the engine was used to lift about 6,120 tonnes of water

**Figure 403.1. Side view of the Winding House, with historic winding house, which houses the East Pit winding engine of Elliot Colliery and around which the Winding House museum is built (© Museums and Heritage Service, CCBC).**

**Figure 403.2. Heart of the Winding House (© Museums and Heritage Service, CCBC).**

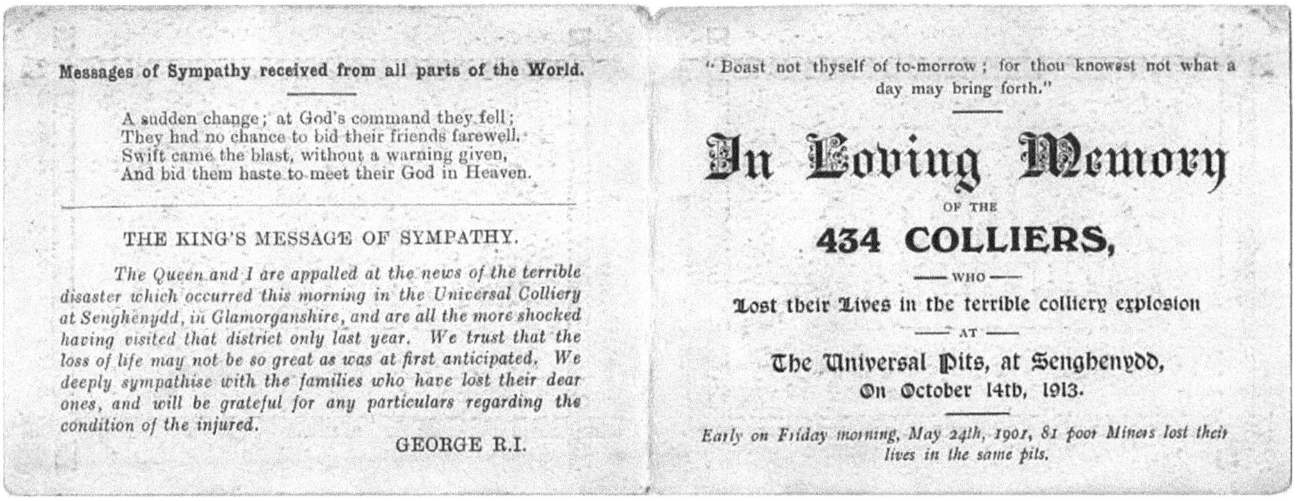

Messages of Sympathy received from all parts of the World.

A sudden change; at God's command they fell;
They had no chance to bid their friends farewell.
Swift came the blast, without a warning given,
And bid them haste to meet their God in Heaven.

THE KING'S MESSAGE OF SYMPATHY.

*The Queen and I are appalled at the news of the terrible disaster which occurred this morning in the Universal Colliery at Senghenydd, in Glamorganshire, and are all the more shocked having visited that district only last year. We trust that the loss of life may not be so great as was at first anticipated. We deeply sympathise with the families who have lost their dear ones, and will be grateful for any particulars regarding the condition of the injured.*

GEORGE R.I.

"Boast not thyself of to-morrow; for thou knowest not what a day may bring forth."

In Loving Memory
OF THE
434 COLLIERS,
WHO
Lost their Lives in the terrible colliery explosion
AT
The Universal Pits, at Senghenydd,
On October 14th, 1913.

*Early on Friday morning, May 24th, 1901, 81 poor Miners lost their lives in the same pits.*

**Figure 403.3. Memorial card printed in the aftermath of the 1913 Senghenydd disaster (© Museums and Heritage Service, CCBC).**

daily. In 1904 high-pressure boilers were installed and the engine was upgraded by the addition of two additional high-pressure cylinders, 71 centimetres in diameter. They were fitted behind and in line with the originals, which were left intact. The initial twin-cylinder engine now became a horizontal-tandem compound engine. The new installation had about 2,000 horsepower, giving faster winding and higher output. Production was about 3,060 tonnes per day. The original parallel drum was replaced at the same time by the unusual lighter semi-spiral drum, known as a diabolo, with a reverse taper. The wind started at the inside rather than the outside of the drum, which is still *in situ*. One cylinder of the winding engine was restored in 2021. The diablo drum is believed to be the only one of its kind surviving in Wales, and possibly the UK.

The pit closed in 1967 and the museum opened in 1998 with an extension for visitor reception and exhibition. The building and its contents were listed as grade II* in 2001. A group of friends and volunteers was formed in 2002. Caerphilly County Borough Council and the New Tredegar regeneration project, which transformed the wasteland that once surrounded the museum, gave a commitment to redevelopment of the site. Following successful bids for £3 million, a newly extended and much larger museum, the Winding House, opened in 2008, rebranded as the county borough museum. The museum occupies an area of about 46 metres by 15 metres, and has two exhibition galleries, the historic winding-engine house, research room and a multi-purpose activities room used for school workshops, museum events and room hire. The museum is owned, operated and funded by Caerphilly Council. There are four members of staff and a small team of volunteers. The museum attracted 8,044 visitors in 2019.

The displays at the museum cover the social history of the local area from prehistory to the present day, thereby placing coal mining in its overall historical context. An interactive touch-screen kiosk displays an animation illustrating the working of the winding engine, which is visible from the gallery through an inter-connecting original window. The displays within the historic winding-engine house itself interpret the history of the local coal-mining industry and of Elliot Colliery in particular. A team of volunteers meet each week to help maintain the engine. The volunteers also run the engine for visitors on the last Saturday of each month or by prior arrangement.

The museum collection includes a wide range of coal-mining artefacts, documents and images relating to Elliot Colliery and other collieries within the area. Particularly important items in the collections include a leather sinker's helmet dating from the early 1900s; the hooter used at Britannia Colliery between 1908 and 1983 to signal the start and end of shifts underground; a memorial card printed after the 1913 Senghenydd disaster (**350**); the National Union of Mineworkers silk banner (see **385**) used on the picket line by miners from Penallta Colliery during the 1984/5 Miners' Strike (**404**); and a series of photographs of mining communities in Bargoed and Aber Bargoed, taken in the mid-1970s by Swedish photographer, Kjell-Åke Andersson (born 1949). These include images taken underground, workers at Bargoed Colliery and family life in the area, forming a unique record of a 1970s South Wales mining community.

*References*


ERIH, *s.n.* New Tredegar

http://www.britishlistedbuildings.co.uk/wa-25495

personal communications, Emma Wilson, museums and heritage officer, 2016; Fran Quantock, lifelong learning assistant, 2021

www.windinghouse.co.uk

your.caerphilly.gov.uk/windinghouse/explore.../history-elliot-colliery

your.caerphilly.gov.uk/windinghouse/explore-history/winding-engine

**404. Woodhorn Museum**

Woodhorn Colliery is situated in the Queen Elizabeth II Country Park to the north-east of Ashington in Northumberland. The colliery was sunk in 1894 by the Ashington Coal Company, although the first coal was not raised until 1898. The first coal-cutting machine was installed in 1903 and by 1914 the colliery employed more than 2,000 workers. An explosion in 1916 killed 13 miners. The colliery declined in the 1960s, because the coal seams were thin and cheaper fuels, such as oil and gas, were challenging coal, so production finished in 1981.

The colliery opened as a museum in 1989, under the auspices of Wansbeck District Council. Following redevelopment, the museum reopened in 2006 with a major new building, the Cutter, with art gallery, exhibition spaces and facilities for Northumberland Archives. Subsequently responsibility for the museum was transferred from Northumberland County Council (the successor body to Wansbeck Council) to the Woodhorn Charitable Trust, a registered charity and company limited by guarantee. The trust receives an annual grant from the county council, with additional funds raised through commercial activities on the site. The museum has about 6,700 square metres of floor space. It employs about 100 staff, with some 90 volunteers, and attracts about 100,000 visitors a year.

The buildings of the colliery are all listed as grade II or II*, and the site as a whole was designated as a scheduled ancient monument in 1999. Original buildings remaining on site include an administrative building; blacksmiths' and joiners' shops, now used for large-scale temporary exhibitions, events and functions; stables; winding-engine house; number 1 heapstead, housing education and function spaces, and number 2 heapstead; capell-fan house, used to interpret the story of the 1916 disaster; pick sharpeners' shop (not open to the public); the Jack engine house, with information about the Hartley Colliery disaster (**375**); the Walker fan house, with access on guided tours; and the Crab engine house, containing artefacts and accessed on guided tours. There are timed demonstrations of the winding engine, and new interpretation and facilities opened in 2016. The 1899/1900 number 2 winding-engine house is the last Victorian engine house in Northumberland still containing a working winder. Similarly the capell fan, a small fast-running fan developed about 1860 and used at Woodhorn between 1900 and 1919, is believed to be the only capell fan still surviving in the UK. The Crab engine remained operational until 1981, making it the last steam-driven winder to work in Northumberland.

**Figure 404.1. Winding wheel at entrance to museum with two headgears to the right and their two winding-engine houses in the centre (© Woodhorn Charitable Trust).**

**Figure 404.2. Reception and display building known as the Cutter (© Woodhorn Charitable Trust).**

**Figure 404.3. Window of winding-engine house with reflection of headgear (© Woodhorn Charitable Trust).**

The two most important elements in the collections are on permanent display. The museum has 17 National Union of Mineworkers' banners from the Northumberland collieries. Five of these are displayed in rotation, being of silk (**385**), hanging above a sloping ramp. The displays deal with the 1984/5 Miners' Strike, led by Arthur Scargill (born 1938) in an attempt to prevent pit closures, including a large photograph of the then-prime minister, Margaret Thatcher (1925-2013), and her description of the miners as 'the enemy within'. The peak of coal production in the UK came in 1913, after which coal usage declined after Winston Churchill (1874-1965; see also **394**), as First Lord of the Admiralty (1911-15), began a programme of building dreadnought ships for the navy. These were fired by oil, sourced mainly from Persia (now Iran), because too much coal would have had to be carried to give sufficient increased speed. So the Conservative Margaret Thatcher in the 1970s completed what the Conservative Winston Churchill had begun in 1912.

In 1934 the Workers' Educational Association set up an art-appreciation class for miners in the village of Ashington, but the pitmen preferred to spend the sessions producing their own works showing life underground and in their community. A purpose-built gallery contains the Spennymoor Settlement Painters (also known as the Ashington Group) collection of 23 works created between 1934 and 1984 by the self-taught artists, who came to be known as the Pitmen Painters. The gallery also contains 'My life as a pitman', 39 works by Oliver Kilbourn (1904-93), one of the Pitmen Painters (see also **386**).

*References*

ERIH, *s.n.* Ashington

https://historicengland.org.uk/listing/the-list/list-entry/1016976

https://museum.wales/articles/1172/The-Miners-Strike-1984-85/

http://www.ashingtongroup.co.uk/home.html

personal communication, Deborah Tate, marketing officer, 2016

showcaves

The_1984-5_Miners__Strike_Resource__hi_res_ (2).pdf

# United States of America

America has the largest coal reserves in the world, estimated at 200 years of usage (Aden, Fridley and Zheng 2008, p. 9), and in 2012 was the second-largest producer of coal in the world, after China. There are mining museums in all the major coalfields (CMMCP 2016, fig. 36.1): the bituminous Appalachian (**407, 417**), although with some anthracite and once considered to be the most productive coalfield in the world; the bituminous Illinois coalfield in the Midwest (**409, 415, 416, 420, 422, 423, 425**); the anthracite and lignite Rocky Mountains coalfield in the west (**412, 431**); and the bituminous Black Warrior and Arkoma coalfield of the south-east, including Alabama (**405, 406, 418, 419, 428**). Coal deposits are also found elsewhere in America, including anthracite and semi-bituminous in Pennsylvania, which runs into the Appalachian coalfield and has the largest deposit of anthracite in America (**410, 424, 426, 427, 429, 430**), and lignite in North Dakota in the Midwest, worked by opencast mining (**408, 411, 413, 414**). In 2016 coal accounted for 24.7 *per cent* of energy production in America, approximately 12 *per cent* of which was exported. All American coal mines are owned and operated by private-sector companies. Coal is produced in 25 states spread across three major coalfields and there are more than 7,500 abandoned underground mines (CMMCP 2016, 36, p. 331). Unlike other countries that have committed to ending coal mining, as of January 2022 America has still declined to set an end date for the coal sector.

Outcroppings on the south bank of the river James in Chesterfield, Virginia, in the south-east, were discovered about 1700 by a hunter who dislodged lumps of coal. The coal seams were identified by French Huguenots. The earliest record of coal being mined in America was of a pit near Mid-Lothian (**419**) in 1737. In the eighteenth and nineteenth centuries there were a number of collieries working in the Richmond coalfield, which was about 38,850 hectares in size. This was North America's first commercial coal-mining venture, representing the earliest systematic exploitation of fossil fuel in America. The main development in mining came from the first half of the nineteenth century.

Foreign miners, initially English, French Huguenot, German, Scottish and Welsh and later from throughout Europe (**417, 423**), were hired for their expertise in planning and supervising the mining. The hard labour was performed by both free unskilled white men and enslaved and free Africans, including boys as young as eight. Many miners in American mines were slaves who were either owned directly by mine owners or rented from plantation owners, who took out insurance policies on their 'property', so that they could be compensated in the event of their death or serious injury. Early mining operators stressed the salutary working conditions and medical facilities when advertising in regional newspapers for rental slaves (Frantel 2008).

Men were often attracted to the mines by the offer of company housing and other facilities, such as medical care, schools and stores. But the large mining companies controlled every aspect of life and could evict miners who went on strike (**406**), as in the UK (**382**). Many American museums deal in detail with this aspect of life on the coalfields as well as with labour resistance, trade-union activity and supportive action by women, and the often-violent reaction by the mine owners. Miners also resisted the introduction of the longwall system until the 1920s (Palmer 2021, p. 7), because they considered it similar to a factory environment with direct supervision and so eroded the control they had over the work process with the pillar-and-stall system (**47**). Conditions continued to be worse for miners into the twentieth century in America than in Europe, and miners, and sometimes their families, died in the attempts to improve conditions and wages (**417, 426, 431**). American mines continue to be hazardous: in 2010 29 non-union miners were killed in West Virginia, one of whom had written a farewell letter to his fiancée because he thought that the conditions were so dangerous (Drexler 2011, p. 244), just as had been described in England in the 1930s (Orwell 1937, p. 40).

### 405. Alabama Mining Museum

The Alabama Mining Museum in Dora is located in a gymnasium built in 1935 by the Works Project Administration, set up to provide jobs and income during the Great Depression (1929-39). Coal mining was the motivating force behind the growth of cities such as Birmingham, Alabama, in the period between 1890 and 1940. The museum was designated by the Alabama State Senate as the Official State Coal Mining Museum of Alabama in 1984. Its aim is to preserve the history of how coal mining contributed to the development of the state.

The collections and displays include mining tools, which demonstrate developments in mining technology from hand-getting to modern machinery, social-history items and a 1900s train with its mining waggons. Information is provided about life in an early mining camp and how large mining companies controlled everything (**406**).

*References*

https://www.birminghamal.org/listings/alabama-mining-museum/

showcaves

### 406. Aldrich Coal Mine Museum

The Aldrich Coal Mine Museum is located in Alabama. Aldrich was originally a separate settlement, named after Truman Aldrich (1848-1932) when he started mining here in 1872, but is now effectively part of Montevallo just to its east. The mine produced high-quality coal for heating and steel production. The mine closed in 1942.

Coal camps were completely self-sufficient with their own churches, company physicians, post office, prison, schools, stores and other necessary facilities. The museum occupies a 1928 building, which was formerly the company store of the Montevallo Coal Mining Company and is listed on the Alabama Register of Landmarks and Heritage. Miners bought everything they required here, such as clothes, food, goods and tools. Goods were bought on credit and paid for by labour, resulting in the miners being totally dependent on the company. Company stores charged about 15 *per cent* more than private stores. This tommy system was effectively the same as what was known in the UK as the truck system, against which laws were passed in 1831, 1887 and 1896, and in the Rhineland in Germany in 1787. The tommy system was especially prevalent in the early United States because there was no national currency, an insufficient supply of coinage and banks often failed during financial crises so their notes became worthless. There were strikes against the practice by the newly formed UMWA (**416**), but there does not seem ever to have been legislation passed against it. Its abolition was not a demand during the major 1902 strike (**416**), which suggests that the major producers were not using it very much, if at all. There were, however, numerous small producers in rural areas, and it is likely that the tommy system persisted in these places both in anthracite, and probably bituminous, mining during the first decade of the twentieth century.

The museum exhibitions cover coal-mining memorabilia, fossils from coal seams and the original sales counter and cash register of the stores. There is an extensive collection of pictures of the mine and the town of Aldrich during its peak coal-mining days. Mining equipment is on display and a simulated 30-metre-long drift, called the Rajah Slope. Outside the building is the only monument to coal miners in Alabama.

*References*

https://en.wikipedia.org/wiki/Aldrich_Coal_Mine_Museum

https://en.wikipedia.org/wiki/Truck_system

https://www.cityofmontevallo.com/AldrichCoalMineMuseum.aspx

https://www.shelbyal.com/Facilities/Facility/Details/Aldrich-Coal-Mine-Museum-19

personal communication, Robert Wolensky, professor emeritus, University of Wisconsin, 2022

showcaves

### 407. Beckley Exhibition Coal Mine

Beckley is located in southern West Virginia, in the Appalachian Mountains. The Beckley Exhibition Coal Mine, which opened in 1962, is situated in the Beckley New River Park. It is funded by the city of Beckley and self-generated revenue. There are 30 members of staff and the mine attracts some 50,000 visitors a year.

The drift mine was worked by the Phillips family between the 1880s and 1910, and has a constant temperature of 14 degrees Celsius. Visitors travel in a mine car, which holds up to 32 people, for a 35-minute trip, led by retired coal miners, along 460 metres of roadway through worked-out areas. The tours start with the history of the mine and mining in the area, then various stopping points illustrate mining methods, ranging from timber roof supports to steel plates and bolts, and early coal-cutting machines.

The reconstructed company general store forms the visitor centre, 1,300 square metres in size. The centre has exhibits on the second floor, including many mining tools, and carbide and safety lamps. Also on the surface is the coal village of about 8,094 square metres, with a number of re-erected buildings that represent what would have been found at a mining camp in the early-twentieth century. A 1920s one-room shanty, which would have provided accommodation for a single miner during the week, was transferred from the Helen Coal Camp in West Virginia. Proper houses were provided by the coal company for rental to miners living on site with their families (**406**).

Beckley exhibits eight reconstructed buildings, which were donated and brought to the site from various coal camps in the area. Most of these were built between 1925 and the late 1940s by the New River Coal Company for their employees. The 1906 three-storey superintendent's house, which was moved from Skelton, has both living accommodation and the superintendent's office. Also displayed are reproductions of the barber's shop, company post office and physician's consulting room.

The school was built in 1925 at Berry Branch in Helen for black children living in the coal camp (there was a separate school for white children). The building originally had two rooms, each for one teacher and about 15 children. The schools were owned by the coal company, although administered by the board of education, and only residents of the camp were allowed to attend. Students learned the basic skills of reading, writing and arithmetic. As numbers in the camp declined, the school was changed to one room. The last one-room coal-camp school only closed in 1965.

The Pemberton Coal Camp church was built in 1921 and was used not only for worship, but also for business and social meetings. It had a basement that was used for church dinners, church gatherings, homecomings and meetings of groups such as the Ladies Aid Society. As the congregation were leaving the church at Christmas, the minister gave a gift of apples, chocolates, nuts, oranges and ribbon candy, paid

**Figure 407.1. Exterior of the Rahall general store, which houses the visitor centre and displays (© Beckley Exhibition Coal Mine).**

**Figure 407.2. Exterior of 1906 superintendent's house moved from Skelton (© Beckley Exhibition Coal Mine).**

**Figure 407.3. Exterior of 1921 church moved from Pemberton Coal Camp (© Beckley Exhibition Coal Mine).**

for by the coal company. This has become an Appalachian tradition which still survives today. Finally there is material on modern mining techniques and the development of clean-coal technology as a source of energy.

*References*

https://beckley.org/coal-mine/

http://www.mountaindiscoveries.com/stories/pdf/fw2002/heritagetrail.pdf

personal communications, Leslie Gray Baker, director of operations, 2016, 2021

showcaves

### 408. Beulah Mine

Beulah Mine is an opencast lignite mine in North Dakota, which opened in 1963 and was extended in 1980. It occupies an area of some 3,640 hectares and produces approximately 6.2 million tonnes of coal a year. The mine provides coal to the nearby 427-megawatt Coyote power station and the 75-megawatt Heskett power station, 120 kilometres away and supplied by railway. In 2009 Beulah won the North Dakota award for excellence in surface coal mining and reclamation. In 2015 the mine received the North Dakota public-service-commission reclamation award for avoiding working wooded areas that were important for landscape diversity and wildlife habitat. It also agreed a new contract to supply approximately 408,000 tonnes of coal a year to the Heskett power station until 31 December 2021.

The mine uses 195 BE shovels and 992 caterpillar loaders, together with a dragline, to work the coal. It operates with two shifts a day, five days a week. The mine provides tours of the site for visitors at all times when the mine is operating, although 24 hours' notice is usually required. It was operated by the Dakota Westmoreland Corporation, but in 2018 the corporation filed for bankruptcy and it has not been possible to obtain copyright permission to use the images of the mine previously suppled for this publication.

*References*

https://businessbankruptcies.com/cases/dakota-westmoreland-corporation

https://visitbeulah.com/play/explore/

https://www.lignite.com/mines-plants/mines/8472/

http://westmoreland.com/location/beulah-mine-north-dakota/

personal communication, B. Weaver, Dakota Westmoreland Corporation, 2012

showcaves

### 409. Big Brutus

Big Brutus, the nickname of a Bucyrus-Erie model 1850B electric shovel (see also **422.2**), is located at West Mineral in south-east Kansas. The shovel stands 49 metres, or 15 storeys, high and could move at 6 metres per minute. It has eight treads, a boom 46 metres long and weighs 4,990 kilograms. The bucket holds 69 cubic metres, or 136 tonnes, which would fill three waggons.

The shovel cost $6.5 million when it was built in 1962 and took 150 railway waggons to move it to the opencast mine. There it dug at a depth of between 6 and 21 metres. By 1974 the shovel was costing $27,000 a month to operate, which meant it was no longer economical. As it was thought to be too large to move, it was left *in situ*. In 1984 the Pittsburg and Midway Coal Mining Company donated the shovel to form the centrepiece of a museum as a tribute to American coal miners and the coal-mining heritage of south-east Kansas. The adjacent void left by the opencast mine is now filled with water and stocked with fish. In 1987 Big Brutus was designated as a regional historic mechanical-engineering landmark by the American Society of Mechanical Engineers, as it was believed to be the second largest electric-powered shovel in the world and the largest still in existence.

A 1920 Page dragline, as used in opencast mining since its invention in 1904, was donated by the Wilkinson Coal Company in 1991. The dragline was originally acquired to dig the Chicago Canal. The adjacent visitors' centre has exhibits, displays that are changed regularly and photographs of both Big Brutus and the history of the area. One prime exhibit is Little Giant, the world's smallest working replica of an early electric-mining shovel. This was constructed over an 11-year period by an enthusiast at a scale of 1 inch (25 millimetres) to 1 foot (305 millimetres) and weighs 317 kilograms. It was made in the 1930s and 1940s and purchased in 1946 by the Pittsburg and Midway Coal Mining Company. This shows exactly how electric-mining shovels operate and was designed to be able to be moved for exhibition away from the site.

*References*

http://the-working-man.com/big-brutus.html

http://www.bigbrutus.org/

showcaves

### 410. Broad Top Area Coal Miners Museum

Robertsdale is situated on Broad Top Mountain in Pennsylvania. The Rockhill Iron and Coal Company, which operated many of the local mines, founded the town in 1872 for its headquarters and as the terminus for the railway. The company owned the houses occupied by the miners, the post office, store, and the water and power systems (**406**). With the development of cheaper opencast mining, the last deep coal mine closed in the early 1950s. The Broad Top Coal Miners Historical Society was formed as a membership organisation to preserve and interpret the coal-mining and railway heritage of south and central Pennsylvania. The society acquired the 1948 Reality theatre and the United Methodist church to accommodate the Broad Top Coal Miners Museum.

The museum has artefacts and information about the Broad Top coalfield, including the history of the two railway systems in the area, ironworking and the mining of semi-bituminous coal. The museum deals with the social history and living conditions in the small hamlets of the region and of the coal miners and their families, together with the trade-union movement and strikes by the miners. Working conditions underground are depicted by a replica of a colliery underground. The museum also has a research library.

*References*

https://en.wikipedia.org/wiki/Robertsdale,_Pennsylvania

http://www.angelfire.com/pa3/btcoal/

### 411. Center Mine

Center Mine is located 6 kilometres south-east of Center in North Dakota. Work began in 1970, producing about 1.4 tonnes a year by opencasting. The mine expanded in 1977, so that each year BNI Coal now works some 81 hectares to supply about 4 million tonnes of lignite solely to the Milton R. Young Generating Station at Center. BNI has a permit to work a further 7,284 hectares, with an estimated 545 million tonnes of reserves. The land is reclaimed each year after mining has been completed. In 2015 BNI Coal acquired a new 4-million-kilogram giant dragline, called Liberty. It took over a month to move the dragline to the mine by road. There are tours of the opencast mine every weekday.

*References*

https://lignite.com/mines-plants/mines/8473/

http://www.bnicoal.com/

showcaves

### 412. Colowyo Mine

Coal mining began in the early-nineteenth century in the Danforth Hills of Colorado. The original miners' settlement of Mount Streeter was destroyed in the late 1970s by the operation of the mine. There are eight seams, between 50 centimetres and 3.65 metres thick, of hard low-sulphur lignite close to the surface. Colowyo Mine works this coal by opencasting, the upper three seams with three front-end loaders and 24 trucks and the lower by two large shovels and three draglines. The site also has seven bulldozers. In 1998 production was 5.3 million tonnes, the largest opencast production in Colorado. Coal is crushed on site, then moved by 45-tonne trucks, mainly to power plants. In 2001 there were still reserves of 151 million tonnes to be worked by opencasting and 100 million tonnes accessible only by deep mining. Reclamation of opencasted areas with native flora has won many awards. Tours of the site are offered for groups.

*References*

https://www.gem.wiki/Colowyo_Mine

https://www.wrcc.osmre.gov/initiatives/colowyo.shtm

showcaves

### 413. Coteau Freedom Mine

Coteau Freedom Mine is located just north of Beulah in North Dakota. It is the largest opencast lignite mine in the United States, producing some 14.5 million tonnes a year. Coteau Properties Company work the site and then reclaim about 243 hectares a year. The mine provides coal to the Great Plains Synfuels plant and the Antelope Valley, Leland Olds and Stanton stations, which are all able to be visited by the public.

Overburden is removed by three Bucyrus-Erie 2570 120-square-metre draglines and a fleet of 181-tonne overburden trucks. It would appear that it is not possible to visit the mine, but it can be seen from the road.

*References*

https://lignite.com/mines-plants/mines/freedom-mine/

https://www.youtube.com/watch?v=v9qKCaClZwo

showcaves

### 414. Falkirk Mine

Falkirk Mine, located near Underwood, North Dakota, produces an average of 7.3 million tonnes of lignite a year by opencasting. The mine is owned by the Falkirk Mining Company, a subsidiary of the North American Coal Corporation. The coal is sold to the nearby Coal Creek power station, which generates electricity for two-thirds of rural Minnesota.

Construction of the mine began in 1975 and deliveries to the power station began in 1978. The mine employs about 350 staff and has some of the largest earth-moving equipment in use today. Tours of the mine are available for the public every weekday during working hours.

*References*

https://lignite.com/mines-plants/mines/8474/

https://www.nacoal.com/operation/falkirk-mine

showcaves

### 415. Harrison Coal and Reclamation Historical Park

The Harrison Coal and Reclamation Historical Park is located in New Athens in Ohio. The organisation was founded in 1992 to ensure the preservation of the heritage of coal and opencast mining in Ohio, by collecting archives and machinery related to the surface-mining industry. Further aims are to educate the public in the importance of mining and the role that coal plays in everyday life.

At present the park is used for the storage and restoration of machinery. The organisation is working towards acquiring 81 to 101 hectares of land nearby, on which to construct a visitor centre and an open-air museum, where the annual old-construction and surface-mining equipment show can also be held each year. The main projects in which the organisation is currently engaged are the conservation of a Marion 111-M crawler dragline, a Marion 7200 walking dragline and an Insley model L dragline.

The park already has an extensive collection of machinery used in opencast mining in Ohio. The earliest item is a hand-loaded underground mine car, built about 1920 and weighing 272 kilograms, which came from the Bill Able Mine in New Athens. It is displayed filled with coal mined on the park site in the 1960s. From the early 1940s there is a horizontal highwall auger, mounted on a 1942 Army 6 by 6 truck. The auger was used to drill blastholes into the wall of the opencast mine above the coal seam. There is also a 1947 Hanson 3/8 cubic shovel, which was used for removing overburden above coal seams and was the first fully revolving shovel produced in America.

The 1950s are represented by a 2.7-tonne Euclid bulldozer blade and a portable air compressor used to power air tools and repair equipment. A 54-tonne bottom dump, made by the Dart Company, was used to move coal to the screens where the coal was loaded into railway waggons. This was the largest hauler when it was produced. An Insley model L dragline, with a 11-metre boom and ⅝-yard (57-centimetre) bucket, dates to 1966 and an 8280 Euclid bulldozer, weighing 33 tonnes and the most powerful of the Euclid Terex shovels with 476 horsepower, dates to 1968. Also from the 1960s is a rotary blasthole drill, used to make holes into which explosives could be placed to blast away the overburden. This weighs 9 tonnes and could drill down to between 46 61 metres.

An S-18 Euclid scraper pan, manufactured in 1975 and weighing about 20 tonnes with a pan holding 14 cubic metres of dirt, was used initially to remove topsoil and later to return it in the reclamation phase. Also from the 1970s is a Komatsu 355 bulldozer blade, used for moving dirt and weighing about 4.5 tonnes. The collection also includes about a further 30 items of heavy equipment.

*References*

https://coalpark.org/

http://www.hcrhp.org/

### 416. John L. Lewis Memorial Museum of Mining and Labor

The John L. Lewis Memorial Museum of Mining and Labor is located in Lucas in the state of Iowa. The museum was set up with grants from various trade unions, members of the local community and other organisations, such as the Central Iowa Building and Trades Council. It is run by a board of 12 and concentrates on the history of the United Mine Workers of America trade union (UMWA) and its president for 40 years, John L. Lewis (1880-1969).

UMWA was founded in 1890, when the Knights of Labor Trade Assembly number 135 and the National Progressive Miners Union merged, operating first in the bituminous coalfields, and shortly afterwards moving into the anthracite coalfields. In 1902 UMWA went on strike in the anthracite coalfield of eastern Pennsylvania for higher wages and shorter hours, both of which were achieved, and recognition of the union, which was not. This was the first labour dispute in which the federal government intervened as a neutral arbitrator. By 1933 the union was recruiting miners throughout America, eventually achieving a membership of 800,000. Its aims were to use the collective bargaining power of the miners to increase safety in the mines and to decrease their reliance on the mining companies and their stores (**406**). Under the leadership of John L. Lewis, UMWA left the American Federation of Labor (AFL) in 1935, when UMWA set up the Congress for Industrial Organizations (CIO), which recruited in the automobile, chemical, electrical-equipment, paint, rubber and steel industries. In 1942 UMWA resigned from the CIO and in 1944 rejoined the AFL. Today UMWA represents coal miners, clean-coal technicians, healthcare workers, manufacturing workers, public employees and truck drivers throughout America and Canada.

John Llewellyn Lewis began his career as a coal miner in Iowa and then in Illinois and was eventually elected as a branch secretary of UMWA. He became president of the union in 1921 and continued in office until 1960. During his presidency UMWA became the largest and most influential trade union in America, as coal was essential to the economy and the defence of the nation. He was regarded by many as autocratic, opposing the 1921 Amazon Army March by women supporting striking miners in south-east Kansas (**420**) and preventing communists standing for union office in 1923. In 1952 his campaigning resulted in the first Federal Mine Safety Act.

The coal-mining area of the museum shows old and modern coal-mining equipment and tools and relevant documents, items and photographs belonging to UMWA. The museum holds the original blueprints for a carbide light, along with its 1911 patent and instructions for its use. The labour-history section has letters, photographs, and the text of speeches and other items relating to Lewis' association with presidents of America and his work with UMWA, as well as other American trade unions, such as those of the workers in the automobile, rubber and steel industries. A cartoon gallery portrays Lewis with a variety of prominent people, including presidents of the United States. A memorial gallery has his letters, personal documents, photographs and other items, together with a life-size bronze statue of Lewis.

There are items from when Lewis lived in Lucas during his coal-mining days. The 40-seat theatre shows videos about coal mining, labour history and Lewis, while the library contains a collection of books, documents and photographs about Lewis, coal mining, labour history and the history of families in Lucas.

*References*

https://coalmininglabormuseum.com

https://en.wikipedia.org/wiki/United_Mine_Workers

http://www.umwa.org/

personal communication, Robert Wolensky (**406**), 2022

### 417. Kentucky Coal Mining Museum and Portal 31 Mine

The Kentucky Coal Mining Museum is situated in Benham in Harlan County, Kentucky. The museum tells the story of coal mining in eastern Kentucky, focusing especially on the company towns of Benham and nearby Lynch. Housing for the workforce was provided by 1,000 company-owned structures (**406**). By the 1940s the mining complex employed more than 4,000 people above and below ground and the population of Benham reached 10,000 at its peak. On 12 February 1923 a world record for coal production in a single nine-hour shift was set when miners, operating 40 shortwall-cutting machines, produced 11,630 tonnes of coal, which filled 256 rail waggons.

All collieries in the area were adit mines and all the coal produced was used for steel production. The mining company was upheld as a model employer, because it provided benefits, churches, education, healthcare, housing, recreation, social services and wages (**406**), but on the other hand the company's police force kept out union organisers and dissuaded miners from joining the union. The resultant strikes in the 1920s and 1930s and the counter-action by the coal companies meant that the county came to be called bloody Harlan.

In 1990 the Tri-City Chamber of Commerce purchased the former company store built in 1923 of concrete and masonry by the local coal company, International Harvester. The museum opened in 1994 with financial assistance from the state of Kentucky. In 2009 the museum began to handle tours in nearby Lynch of the Portal 31 Mine, named from its entrance. The mine produced some 0.98 million tonnes a year, from a seam 1.8 metres thick.

Outside the museum are a 1940s electric locomotive and a 1.814-tonne block of coal, cut specially for the museum. As well as the company store, the displays also include information about a miner's home, churches, company hospital, post office, school, theatre and the technology involved in coal mining. In the 1920s the United States Coal and Coke Company operated the world's largest coal-preparation plant, with a capacity of 13,608 tonnes, and the displays include a working scale model of a preparation plant. The collection belonging to the country-music singer, Loretta Lynn (born 1932), is on display on the third floor, entitled 'Coal miner's daughter'. A photographic exhibition deals with the stories of the many families, who came to work the coal in the region in the 1920s and 1930s, of 38 different nationalities, especially Czech, English, Irish, Italian, Polish, Scottish, Spanish and Welsh.

There are 30-minute tours offered of the Portal 31 Mine, which opened in 1917. The displays cover mining from the 1910s to the present day, depicted with animatronics at eight different stopping points. Material on display includes a scene of a miner making a new cut for ventilation, an electric locomotive and heavy mining machinery. Visitors are taken around the fully automated tour by train. Although no longer a working mine, the site is used by universities to train mining-engineering students. The museum website gives details of the development stage, which cost $2.5 million, and took decades to complete. Roof-bolting was used to secure the top of the tunnel, the roof and tunnel walls were covered with sealant and unused roadways were sealed to keep out methane. The website provides detailed plans as it is now, with the machines in place. Also on the site are the pithead baths, a 1920s lamproom, where mining tools are displayed and, at the train depot, what is claimed was the largest coal-loading system in the world when constructed.

*References*

http://www.portal31.org/

showcaves, *s.n.n.* Kentucky Coal Mining Museum; Portal 31

### 418. Madrid Old Coal Town Museum (formerly Old Coal Mine Museum)

The Old Coal Mine Museum opened some 50 years ago in Madrid, New Mexico, with displays of coal-mining equipment and objects from the coal-mining town. Coal mining began in Madrid around 1835. The company owned all the housing, hotel, stores and tavern (**406**), which was built about 1895, burned down in 1944 and rebuilt in 1947. In the 1930s Madrid was a multi-ethnic community and had over 3,000 residents. It was one of the industrial towns that was considered to be the most state-of-the-art in the south-west. Most of the mines closed in the 1950s as a result of a declining coal market and Madrid became a ghost town from about 1953 to 1973, but in the 1970s newcomers began to move into Madrid and to renovate the houses.

By that time many of the museum buildings and objects were in bad condition and did not meet the expectations of modern visitors. In 2007 the museum purchased the Mine Shaft tavern and began the process of consolidating and restructuring. Substantial changes have been made to the Madrid Old Coal Town Museum since 2009, with the focus on archiving, cataloguing, collecting, preserving and development of exhibits to a modern standard. The museum now consists of three buildings over an area of about 4,047 square metres. It employs one or two people on a seasonal basis and is in the process of obtaining non-profit-making status. Much of the most recent effort has been in digitising images, which will be used in future displays.

The museum has everything that was left when the coal company moved out and that the company could not use or take to another operation. Most of the objects in the museum's collections are on display. The most important object in the collections is the Richmond Locomotive Works engine number 769, which was built in 1900. It is one of only a handful of Santa Fe steam locomotives still in existence that were not built at the Baldwin Locomotive Works. The locomotive began life as Santa Fe Pacific Railroad 266, the first of a group of 20 2-8-0s built for this subsidiary of the Atchison, Topeka and Santa Fe railway company (AT & SF). In 1902 the locomotive became AT & SF 3045. Sometime later it was renumbered as 769 and sold to the Albuquerque and Los Cerrillos Coal Company in 1950. Along with locomotives numbers 870 and 874, 769 was sent to work at the coal mine at Madrid, located at the end of a branch line extending south from the Santa Fe railway at Waldo. The coal company never renumbered 769, and it spent its post-Santa Fe career in full AT & SF livery. When the mine shut, 769 was abandoned on site and left to rust for 20 years, until it was almost completely restored by John Coker. It is now considered to be the most complete non-operational steam locomotive in the country. The engine house that once housed 769 has been converted into the Engine House theatre for melodrama productions and is also used today for community events.

*References*

https://www.themineshafttavern.com/

http://www.cityprofile.com/new-mexico/old-coal-mine-museum.html

personal communication, Melinda Bon'ewell, joint owner, 2014

### 419. Mid-Lothian Mines Park

The Mid-Lothian Mines Park is located in Chesterfield, Virginia, where coal was first worked in America. The Mid-Lothian Coal Mining Company was chartered in 1835, working with four separate shafts where the bituminous coal seam was 231 metres deep and approximately 11 metres thick. The Mid-Lothian shaft was 220 metres deep, worked by 150 men and boys and 25 mules. A preacher, Jeremiah B. Jeter (1802-80), gave a sermon underground in 1843. The same year an author, Henry Howe (1816-93), was allowed to make an underground visit, about which he published an account in 1845. The Mid-Lothian shaft was renamed as the Pump shaft when a 500-horsepower Cornish pumping engine was installed in 1858 to recover operations after a methane explosion in 1855 killed 55 men (35 black and 20 white) and subsequent flooding. The coal was shipped to Baltimore, New York and Washington for use in blacksmiths' forges, in the production of gas and coke, and in the manufacture of chemicals, copper, glass and iron. Many men were killed in disasters in these shafts; for example, in 1882 32 miners were killed in an explosion at the Grove shaft. In the 1920s the Murphy Coal Corporation attempted to pump out and reopen the Grove shaft, but it still proved unprofitable and closed in 1923.

The Mid-Lothian Mines Park opened in 2004, with 18 hectares of woodland surrounding the remains of the 1870 main building of the mine. The granite building stands two storeys high and housed the pumps, steam boilers, ventilating fans and winding equipment. A replica headgear, constructed from 14.5 tonnes of wood, has been erected over Middle Shaft Mine in the same location as the 1836 original.

The Mid-Lothian Mines and Rail Roads Foundation board of directors was established in 2014 to build on the research that the foundation had been carrying out. In 2015 the foundation signed an agreement for maintenance and lease with Chesterfield County, which meant that the foundation could begin fundraising for its planned development. As well as consolidating the building remains, it is intended to build three structures, totalling 1,115 square metres, to employ 10 full-time staff, along with between six and 12 volunteers, and to attract about 100,000 visitors a year. These plans are currently on hold due to lack of funding.

Figure 419.1. Amphitheatre at Mid-Lothian Mines Park (© Mid-Lothian Mines and Rail Roads Foundation).

Figure 419.2. Reconstructed headgear in same location as 1836 original over the Middle Shaft Mine (© Mid-Lothian Mines and Rail Roads Foundation).

**Figure 419.3. Remains of 1870 ventilation building of Grove Shaft Mine (© Mid-Lothian Mines and Rail Roads Foundation).**

*References*

https://en.wikipedia.org/wiki/History_of_unfree_labor_in_the_United_States

personal communications, Robert P Jones, executive director, Mid-Lothian Mines and Rail Roads Foundation, 2104, 2015; Tom Garner, president, 2021

www.midlomines.org

**420. Miners Hall Museum**

The Miners Hall Museum is located in Franklin, Kansas. The museum is managed by a board of trustees and is funded by grants and by private and business donations. The museum employs three members of staff with about 50 volunteers and attracts more than 3,500 visitors each year. It opened in 2012 and has 465 square metres of interior displays, plus a large outdoor area of exhibits.

The mining of bituminous coal began in Kansas in the 1850s, and the first deep-shaft mine was sunk in Scammon in 1874. Production peaked in 1917/18, but continued until 2020. The aim of the museum is to preserve and display authentic materials and artefacts that document the history of coal mining and its impact on south-east Kansas. In particular the museum records the history of the immigrants who came to the area to work in the coal-mining industry, many of whose descendants still live in south-east Kansas. Subjects covered include the countries from which they came, their lives in the coal camps, education, living conditions, working conditions, and the social and religious lives of the families.

The first items taken into the collection of the museum included carbide cap lamps, notebooks, a canvas hat with lamp from the first half of the twentieth century and an aluminium dinner bucket from the 1910s. The displays include a set of shotfiring tools, such as rods for cleaning out holes before shotfiring and stemming, a selection of drill steels to use with a ratchet drill and a drill stand. The collections include a number of early photographs; for example, of the coal camp of Arma, of miners in a low seam at Umbria Mine near Mulberry, Kansas, and of a family (three older men, two male adolescents and a boy younger than 10) who all worked together in a coal mine.

In December 1921 thousands of women, mostly immigrants from south-east Europe, marched in south-east Kansas in support of better working conditions for their husbands and male relatives and of the striking coal miners, who were protesting against what they considered to be unfair labour practices. The women's march halted work in the mines for three days, but it was undertaken without the support of UMWA (**416**). The UMWA president, John Lewis, supported the mine owners, who invoked the 1920 Kansas Industrial Relations Act, which banned actions such as strikes and picketing. The march began from what is today the site of the Miners Hall Museum. Photographs on display show the site from which the march set out and the departure on the march of just a portion of the thousands of women.

*References*

http://www.amazonarmy.com/

http://www.minershallmuseum.com/

personal communications, Phyllis Bitner, secretary/treasurer, Miners Hall Museum Foundation, 2014; Linda Roberts, trustee, 2021

**421. Museum of Science and Industry, Chicago**

The Museum of Science and Industry, Chicago, is located in Illinois, in the eastern Midwest. It occupies the building that was originally the palace of fine arts at the 1893 Chicago World Fair, held to celebrate 400 years since the arrival of Christopher Colombus (1451-1506) in America in 1492. Designed by Charles Atwood (1849-95), it was more substantial than the other buildings at the fair, having a brick substructure under its plaster façade. The president of Sears, Roebuck and Company donated some $5 million and persuaded fellow members of the Commercial Club of Chicago to contribute to the cost of converting the palace of fine arts into a museum of science and industry. The facade of the building was reclad in limestone and the interior was remodelled in late-art-deco style. The museum opened in three stages, the first in 1933 during another Chicago World Fair held to celebrate the hundredth anniversary of the city, and the final one in 1940.

The simulated coal mine was opened in 1933 in the first phase of development. It is located on the main level of the museum's central pavilion, and was the first interactive experience created at the museum. It is a reconstruction of a working deep-shaft bituminous-coal mine. The replica mine was fitted out with original machinery and equipment transferred in 1933 from a mine known as Old Ben 17 in Lake Creek Precinct, Williamson County, Illinois (this is not related to Old Ben 1 in Lynnville, which is 207 kilometres away: **422**). The simulated mine is intended to demonstrate to visitors the development of coal-mining technology. Tours have to be pre-booked and are undertaken in groups, with explanations from a guide. The underground is reached by a slow ride accompanied by comments from the guide.

The first scene is coal being loaded into waggons, where it is weighed. It is moved by train into a skip, which takes it to the surface. Water in the mine drains into a sump at the bottom of the shaft, from where it is pumped to the surface. The next stage of the tour deals with the safety room, where testing for the presence of methane is explained. A ride on an underground train is accompanied by an explanation and demonstration of roof bolting. Simulated headings contain examples of different types of mining through the ages, starting with pillar-and-stall working (**47**) in which the roof is supported by wooden chocks. A guide demonstrates the use of an undercutting machine with simulated coal seams in the background. Next on display are a face-drilling machine and an automated loading machine. A geology crosscut illustrates fossils embedded in shale above a coal seam. A simulated longwall face is used to demonstrate modern mining technology. Back 'above ground' there is a reconstructed control room, with extensive banks of equipment to monitor conditions underground, such as levels of different types of gas, and computers recording loads of coal coming out of the mine, either up the drift or in tubs in the cage.

*References*

https://en.wikipedia.org/wiki/Museum_of_Science_and_Industry_(Chicago)

https://www.msichicago.org/explore/whats-here/exhibits/coal-mine/

showcaves

**422. Museum of the Coal Industry**

The Museum of the Coal Industry is situated in the city park of Lynnville in Warrick County in the state of Indiana. The museum was founded in 1985, with the goal of educating the public in the history and heritage of the coal-mining industry in Indiana, and in Warrick County in particular. The museum is administered by seven board members and the curator, with support from several volunteers on the site. Admission to the museum is free. To support its operation, the museum depends on donations, mainly from coal miners, and three coal miners' breakfasts each year. The museum attracts about 2,000 visitors a year.

The site of the museum was originally mined for coal, and so contributes to the museum's aim by illustrating how land can be reclaimed after mining. The museum has a frontage measuring almost 1.6 kilometres and opened all its facilities in 2000. The displays are housed in a building measuring 17 metres by 12 metres, which is full of mining items and tools. There is a montage of photographs of coal-stripping shovels that worked at the opencast Enos Mine, which was later renamed as Old Ben 1 (this is not related to Old Ben 17: **421**). A separate photograph, taken in 1921 when George Enos started the Enos Coal Mining Company, shows the Marion 300M machine with its crew, including a fireman, groundman, oilman, runner and seven trackmen. The Marion 300M was one of four stripping shovels at Enos Mine and was photographed as it was about to begin work removing the overburden in order to reach the coal. The library has books and other items on coal mining, including a time book dating to 1901 from the Big Four Coal Company.

The first major and rare item in the collections is a 50-pound (23-kilogram) coal loader, made in 1928 and used at Enos Mine between 1926 and 1969. This was the first totally electric machine built by the Bucyrus

**Figure 422.1. 23-kilogram coal loader used at the Enos Coal Mine, 1926 to 1969; made in 1928 as the first totally electric machine built by the Bucyrus Erie Company (© Museum of the Coal Industry).**

**Figure 422.2. First electric rope shovel built by the Bucyrus Erie Company; currently under conservation (© Museum of the Coal Industry).**

Erie Company in Evansville, 32 kilometres to the south (see also **409**). The second is the Tinker Bell locomotive, donated by Peabody Energy in 2011. This is an EMD SW1 surface locomotive, built in 1940 and used at the Peabody Lynnville Mine between 1968 and 1999. It would have been used for shunting the full or empty coal waggons that made up coal trains in the railway sidings, and also for moving waggons of materials on site. The locomotive has been stripped and repainted in its original livery. There is an outside walkway, with various pieces of mining equipment displayed alongside it, ranging from the Tinker Bell locomotive to some underground water pumps.

A joint company house (with four rooms) and company store (one room), 6 metres wide and 19 metres long, has been built at the museum, using volunteer labour and donated materials, such as rough-cut timber. This illustrates how people lived between 1880 and 1930 in homes owned by the company (**406**). The building is elevated on stilts, as was traditional. The rooms all have authentic furnishings.

Conservation is being carried out on the first electric rope shovel built by the Bucyrus Erie Company; most rope shovels of similar date were steam-powered rather than electric. The museum also hopes to add an underground experience when funds allow.

*References*

http://lynnvillecoalmuseum.org/index.html

personal communication, Aja Mason, curator, 2016

### 423. Novinger Coal Miners Museum

Novinger is located in Missouri, where coal mining began in 1890. At its peak, between the late 1880s and the early 1900s, Novinger was the leading coal centre in Missouri, with 27 coal mines and some 5,000 inhabitants. Most of these were originally of German descent from Pennsylvania, and later workers came from Croatia, France, Ireland, Italy, Poland and the UK. The mining payroll was more than $50,000 a month. Novinger was at the intersection of two railways, one running north-south and the other east-west. Several trains a day, loaded with tonnes of coal, left Novinger every day for markets all over the Midwest.

The area began to boom with many square, four-room miners' houses being built. This brought the need for more businesses, causing many brick buildings, with one or two-storeys, to be erected to house them. By 1902 Novinger had gas streetlights and a few years later pedestrian pavements were added to the downtown area. By 1909 some mines were working shifts around the clock. It was said the miners' pit lamps resembled fireflies in the dark as they travelled to and from work. The Novinger coalfield reached peak production in 1918, with 715,800 tonnes.

The mines were unionised and strikes by the miners were frequent. In the 1930s many houses were moved intact into the town from the mining camps. The mines were especially active between 1914 and 1945, after which they became unprofitable and nearly all had closed by the early 1950s. Billy Creek Coal Mine, the last deep-shaft mine in Missouri, closed in 1966.

By 1930 Novinger's population had dropped to 846 and the population today is only 547. The concept of preserving the city's heritage was agreed when the city started planning for its centennial celebration in 1976. Novinger Planned Progress, set up in the late 1950s to do community projects, took on responsibility for developing the Novinger Coal Miners Museum to preserve memories of Novinger's coal-mining community. In 1971 the organisation purchased the Rock building (named after the Rock family who ran a butcher's shop here from 1924), and opened it as their first museum.

More buildings were acquired as the collections grew. The Rock building was redisplayed as a replica of an early butcher's shop, along with a reproduction of Gatlin millinery, a hat shop from 1905 to 1940, and the office of an accountant, Charles Gaspen, who provided book-keeping services to a number of the mines in the area between 1924 and 1966. The present museum opened in 1990 and occupies an area of about 300 square metres. It is run by about 10 volunteers and is funded by donations. The museum is visited by about 200 visitors a year, depending on what is happening in the area to attract people to the town. It is open by appointment.

The museum deals with the history of coal mining in northern Missouri from about 1900 to the 1950s. Displays include a simulated coal mine and a reconstructed blacksmith's forge, a model of a screening and rail-loading plant, closely integrated with the actual mine headgear and many other coal-related items. The museum also features four store-front exhibits containing artefacts from various former businesses and many objects connected to the community's history. There is a library for genealogical research, which includes material relating to Novinger's coal-mining past.

*References*

https://www.visitmo.com/novinger-coal-miners-museum.aspx

http://www.adairchs.org/NovingerRenewal/Novinger/coal-miner.htm

personal communication, Glenna Daniels-Young, volunteer, 2016

### 424. Number 9 Mine and Museum

Anthracite mining began in the early-nineteenth century in the region around Lansford. The Lehigh Coal and Navigational Company (LCNC) played a major role in the development of the eastern end of the southern Pennsylvania coalfield. Planning by the company for Number 9 Mine began in 1845, but flooding thwarted the first two attempts to sink the mine. It was not until 1856 that the coal seam was reached successfully. The local anthracite is often referred to as the black diamonds, because it is a very high-grade, quite rare, coal which is almost pure carbon and essential for certain industrial processes, such as the production of steel.

The mine was one of the most productive in the area and at its peak employed between 450 and 500 men. In 1931 the mine entrance and several hundred metres of the main roadway were rebuilt with steel arched girders and reinforced concrete in an effort at modernisation to extend the mine's useful life. The mine finally closed in 1972,

**Figure 423.1. Exterior of Novinger Coal Miners Museum (© Novinger Planned Progress, Inc).**

**Figure 423.2. Model of a screening and rail-loading plant, closely integrated with the actual mine headgear (© Novinger Planned Progress, Inc).**

**Figure 424.1. Exterior of Number 9 Mine and Museum, using the original pithead baths, known as the 'wash shanty' (© Panther Creek Valley Foundation).**

**Figure 424.2. (© Panther Creek Valley Foundation).**

at which time it was claimed by UMWA (**416**) to be the 'world's oldest continuously-operated deep anthracite-coal mine'. This may be an exaggeration, but it would certainly appear to have been the longest working anthracite mine in America.

The Panther Creek Valley Foundation was established in 1990 as a not-for-profit organisation to preserve the mine and enable the public to see the underground workings. The museum opened in 1992 and the first underground tours began in 2002. There are 13 officers and directors of the foundation and up to 15 volunteers assist each year. There is one full-time employee and a rotating staff of between 10 and 12, with three to four on duty each day. The museum attracts over 10,000 visitors annually. It is funded by admission charges, donations, gift-shop sales and membership dues, plus grants when they are available.

The museum deals with the story of the anthracite miners of north-eastern Pennsylvania in the nineteenth and early-twentieth centuries. It is located in the original pithead baths, known as the 'wash shanty'. The way that the miners' clothes were stored on hooks that were hoisted up to the ceiling is the same as the system used in continental Europe. A wide range of mining tools and equipment are displayed in the engineering workshops, as well as two large models of coal-preparation plants. One of these was commissioned by the company's advertising department in 1952 and shown at coal-trade shows across America and Canada in the early 1950s. Models of a boy and a mule illustrate how tubs of coal and equipment were moved underground. Oxen were used in some of the anthracite mines in the region before the widespread adoption of mules, but the LCNC used only mules throughout the life of their mines until their closure in the 1950s. Mules had the advantage that they could go into areas where it was dangerous to use electricity for powering mine cars. The museum has a death certificate for a mule killed in the mine and two leather mule caps that were used to protect the mules' heads after overhead electric power lines were introduced underground.

The museum also deals with social history, such as reproductions of a typical kitchen in a miner's home and a coal-company store (**406**). Also on display is a small steam locomotive, Lokie (*cf.* **427**), which was built in 1923 and operated on the narrow-gauge line connecting the Nesquehoning, Lansford and Coaldale mines. It has a gauge of 42 inches (1.1 metres), the standard for the LCNC mines. It was sold in 1954 and acquired by the museum in 2009. The museum has many maps, pictures and tools from the earliest dates of anthracite mining up to the closure of the mine.

Visitors enter the mine by railway, travelling some 488 metres into the hillside, where the temperature is between 9 and 12 degrees Celsius all year. Visitors then do a 183-metre guided walk through the mine. There are two cages at the bottom of the 275-metre-deep shaft that brought coal up from lower levels and a narrow roadway between the first and the second levels, 46 metres below the first. This was known as the muleway, as it was where the mules were brought up and down as needed, and was also a secondary means of escape for miners to get back from the lower levels to the first level if the cages were not operating. A shaft driven in the late 1880s goes straight up about 18.3 metres to the monkey, which is the airway just below the workings. The original underground first-aid room, constructed about 1930, was used for rapid treatment of injured miners before they were taken back to the surface.

*References*

https://no9minemuseum.wixsite.com/museum

personal communications, David Kuchta, president, Panther Creek Valley Foundation, 2016; Zachary Petroski, president, 2021

showcaves

### 425. O'Gara Mine Number 12

The winding unit of O'Gara Mine Number 12 is preserved in Muddy in the township of Harrisburg in Saline County in the state of Illinois. The Harrisburg coalfield covers 200 square kilometres of land. The area was originally dependent on cotton and tobacco growing, but in the Reconstruction Era (1865-77) after the American Civil War (1861-5), these were replaced by a number of industries, in particular coal mining. The first mine was sunk in 1854 south-east of Harrisburg, the coal being transported by waggon for local use. The first railway, the Cairo and Vincennes, was completed in 1872 and enabled coal to be moved further and more easily. From 1890 Harrisburg became one of the leading centres for coal mining in Midwestern America. There were three leading mining companies in the area, each with its own railway yards. One of these companies was set up by Thomas O'Gara of Chicago (born about 1865), who operated 12 mines in Saline County, which employed 6,000 men and produced 7 million tonnes each year.

Large numbers of immigrants moved from England, Wales and eastern Europe to work in the mines in the early-twentieth century. They established a number of small satellite villages around Harrisburg, including the village of Muddy. In 1923 the O'Gara Company erected a large headquarters building in Harrisburg, where the coal industry continued to thrive, even during the Great Depression (1929-39).

Harrisburg is located on the south bank of the river Saline. Muddy was built on the opposite, north, bank of the river to house miners working in O'Gara Mine Number 12, which began operating in 1923. Photographs of this mine when it was operating are housed in the Coal Mine History Room of the Saline County Museum in Harrisburg. In early 1937 a catastrophic flood along an 885-kilometre stretch of the river Ohio, between Pittsburgh and Cairo, left 385 people dead and a million homeless. In Harrisburg alone 80 *per cent* of the city was flooded and 4,000 people were left homeless. As well as homes, many mines were submerged. Although the state of Illinois paid to pump 11 billion litres of floodwater out of Sahara Mine Number 3, it was not economic to drain the majority of the flooded mines, thereby leaving the local economy devastated. O'Gara Mine Number 12 was one of those that had to cease production in 1937.

The only surviving structure from the mine itself is the reinforced-concrete winding unit. The structure resembling a headgear at the top of the winding unit is actually for raising tubs to the uppermost level, so the coal would then flow down through the various sorting screens and into the waiting waggon. It is reminiscent of locomotive coaling towers built in the UK in the inter-war years (1918-39), except that there would have been two tubs counterbalanced in the same manner as two colliery cages, so that the winding unit was dealing only with the weight of coal. The winding unit must have been adjacent to a screening and rail-loading plant, because in the earliest photographs there are two small winding wheels on the top with ropes going down behind the building. The hoisting must have been done at speed or else with very large tubs, as otherwise there would have been a bottleneck in the flow of coal.

The winding unit bears the date of 1923. This makes it one of the oldest surviving reinforced-concrete mine structures in northern America, the other being the headgear dated 1913 at Morden Mine (**52**), which is in worse condition. Concrete was being used in the United States in the nineteenth century, but not on this sort of scale except for dams and coastal-defence forts. Timber continued to be used well into the twentieth century. So the use of concrete here suggests that perhaps the supply of timber of the strength and size required was not readily available locally and so the cost of transport made the use of concrete a competitive alternative. Alternatively the company may have been making an architectural statement, as is perhaps suggested by the detailing, such as the inset date and name, which is reminiscent of modernist architecture.

In 1944 John Molinarolo Senior purchased the mine site, together with 45 company homes, a number of which still survive. The winding unit also still survives on private land, which was owned by John Molinarolo Junior (1926-2020) of Flint Hill Vineyards and Winery, who ensured its preservation. The public can visit the site with permission from the Village Board. Surprisingly the winding unit does not have any statutory protection and it is not registered in the section for Saline County of the United States National Register of Historic Places.

*References*

https://en.wikipedia.org/wiki/Harrisburg,_Illinois

https://en.wikipedia.org/wiki/Muddy,_Illinois

https://en.wikipedia.org/wiki/Ohio_River_flood_of_1937

https://towns-and-nature.blogspot.com/2017/04/muddy-il-ogara-coal-mine-12s-tipple-is.html

### 426. Pennsylvania Anthracite Heritage Museum, Eckley Miners' Village and Lackawanna Coal Mine Tour

The Pennsylvania Anthracite Museum complex is made up of the Pennsylvania Anthracite Heritage Museum and the Lackawanna Coal Mine Tour, which are situated in McDade Park, Scranton, together with the Eckley Miners' Village. The complex is owned and operated by the Historical and Museum Commission of the Commonwealth of Pennsylvania and is supported by the Anthracite Heritage Museum and Iron Furnaces Associates, and the Eckley Miners' Village Associates. Both of these are not-for-profit community-based organisations.

The Anthracite Heritage Museum preserves the history of hard-coal mining and related industries in north-eastern Pennsylvania, together with the story of the immigrants who came from some 38 different countries in Europe to work as miners (**417, 423**). Some were specialist workers and others uneducated peasants with no industrial experience. Working and social life in the factories, mills and mines is explained with, for example, a reconstructed church, kitchen and pub (**406**). The museum has a library, with books, documents, ephemera, images, maps and periodicals. The museum also runs a four-day summer camp, 'young curators in our community', to introduce children to the heritage and material culture of the region, with visits to local sites conducted by professional staff and volunteers.

The Lackawanna Coal Mine Tour is situated in three separately worked seams of an anthracite adit mine known as Slope 190, which opened in 1860 and closed in 1966. The local anthracite is a very high-grade coal (**424**). After watching an explanatory video in the interpretation centre, visitors take about four minutes to descend in a mine car to the Clark seam, at a depth of 91 metres. The seams here are known as veins, a term more generally used in Europe for lodes in metalliferous mining. There they have an hour-long guided walking tour of about 805 metres. The temperature in the mine is 12 degrees Celsius all year. Visitors stop at various points on the main roadway of the mine, then move into the Dunmore 1 seam by a rock-cut roadway, for explanations about transport and mining in a fault. The tour then stops at what is described as the peg shanty to learn about the role of the deputy, known here as the fire boss. A peg shanty was usually above ground near the entrance to the shaft or drift, where the mineworkers would peg in, or place their tally (known here as a chit), on a hook indicating that they were underground, a role performed in the UK by the banksman. When they returned to the surface, they had to peg out, or move the tally to the out section of the peg board.

In the Dunmore 2 seam guides explain about working in thin seams, the second means of egress from the mine, the role of air doors in ventilation and the work of mule boys and door boys (known as nippers), who opened and closed the ventilation doors. In 1900 there were 25,000 boys under the age of 16 working in Pennsylvanian mines. Boys under 14 were forbidden to work underground, but an investigation by the National Child Labor Committee in 1906 found over 150 boys as young as nine working underground in one small town, with frequent deaths. The tour finishes with mining at the time of the closure of the colliery.

Guided tours of the Eckley Miners' Village explain the history of anthracite mining and the associated settlements and their residents. Established in 1854, Eckley is an example of a planned nineteenth-century town built by the mining company close to the mines (**406**). In 1969 Eckley was chosen, because its 1870s appearance was so well preserved that only the television aerials and power lines had to be removed, for Paramount Studios to film 'The Molly Maguires'. This was a secret organisation in nineteenth-century Pennsylvania of Irish coal miners who fought the mine owners over safety concerns, resulting in the execution of 20 miners in 1877/8. The making of the film saved the town from demolition and led to its preservation under the auspices of the Historical and Museum Commission.

A visit begins in the visitor centre, which opened in 1975, with a 17-minute film and the exhibition of tools and pictures of the working and social life of the mining families. The oldest buildings in the village, dating back to 1854, are the labourers' dwellings, which were double structures with each half housing up to 15 people. Boarders slept in the two rooms upstairs and the family in one room downstairs. The kitchens downstairs are still equipped with the large coal stoves typical of the nineteenth century, as well as outhouses and cellars for storing root crops like turnips. There are also three houses of slate pickers, who were the lowest paid of the workers, so could only afford to live in tiny houses with two rooms downstairs and a sleeping loft upstairs. They were usually boys or older or disabled men, who picked out slate from the coal in the screens building (*cf.* **378**), known here as the breaker. The breaker on display on the site was built in 1970 by Paramount Studios. Buildings in the village were often coloured red or black, because those were the least expensive paint colours available (*cf.* the blue colour of the screens building at **386**).

The Immaculate Conception church was built in 1861 for the Catholic Irish miners. It was deconsecrated in the 1950s and restored to its 1920s appearance. The Saint James Episcopal church, built from wooden planks, dates from 1859 and served the English, German and Welsh Episcopalians. The 1946 Eckley Sports and Social Club was donated by the mining company for use as a community centre and is still in use. The original company store was built in 1857 and demolished in the 1940s (**406**); the present replica was built by Paramount Studios for their film.

*References*

http://eckleyminersvillage.com/

https://en.wikipedia.org/wiki/The_Molly_Maguires_(film)

https://morningsonmaplestreet.com/2014/11/26/joseph-puma-page-one/

https://www.youtube.com/watch?v=Rrpl2VpQMgc

http://www.anthracitemuseum.org/

http://www.undergroundminers.com/slope190.html

personal communication, Robert Wolensky (**406**), 2021

showcaves

### 427. Pioneer Coal Mine Tunnel

Pioneer Colliery, in Ashford in eastern Pennsylvania, was constructed as a horizontal adit mine, running 550 metres horizontally into the side of Mahanoy Mountain. It was dug specifically to drain water downwards, after the first opencast mines in the area filled with water when only about 10 to 15 metres deep. The local anthracite is a very high-grade coal (**424**).

In 1911 the Pioneer Tunnel Mine was being operated by the Philadelphia and Reading Coal and Iron Company. Most of the inhabitants of Ashford in the early-twentieth century earned their living from coal mining. As a result the town gradually declined after the closure of the mine in 1931, so that it is still virtually the same as it was a century ago. The Ashland Community Enterprises was set up as a not-for-profit organisation to develop and operate a tourist mine. After renovation work, the Pioneer Coal Mine Tunnel reopened in 1963, and now attracts about 40,000 visitors a year. Some of the tunnel had to be retimbered, but the rest had been cut through solid rock, which meant that there the roof did not need supporting.

Visitors are taken 550 metres into the tunnel on battery-powered underground mine cars. They then switch to walking for the 35-minute tour, guided by former miners. The guides explain how deep mining operates now and in the past, and show how the seams of coal lie in strata interspersed by rock. The temperature underground is a constant 10 degrees Celsius.

A further 30-minute tour is available on the surface on a railway using a narrow-gauge 27-tonne saddle-tank steam locomotive, known as the Henry Clay, built in the 1920s. Known as a lokie (*cf.* **424**), these small locomotives, of which only a few are left in existence, measured only 7 metres in length, 2.4 metres in width, 3.7 metres in height and had 10-metre drivers. The route goes 915 metres around the side of Mahanoy Mountain to the site of the opencast Mammoth Stripping Mine; this worked the thick Mammoth seam where it outcropped, leaving behind a rock face 50 metres high.

As well as the panoramic view over the town, visitors also see an example of a bootleg hole, dug surreptitiously by miners to obtain coal for themselves for heating or to sell. Smoke can also be seen rising from the ground at a number of places from the Centralia Mine fire. It is not certain whether it is just the Buck Mountain seam which is on fire, or whether the many roadways underground have allowed the fire to spread into seams above that. Attempts to extinguish it have been unsuccessful and it may continue to burn for hundreds more years.

*References*

http://www.fieldtrip.com/pa/78753850.htm

showcaves

www.pioneertunnel.com

**428. Pocahontas Exhibition Coal Mine**

Pocahontas is located in Virginia. It was named after the daughter of Chief Powhatan (1545-1618), Pocahontas (about 1596-1617), who lived in the seventeenth-century Jamestown settlement of Virginia. She married an English settler, John Rolfe (1585-1622), and died and was buried in Gravesend in England. The Pocahontas Exhibition Mine opened in 1882 and was the first colliery in the Pocahontas coalfield. It worked Pocahontas number 3 seam, which was 4.3 metres thick and produced some 40 million tonnes of coal. It covered an area of 78 square kilometres. Bituminous Pocahontas coal was used to power the ships of the United States navy, because it did not produce smoke when burnt, thus concealing the presence of the ships from the enemy when at a distance. Pocahontas coal was also especially good for coking, but eventually on-site coking ceased and only three beehive coking ovens now survive.

The Pocahontas Exhibition Coal Mine and associated museum were opened to the public in 1938, the first tourist mine in America. Until the mid-1950s the visitors were able to drive their cars through the mine, entering through the fan drift and leaving through the main adit. It was found, however, that exhaust fumes were damaging the roof of the mine, so tours are now undertaken only on foot, led by guides. Several roadways, 290 metres in length, were maintained for the tour when the mine closed in 1955. The mine and the museum were declared a national historic monument in 1994.

The tour begins with mining in the late 1800s, when coal was worked with pick and shovel. This is followed by demonstrating the old method of undercutting by machine, then blasting down the overhang for the coal to be loaded onto mine cars. Finally modern methods of mining are explained.

*References*

dmme.virginia.gov/dmme/dapocahontas.shtml

https://en.wikipedia.org/wiki/Pocahontas_Exhibition_Coal_Mine

showcaves

**429. Seldom Seen Tourist Coal Mine**

The Seldom Seen Tourist Coal Mine is located in North Patton in Pennsylvania. The colliery opened as a drift mine in 1902. Throughout its history it was always a small family-operated mine, as was common in this area. In 1933 the adit roadway was extended right through the hillside. Where it came out on the other side is now the tourist entrance. Generally motive power came from mules, and the only electrical equipment was the pump to keep the mine dry. In 1943 a shortwall coal cutter was installed, along with a battery-operated locomotive. A conveyor belt was added in the late 1940s. The mine reached its peak production of 30,400 tonnes, with 37 employees, in 1951 to meet the increased demand during the Korean War (1950-3). In 1963 only 703 tonnes were produced, so the mine closed and the top of the hill was removed by opencasting. The same year the underground mine was sold to E.J. Haluska, who opened it to the public as the Seldom Seen Valley Mine. In 1973 rising petrol prices meant that it cost much more for visitors to reach the site, so their numbers dropped and the mine was sold. The tourist mine closed in 1990 but reopened in 1993, run by volunteers operating as members of a not-for-profit organisation.

As the mine went straight from being a productive colliery to a tourist attraction, all the equipment from when it was operating remained underground, so it is more complete than many other underground museums. Visitors are taken by the battery-powered former mine train to the coalface, where the guides explain about the use of mules for haulage and the pillar-and-stall method of mining (**47**). Also on display are the conveyor belt, pumps and shortwall cutter. The conveyor belt, loaded with coal, and the cutting equipment are operated to show how noisy the working environment was.

In 1901 three men were charged with being part of a gang of six who broke into a house and robbed the occupants of a considerable amount of money and jewellery. As the robbers were masked, they could not be positively identified, so the charged men were acquitted. The three men were never seen again. Legend says that they and their colleagues buried their haul in the Seldom Seen Mine, but were themselves buried later by a roof fall when trying to collect it. Neither their bodies nor the loot have ever been found, but the mine is reputed still to be haunted by their ghosts.

*References*

https://www.onlyinyourstate.com/pennsylvania/pittsburgh/historic-coal-mine-near-pgh/

https://www.visitpa.com/region/alleghenies/seldom-seen-valley-tourist-coal-mine

showcaves

**430. Tour-Ed Mine**

The Tour-Ed Mine is located in Tarentum in Pennsylvania. It operated from 1859 until 1969 as the Avenue Mine, and opened to the public in 1970. The mine is operated by a not-for-profit organisation and is set in the 5-hectare Ira Wood Park.

Visitors are first given a short talk about mining and safety by one of the guides, all of whom are former coal miners. The 30-minute tour shows what it was like to be a coal miner in the 1860s and in modern times. The mine is accessed by battery-operated locomotive, with groups of about 25 visitors, and then the tour continues on foot. The form of mining demonstrated is pillar-and-stall (**47**), with pillars 18 metres long and 12 metres wide. There is a mining museum in a chamber underground, with equipment ranging from the 1850s, such as pick and shovel, through to pneumatic drills, a 1930s conveyor belt and a 1940s loader. Much of the mining equipment is still operated for visitors. The temperature underground is a constant 11 to 13 degrees Celsius. More than a million visits have been made to the mine since opening. The mine has been used a number of times as a set for films because, as well as the underground, a typical 1850s home of a coal miner and his family has been recreated on the site.

*References*

http://www.tour-edmine.com/

showcaves

### 431. Walsenburg Mining Museum

Walsenburg is located in Colorado. It attracted many German settlers and the name of the growing town was changed from the Spanish *La Plaza de los Leonas* (the lions' plaza) to Walsenburg. It was named after Fred Walsen (1841-1906), who came to America in 1859 from Prussia in Germany and in 1870 built a large store here. He was also responsible for opening the first coal mine in the area in 1876 and several more subsequently, after which coal mining became the main industry of the area. It is estimated that about 450 million tonnes of coal were mined here, leading to Walsenburg being known as 'the city built on coal'. Walsenburg first became well known nationally as the home of Bob Ford (1862-92), who opened a saloon and gambling den in Walsenburg about 1890 and who killed his gang leader, the outlaw Jesse James (1847-82).

In 1912 the death rate in Colorado mines was 7.055 per 1,000 employees, compared to a national rate at that time of 3.15, because explosions and roof falls were common, and with 0.011 per 1,000 employees in American mines in 2016. The poor pay and harsh conditions under which miners worked in Colorado led to many strikes and to violent conflict in 1914 between the miners and the mine operators. Accounts of the numbers of those killed differ. It was claimed that in the Walsenburg area in just 10 days 54 men, women and children were killed during the dispute. Walsenburg was brought to the attention of a wider audience in 1944 with the lyrics by the folk singer, Woody Guthrie (1912-67), about the 1914 Ludlow Massacre. This relates to the deaths of about 24 people during the strike by 1,200 miners. The dead included four women and 11 children, who suffocated when the tent above them was set on fire. One verse says:

> Our women from Trinidad they hauled some potatoes,
> Up to Walsenburg in a little cart.
> They sold their potatoes and brought some guns back,
> And they put a gun in every hand.

Trinidad is 61 kilometres south-east of Walsenburg, which in turn is 40 kilometres north-west of Ludlow.

The Huerfano County Historical Society took over the 1896 jail building in Walsenburg and has converted it into the Walsenburg Mining Museum. This deals with the stories of Bob Ford, the local mining camps, and the mining strikes and disputes. One character featured is 'Mother' Mary Jones (1837-1930), who was a union organiser and a leading figure in strikes between 1871 and the 1920s, including organising resistance by the wives and children of miners. In 1902 she was described at one of her trials as 'the most dangerous woman in America'. The jail cells display coal-mining equipment. Outside the building is a bronze statue of a coal miner on a plinth inscribed with the names of local coal miners.

*References*

http://en.wikipedia.org/wiki/Walsenburg,_Colorado

https://arlweb.msha.gov/stats/centurystats/coalstats.asp

https://en.wikipedia.org/wiki/Ludlow_Massacre#Massacre

https://en.wikipedia.org/wiki/Mary_Harris_Jones

http://woodyguthrie.org/Lyrics/Ludlow_Massacre.htm

http://www.huerfanohistory.org/contacts.html

# Vietnam

Vietnam has two major coalfields: anthracite is found in the Quảng Ninh coalfield and sub-bituminous coal in the Red River Delta, both on the north-eastern coast. Exploitation of Quảng Ninh coal began in 1839 (Bui and Lê 2011), but the full extent of the coalfield was only discovered by the French in 1881-2. During the 1870 Franco-Prussian war, the UK refused to supply the belligerents, meaning that France was cut off from coal supplies from both the German Ruhr and Britain, so securing a reliable source of coal became essential. France conquered Vietnam, known as French Indo-China, between 1883 and 1885, primarily to appropriate its coal resources to supply the French navy (Campagne 2020). Tonkin became one of the largest coal exporters in the world, and large-scale coal mining in the 1920s created enormous opencast voids, while the demand for pit props led to illicit logging which denuded forests (Nguyen 2021). A general strike took place of more than 30,000 miners demanding higher wages on 12 November 1936, which later became the date for the annual celebration of the miners. The Red River Delta coal was only discovered in 1960 during a search for gas and oil (Mijał 2018): there is still major local opposition to exploitation because of fears of pollution and because this is the area of the largest rice fields in Vietnam (Bui and Lê 2011).

Vietnam ranks twentieth in world coal reserves. The state-owned mining company, Vinacomin, oversees 95 *per cent* of production and foreign companies have to sign a production-sharing contract. This has proved a major disincentive to foreign investment, even though the mining industry needs foreign technology and expertise to install mechanisation (CMMCP 2020, 37), especially as the Quảng Ninh field has many faults (Mijał 2018, p. 278). The coal seams dip steeply, making half suitable for surface mining while the other half must be mined underground. Vietnam's electricity generation in 2017 was 37.5 *per cent* from coal. Coal production decreased by 10.6 *per cent* between 2007 and 2017, while consumption increased by 348 *per cent*, mainly supplied from China (CMMCP 2020, 37). Vietnam has reserves of 7.9 billion tonnes of anthracite and 35.7 billion tonnes of bituminous coal; the master plan for the coal industry aims at output of 59 to 64 million tonnes by 2025 and 59 to 68 million tonnes by 2030 (Mijał 2018, pp. 275, 277).

### 432. Quốc Tế Bảo Tàng Than (Quang Ninh Museum)

The Quốc Tế Bảo Tàng Than first opened in 1960. The name translates as International Museum, but it is usually known in English as Quang Ninh Museum. The new museum building was designed by the Spanish architect, Salavador Perez Arroyo (born 1945), to look like a giant mirror and opened in 2013. There are no museums in Vietnam specifically dedicated to coal mining. The museum has three floors of displays, with coal mining forming the bulk of those on the third floor.

Almost 80 *per cent* of Quảng Ninh province is mountainous, with forest, mineral and water resources. The displays deal with the role of Quang Ninh Province as the producer of almost 90 *per cent* of coal in Vietnam and the role of coal mining in the development of the province. Original images and objects outline the history of coal mining in Quang Ninh province, including a model of the roadways of an underground mine and examples of opencast mines. Outdoors there is a block of anthracite coal, weighing about 25 tonnes. The Vietnam Centre for Cultural Heritage Research and Promotion supported the Quốc Tế Bảo Tàng Than with research and design for the coal-mining exhibition.

*References*

https://halongtourism.info/quang-ninh-museum/

http://www.baotangquangninh.vn/

personal communications, Nguyen Hai Ninh, vice-head, Museum Management Bureau, Department of Cultural Heritage, Ministry of Culture, Sports and Tourism, Government of Vietnam, 2013, 2014

# Concordance

This concordance deals with the major aspects that are found at a large number of sites. A number of items are not included either in the concordance or in the introduction, such as staffing and visitor numbers, but such information can be found by going through the entries. The references in the concordance are to information to be found at each museum, not in each entry; for example, virtually all entries have information about coal-mining history, but all references in the concordance under 'coal-mining history' refer to those museums that specifically cover coal-mining history.

The entries are arranged in alphabetical order by topic, not in order of importance. As far as possible, this concordance is up to date as of 2022. All institutions which responded to requests for information were asked to state which items were to be found at their institution. This means that in some cases the presence of a particular item, especially topics such as visitor facilities, may not be covered in the actual textual entry for that place. Conversely items may not be included that are present at sites that have not responded to the request for information.

Some items held by a particular museum, such as large machinery or geological collections, also may not be on public display and may instead be in store. Those intending visiting a particular institution, particularly if wishing to consult the archives or library, should contact that museum and its website in advance rather than relying solely on information in this publication.

# Bibliography


Aden, N.T., Fridley, D.G., and Zheng, N., 2008. *Outlook and challenges for Chinese coal* (Berkeley, USA).

Agricola, G., 1556. *De re metallica*, trans. and ed. H.C. Hoover and L.H. Hoover (New York, USA, 1950).

Alexander, P., 2008. 'Challenging cheap-labour theory: Transvaal and Natal coal miners, *ca* 1890-1950', *Labour History*, 49.1, pp. 47-70.

Andrews, B.J., 2011. *Coal mines of NSW* (Sydney, Australia).

Andrews, D.A., and Hi, C.C., 1956. *Coalfields of the Republic of Korea*, Bulletin of the Korea Geological Survey, 1041 (Washington, USA).

Arnot, R. Page, 1949-61. *A history of the Miners' Federation of Great Britain*, 3 vols (London).

Ashworth, W., 1986. *The history of the British coal industry; V: 1946-1982: the nationalized industry* (Oxford).

Austin, D. (ed.), 1982. *Boldon Book: Northumberland and Durham*, Domesday Book supplementary volume, 35 (Chichester).

Avango, D., Hacquebord, L., Aalders, Y., *et al*, 2011. 'Between markets and geo-politics: natural resource exploitation on Spitzbergen from 1600 to the present day', *Polar Record*, 47.1, pp. 29-39.

Baiyewu-Teru, A., 2015. 'The history of coal in Nigeria' (published on-line at https://ng.boell.org/en/2015/12/03/history-history-coal-nigeria).

Battiscombe, G., 1974. *Shaftesbury: a biography of the seventh earl, 1801-1885* (London).

Benoit, P., and Verna, C. (ed.), 1999. *Le charbon de terre en Europe occidentale avant l'usage industriel du coke* (Turnhout, Belgium).

Berckmans, P. [2016]. *Le Borinage, 1781-2014: naissance et développement de la culture et du paysage industriels dans le plus ancient basin houiller du continent européen*, Les Dossiers de L'Institut du Patrimoine Wallon, 17 (Namur, Belgium).

Besson, J-C., Malolepszy, V., Gigault, L., *et al*, 2019. *La mine en France; histoire industrielle et sociale: actes du colloque du 20 juin 2018 organisé à l'occasion de la parution de l'ouvrage collectif*, Comité d'Histoire Mémoire, 27 (Ars-sur-Moselle, France).

Bui, X.N., and Lê, Q.T., 2011. 'Coal mining industry in Vietnam', *Proceedings of the International Symposium on Earth Science and Technology* (Fukuoka, Japan), pp. 155-9.

Bulearcă, M., Neagua, C., Simaa, C., and Mărguş, D., 2014. 'The Romanian extractive industry, from its earliest beginnings up to year 1948', *Procedia Economics and Finance*, 8, pp. 106-12.

Burton, W. Donald, 2014. *Coal-mining women in Japan: heavy burdens* (London).

Busson, F., 1906. 'Die Abbaumethoden im Voitsber-Köflacher Braunkohlenrevier', *Berg- und Hüttenmännisches Jahrbuch*, 54, pp. 99-164.

Butler, J., 2021. 'A coal mine for every wildfire', *London Review of Books*, 43, number 22 (18 November), pp. 9-12.

Campagne, A., 2020. 'French energy imperialism in Vietnam and the conquest of Tonkin (1873-1885)', *Journal of Energy History/Revue d'Histoire de l'Energie*, 3 (published on-line at energyhistory.eu/en/node/218).

Cao, D., Wang, A., Ning, S., *et al*, 2020. 'Coalfield structure and structural controls on coal in China', *International Journal of Coal Science & Technology*, 7, pp. 220–39.

Carrington, D., 2021. 'More aid goes to fossil fuels than to cutting air pollution', *The Guardian* (7 September), p. 11.

Chatterjee, J.S., 2020. 'The politics of denial: a history of diseases in coal mines of colonial India', in [Debrabant] 2020, pp. 190-3.

Chen, S.P., 1993. 'Coal as an energy resource in Malaysia', *Bulletin of the Geological Survey of Malaysia*, 33, pp. 399-410.

Cherchi, S., 2011. 'The town of Carbonia and life after the mine', in [Debrabant] 2011, pp. 239-41.

Church, R., 1986. *The history of the British coal industry; III: 1830-1913: Victorian pre-eminence* (Oxford).

Clayton, A.K., 1962. 'The Newcomen-type engine at Elsecar', *Transactions of the Newcomen Society*, 35, pp. 97-108.

Cliff, N. (trans.), 2016. *Marco Polo: the travels* (London).

CMMCP 2016. *Coal mine methane profiles*, published on-line by the Global Methane Initiative.

CMMCP 2020. *Coal mine methane profiles*, published on-line by the Global Methane Initiative.

Cunningham, M., van Uffelen, L., and Chambers, M., 2019. 'The changing global market for Australian coal', *September 2019 Bulletin of the Reserve Bank* (published on-line at https://www.rba.gov.au/

publications/bulletin/2019/sep/the-changing-global-market-for-australian-coal.html).

[Debrabant, V., ed.], 2006. *10 Mars 1906 la catastrophe des mines de Courrières . . . et après?* (Lewarde, France).

[Debrabant, V., ed.], 2009. *Les paysages de la mine, un patrimoine contesté?* (Lewarde, France).

[Debrabant, V., ed.], 2011. *Le statut social des mineurs de charbon dans le monde: the social status of coal miners in the world* (Lewarde, France).

[Debrabant, V., ed.], 2013. *Des machines et des hommes: machinery and men* (Lewarde, France).

[Debrabant, V., ed.], 2016. *Charbon et conflits dans le monde: coal and war throughout the world* (Lewarde, France).

[Debrabant, V., ed.], 2020. *Santé à la mine: acteurs et systems de soins* (Lewarde, France).

[Debrabant, V., and Dumont, G.], 2014. *Le basin minier au cœur des conflits* (Lewarde, France).

de Dinechin, P., 2001. *Identidad y reconversion en las ciudades carboniferous de Lota y Coronel-Chile* (Santiago, Chile).

Delaney, J., 1998. *A history of the Greta coal measures, 1861-1998* (Newcastle, Australia).

Delbroek, B., 2011. 'The social status of coalminers: the miners of Belgian Limburg', in [Debrabant] 2011, pp. 221-4.

Dias, P. Alves, Kanellopoulos, K., Medarac, H., *et al*, 2019. *EU coal regions: opportunities and challenges ahead*, JRC Science for Policy Report (Luxembourg City, Luxembourg).

Drexler, J., 2011. 'ICEM and the mining industry', in [Debrabant] 2011, pp. 243-5.

Dumont, G., 2011. 'The identity and social status of coal miners in France, 1720-1880', in *ibid.*, pp. 191-6.

Elliott, B., 2006. *South Yorkshire mining disasters; I: the nineteenth century* (Barnsley).

ERIH European Route of Industrial Heritage (published on-line at http://www.erih.net/european-theme-routes/mining.html).

Escudier, J-L., 2006. 'De l'accident du travail à la catastrophe minière: une affaire d'État', in [Debrabant] 2006, pp. 299-309.

euracoal. published on-line at https://euracoal.eu/info/country-profiles/other-eu-energy-community/.

Farrenkopf, M., 2016. 'Coal and the Third Reich's economic policy in wartime', in [Debrabant] 2016, pp. 214-17.

Fejér, L., 1989. 'History of hard and soft coal exploration in Hungary till 1945', in *History of mineral exploration in Hungary until 1945*, ed. G. Csíky and Gy. Vitális, Annals of the History of Hungarian Geology, special issue, 2 (Budapest, Hungary), pp. 31-4.

Fernández, S., 2015. 'Pasao y presente de la lucha minera en España', *¡Unidady Lucha!: Órgano del Comité Central del Partido Comunista de los Pueblos de España.*

Flinn, M., and Stoker, D., 1984. *The history of the British coal industry; II: 1700-1830: the Industrial Revolution* (Oxford).

Fontaine, M., 2016. 'Introduction', in [Debrabant] 2016, pp. 201-12.

Forti, A., 2011. 'Mining in Belgium: traditions, legislation and the art of compromise', in [Debrabant] 2011, pp. 199-203.

Forti, A., 2016. 'The two world wars and the Walloon coal industry: a time bomb', in [Debrabant] 2016, pp. 265-9.

Frantel, N.C., 2008. *Chesterfield County, Virginia, uncovered: the records of death and slave insurance records for the coal mining industry,1810-1895* (Westminster, USA).

Gales, B., and Hölsgens, R., 2017. 'Coal transition in the Netherlands' (published on-line at https://www.researchgate.net/publication/331088234_Coal_Transitions_in_the_Netherlands_An_historical_case_study_for_the_Project_Coal_Transitions_Research_and_Dialogue_on_the_Future_of_Coal).

Goodchild, J., 2000. *Images of England: West Yorkshire coalfield* (Stroud).

Goodman, R., 2020. *The domestic revolution: how the introduction of coal into our homes changed everything* (London).

Gray, M., and Sundaresan, S., 2020. *How to waste over half a trillion dollars: the economic implications of deflationary renewable energy for coal power investments* (London).

Guiollard, P-C., 2013. '1918 and 1946: the two stages of mine reorganisation in the Nord and Pas-de-Calais regions that were factors of innovation', in [Debrabant] 2017, pp. 176-9.

Hartley, R.F., 1994. 'The Tudor miners of Coleorton, Leicestershire', *Bulletin of the Peak Mines Historical Society: Mining before Powder*, 12.3, pp. 91-101.

Hatcher, J., 1993. *The history of the British coal industry; I: before 1700: towards the age of coal* (Oxford).

Heidecker, K., 2010. *The divorce of Lothar II: Christian marriage and political power in the Carolingian world*, trans. T. M. Guest (Ithaca, USA).

Henriksson, H., 1996. 'Bland gruvpigor och gruvdrängar; Svenskt gruvarbete under 1700- och 1800-talen', *Dædalus: Tekniska Museets Årsbok*, pp. 165-204.

Hobbs, H., 2017. 'Bosnia: a longwall mine of potential', *World Coal* (6 July; published on-line at https://www.worldcoal.com).

Horne, D., 1986. *The lucky country*, 3rd ed. (Ringwood, Australia).

Houp, R., 2011. 'The Coal Pact, or how to end a sixty-year battle for coal', in [Debrabant] 2011, pp. 241-2.

Hui, M., 2021. 'China's boycott of Australia has redirected global flows of coal' (published on-line at https://qz.com/2051594/chinas-boycott-of-australian-coal-has-been-good-for-india/).

Islam, S.E., 2013. 'Coal: an Indian narrative', in [Debrabant] 2013, pp. 193-7.

Jackson, J., 2012. *1912-2012: 100 years of Wolf safety lamps* (Sheffield).

Jaskó, S., 1989. 'History of lignite exploration in Hungary', in *History of mineral exploration in Hungary until 1945*, ed. G. Csíky and Gy. Vitális, Annals of the History of Hungarian Geology, special issue, 2 (Budapest, Hungary), pp. 35-40.

Jurkiewicz, J.G., 2013. 'The Glowna Kluczowa Sztolna Dzedziczna gallery: a unique example of a 19th century civil engineering project', in [Debrabant] 2013, pp. 191-3.

Kaur, A., 1990. 'Hewers and haulers: a history of coal miners and coal mining in Malaysia', *Modern Asian Studies*, 24.1, pp. 75-113.

Khaitan, U., 2020. 'Women beneath the surface: coal and the colonial state in India during the Second World War', *War & Society*, 39.3, pp. 171-88.

Kopp, H., 2015. 'Montangeschichte des Kohlebergbau Fohnsdorf', in *Res montanarum Fachzeitschrift des Montanhistorischen Vereins Österreich*, 54, pp. 1-10.

Kotler, M.L., 2021. 'UNESCO and Japan's rewriting of history', *The Diplomat* (13 September).

Labuda, J., 2009. 'L'influence de l'industrie manière et sidérugique sur le paysage de la region de Banská Štiavnica', in [Debrabant] 2009, pp. 90-7.

Llorens, J.P., 2012. 'El patrimonio minero del Franquismo y su incidencia en el paisaje industrial Catalán' (published on-line at https://www.um.es/hisminas/wp-content/uploads/2012/06/Texto-completo2.pdf).

Lim, T.W., 2017. *Energy transitions in Japan and China: mine closures, rail developments, and energy narratives* (London).

Lucarelli, B., 2010. 'The history and future of Indonesia's coal industry: impact of politics and regulatory framework on industry structure and performance', *Program on Energy and Sustainable Development Working Paper,* 93 (Stanford, USA).

Lukasczyk, C., 2000. 'Der Österreichische Kohlenbergbau im 20. Jarhundert' (lecture to the Montanhistorischen Verein für Österreich, 28-30 September, Leoben, Austria).

Luyten, M., 2018. *Het geluk van Limburg* (Amsterdam, Netherlands).

Maja-Pearce, A., 2022. 'Short cuts', *London Review of Books*, 44, number 1 (6 January), pp. 32-3.

[Malolepszy, V., ed.], 2020. *Santé à la mine: health down in the mine* (Lewarde, France).

Martínez, I.D., 2020. 'From mobilization to the creation of the instituto nacional de Silicosis (National Silicosis Institute) of Asturias, Spain: the development of working safety in the mining sector during Francoism', in [Debrabant] 2020, pp. 216-18.

McFadden, P., 2016. 'Mining and warfare in South Africa – Anglo Boer war 1898-1902 to the Great War 1914-1918', in [Debrabant] 2016, pp. 259-62.

McIvor, A., and Johnston, R., 2007. *Miners' lung: a history of dust disease in British coal mining* (Aldershot).

Medunic, G., Radenović, A., Bairamović, M., and Švec, M., 2016. 'Once grand, now forgotten: what do we know about the super high-organic-sulphur Raša coal?', *Rudarsko-geološko-naftni zbornik*, 31, pp. 27-45.

Mijał, W., 2018. 'Coal mining and coal preparation in Vietnam', *Journal of the Polish Mineral Engineering Society*, pp. 275-86.

Moorhouse, S.A., 1981. 'The rural medieval landscape', in *West Yorkshire: an archaeological survey to A.D. 1500*, ed. M.L. Faull and S.A. Moorhouse (Wakefield), pp. 581-850.

Mount, F., 2021. 'Short cuts', *London Review of Books*, 43, number 13 (1 July), pp. 18-19.

Murray, A., 2011. 'Voices from the shadows', in [Debrabant] 2011, pp. 209-14.

Murray, A., 2013. 'The dark side of the boom', in [Debrabant] 2013, pp. 202-6.

Musa, M.G., 2011. 'These archive collections which recount the history of the mining town of Carbonia and its miners', in [Debrabant] 2011, pp. 230-5.

Musa, M.G., 2016. 'Coalmining in the Sulcis area during the First World War', in [Debrabant] 2016, pp. 212-14.

Nguyen, T.L., 2021. 'Coal mining, forest management, and deforestation in French colonial Vietnam', *Environmental History*, 26.2, pp. 255-77.

Oldfather, C.H. (trans.), 1939. *Diodorus Siculus: The library of history; III* (Cambridge).

Orwell, G., 1937. *The road to Wigan Pier* (London; penguin edition, 1989).

Page, G.T., and Shapiro, L., 2006. 'Coal in Canada' (published on-line by the Canadian Encyclopedia at https://www.thecanadianencyclopedia.ca/en/article/coal; updated 2018 by N. Block and N. Baker).

Palmer, D., 2021. 'Japan's World Heritage Miike Coal Mine – where prisoners-of-war worked "like slaves"', *The Asia-Pacific Journal: Japan Focus*, 19.13, number 1, pp. 1-20.

[Paris, A., ed.], 2002. *La conservation du patrimoine technique et industriel* (Lewarde).

Ponty, J., 2016. 'Coal: a bone of contention in central and eastern Europe in the 20th century', in [Debrabant] 2016, pp. 222-4.

Preece, R., 2007. 'Brève étude des rapports de sécurité et de legislation dans les houillères britanniques au XIXe et au début du XXe siècle', in [Debrabant] 2007, pp. 163-7.

Preece, R., 2016. 'Courage, conflict and commemoration – the contribution of British coal miners during two world wars', in [Debrabant] 2016, pp. 251-3.

Preece, R., Roberts, I., and Rose, M., 2009. 'Investigation of a miner's cottage at Caphouse Colliery', *Forum: the annual newsletter of CBA Yorkshire*, pp. 15-17.

Przigoda, S., 2016. 'German coal mines during the First World War', in [Debrabant] 2016, pp. 244-7.

Rackham, H. (trans.), 1952. *Pliny: Natural history; IX: books 33-35*, Loeb Classical Library, 394 (Harvard).

Ranestad, K., and Sharp, P.R., 2020. 'Success through failure? Four centuries of searching for Danish coal', *European Historical Economics Society Working Paper*, 183, pp. 1-25.

Ratanasthien, B., 2011. 'Coal deposits', in *The geology of Thailand*, ed. M.F. Ridd, A.J. Barber and M.J. Crow (London).

Ratshomo, K., and Nembahe, R., [2016]. *South African coal sector report* (Pretoria, South Africa).

Rice, G.S., 1910. *The explosibility of coal dust*, United States Geological Survey Bulletin, 425 (Washington, USA).

Rodríguez-Valencia, C., 2011. 'Spanish women's work at the coal face', in [Debrabant] 2011, pp. 238-9.

Roijen, E., 2011. 'The use of an open-loop heat-pump system to heat social housing and an office building throughout the town of Heerlen, the Netherlands', in *Heat-pump technology using minewater*, ed. M.L. Faull, National Coal Mining Museum for England Publications, 13 (Overton), pp. 15-25.

Romano, R.B., and Papastefanaki, L., 2020. 'Women and gender in the mines: challenging masculinity through history: an introduction', *International Review of Social History*, 65.2, pp. 191-230.

Salway, G., 1999. *The British Coal Collection*, National Coal Mining Museum for England Publications, 2 (Overton).

Saunders, N.J. (ed.), 2004. *Matters of conflict: material culture, memory and the First World War* (Abingdon).

Sevük, I.H., 1942. *Yurttan yazilar* (Istanbul, Turkey).

Sherwood, A., and Phillips, J., 2021. 'Coal and coal mining', *Te Ara – the encyclopedia of New Zealand* (published on-line at http://www.TeAra.govt.nz/en/coal-and-coal-mining).

showcaves published on-line at http://www.showcaves.com/english/explain/Index/index.html.

Slotta, D., and Kierdorf, A., 2016. *Der Förderturm Camphausen IV*, Historische Wahrzeichen der Ingenieurbaukunst in Deutschland, 18 (Berlin, Germany).

Slotta, R., 2011. 'Social conditions in the German mining industry', in [Debrabant] 2011, pp. 188-90.

Sombathy, E., Kúšik, D., and Mižák, J., 2018. 'Slovak Mining Road', *Slovak Geological Magazine*, 18.1, pp. 83-106.

Supple, B., 1987. *The history of the British coal industry; IV: 1914-1946, the political economy of decline* (Oxford).

Swinbank., E., 2011. '17th and 18th century coal miners in Scotland "everything that meets the eye or the ear tells of slavish labour"', in [Debrabant] 2011, pp. 226-9.

Szemán, A., 1994. 'Georgius Agricola de re metallica c. művének újabb magyar vonatkozású metszetei¬ről', in *Agricola Évszázada; publikációk a magyarországi ásványkincsek történetéből*, 4 (Miskolc, Hungary), pp. 89-104.

Szewczyk, K., 2009. 'Les changements du paysage naturel dans la vallé de Walbrzych après la fermeture des mines', in [Debrabant] 2009, pp. 176-83.

Szmidtke, Z., 2011. 'Bernard Bugdol: from hewer to mine director', in [Debrabant] 2011, pp. 224-6.

Takashi, M., 2017. 'Convict labor and its commemoration: the Mitsui Miike coal mine experience', *The Asia-Pacific Journal: Japan Focus*, 15.1, number 3, pp. 1-14.

Thee, K.W., 2013. 'The Indonesian economy during the Japanese occupation', *Masyarakat Indonesia*, 39.2, pp. 327-40.

Thomas, W. Gerwyn, 1986. *Welsh coal mines* (Cardiff).

Thompson, N., 2021. 'Japan's "Battleship Island" hides history of wartime forced labor', *Global Voices* (21 August).

Tierney, J.J., 1960. 'The Celtic ethnography of Posidonius', *Proceedings of the Royal Irish Academy: Archaeology, Culture, History, Literature*, 60, pp. 189-272.

Travis, J.R., 2008. *Coal in Roman Britain*, BAR British Series, 468 (Oxford).

Trueman, E., 1954. *The coalfields of Great Britain* (London).

Turner, S.A.M., 2021. *The Miners' Welfare Fund 1921-1952: the greatest piece of social reform of its time* (London).

Verde, M.Á.P., Soto, Á/P/M., and Gómez, J.J.G., 2020. 'Female workers in the Spanish mines, 1860-1936', *International Review of Social History*, 65.2, pp. 233-65.

Vitális, I., 1939. *Coal occurrences in Hungary* (Sopron, Hungary).

Vrankar, L., 2019. Slovenia and its coal regions' (published on-line at https://ec.europa.eu/energy/sites/ener/files/documents/1.5_vrankar_plenary.pdf).

Vujić, S., 2014. 'The history of Serbian mining', in *Serbian mining and geology in the second half of XX century*, S. Vujić, A. Grubić *et al* (Belgrade, Serbia), pp. 1-38.

Xue, Y., 2011. 'Coal mine ballads: a true picture of the labor of miners', in [Debrabant] 2011, pp. 203-9.

Xue, Y., 2013. 'From tradition to modernity: the development of coal extraction methods and techniques in China', in [Debrabant] 2013, pp. 186-90.

Xue, Y., 2016. 'The Japanese occupation of China's mines in the modern era (1895-1945)', in [Debrabant] 2016, pp. 217-21.

Zhen-Hui, L., and Yung-Jaan, L., 2015. 'A method for development of ecomuseums in Taiwan', *Sustainability*, 7.10, pp. 13249-69.

www.ingramcontent.com/pod-product-compliance
Ingram Content Group UK Ltd.
Pitfield, Milton Keynes, MK11 3LW, UK
UKHW061843041026
15591UKWH00015B/887

9 781407 358123